Javier Roa

Regularization in Orbital Mechanics

De Gruyter Studies in Mathematical Physics

Volume 42

Javier Roa

Regularization in Orbital Mechanics

Theory and Practice

DE GRUYTER

Mathematics Subject Classification 2010
35-02, 65-02, 65C30, 65C05, 65N35, 65N75, 65N80

Author
Javier Roa
Technical University of Madrid
Madrid, E-28040
Spain

ISBN 978-3-11-055855-5
e-ISBN (PDF) 978-3-11-055912-5
e-ISBN (EPUB) 978-3-11-055862-3
Set-ISBN 978-3-11-055913-2
ISSN 2194-3532

Library of Congress Cataloging-in-Publication Data
A CIP catalog record for this book has been applied for at the Library of Congress.

Bibliographic information published by the Deutsche Nationalbibliothek
The Deutsche Nationalbibliothek lists this publication in the Deutsche Nationalbibliografie; detailed bibliographic data are available on the Internet at http://dnb.dnb.de.

Typesetting: le-tex publishing services GmbH, Leipzig
Printing and binding: CPI books GmbH, Leck
♾ Printed on acid-free paper
Printed in Germany

www.degruyter.com

To my family

Foreword

It has been almost 50 years since Eduard Stiefel and Gerhard Scheifele published their celebrated book "Linear and Regular Celestial Mechanics", which brought the topic of regularising the equations of orbital motion to the forefront of research in celestial mechanics and astrodynamics. Surprisingly, no detailed monograph on this topic has been published ever since. The present book is the long-anticipated return of the theory of regularisation, providing a fresh and updated review of the subject. Despite its theoretical nature, the book includes practical applications which prove the great potential of using regularised equations of motion, and which make this monograph an essential addition to the libraries of both theoreticians and practitioners.

Regularisation is a broad topic that touches many fields of classical and celestial mechanics. Besides explaining this topic in depth and offering practical examples, this book also sheds light on the elegant connections of this topic with other fields such as topology, abstract algebra, or number theory. The reader will find references to canonical transformations, ideal frames, quaternions, or relativity, among many others.

The introductory Chapter 1 gives a broad-brush description of the present challenges in space exploration and of the ensuing problems in orbit computation.

In Chapter 2, the author provides a comprehensive review of the various techniques that can be applied to regularise and stabilise the equations of orbital motion. Before diving into theoretical details in the subsequent chapters, here the author convinces the reader of the power of the techniques to be discussed in the book.

While pursuing an improved performance from a computational perspective, the author also addresses the mathematics underlying those transformations, and explains how regularising techniques can lead to new analytic results. For example, in Chapter 3, the reader will find an original approach to the Kustaanheimo–Stiefel (KS) transformation, focusing on this transformation's link to the Hopf fibration. The topological aspect of this link is elucidated at length, serving as a foundation to analyse the evolution of numerical errors in problems with strong perturbations. The chapter contains new results, like the concept of manifold of solutions and topological stability, as well as the derivation of the gauge-generalised non-osculating orbital elements in KS space.

As an example of a set of alternative orbital elements, the Dromo formulation is presented in Chapter 4. This formulation provides a convenient framework to apply various regularisation techniques for improving the behaviour of the equations of motion. It also renders the author an opportunity to recover the concept of ideal frames. After analysing the KS and Dromo transformations in detail, the author shows how regularisation techniques can be applied to specific problems, like the propagation of flyby trajectories. In particular, Chapter 5 discusses the connection between Minkowski's space-time and the geometry of hyperbolic orbits.

https://doi.org/10.1515/9783110559125-201

Regularisation implies a combined use of advanced analytic techniques and highly efficient computation schemes. Accordingly, in this monograph numerical integration methods receive the attention they deserve. The author builds a dedicated software tool to compare the performance of different formulations and integrators, and Chapter 6 defines a systematic way of benchmarking the numerical fitness of various methods.

The second part of the book, which starts with Chapter 7, applies regularisation to real-world problems, yielding novel results in the areas of relative motion and low-thrust trajectory design. Chapter 7 presents the theory of asynchronous relative motion, a theory that takes its origin from the use of a generic transformation to replace the physical time with a fictitious time. The result is a new paradigm for approaching relative motion using differential orbital elements. This approach improves the accuracy of orbit propagation, as compared with the classical linearised solutions. The power of this technique resides in the fact that it can be easily applied to the already known solutions. So the reader will find improved versions of the Clohessy–Wiltshire and the Yamanaka–Ankersen solutions. In addition, fully non-singular and universal (valid for any eccentricity) solutions are derived using regular sets of elements.

Application of time transformations to the continuous-thrust problem leads to a new class of analytic solutions, termed generalised logarithmic spirals. This beautiful family of trajectories admits two integrals of motion, strikingly similar to the conservation of the energy and angular momentum in Kepler's problem. The new family of solutions is a useful addition to the field of preliminary low-thrust mission design, since designing trajectories with continuous thrust can be approached with the classical methods, with Keplerian arcs brought in. Along these lines, Chapter 10 poses the spiral Lambert problem and finds an equivalent, in the low-thrust realm, to many of the properties of the original Lambert problem. Then, Chapter 11 presents examples of application of this approach to actual mission design. Finally, Chapter 12 surprises the reader by deriving even more families of analytic solutions that admit conservation laws, also providing unexpected parallels with some cosmological concepts. These chapters will be most enjoyable to geometricians and mission designers.

All in all, this wonderful monograph provides a broad and up-to-date picture of regularisation theories, as well as enlightening examples of application of those theories to practical problems. I would recommend it to any graduate student or researcher working in the area of celestial mechanics and astrodynamics.

Michael Efroimsky, Ph.D.
Astronomer,
US Naval Observatory,
Washington DC

Contents

Part III: **Appendices**

1 Introduction. Current challenges in space exploration

Space programs have undergone profound changes, from economic restructuring to the implementation of novel mission architectures. During the Cold War, space projects enjoyed an apparently unlimited funding that led to unprecedented technical and scientific breakthroughs in just a few years. In fact, between 1958 and 1966, when it reached its historical maximum, the budget of NASA rose from 0.1% to 4.4% of the total US federal budget. However, after the Apollo program this figure has been going down progressively until reaching a steady 0.5%, which has remained almost constant in the last decade. In order to maintain the highest scientific standards for a given budget, the efficiency of the mission concepts must be maximized.

From a scientific perspective, the exciting goals set for the future require pushing the boundaries of the current technology levels, mission architectures, and even changing the paradigm of space exploration in general. In the coming decades we may see probes investigating the subsurface oceans of the icy moons, asteroid rocks brought to Earth, lunar bases, and even humans walking on the surface of Mars. The way to go is more or less clear, but much work needs to be done to reach such objectives.

Leaving the Earth is one of the most expensive phases of a mission, with the launch costs becoming as high as 30–40% of the total cost. It is also the most violent part, and malfunctions at this stage are catastrophic. Once in orbit, the probe still needs to travel to its final destination. The amount of propellant that remains in the spacecraft after reaching its target orbit determines the mission lifetime. Planning an efficient space travel requires high-fidelity physical models to ensure that the probe will follow an orbit that is at least close to the optimized nominal trajectory. A poor model will lead to unexpected maneuvers to correct the course, reducing the amount of fuel available for nominal operations and, consequently, shortening the duration of the mission.

1.1 Accessible space

An exciting advance in the last decade has been the rise of private ventures to satisfy the increasing demand for efficient launch vehicles. With the end of the Space Shuttle Program approaching, NASA needed to guarantee its launch capabilities to supply the International Space Station (ISS). In December 2008, two privately owned companies, SpaceX and Orbital Sciences (currently Orbital ATK), were awarded Commercial Resupply Services contracts to conduct unmanned cargo launches to the ISS. Russia had been in charge of this task for more than 20 years, with over 50 launches. On

https://doi.org/10.1515/9783110559125-001

May 22, 2012, Elon Musk's SpaceX became the first company to send commercial cargo to the ISS, followed by Orbital Sciences' Antares rocket launch in September 2013. In 2014, NASA awarded $4.2B and $2.6B contracts to Boeing and SpaceX, respectively, in order to regain the capability of launching astronauts to the ISS from US soil. Boeing is developing the Orion capsule, whereas SpaceX is focusing on the crewed version of the Dragon capsule. In the future, the Orion capsule would have sent astronauts to an asteroid as part of the Asteroid Redirect Mission (ARM), whereas the Dragon capsule will evolve into the Red Dragon, designed to carry humans to Mars.

These initiatives have improved the competitiveness between launch providers significantly, resulting in reductions of the cost of the launchers. Table 1.1 compares the cost-per-kilogram of launches to low Earth orbit (LEO) with different vehicles from various providers. It is worth noting that the ultimate goal of the Space Shuttle Program was to reduce launch costs thanks to partially reusing the launch vehicle. Several factors like design changes or high maintenance requirements meant that the final cost was around 18,000 USD/kg to LEO, much higher than Russia's Proton. This factor, together with safety concerns, resulted in the cancelation of the program in 2011. SpaceX followed a different approach toward reusability: the first stage of the rocket lands autonomously and, after refueling, it will be ready for the next launch. Given the success of Falcon 9, the future Falcon Heavy will eventually reduce the launch cost down to 2500 USD/kg, the most affordable option to date.

Tab. 1.1: Launch cost to LEO (in USD/kg).

Ariane 5	Delta IV-H	Falcon 9	Falcon Heavy	Vega	Proton M
10,500	13,800	4100	2500	15,600	4300

A direct consequence of the launch cost reductions is space becoming more accessible. This factor, combined with the advances in miniaturization, has led to the flourishing of cubesats. Cubesats are smaller, cheaper, and easier to build than regular spacecraft and have many applications, ranging from merely university experiments to astronomical observations, remote sensing, and communications. Their reduced size and mass make them perfect candidates for secondary payloads, or even for exploring alternative deployment strategies. The NanoRacks CubeSat Deployer aboard the ISS is capable of deploying 6U cubesats, which can be sent as cargo with the supply capsules. Another popular concept is the use of rocoons, the combination of a rocket and a balloon. The Spanish company Zero 2 Infinity has already conducted successful experiments based on this concept. The rapid development in these areas has turned small satellites into key players in the space industry.

1.2 Distributed space systems

The full potential of miniaturized spacecraft relies on operating several satellites that perform cooperative tasks, forming a distributed space system. The relative motion between small spacecraft (down to the pico-scale, weighing around 100 g) has received much attention for industrial applications in recent years. Some advanced concepts promoted by DARPA (Defense Advanced Research Projects Agency), among others, even consider the use of spacecraft swarms, with tens of thousands of spacecraft.

Formation flying introduces a new paradigm of space mission design, and it is rapidly replacing monolithic solutions in many scenarios. Even the startup world is taking advantage of the wide range of possibilities provided by this concept. The San Francisco-based Planet Labs, for example, currently operates a constellation of almost a hundred 4 kg-spacecraft, which image the Earth continuously.

But the potential of distributed space systems is not merely economic. It also opens a whole new world of possibilities from a scientific and operational perspective. The German TanDEM-X formation flying mission (launched in 2010) served as a proof concept, and it also generated a high-accuracy digital elevation model of the Earth using synthetic aperture radar. The Gravity Recovery and Climate Experiment mission (GRACE) consists in two spacecraft flying in formation taking precise measurements of their relative states. This data allows scientists to generate the most precise gravity model of the Earth to date. This same concept was exploited by the Gravity Recovery and Interior Laboratory (GRAIL), which mapped the Moon's gravity field. In February 2016, the LIGO team announced the detection of gravitational waves for the first time, confirming Einstein's theory of general relativity. To extend these pioneering results the evolved Laser Interferometer Space Antenna (eLISA), consisting in three spacecraft in wide formation, will be launched in the 2030s to detect more accurately the ripples in space-time. LISA Pathfinder was launched in 2015 to test the key technologies required by eLISA, in particular the formation-keeping capabilities.

When designing a telescope, the aperture of the instrument is determined by the diameter of the main mirror. This is limited by obvious practical constraints, like the diameter of the launch vehicle. For instance, the diameter of the main mirror aboard the Hubble Space Telescope is 2.4 m, just enough to fit in the space shuttle cargo bay. The James Webb Space Telescope (JWST) will replace Hubble in 2018, and the new observatory mounts a mirror of 6.5 m in diameter. The mirror cannot fit inside the Ariane 5 fairing, so the engineering team opted for a foldable mirror composed of hexagons. Although more flexible, this solution is still limited by the launcher and future concepts explore the use of multiple satellites. The Terrestrial Planet Finder concept involved multiple small infrared telescopes flying in precise formation, simulating an unprecedentedly large aperture observatory. In addition, having multiple collectors allows the astronomers to apply sophisticated reduction techniques in the data pipeline to subtract bright stars. Unfortunately, this mission was canceled in 2011 due to budget issues.

The Kepler space telescope has discovered up to 2335 confirmed exoplanets, improving our knowledge about extrasolar worlds and unveiling astonishing examples of bizarre planetary systems. The telescope's only instrument is a photometer, which transmits lightcurves of almost 150,000 stars back to Earth. The search for exoplanets is based on the transit method: if the apparent brightness of a star decreases periodically, this could mean that a planet is orbiting it. This method has some disadvantages, like the high number of false positives (around 40% for Kepler) and the fact that the only planets that can be detected are those in edge-on orbits.

An interesting alternative to transit detection is the direct imaging of the planets. The telescope observes the planets themselves, instead of focusing on the central star. The challenge is that the stars are much brighter than the planets, and light reflected on them is typically lost. To solve this problem, coronographs can be placed between the telescope and the source to block the light of the star, and then the telescope is pointed to the planets. The most flexible solution is to design an occulter that flies in formation with the telescope, and blocks the light accordingly. NASA's WFIRST mission will include this technology and aim for direct observations of exoplanets. The required precision in both the relative positioning and pointing of the spacecraft make formation flying for astronomical applications one of the major challenges in future missions.

Relative motion between spacecraft will soon have interplanetary applications. The InSight mission incorporates two cubesats, which will become the first to fly in deep space. Provided with both UHF and X-band antennas, the Mars Cube One (MarCO) will serve as a communication relay for InSight, specially during entry, descent, and landing. The cubesats will follow their own orbits to Mars, and this poses an important challenge in regards to the navigation and operation of multiple spacecraft. Even the design team of the Europa mission, the latest of NASA's flagship program, is considering the use of two cooperative spacecraft. After reaching Jupiter's moon Europa along a low-energy transfer, the main spacecraft will deploy a lander that will maintain a communication link with the mothership until it completes its experiments on the surface of Europa. These are just some applications showing the relevance of distributed space systems in future mission concepts.

1.3 Efficient orbit transfers

Advances in launch systems have cut down the cost of leaving the Earth, and the use of spacecraft formations reduces the unitary cost of the craft and improves the scientific capabilities. But, once in space, the problem of reaching the final orbit still remains. A mission comes to an end when the spacecraft runs out of fuel. Thus, the propellant expenditures during the cruise phase need to be minimized, so the craft has as much fuel as possible when starting its nominal operations.

Interplanetary travels are the best examples of the need for smart transfer strategies. Finding the adequate launch window is one of the key elements. For example, a 154-day transfer to Mars launched on April 18, 2018 will require a mass of propellant that is 55% of the total mass of the spacecraft. However, if the launch is delayed by only 20 days, the propellant fraction will rise to 93% to keep the same time of flight, because of the configuration of the planets. Trajectories are usually very sensitive to changes, which complicates the design process. This is particularly critical during the preliminary phases, in which there are many iterations between the different engineering and scientific teams.

1.3.1 Low-thrust missions

The specific impulse (I_{sp}) of a propulsion system can be regarded as a measure of its efficiency: the higher the specific impulse, the less propellant is needed to exert the same change in the velocity of the spacecraft. Table 1.2 compares the specific impulse and maximum thrust of different propulsion systems. Solid rocket boosters deliver high thrust levels with low specific impulses, and are good choices for launch vehicles. Once ignited, the combustion cannot be stopped. Rockets with liquid fuel (bipropellant) can be switched on and off, and they have been the preferred choice for space maneuvers. New concepts based on electric propulsion, like ion engines and the variable specific impulse magnetoplasma rocket (VASIMR), increase the specific impulse significantly, at the cost of producing smaller thrust forces.

A typical space maneuver consists in switching on a chemical engine for a few minutes, which yields a change of velocity that is almost instantaneous given the total duration of the mission. But its low specific impulse makes it less efficient than electric propulsion systems. For these reason, missions like Deep Space 1, Hayabusa, or Dawn used the latter instead of the usual chemical thrusters. Given the low thrust levels of ion engines, the thruster needs to be on for months instead of minutes, and the spacecraft follows a spiral trajectory that slowly takes it to its final orbit. The mass of propellant can be reduced significantly, at the cost of increasing the time of flight.

On the industrial side, low-thrust electric propulsion has great potential for placing geostationary satellites into orbit. The high altitude of the geostationary orbit

Tab. 1.2: Specific impulse and thrust of different propulsion systems.

	Solid rocket (STS booster)	**Biprop. rocket (RS-25)**	**Ion engine (NEXT)**	**VASIMR (VX-50)**
I_{sp} [s]	250	350–450	4000	5000
Thrust [N]	13,800,000	1,900,000	0.25	0.50

(GEO), which is almost 36,000 km, makes it impossible for a rocket to insert a satellite directly into this orbit. The rocket puts the spacecraft into a highly elliptical orbit with its apogee at GEO altitude, and then the satellite's engines are used to circularize the orbit. The ABS-3A geostationary satellite (Boeing 702SP bus) was launched in March 2015 and used a revolutionary transfer strategy, as it was the first GEO satellite using electric propulsion. The spacecraft spiraled for over six months until it reached its operational orbit. Less propellant was spent in this phase and, as a result, there is more fuel available for the required station-keeping maneuvers, increasing its operational life. These corrective maneuvers compensate the drift of the satellites at GEO due to the Earth's nonspherical gravity field.

The design of a transfer using chemical propulsion consists in finding the optimal distribution of the maneuvers along the orbit, and in characterizing such maneuvers. When using electric thrusters, the engine is on for almost the entire duration of the transfer. The steering and throttling of the engine needs to be determined, and this process turns out to be a complex optimization problem.

1.3.2 Solar sailing

The Russian space pioneer Konstantin Tsiolkovsky envisioned potential alternatives to rockets like the use of solar sails that, by the effect of the solar radiation pressure, accelerate the spacecraft without spending fuel. The concept has been around for decades, but it involves some technical difficulties that have hindered its implementation in actual missions. In 2010, the Japan Aerospace Exploration Agency (JAXA) launched IKAROS, the first successful interplanetary mission provided with a solar sail. The spacecraft deployed a 200 m^2 sail that propelled it to Venus. Later that year, NASA launched the NanoSail-D2 cubesat equipped with a 10 m^2 solar sail. The LightSail cubesat, launched in 2015, was promoted by the Planetary Society.

The dynamics of spacecraft using solar sails are similar to the motion of probes with low-thrust electric propulsion systems. The force due to the solar radiation is small in magnitude, and it is accelerating the spacecraft continuously. This acceleration can be controlled by changing the attitude of the sail. The techniques used for low-thrust trajectory optimization can be extrapolated to the design of solar sail missions. The advantage of the latter is the fact that no propellant is required.

1.3.3 Gravity-assist trajectories

In 1966, Gary Flandro published an interesting discovery he made during a summer he spent at the Jet Propulsion Laboratory (JPL). Between 1977 and 1978, Jupiter, Saturn, Uranus, and Neptune would be aligned in such a way that a flyby at Jupiter would send a spacecraft to Saturn, a second flyby at Saturn would aim the trajectory to Uranus,

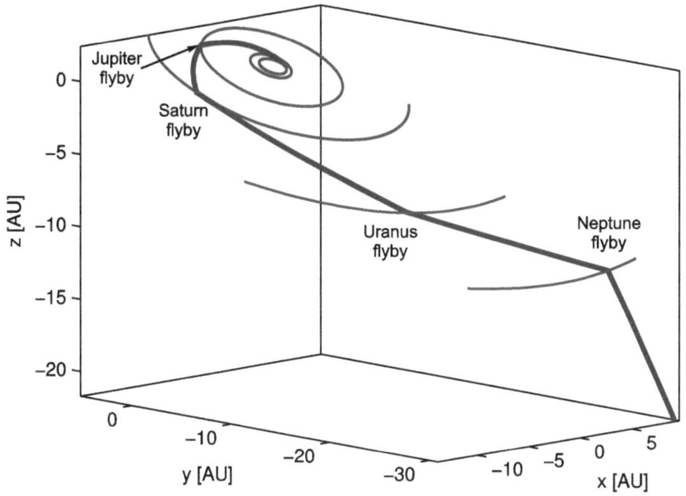

Fig. 1.1: The orbit of Voyager 2 from 08/21/1977 (launch) to 12/01/1992, represented in the ICRF/J2000 frame.

and a third flyby would send the probe toward Neptune. He coined the term Planetary Grand Tour, and this sequence was exploited during the design of the trajectory of Voyager 2 (Fig. 1.1). During the close approach there is an energy exchange between the probe and the planet that is flown by. Because of the difference in their masses the change in the orbit of the planet is negligible, but the spacecraft experiences an important change in its velocity. Although Pioneer 10 was the first mission to use this technique, it showed all its potential with the Voyager program.

Gravity-assist maneuvers have been widely used since the 1970s. In fact, most interplanetary trajectories involve flybys of intermediate planets in order to reduce the use of its engines. This makes the design process even more complicated for the mission analyst: determining the optimal trajectory is not about finding the most direct way, but rather to come up with the best sequence of flybys, dates, impulsive maneuvers, etc. The preliminary design becomes a counterintuitive task, with millions of possible combinations and solutions. Adequate analytical and numerical tools need to be implemented as the mission goals become more ambitious, specially if the trajectory involves not only planetary flybys but also arcs with low-thrust electric propulsion, which need to be optimized as well.

1.4 Orbit propagation

The Pioneer 10 and 11 spacecraft were the first to visit Jupiter and Saturn, respectively, in 1973 and 1979. After reaching the two gas giants, they initiated their journey to interstellar space. But the spacecraft were not moving as they were expected to. An anomalous acceleration was reducing their speed faster than the models predicted. This phenomenon, known as the Pioneer anomaly, puzzled the scientific community for many years. In fact, it was not until 2012 when scientists agreed that the most likely cause was the effect of the emission of thermal radiation from the spacecraft. The magnitude of the anomalous acceleration was just of the order of 10^{-10} m/s^2, which shows how extremely accurate force models and simulations need to be.

A careful analysis of Doppler data taken during the geocentric flybys of Galileo (December 8, 1990), NEAR (January 23, 1998), and Rosetta (March 4, 2005) revealed an unexpected change in their velocities, of the order of millimeters per second. The origin of this energy gain remains a mystery, and has motivated many theories: from the existence of dark matter halos around the Earth to relativistic effects, issues with the sensors at the DSN stations, propagation errors... These anomalies are, in the end, unexpected differences between the motion predicted by the models and the actual measurements.

The flyby anomaly and the Pioneer anomaly are famous examples, but discrepancies between the computed trajectory and the actual measurements always appear. In fact, measuring these differences can provide valuable scientific data. For example, the Juno science team will study the structure of Jupiter's interior and winds based on the deviations between the true and predicted orbits, which are obtained with numerical simulations. Numerical errors are always present, and minimizing this error is a critical task because errors in the propagation may lead to incorrect scientific conclusions.

From an operational point of view, accurate and reliable orbit propagators are used to compute the nominal trajectory. The navigation and control teams will be in charge of making sure that the spacecraft stays on its design course. If the nominal orbit is constructed with a poor force model, adjusting maneuvers will be constantly needed, because the probe will never follow the predicted path. This is even more critical when multiple flybys are present. A close encounter has an amplifying effect on the numerical error. Therefore, the propagation must be as accurate as possible for the trajectory solution to be trusted. With examples like the Cassini mission, which has completed 120 flybys of Titan, 22 of Enceladus, and others of Iapetus, Rhea and Dione, ensuring the accuracy of the propagation becomes a complicated problem. Figure 1.2 depicts the orbit of Cassini during its first 18 months of operation, and shows the complexity of the trajectory.

Every effort toward more efficient and accurate propagation methods benefits almost every application in celestial mechanics. Trajectory optimization problems are interesting examples. In practice, optimization algorithms will be given a cost function to minimize. This cost function typically includes orbit propagations. Since the function will be evaluated thousands of times, the runtime needs to be as short as possible. In addition, numerical errors will make the optimizer converge to erroneous solutions, or it simply will not converge to any solution. Conversely, the relative motion between spacecraft is sensitive to the differential effect of orbital perturbations. Nu-

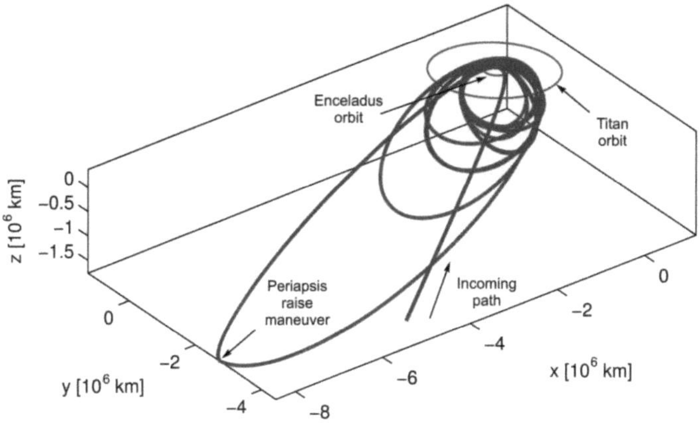

Fig. 1.2: The orbit of Cassini in the ICRF/J2000 frame centered at Saturn (06/20/2004–12/01/2005).

merical models are required for high-precision formation flying applications, which again involve numerical propagations.

1.5 The aim of the present book

The physics that govern motion in space have not changed since the time Kepler formulated his celebrated laws. It is only our perception, our models, that have evolved. And depending on how a physical problem is described using a mathematical model, different conclusions may be derived. This book focuses on the theory of regularization, which was born in an attempt to eliminate the singularities in the equations of orbital motion. This theory can be regarded as a collection of mathematical and dynamical contrivances that provide a more adequate description of the dynamics. Following Marcel Proust's quote

> "The only true voyage of discovery, the only fountain of Eternal Youth, would be not to visit strange lands but to possess other eyes [...],"

this book takes regularized formulations as "new eyes" and approaches three of the main challenges in modern astrodynamics: high-performance orbit propagation, formation flying, and low-thrust mission design. The use of an unconventional formulation of the dynamics might be advantageous in certain scenarios, presented in the following chapters.

The main application of regularization for the last 60 years has been the development of improved schemes for numerical integration. By recovering the basics of the theory, this book shows how regularization can be applied systematically to solve other problems of practical interest, as well as how to exploit all the advantages of these methods for conducting numerical simulations.

The first part of the book is of theoretical nature, and focuses on regularization itself. Chapter 2 introduces the theory, justifying why it is worth seeking alternative representations of the dynamics. The different methods and techniques that lead to regularization are presented, together with their advantages. Chapter 3 is focused on a specific formulation, the Kustaanheimo–Stiefel regularization (KS for short). In particular, the concept of topological stability is developed resulting in an alternative Lyapunov indicator for characterizing chaotic regimes. The next chapter, Chap. 4, analyzes the Dromo formulation as a representative example and discusses some practical aspects. In Chap. 5 we will show how regularization can be used to derive formulations to solve specific problems. The last chapter of the first part, Chap. 6, compares the numerical performance of a collection of regularized formulations.

The second part of the book includes the applications of regularization to spacecraft relative motion. Chapter 7 presents the theory of asynchronous relative motion, an alternative way to formulate the relative dynamics in space. In Chap. 8, this same problem is solved using different regularized formulations, each presenting particular advantages. The third part of the book applies regularization to low-thrust mission design. The search for more convenient descriptions of the problem of low-thrust transfers led to the discovery of a new family of spiral trajectories, called generalized logarithmic spirals. The definition and main properties of the orbits can be found in Chap. 9. In Chap. 10, the spiral Lambert problem is solved using this new family of spirals, and Chap. 11 presents a complete strategy for designing three-dimensional low-thrust transfers. These results can be generalized to define other families of new orbits, with different physical interpretations ranging from solar sailing to representing the Schwarzschild geodesics (Chap. 12).

Part I: **Regularization**

"Read Euler, read Euler, he is the master of us all."

– Pierre Simon Laplace

2 Theoretical aspects of regularization

Phillip Herbert Cowell (1870–1949) lived during an exciting period of dynamical astronomy, and was one of the various luminaries who contributed to the theory of the motion of the Moon. As second chief assistant at the Royal Observatory at Greenwich, he became intrigued by the discrepancies between the observed trajectory of the Moon and the predictions in Hansen's tables. He corrected some coefficients of the periodic terms predicted by Hansen, and introduced new terms to account for long-period dynamics. With powerful perturbation theories at hand, he then focused his attention on the imminent return of Halley's comet (it approached the Earth in 1910). Andrew C. Crommelin noticed in 1906 a difference of almost three years in the predictions of the date of the approaching return of the comet and, being Cowell's colleague at the Greenwich Observatory, Crommelin suggested he recalculate the orbit.

In 1892, 1904, and 1905 three additional moons were discovered orbiting Jupiter. Astronomers were struck by these findings, because for almost 400 years the Galilean moons were thought to be the only Jovian satellites. But on the night of February 28, 1908, an even more surprising discovery was made. A new moon was observed with a period that seemed tens of times larger than that of the Galilean moons. A direct orbit with this characteristic could not be stable according to Laplace's classical theory, so Crommelin made a revolutionary suggestion: the orbit might be retrograde.[1] He then teamed up with Cowell to explore the rare motion of this object. The perturbation from the Sun's attraction was so large (between six and ten percent of the attraction from Jupiter), that they did not even consider relying on the analytical theories that worked so well for the Moon. Instead, they decided to integrate the equations of motion in Cartesian coordinates by numerical quadrature. The method presented by Cowell and Crommelin (1908) was later referred to as Cowell's method, and it is possibly the most common method for propagating orbits. It reduces to integrating the system of differential equations

$$\frac{\mathrm{d}^2\mathbf{r}}{\mathrm{d}t^2} + \frac{\mu}{r^3}\mathbf{r} = \mathbf{a}_p \tag{2.1}$$

in which $\mathbf{r} = [x, y, z]^\top$ is the position vector of the particle with respect to the central body in an inertial frame, μ is the gravitational parameter, and $\mathbf{a}_p$ are external perturbations.

Cowell himself was well aware of the power of this method and he soon applied it to the propagation of the orbit of Halley's comet (Cowell and Crommelin, 1910). It is interesting to note that, in the mean time, Tullio Levi-Civita was looking for alternatives

---

1 In 1975 this moon was called Pasiphae, and its orbit is indeed retrograde. Its orbital period of 764.1 days is significantly longer than the period of Io (1.8 days), Europa (3.5 days), Ganymede (7.1 days), and Callisto (16.7 days). The moon was discovered at the Royal Greenwich Observatory by astronomer Philibert J. Melotte, a colleague of Cowell and Crommelin.

https://doi.org/10.1515/9783110559125-002

to the use of Cartesian coordinates in the problem of three bodies (Levi-Civita, 1903, 1904, 1920). In the N-body problem collisions may occur, meaning that the denominator in Eq. (2.1) will vanish. Thus, the equations of motion become *singular* and fail to reproduce the dynamics. An entire new branch of celestial mechanics was born from the pioneering studies by Levi-Civita and Peter Hansen about the three-body problem: the theory of regularization. In Sect. 2.4 we will discuss early contributions by Laplace.

Regularization is a collection of mathematical and dynamical transformations that seek a more convenient formulation of the equations of motion. In the second half of the 19th century and the first half of the 20th century regularization was mostly applied in the context of the three-body problem. With the rise of computers and the Space Age in the 1960s and 1970s, the numerical integration of the equations of orbital motion received renewed interest. Regularization was no longer a mere theoretical artifact, for scientists like Joachim Baumgarte, Eduard Stiefel, or André Deprit proved that regularized schemes were better suited for numerical integration. Different formulations were published in those years with operational applications.

This chapter is devoted to presenting the foundations of regularization, together with more detailed historical notions. In particular, the main techniques used to regularize and stabilize the equations of motion will be analyzed. Section 2.1 deals with the question of whether using sets of variables that are different from the Cartesian ones is worthwhile or not. The Sundman time transformation, one of the best known regularizing transformations, is discussed in Sect. 2.2. Sections 2.3–2.5 present different techniques for improving the overall performance of the integration, including the stabilization of the equations of motion by embedding first integrals, or the use of orbital elements. Canonical transformations are briefly discussed in Sect. 2.6, and the concept of gauge freedom in celestial mechanics, developed by Efroimsky and Goldreich (2003), is reviewed in Sect. 2.7.

2.1 Why bother?

Cowell's method presents a series of issues related to its numerical behavior and the form of the equations themselves. A singularity arises from the form of the specific potential of a particle orbiting a central mass with gravitational parameter $\mu = Gm$,

$$\mathcal{V}(r) = -\frac{\mu}{r}$$

Indeed, at the origin of the potential well ($r \to 0$) the function $\mathcal{V}(r)$ takes infinite values. This is not only a theoretical singularity, because during close encounters and deep flybys the relative separation decreases rapidly.

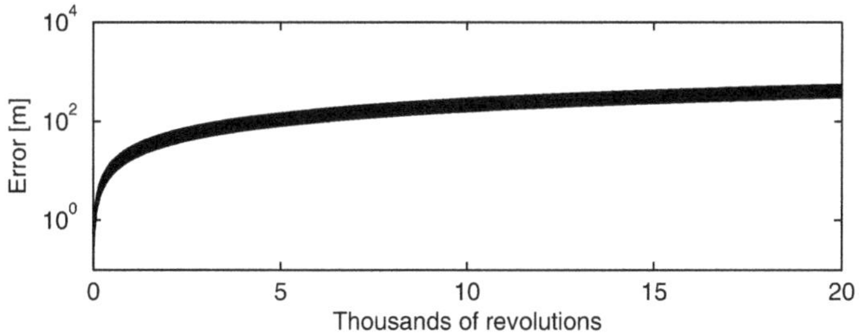

Fig. 2.1: Evolution of the propagation error of a Keplerian orbit. The orbit is integrated with a variable step-size Störmer–Cowell integrator of order nine with a tolerance of 10^{-14}.

This singularity is even more critical in the context of the N-body problem, where the potential takes the form

$$\mathcal{V}(r) = -G \sum_{i<j} \frac{m_i m_j}{||\mathbf{r}_i - \mathbf{r}_j||}$$

and becomes singular when any pair of particles (i, j) collide. Collisions in N-body simulations of stellar systems are usual, and the singularity needs to be overcome. We refer to Aarseth (2003, chap. 5) for a more detailed discussion about the need for regularization in N-body systems.

The singularity also affects the velocity. The local circular velocity of a particle orbiting a central mass is simply

$$v = \sqrt{\frac{\mu}{r}}$$

It increases as the orbit reduces its size. Thus, in the limit case when the orbit collapses to the origin ($r \to 0$) the corresponding orbital velocity becomes infinite. This phenomenon can also be understood in view of the conservation of the angular momentum, $h = rv$ in a circular orbit. As the radial distance goes to zero, the velocity grows rapidly to compensate the angular momentum.

Another issue related to the form of Eq. (2.1) is its behavior when integrated numerically. If one solves these differential equations in the Keplerian case ($\mathbf{a}_p = 0$), which is the simplest case possible, the evolution of the error is that shown in Fig. 2.1. It is computed as the difference between the solution propagated numerically and the analytic solution. Even when integrated with the Störmer–Cowell integrator and a fine tolerance (10^{-14}), the error grows exponentially in time, which yields a separation of about 500 m after 20,000 revolutions. The example corresponds to a geocentric orbit with $e = 0.3$ and $a = 8000$ km.

The exponential growth of the error in Cowell's method can be understood by studying the relative dynamics between two neighboring trajectories. Let $\mathbf{r}^*$ define a

Keplerian orbit, initially separated by a distance $\delta\mathbf{r}_0$ from a reference solution $\mathbf{r}$. The relative motion is governed by the equation

$$\ddot{\mathbf{r}}^* - \ddot{\mathbf{r}} = -\frac{\mu}{r^{*3}}\mathbf{r}^* + \frac{\mu}{r^3}\mathbf{r}$$

Under the assumption that the separation is small, i.e., $||\delta\mathbf{r}_0|| \ll ||\mathbf{r}_0||$, the dynamics can be linearized about the reference trajectory. Retaining only the first-order terms the problem reduces to

$$\ddot{\mathbf{r}}^* - \ddot{\mathbf{r}} \approx -\frac{\mu}{r^5}\left[r^2\,\mathsf{I} - 3(\mathbf{r}\otimes\mathbf{r})\right]\circ\delta\mathbf{r} = \mathsf{A}(\mathbf{r})\,\delta\mathbf{r}$$

Here, I is the second-order unit tensor (represented by the identity matrix), $\otimes$ is the dyadic product, and $\circ$ denotes contraction.[2] This second-order dynamical system can be reduced to a first-order system when introducing the state vector $\delta\mathbf{x}$ that contains both the relative position and velocity vectors, $\delta\mathbf{r}$ and $\delta\mathbf{v}$:

$$\frac{\mathrm{d}(\delta\mathbf{x})}{\mathrm{d}t} = \mathsf{L}(\mathbf{x})\,\delta\mathbf{x}$$

The linear operator $\mathsf{L}(\mathbf{x})$ is

$$\mathsf{L}(\mathbf{x}) = \begin{bmatrix} \mathbf{0} & \mathsf{I} \\ \mathsf{A}(\mathbf{x}) & \mathbf{0} \end{bmatrix}$$

and models the linear dynamics of the system. The spectrum of the linear operator is

$$\lambda_1 = -\sqrt{\frac{2\mu}{r^3}}, \quad \lambda_2 = +\sqrt{\frac{2\mu}{r^3}}, \quad \lambda_{3,4} = -\lambda_{5,6} = \mathrm{i}\sqrt{\frac{\mu}{r^3}} \tag{2.2}$$

Four eigenvalues are purely imaginary and yield oscillatory modes. There are two eigenvalues with nonzero real parts, λ_1 and λ_2; the real part of λ_1 is negative, whereas the real part of λ_2 is positive. This yields an exponential growth of the separation, according to the law $\mathrm{e}^{\lambda_2 t}$. The system is unstable in the sense of Lyapunov. See Bond (1982) for a connection with the stability of the numerical integration methods. The stable and unstable manifolds associated to the resulting saddle configuration relate

2 Under this notation it is:

$$(\mathbf{a}\otimes\mathbf{b})\circ\mathbf{c} = \mathbf{a}\,(\mathbf{b}\cdot\mathbf{c})$$

The dyadic product of two rank-one tensors (vectors) results in a rank-two tensor (matrix),

$$\mathbf{a}\otimes\mathbf{b} = \begin{bmatrix} a_1b_1 & \cdots & a_1b_n \\ \vdots & \ddots & \vdots \\ a_nb_1 & \cdots & a_nb_n \end{bmatrix}$$

It is often written in matrix form as $\mathbf{a}\,\mathbf{b}^\top$, with $\mathbf{a}$ and $\mathbf{b}$ column vectors.

to the right eigenvectors $\mathbf{v}_1$ and $\mathbf{v}_2$, associated to λ_1 and λ_2, respectively. If the components of the position vector in an inertial frame $\mathcal{I}$ are $[x, y, z]$, the eigenvectors $\mathbf{v}_1$ and $\mathbf{v}_2$ read

$$\mathbf{v}_1 = [-kx, -ky, -kz, x, y, z]^\top$$
$$\mathbf{v}_2 = [+kx, +ky, +kz, x, y, z]^\top$$

with $k = r^{3/2}/\sqrt{2\mu}$.

The referenced instability is closely related to the choice of the independent variable. Consider two circular orbits of radii r and $r + \delta r$. If the relative separation is initially small, $\delta r \ll r$, the evolution of the angular separation between the particles $\delta\vartheta$ can be visualized thanks to a simple series expansion:

$$\delta\vartheta = \left[\sqrt{\frac{\mu}{(r+\delta r)^3}} - \sqrt{\frac{\mu}{r^3}}\right] t = -\frac{3}{2}\sqrt{\frac{\mu}{r^5}}\, t\, \delta r + O(\delta r^2) \tag{2.3}$$

It grows secularly in time (the motion is Lyapunov unstable). However, the separation between the orbits, δr, is constant, and remains small. That is, the orbital motion is Poincaré (or orbitally/structurally) stable. Let Σ be a Poincaré section that is transversal to the flow and let $\mathbf{p}_1, \mathbf{p}_2, \ldots$ be the consecutive intersections of the orbit with Σ. The Poincaré map $\mathscr{P}: \Sigma \mapsto \Sigma$ is such that

$$\mathbf{p}_{n+1} = \mathscr{P}(\mathbf{p}_n)$$

and for a periodic orbit $\mathbf{p}_i = \mathbf{p}_j \equiv \mathbf{p}$ or $\mathscr{P}^n(\mathbf{p}) = \mathbf{p}$. When the separation between the orbits is measured at a given time t it is defined as $||\mathbf{r}^*(t) - \mathbf{r}(t)||$, and grows exponentially. However, when the separation is measured at a value of the true anomaly (which is equivalent to analyzing the separation on a Poincaré section) it is $||\mathbf{p}^* - \mathbf{p}||$, and remains constant. This suggests that the independent variable plays a critical role when formulating orbital motion.

Szebehely (1967, pp. 233–234) presents an elegant and unified treatment of the Lyapunov and structural stabilities by introducing the concepts of *isochronous* and *normal* correspondence. The former yields a synchronization in physical time and therefore Lyapunov stability, whereas the latter relates to the synchronism in an angular variable and involves structural stability.

2.2 The Sundman transformation

The three-body problem is the simplest among the N-body systems, and collisions appear naturally. The Finnish astronomer Karl F. Sundman worked extensively on this problem and was well aware of the importance of regularizing the equations of motion.

He introduced the time transformation[3]

$$dt = r\,ds \tag{2.4}$$

that was later named after him. The fictitious time s replaces the physical time t as the independent variable, and behaves as an angle. For historical reasons we shall mention that Levi-Civita (1906) and Bohlin (1911) also considered this transformation when regularizing the restricted three-body problem. The latter worked directly with the eccentric anomaly instead of considering it a fictitious time.

The derivatives with respect to physical time transform into

$$\frac{d}{dt} = \frac{1}{r}\frac{d}{ds} \quad \text{and} \quad \frac{d^2}{dt^2} = \frac{1}{r^3}\left(r\frac{d^2}{ds^2} - \frac{dr}{ds}\frac{d}{ds}\right)$$

The typical notation for the derivatives with respect to fictitious time is the prime $\square'$, whereas dots $\dot{\square}$ refer to derivatives with respect to physical time. If Eq. (2.1) is rewritten in terms of the fictitious time it transforms into

$$\frac{d^2\mathbf{r}}{ds^2} - \frac{1}{r}\frac{dr}{ds}\frac{d\mathbf{r}}{ds} + \frac{\mu}{r}\mathbf{r} = r^2\mathbf{a}_p \tag{2.5}$$

This equation still presents problems for $r \to 0$ although the order of the singularity has been reduced, changing the denominator from r^3 to r. Additional transformations are required to achieve the complete regularization of the system.

The Sundman transformation has a beneficial effect in the numerical integration process. It yields what Jörg Waldvogel calls “the slow motion movie” (Waldvogel, 2007); if the orbital motion is discretized using a fixed step in physical time, most of the points will be gathered around the apoapsis of the ellipse. Unfortunately, at periapsis, where the dynamics are faster and require a finer step size, there will be fewer points. This is shown by Kepler’s second law: given a differential dt the particle sweeps a differential of area which is constant,

$$dA = \frac{1}{2}r^2\,d\vartheta \implies d\vartheta \propto \frac{1}{r^2}$$

and $d\vartheta$ is angular separation in true anomaly between consecutive points. When the particle is closer to the origin (periapsis) the angular separation grows. When r is

3 We refer to the memoir that the mathematician Magnus G. Mittag-Leffler invited Sundman to write on the subject in 1913 (Sundman, 1913). The results presented in that paper were a collection of previous results obtained by the author and published in different works, mainly in *Recherches sur le problème des trois corps* and *Nouvelles recherches sur le problème des trois corps*. These two papers appeared in *Acta Societatis Scientiarum Fennicae*, volumes 34 (1907) and 35 (1909), respectively. Although in this chapter we only recover the time transformation, we must emphasize that the most memorable result presented in Sundman’s *Mémoire* is the conclusion that all solutions to the regularized three-body problem can be represented by convergent series. The time transformation from Sundman (1913, eq. 71) yields a new form of the canonical system; Levi-Civita (1920, §2) referred to that transformation as the Darboux–Sundman transformation.

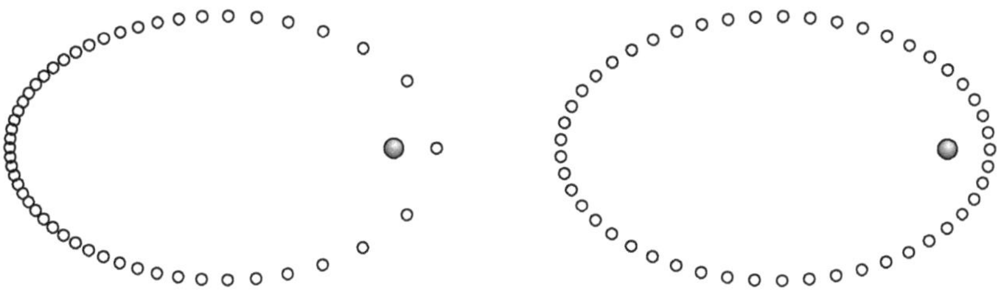

(a) Fixed step in t (b) Fixed step in s

Fig. 2.2: Discretization of a highly elliptical orbit ($e = 0.8$) using the physical and the fictitious times and the same number of points.

large (apoapsis) the points are close to each other, and yield the distribution shown in Fig. 2.2a. Conversely, differentiating Kepler's equation yields

$$dt = \alpha r \, dE \tag{2.6}$$

with E the eccentric anomaly and α a constant depending on the semimajor axis and the mean motion. Identifying Eqs. (2.4) and (2.6) shows that the fictitious time s is in fact equivalent to the eccentric anomaly E, except for a constant factor affecting its time evolution. The discretization of the orbit with fixed steps in fictitious time (equivalent to E) is shown in Fig. 2.2b.

The examples in Fig. 2.2 have been constructed with the same number of points, meaning that the integrator will take the same number of steps. However, using the discretization in physical time (Fig. 2.2a) yields very large steps during periapsis passage. This introduces important numerical errors because the integration fails to capture the rapid changes in the states. On the contrary, a discretization in fictitious time (Fig. 2.2b) leads to a finer representation of the dynamics around periapsis, and the result will be more accurate. At apoapsis, where the dynamics are much slower, the wider step size of Fig. 2.2b reduces the number of unnecessary steps, speeding up the integration.

When propagating an orbit from t_0 to t_f, the final value of the fictitious time, which makes $t(s_f) = t_f$, is usually not known. In order to stop the integration at the correct time t_f, the integrator must incorporate a root-finding algorithm to find the correct stopping value, by interpolating the integration of the Sundman transformation until $t = t_f$.

2.2.1 Generalized Sundman transformation

The time transformation in Eq. (2.4) is often generalized to

$$dt = \alpha r^n \, ds \tag{2.7}$$

where α is an arbitrary constant and n defines the order of the transformation. Different values of n have been tested in the past century. Before diving into the possible transformations we shall first borrow a theorem from Stiefel and Scheifele (1971, p. 78):

Theorem 1. *(Stiefel and Scheifele, 1971, Thm. 3) Given the analytic step-size adaption in Eq.* (2.7), *a full regularization of the equations of motion* (2.1) *can only be achieved for* $n < 3/2$.

The original transformation by Sundman corresponds to $n = 1$ and, according to this theorem, the complete regularization of the equations of motion is possible. If the constant parameter α takes the value $\alpha = (a/\mu)^{1/2}$, the fictitious time becomes the eccentric anomaly. Many regularization methods are based on the transformation $n = 1$. Some examples are the formulations by Levi-Civita (1904, 1906, 1920) and later extended by Kustaanheimo and Stiefel (1965), or alternative methods by Sperling (1961), Burdet (1967), Janin (1974), Bond and Gottlieb (1989), Bond (1990), Roa and Peláez (2015d), and Baù et al. (2015) among others. Interestingly, Pellegrini et al. (2014) proved that the exponent $n = 1$ might yield more stable series solutions to orbital motion when compared to other values of n. The reason is that, being the fictitious time equivalent to the eccentric anomaly, the orbit is discretized in uniformly distributed points with a homogeneous geometry.

The case $n = 3/2$ was explored by Nacozy (1977), who coined the term *intermediate anomaly* for referring to the fictitious time in this case. The constant factor α reads $\alpha = \mu^{-1/2}(1 + e^2)^{1/2}$. The intermediate anomaly relates to the true anomaly by virtue of

$$s = ek^2\,\mathrm{F}(\vartheta/2, k)$$

with $k^2 = 2e/(1 + e)$ and $\mathrm{F}(\vartheta/2, k)$ the incomplete elliptic integral of the first kind. Nacozy's discovery was motivated by the concept of the partial anomalies proposed by Hansen (1853), a method that he had already applied to determine the orbit of comet Encke (Nacozy, 1969). In practice, the value of n can be tailored to optimize the performance of the Sundman transformation when applied in different scenarios (Velez and Hilinski, 1978).

Hansen (1853) derived a new general perturbation method for propagating cometary orbits. As the director of the Seeberg Observatory he was concerned about the accuracy of the computations. The orbits of comets are typically elliptical and cross the circular orbits of the planets. Consequently, the planetary perturbations upon the comet vary in different time scales and involve many frequencies. This phenomenon complicates the development of series solutions because they depend heavily on the natural frequencies. In order to account for the different time scales, he introduced the *inferior* and *superior* partial anomalies related to the quadrants of the ellipse. In Hansen (1853, §3, p. 57) a detailed discussion of the possible situations regarding the in-orbit position of the comet can be found. Later, Glydén (1870) improved the convergence of the series solution by introducing elliptic functions. A review of the methods can be found in a dedicated publication by Baillaud (1876).

In the case $n = 2$ and $\alpha = 1/h$, the fictitious time evolves like the true anomaly. There are a number of regularized theories based on this time transformation, for example those from Burdet (1969); Vitins (1978); Palacios and Calvo (1996), Peláez et al. (2007), and Urrutxua et al. (2015b). It is interesting to note that Burdet (1969) introduced the terms *central* and *focal method*, referring to the regularization with $n = 1$ and $n = 2$, respectively. The names come from the fact that the eccentric anomaly rotates around the center of the conic section, whereas the true anomaly is referred to the focus. Vitins (1978) and Peláez et al. (2007) exploited the gyroscopic representation of the orbital plane, later taking advantage of the theory of ideal frames developed by Hansen (1857). We refer to Jochim (2012) for a translation of the most important parts of the original manuscript, as well as for a detailed analysis of the fundamental results. It is worth noting that this transformation was already applied by Laplace (1799, book II, chap. 2, p. 150) when deriving his theory of orbital motion.

The time transformation in Eq. (2.7) can be written in its most general form as

$$\mathrm{d}t = g(r)\,\mathrm{d}s \tag{2.8}$$

where $g(r) \neq 0$ is an arbitrary function. Many different forms of the function $g(r)$ have been explored, specially for regularizing N-body systems (Sect. 2.2.2). For example, Cid et al. (1983) addressed the dynamics of particles moving in a central force field governed by the potential

$$\mathcal{V}(r) = \sum_i \frac{a_i}{r^i}$$

and introduced a time transformation with

$$g_n(r) = \frac{r^{n/2}}{\sqrt{r^{n-2} + b_1 r^{n-3} + \ldots + b_{n-2}}}$$

in order to achieve the full regularization of the Hamiltonian. Ferrándiz et al. (1987) unified the transformation $g_3(r)$ and the one proposed by Belen'kii (1981), $g(r) = r^{3/2}(1 + br)^{-1/2}$, by introducing the time transformation:

$$\mathrm{d}t = \frac{r^{3/2}}{\sqrt{b_0 + b_1 r}}\,\mathrm{d}s$$

A similar equation played a key role in the regularization and linearization (in Stiefel and Scheifele's sense) of the radial intermediary introduced by Cid et al. (1986).

The last section of the celebrated work by Kustaanheimo and Stiefel (1965) has not received much attention in the past; this section is devoted to the formal treatment of nonconservative perturbations in KS formalism. They suggested the use of a generic time transformation, in an attempt to find an adequate form of $g(r)$ so that the angular frequency of the harmonic oscillator is preserved.

In their book, Stiefel and Scheifele (1971, §19) introduced a new time variable that relates to the *total* energy of the particle. Thus, disturbing potentials are introduced in the time transformation improving the stability of the integration. Baumgarte and Stiefel (1974) published a dedicated analysis of other possible transformations and discussed their advantages.

Satō (1998) published an ingenuous regularization of collisional orbits by representing Keplerian motion in the projective space. Thanks to this construction he arrived at the concept of the projective anomaly, a new parameterization of orbital motion that successfully regularizes the problem when $r \to 0$. By connecting the projective anomaly with the eccentric anomaly, Satō introduced the generalized anomaly. This variable includes not only the eccentric and projective anomaly as special cases, but also the true anomaly. Kopeikin et al. (2011, §1.3.5 and §1.3.6) reviewed this concept in detail, and took advantage of its properties for exploring the gauge freedom of orbital motion (Sect. 2.7).

2.2.2 Time transformations in *N*-body systems

It was Sundman (1913, §11) himself that was one of the first to realize that the transformation in Eq. (2.4) does not work well when there are more than two bodies. For an N-body system r is not unique, because there are $N(N-1)/2$ relative distances r_{ij} between the particles. He focused on the three-body problem and defined an alternative transformation in terms of the regularizing function

$$g(r_{ij}) = \left(1 - \mathrm{e}^{-r_{12}/\ell}\right)\left(1 - \mathrm{e}^{-r_{13}/\ell}\right)\left(1 - \mathrm{e}^{-r_{23}/\ell}\right)$$

Here ℓ is a constant depending on the physical properties of the system.

Finding an adequate transformation of the independent variable is a critical part of the regularizing strategy. As shown by Sundman, when there are more than two bodies it is more convenient to define a transformation involving the relative states of all the bodies. Aarseth and Zare (1974) arrived at a remarkable regularization of the three-body problem by introducing the KS transformation in order to model the dynamics of a single binary. Binary collisions are completely regularized, and the integration of triple encounters is well behaved numerically. They found a regular form of the Hamiltonian in KS variables relying on the time transformation

$$\mathrm{d}t = \alpha r_{ik} r_{jk}\,\mathrm{d}s \tag{2.9}$$

Under this notation the regularized double encounters are $r_{ik} \to 0$ and $r_{jk} \to 0$. Later that year, Heggie (1974) advanced on Aarseth and Zare's method by generalizing the regularization to any pair of particles, and then extending the formulation to generic

N-body systems. For that he considered the new time transformation

$$dt = \prod_{i<j} r_{ij}\, ds$$

that involves the relative separation between all the particles. In the particular case $N = 3$ he explored an alternative transformation

$$dt = \frac{r_{12} r_{13} r_{23}}{(r_{12} + r_{13} + r_{23})^{3/2}}\, ds$$

and found slight improvements in the accuracy and numerical performance. The special role of the exponent $3/2$, which later led to the intermediate anomaly (Nacozy, 1977), will be discussed in more detail in Sect. 2.3. It is again remarkable that a transformation similar to the one in Eq. (2.9) can be found in a note by Levi-Civita.[4]

An important breakthrough in the definition of time transformations was the introduction of conserved quantities such as the energy or the Hamiltonian in the regularizing function g. Zare and Szebehely (1975) approached the problem from a rigorous perspective looking for values of g that make

$$\sum_i p_i q_i = \int g(2\mathcal{T} - \mathcal{V})\, ds$$

integrable, where q_i and p_i are the coordinates and the momenta, and $\mathcal{T}$ is the kinetic energy. Noting that $2\mathcal{T} - \mathcal{V} = \mathcal{H} + \mathcal{T} = 2\mathcal{H} + \mathcal{V} = (3\mathcal{H} + \mathcal{L})/2$, with $\mathcal{L}$ the Lagrangian, the previous expression transforms into

$$\begin{aligned}
\sum_i p_i q_i &= \int g(\mathcal{H} + \mathcal{T})\, ds &&= \mathcal{H}(t - t_0) + \int g\mathcal{T}\, ds \\
\sum_i p_i q_i &= \int g(2\mathcal{H} + \mathcal{V})\, ds &&= 2\,\mathcal{H}(t - t_0) + \int g\mathcal{V}\, ds \\
\sum_i p_i q_i &= \int g(3\mathcal{H} + \mathcal{L})/2\, ds &&= \frac{3}{2}\mathcal{H}(t - t_0) + \frac{1}{2}\int g\mathcal{L}\, ds
\end{aligned}$$

The form of the last term suggests the use of the regularizing functions $g = 1/\mathcal{T}$, $g = 1/\mathcal{V}$, or $g = 1/\mathcal{L}$ in order to arrive at the trivial integrals $\int ds$. Recall that the use of the potential function in the regularizing process had already been explored by Levi-Civita (1903, p. 66) and Stiefel and Scheifele (1971, §19). The transformation $g = 1/\mathcal{L}$ was introduced into Heggie's method by Mikkola (1985). Thanks to this modified transformation, the referenced method now allows to handle simultaneous binary collisions. In addition, the Hamiltonian adopts a more compact form that makes its differentiation with respect to canonical coordinates and momenta simpler. The re-

4 Levi-Civita, T. “Sul problema piano dei tre corpi. Caso limite in cui una delle masse è infinitesima”, *Rendiconti del Lincei*, 1915, 24, p. 559.

sulting method is easier to implement than the original version by Heggie. The transformation $g = 1/\mathscr{L}$ has a remarkable implication: the fictitious time is defined by

$$s = \int_{t_0}^{t} \mathscr{L}\,dt$$

and it is equivalent to the action integral S. Thus, the action is used as the new independent variable. The principle of least action states that $\delta S = 0$, and therefore $\delta s = 0$. The path followed by the particle is the one that minimizes the fictitious time between the departure and arrival points.

2.3 Stabilization of the equations of motion

Baumgarte (1972b) proved in a seminal work that embedding conserved quantities in the system of differential equations to be integrated improves the stability of the integration process significantly. Stability in this case refers to how well the integrals of motion or dynamical constraints are conserved. This technique comes from noting that the numerical integration scheme has in general no information about the conservation laws. Thus, if they are introduced in the system they behave as control terms and reduce the error-growth rate. This technique proves useful not only in the Keplerian case, but also in the presence of conservative perturbations where the total energy is conserved (Baumgarte, 1972a).

Stabilization can be understood from two different perspectives, as Velez (1974) observed: the equations can be stabilized *numerically* and *analytically*. The techniques aiming for one or the other type of stabilization are similar, and reduce to:
1. Transforming the independent variable by introducing a fictitious time.
2. Linearizing the equation of motion in order to transform Eq. (2.1) into a linear oscillator.
3. Embedding the energy equation into the system of equations.

An alternative to steps 2 and 3 is to integrate sets of elements instead of coordinates (Sect. 2.5 is devoted to the use of elements in orbit propagation).

In the 1970s, Dr. Velez's group did a lot of work on numerical integration of orbits at Goddard Space Flight Center, while developing the Goddard Trajectory Determination System (GTDS, see Wagner and Velez, 1972). For example, Velez (1967) presented an innovative predictor-corrector algorithm to integrate multirevolution orbits more efficiently. Velez and Hilinski (1978) analyzed the impact of time transformations on the truncation errors when integrating Cowell's method through a number of numerical experiments. The question about the stability of the numerical integration was further developed by Velez (1975).

The system of equations (2.1) is unstable in the sense of Lyapunov. Under an adequate transformation these equations may be transformed to a harmonic oscillator, which is stable (Sect. 2.4 has a detailed analysis of the linearization of the equations). The Lyapunov instability of the system mainly affects the propagation of the truncation error and is reduced when introducing a fictitious time. Recall the example in Eq. (2.3), which proved that the angular separation between two neighboring particles grows in time proportionally to $-(3/2)r^{-5/2}\mu^{1/2}$. Considering now a Sundman transformation of the form $\mathrm{d}t = r\,\mathrm{d}s$, the separation between the particles is governed by the equation

$$\delta\vartheta_s = \left[(r+\delta r)\sqrt{\frac{\mu}{(r+\delta r)^3}} - r\sqrt{\frac{\mu}{r^3}}\right] s = -\frac{1}{2}\sqrt{\frac{\mu}{r^3}}\, s\,\delta r + \mathcal{O}(\delta r^2)$$

In a normalized example ($\mu = 1$ and $r = 1$) it is simply $\delta\vartheta_s = -(s/2)\,\delta r$, whereas the separation given in Eq. (2.3) evolves as $\delta\vartheta_t = -(3t/2)\,\delta r$. The fact that $\delta\vartheta_s/\delta\vartheta_t = 1/3$ is what motivated Baumgarte's statement: the introduction of a fictitious time $\mathrm{d}t = r\,\mathrm{d}s$ reduces the Lyapunov instability of orbital motion by a factor three (Baumgarte, 1972a). A partial analytic stabilization is achieved. Clearly, when the true anomaly is the independent variable $\delta\vartheta$ does not change, meaning that for $\mathrm{d}t = r^2\,\mathrm{d}s$ the system is completely stabilized in Lyapunov's sense.

Although orbital motion referred to the fictitious time is stabilized, the time transformation needs to be integrated to provide the physical time:

$$t - t_0 = \int g(s)\,\mathrm{d}s$$

and reducing the Lyapunov instability might not improve the numerical conditioning of this integral. The error in the propagation of the time variable is coupled with the total error of the trajectory, and it is difficult to assess its impact in the most general case. An interesting result from Velez (1974) in the case of an orbit perturbed by the J_2 gravity terms of the central body is that the error growth rate is minimum for $n \sim 1.5$, although the system is not strictly stabilized.

The stabilizing effect of embedding first integrals in the equations of motion can be seen in the following example. Let us recover the equations of motion in terms of the fictitious time – Eq. (2.5) –,

$$\mathbf{r}'' - \frac{r'}{r}\mathbf{r}' + \frac{\mu}{r}\mathbf{r} = r^2\mathbf{a}_p \tag{2.10}$$

The equation of the Keplerian energy can be understood as a control term; multiplying this integral of motion by **r** renders

$$\left[\frac{(\mathbf{r}'\cdot\mathbf{r}')}{2r^2} - \frac{\mu}{r} - \mathcal{E}_k\right]\mathbf{r} = 0$$

This control term appears recursively in the literature and it is sometimes referred to as the Poincaré term (Baumgarte and Stiefel, 1974; Janin, 1974). Adding the previous two equations yields the controlled equations of motion

$$\mathbf{r}'' - \frac{r'}{r}\mathbf{r}' + \left[\frac{(\mathbf{r}'\cdot\mathbf{r}')}{2r^2} - \mathcal{E}_k\right]\mathbf{r} = r^2\mathbf{a}_p \tag{2.11}$$

Figure 2.3 shows the difference in accuracy between integrating Eq. (2.10) and Eq. (2.11), which includes the control term. In order to isolate the effect of the control term, the nominal orbit is considered circular. The particle orbits around the Earth and the orbital radius is 8000 km. The equations of motion are integrated using a RKF 5(4) scheme, with absolute and relative tolerances $\varepsilon = 10^{-11}$. The error growth rate of the controlled equations is smaller than the uncontrolled case. As a result, after for example 5000 revolutions, the error in position is reduced by about three orders of magnitude, from about 500 m down to 0.5 m.

Using the energy equation as a stabilizing mechanism is particularly powerful when dealing with conservative perturbations. The perturbing acceleration derives from a disturbing potential

$$\mathbf{a}_p = -\nabla \mathcal{V}_p$$

The total energy of the system is constant and reads

$$\mathcal{E} = \mathcal{E}_k + \mathcal{V}_p$$

Thus, the control term can be referred to the total energy,

$$\left[\frac{(\mathbf{r}' \cdot \mathbf{r}')}{2r^2} - \frac{\mu}{r} - \mathcal{E} + \mathcal{V}_p\right] \mathbf{r} = 0$$

so the controlled system adopts the form

$$\mathbf{r}'' - \frac{r'}{r}\mathbf{r}' + \left[\frac{(\mathbf{r}' \cdot \mathbf{r}')}{2r^2} - \mathcal{E} + \mathcal{V}_p\right] \mathbf{r} = r^2 \mathbf{a}_p$$

Thanks to using the total energy instead of the Keplerian energy, the constant of motion $\mathcal{E}$ is embedded in the system. This technique is widely applied in regularization theories. It is advantageous not only from the numerical perspective, as it might also simplify analytic formulations. In Sect. 2.2 a number of time transformations involving the potential or the energy were discussed. For instance, using the eccentric anomaly as the fictitious time, for a Keplerian orbit it is

$$\frac{\mathrm{d}t}{\mathrm{d}E} = \frac{r}{\sqrt{-2\mathcal{E}_k}}$$

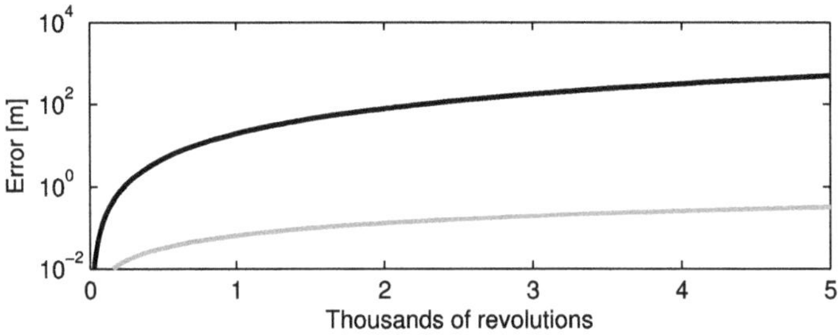

Fig. 2.3: Error in position for the integration of the uncontrolled equations (Eq. (2.10), in black) compared to the controlled equations (Eq. (2.11), in gray).

In the presence of conservative perturbations it is possible to replace the Keplerian energy by the total energy, defining a *generalized* eccentric anomaly, φ. It is the variable that evolves in time like

$$\frac{\mathrm{d}\varphi}{\mathrm{d}t} = \frac{\sqrt{-2\mathcal{E}}}{r}$$

and $\varphi(0) = E_0$. The element formulations proposed by Stiefel and Scheifele (1971, §19) and Baù et al. (2015), for example, are based on this construction.

2.4 Linearization

The linearization of the equations of motion is crucial in the regularization procedure. In this context, linearization does not refer to expansions for small values of a given parameter. It means transforming a system of nonlinear equations into a system of linear equations, without neglecting any terms in the equations. The book by Stiefel and Scheifele (1971) and the essay by Deprit et al. (1994) are two of the most complete works on this particular subject.

The first approach to the linear form of the equations of motion is possibly due to Laplace. We credit Deprit et al. (1994) for writing the original form of the equations by Laplace in a more friendly way.[5] Let $\mathcal{C} = \{\mathbf{u}_\rho, \mathbf{u}_\theta, \mathbf{u}_z\}$ be a rotating frame with $\mathbf{u}_z$ normal to the reference plane, $\mathbf{u}_\rho$ defined by the projection of the position vector onto the reference plane, and $\mathbf{u}_\theta$ completing a dextral frame, $\mathbf{u}_\theta = \mathbf{u}_z \times \mathbf{u}_r$. Figure 2.4 depicts the geometry of the problem. The projected radius is $\rho = \mathbf{r} \cdot \mathbf{u}_\rho$, the projected angular momentum reads $\lambda = \mathbf{h} \cdot \mathbf{u}_z$, and the transversal component of the motion is $z = \mathbf{r} \cdot \mathbf{u}_z$. Introducing the variables $\varrho = 1/\rho$, $\zeta = z/\rho$ and the transformation $\mathrm{d}s = \lambda \varrho^2\,\mathrm{d}t$, the equations of orbital motion take the form:

$$\begin{aligned}
\frac{\mathrm{d}^2\varrho}{\mathrm{d}s^2} + \varrho &= -\varrho\,(\mathbf{a}_p \cdot \mathbf{u}_\rho) - \frac{\mathrm{d}\varrho}{\mathrm{d}s}(\mathbf{a}_p \cdot \mathbf{u}_\theta) \\
\frac{\mathrm{d}^2\zeta}{\mathrm{d}s^2} + \zeta &= -\zeta\,(\mathbf{a}_p \cdot \mathbf{u}_\rho) - \frac{\mathrm{d}\zeta}{\mathrm{d}s}(\mathbf{a}_p \cdot \mathbf{u}_\theta) + (\mathbf{a}_p \cdot \mathbf{u}_z) \\
\frac{\mathrm{d}\lambda}{\mathrm{d}s} &= \lambda\,(\mathbf{a}_p \cdot \mathbf{u}_\theta)
\end{aligned}$$

The resulting differential equations are linear, and explicitly decoupled. In addition, if there are no external perturbations acting on the system ($\mathbf{a}_p = 0$) then the previous equations reduce to harmonic oscillators of unit frequency.

---

5 The system (K) presented by Laplace (1799, p. 151) is written in scalar form and the components of the perturbing acceleration are represented by the derivatives of a force function Q. The time transformation introducing a new independent variable is rather complicated since it involves the integral of a perturbing term.

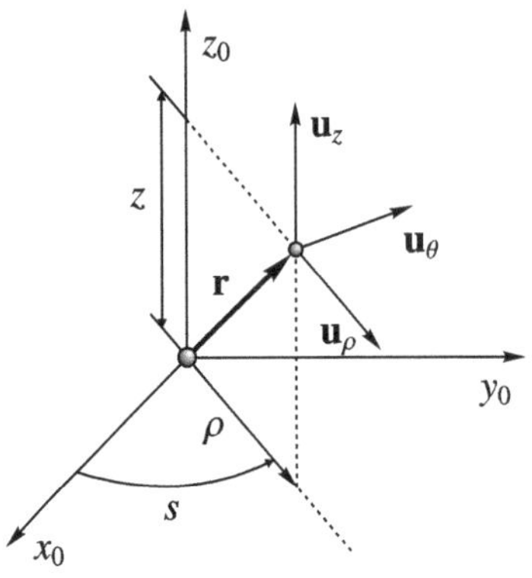

Fig. 2.4: Geometrical decomposition of Laplace's linearization.

2.4.1 Levi-Civita variables

Tullio Levi-Civita was particularly interested in the problem of regularizing collisions between pairs of bodies. In 1906 he suggested the introduction of a new set of coordinates for defining the planar motion, by means of his celebrated transformation in the complex plane (Levi-Civita, 1906):

$$x + \mathrm{i}y = (\xi + \mathrm{i}\eta)^2 \tag{2.12}$$

which provides

$$x = \xi^2 - \eta^2 \qquad \text{and} \qquad y = 2\xi\eta$$

and therefore

$$x^2 + y^2 = (\xi^2 + \eta^2)^2 \tag{2.13}$$

He achieved a global regularization of the motion when introducing a fictitious time, $\mathrm{d}t = r\,\mathrm{d}s$. Special attention was paid to the canonicity of the transformed system. His regularization scheme can be applied to the general perturbed two-body problem. Writing $\mathbf{r} = [x, y]^\top$ and $\mathbf{u} = [\xi, \eta]^\top$ the Levi-Civita transformation $\mathscr{L}: \mathbf{u} \mapsto \mathbf{r}$ reduces to

$$\mathbf{r} = \mathsf{L}(\mathbf{u})\,\mathbf{u} \qquad \text{with} \qquad \mathsf{L}(\mathbf{u}) = \begin{bmatrix} \xi & -\eta \\ \eta & \xi \end{bmatrix}$$

Matrix $\mathsf{L}(\mathbf{u})$ is the Levi-Civita matrix. Applying this transformation to Eq. (2.5) yields:

$$\frac{\mathrm{d}^2\mathbf{u}}{\mathrm{d}s^2} + \omega^2\,\mathbf{u} = \frac{r}{2}\mathsf{L}^\top(\mathbf{u})\,\mathbf{a}_p \tag{2.14}$$

It is the equation of an oscillator of frequency $\omega^2 = -\mathcal{E}/2$, with $\mathcal{E}$ the energy of the system, forced by the external perturbations:

$$\frac{\mathrm{d}\omega^2}{\mathrm{d}s} = -[\mathsf{L}(\mathbf{u})\,\mathbf{u}'] \cdot \mathbf{a}_p \tag{2.15}$$

The derivation of Eq. (2.14) and some interesting properties of the Levi-Civita matrix can be found in Chap. 3, and also in the dedicated monograph by Stiefel and Scheifele

(1971, §8). In the Keplerian case $\mathbf{a}_p = 0$ and Eq. (2.14) reduces to a harmonic oscillator. The differential equations are linear and decoupled, an important improvement with respect to Eq. (2.1). In the latter the errors in one variable propagate rapidly to the others. In addition, Eq. (2.14) has effectively embedded the energy. Its evolution can be integrated separately in order to take advantage of the stabilization mechanism described by Baumgarte (1972b). Figure 2.5 compares the error growth rate of Cowell's method (already shown in Fig. 2.1) with that of Eq. (2.14). The dynamical instability has been weakened, as expected, and the final error in position is reduced by two orders of magnitude. However, the error still grows in time even for this simple case because of the instability of the time transformation.

Volume XXIV of *Rendiconti del Lincei*, published in 1915, includes a total of five notes by Levi-Civita. In the first one, entitled "Sulla regolarizzazione del problema piano dei tre corpi" and presented to the Academy in July, he recovered the transformation in Eq. (2.12) and refined the regularization. The second note dives into more profound concepts related to the dynamics of a special type of holonomic system. The third (November 21) and fourth (December 5) notes focus on the canonical formalism of the three-body problem regularized by means of Eq. (2.12). Finally, his fifth note "Sul problema piano dei tre corpi. Caso limite in cui una delle masse è infinitesima" focuses on the restricted problem and explores more general forms of his transformation, namely

$$x + \mathrm{i}y = f(\xi + \mathrm{i}\eta)$$

He even introduced a set of elliptic coordinates thanks to the transformation $x + \mathrm{i}y = r/2 \cosh(\xi + \mathrm{i}\eta)$. Most of these results were collected in his renowned 1920 memoir. This memoir starts by reviewing the previous efforts by the author toward regularization of the three-body problem, and discusses the work done by P. Painlevé, R. Ball, and K. Sundman. It is interesting to note that Levi-Civita pointed out that Karl Weierstrass, famous for his work on elliptic functions, also contributed to the regularization of the

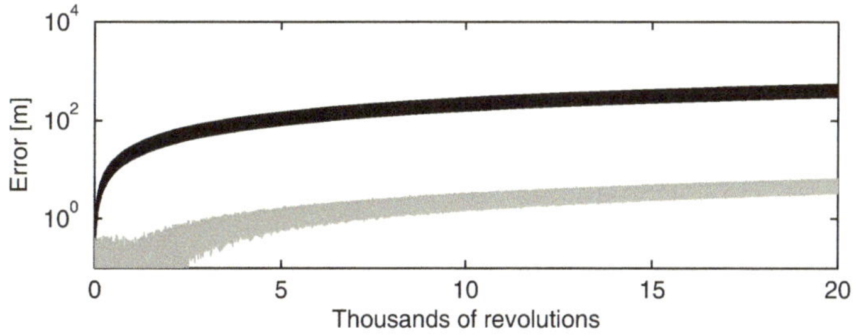

Fig. 2.5: Evolution of the propagation error of the Keplerian orbit discussed in Fig. 2.1. The black line corresponds to the integration of Eq. (2.1), and the gray line shows the accuracy of Eq. (2.14). The latter is integrated with a variable order and variable step-size Adams scheme, setting the tolerance to 10^{-13}.

N-body problem. Indeed, in a letter to Sophie Kowalevski[6], he briefly discussed the topic and hoped to advance on it in the future.

There is an intriguing phenomenon related to the transformation in Eq. (2.12): transformations of this type exist only for dimensions 1, 2, 4, and 8, as proved by Hurwitz (1898) and later by Adams (1960). Thus, is it possible to extend Levi-Civita's regularization to three-dimensions? The answer to this problem required the joint efforts of Paul Kustaanheimo and Eduard Stiefel (Kustaanheimo and Stiefel, 1965). We save the detailed explanation of the KS transformation for Chapter 3, where the need for an extra dimension is explained.

Transformations in the complex plane were also exploited by George D. Birkhoff. In 1915 he published a novel regularization of the restricted three-body problem. First, he sets the origin of the system not at the barycenter (the common practice) but at the midpoint between the primaries. By introducing the complex variable $\mathrm{q} = x + \mathrm{i}y + 1/2 - \nu$ (with ν the mass ratio of the primary bodies) he succeeded in reducing the problem to

$$\ddot{\mathrm{q}} + 2\mathrm{i}\dot{\mathrm{q}} = \nabla_{\mathrm{q}} \mathcal{V}(\mathrm{q})$$

A concise summary of the exposition by Birkhoff (1915) was provided by Szebehely (1967, pp. 97–103). Birkhoff acknowledges that a few months before the publication of his paper Levi-Civita referred him to a previous global regularization admitting double collisions: it was the Thiele–Burrau regularization.[7] Other transformations like the Lemaître regularization are based on similar concepts, and were beautifully unified by Szebehely (1967, §3.8).

2.4.2 Cartesian coordinates

Linearization can also be achieved relying on the usual position and velocity vectors, as Laplace's results suggest. A major breakthrough was that of Hans Sperling and Claude Burdet, who were able to achieve the linear and regular form of the equations of motion (Sperling, 1961; Burdet, 1967). Sperling embedded the integral of motion of the energy and transformed the time variable, seeking a generalized solution to Kepler's problem. Burdet, who was Eduard Stiefel's doctoral student and was deeply influenced by his work in regularization, followed a similar approach but he also introduced the eccentricity vector in the formulation. We shall now derive this formulation

6 The letter dated December 16, 1874, can be found in Weierstrass' biography by Mittag-Leffler (1912, p. 30).

7 Thiele (1895) presented a transformation involving a complex transformation of the form $\xi + \mathrm{i}\eta = \cos(E + \mathrm{i}F)$, with ξ and η relating to the coordinates (x, y), and E and F defining the relative orientation of the bodies. This solution assumed that the primaries had equal masses, as it emanated from the Euler problem of three-bodies. In 1906 C. Burrau extended the method in order to account for cases different from $\nu = 1/2$. It is remarkable that Thiele already made use of the time transformation $\mathrm{d}t = r_{12} r_{13}\,\mathrm{d}s$ that would be adopted later by Levi-Civita.

as an illustrative example of the steps toward regularization described in Sect. 2.3: introducing a fictitious time, linearizing the nonlinear equations, and embedding integrals of motion in the formulation.

First, the integrals of motion of Kepler's problem are written in terms of the fictitious time, defined by $\mathrm{d}t = r\,\mathrm{d}s$. The energy takes the form

$$\mathcal{E} = \frac{v^2}{2} - \frac{\mu}{r} = \frac{(\mathbf{r}' \cdot \mathbf{r}')}{2r^2} - \frac{\mu}{r} \tag{2.16}$$

Similarly, the Laplace–Runge–Lenz vector is defined as

$$\mathbf{e} = \frac{\mathbf{v} \times \mathbf{h}}{\mu} - \frac{\mathbf{r}}{r} \implies \mu\mathbf{e} = \frac{1}{r^2}\left[(\mathbf{r}' \cdot \mathbf{r}')\,\mathbf{r} - (\mathbf{r}' \cdot \mathbf{r})\,\mathbf{r}'\right] - \frac{\mu}{r}\mathbf{r} \tag{2.17}$$

Let us now recover Eq. (2.5) and write it in compact form:

$$\mathbf{r}'' - \frac{(\mathbf{r} \cdot \mathbf{r}')}{r^2}\mathbf{r}' + \frac{\mu}{r}\mathbf{r} = r^2\mathbf{a}_p$$

The second term can be solved from Eq. (2.17), and when introduced in the previous expression yields

$$\mathbf{r}'' - 2\left[\frac{(\mathbf{r}' \cdot \mathbf{r}')}{2r^2} - \frac{\mu}{r}\right]\mathbf{r} = r^2\mathbf{a}_p - \mu\mathbf{e}$$

The terms in brackets correspond to the definition of the energy given in Eq. (2.16), meaning that

$$\mathbf{r}'' - 2\mathcal{E}\mathbf{r} = r^2\mathbf{a}_p - \mu\mathbf{e} \tag{2.18}$$

This equation needs to be integrated together with the Sundman transformation in order to compute the physical time. To achieve the stabilization described by Baumgarte (1972b) one should also integrate the evolution of the Laplace–Runge–Lenz vector,

$$\mu\mathbf{e}' = 2(\mathbf{r}' \cdot \mathbf{a}_p)\,\mathbf{r} - (\mathbf{r} \cdot \mathbf{a}_p)\,\mathbf{r}' - (\mathbf{r}' \cdot \mathbf{r})\,\mathbf{a}_p$$

together with

$$\mathcal{E}' = (\mathbf{r}' \cdot \mathbf{a}_p)$$

Practical comments about the method can be found in the book by Bond and Allman (1996, §9.3). Baumgarte (1972a) and Janin (1974) followed almost the same procedure, although they did not introduce the eccentricity vector.

2.4.3 Universal solutions

Transforming the equations of orbital motion to an oscillator presents additional advantages apart from regularization and numerical stability. In the absence of perturbations the angular frequency of the oscillator is defined in terms of the Keplerian energy $\mathcal{E}_k$, which is constant. The sign of the frequency squared is given by the sign of the energy, or equivalently by the type of orbit (elliptic, parabolic, and hyperbolic). Because Eq. (2.14) reduces to

$$\mathbf{u}'' + \omega^2\mathbf{u} = 0$$

with $\omega^2 = \mathcal{E}_k/2$ constant, this equation is integrated easily and the solution depends on the type of orbit:

$$\mathbf{u}(s) = \begin{cases} \mathbf{u}_0\cos(\omega s) + (\mathbf{u}_0'/\omega)\sin(\omega s)\,, & \text{Elliptic, } \omega^2 > 0 \\ \mathbf{u}_0 + \mathbf{u}_0'\,s\,, & \text{Parabolic, } \omega^2 = 0 \\ \mathbf{u}_0\cosh(\omega s) + (\mathbf{u}_0'/\omega)\sinh(\omega s)\,, & \text{Hyperbolic, } \omega^2 < 0 \end{cases}$$

The solution can be unified by introducing the Stumpff functions $\mathfrak{C}_k(z)$, with k their degree and $z = (\omega s/2)^2$ their argument. Namely

$$\mathbf{u}(s) = \mathbf{u}_0\mathfrak{C}_0(z) + s\,\mathbf{u}_0'\mathfrak{C}_1(z)$$

A dedicated analysis of the Stumpff functions including some aspects on their implementation can be found in Appendix C .

Equation (2.18) can be solved in a similar way for the case $\mathbf{a}_p = 0$, resulting in

$$\mathbf{r}(s) = \mathbf{r}_0\mathfrak{C}_0(4z) + s\,r_0\mathfrak{C}_1(4z)\,\mathbf{v}_0 - \mu s^2\mathfrak{C}_2(4z)\,\mathbf{e} \tag{2.19}$$

It provides a universal vectorial solution to Kepler's problem (see the work by Condurache and Martinuşi, 2007a, for a thorough study). In addition, the Sundman transformation $\mathrm{d}t = r\,\mathrm{d}s$ can be integrated in closed form and yields the universal form of Kepler's equation

$$t - t_0 = s r_0\mathfrak{C}_1(4z) + s^2 r_0'\mathfrak{C}_2(4z) + \mu s^3\mathfrak{C}_3(4z) \tag{2.20}$$

This equation avoids the usual complications related to replacing the eccentric anomaly by the hyperbolic anomaly, or switching to Barker's equation. As an exercise we shall prove that this equation indeed reduces to Barker's equation in the parabolic case: the energy vanishes, meaning that $\omega = 0$; provided that $\mathfrak{C}_k(0) = 1/k!$ – see Eq. (C.1) –, and assuming that the particle departs from periapsis ($r_0 = r_p$) it follows that

$$t - t_p = \frac{s}{6}\left(6r_p + s^2\right) \tag{2.21}$$

The identity $r_0' = \mathbf{r}_0' \cdot \mathbf{r}_0/r_0 = \mathbf{r}_0 \cdot \mathbf{v}_0 = 0$ holds because of being computed at periapsis. The time of periapsis passage is t_p. The fictitious time relates to the true anomaly thanks to

$$\frac{\mathrm{d}\vartheta}{\mathrm{d}t} = \frac{h}{r^2} \implies \frac{\mathrm{d}\vartheta}{\mathrm{d}s} = \frac{h}{r}$$

Inverting this expression and integrating from periapsis to ϑ it is

$$s = (h/\mu)\tan\frac{\vartheta}{2}$$

Introducing the semilatus rectum p one can write $r_p = p/2$ and $h = (p\mu)^{1/2}$, and Eq. (2.21) finally becomes

$$t - t_p = \frac{1}{6}\sqrt{\frac{p^3}{\mu}}\left(3\tan\frac{\vartheta}{2} + \tan^3\frac{\vartheta}{2}\right)$$

which is, indeed, Barker's equation (Battin, 1999, p. 150).

2.5 Sets of orbital elements

Elements are the integration constants in orbital motion. Stiefel and Scheifele (1971, p. 83) extended the concept of element by stating that an element is any variable that remains constant or grows linearly with the independent variable in the Keplerian case. Under this definition the mean anomaly, for example, is considered an element.

Consider the set of classical elements $\mathbf{oe} = \{a, e, i, \omega, \Omega, M_0\}$, namely the semi-major axis, eccentricity, inclination, argument of periapsis, right ascension of the ascending node, and (initial) mean anomaly. Their time evolution is governed by the Gauss form of Lagrange's planetary equations:

$$\frac{da}{dt} = \frac{2a^2}{pr}\left[re\sin\vartheta\,(\tilde{\mathbf{a}}_p\cdot\mathbf{i}) + p\,(\tilde{\mathbf{a}}_p\cdot\mathbf{j})\right] \tag{2.22}$$

$$\frac{de}{dt} = \sin\vartheta\,(\tilde{\mathbf{a}}_p\cdot\mathbf{i}) + \frac{1}{p}\left[(p+r)\cos\vartheta + re\right](\tilde{\mathbf{a}}_p\cdot\mathbf{j}) \tag{2.23}$$

$$\frac{di}{dt} = \frac{r}{p}\cos(\vartheta+\omega)\,(\tilde{\mathbf{a}}_p\cdot\mathbf{k}) \tag{2.24}$$

$$\frac{d\omega}{dt} = -\frac{1}{e}\cos\vartheta\,(\tilde{\mathbf{a}}_p\cdot\mathbf{i}) + \frac{p+r}{ep}\sin\vartheta\,(\tilde{\mathbf{a}}_p\cdot\mathbf{j}) - \frac{r}{p}\cot i\sin(\theta+\omega)\,(\tilde{\mathbf{a}}_p\cdot\mathbf{k}) \tag{2.25}$$

$$\frac{d\Omega}{dt} = \frac{r}{p}\frac{\sin(\vartheta+\omega)}{\sin i}(\tilde{\mathbf{a}}_p\cdot\mathbf{k}) \tag{2.26}$$

$$\frac{dM_0}{dt} = \frac{\sqrt{1-e^2}}{ep}\left[(p\cos\vartheta - 2re)(\tilde{\mathbf{a}}_p\cdot\mathbf{i}) - (p+r)\sin\vartheta(\tilde{\mathbf{a}}_p\cdot\mathbf{j})\right] \tag{2.27}$$

in which p is the semilatus rectum of the orbit. The basis $\mathcal{L} = \{\mathbf{i}, \mathbf{j}, \mathbf{k}\}$ is the orbital frame, defined by $\mathbf{i} = \mathbf{r}/r$, $\mathbf{k} = \mathbf{h}/h$ and $\mathbf{j} = \mathbf{k}\times\mathbf{i}$. The modified perturbing term reads

$$\tilde{\mathbf{a}}_p = (h/\mu)\,\mathbf{a}_p$$

Equations (2.22–2.27) can be integrated numerically in lieu of Eq. (2.1), obtaining the osculating classical elements at each integration step. The state vector is then solved analytically from the dynamics in a Keplerian orbit. When there are no perturbations acting on the system Eq. (2.1) reduces to

$$\frac{d^2\mathbf{r}}{dt^2} = -\frac{\mu}{r^3}\mathbf{r} \tag{2.28}$$

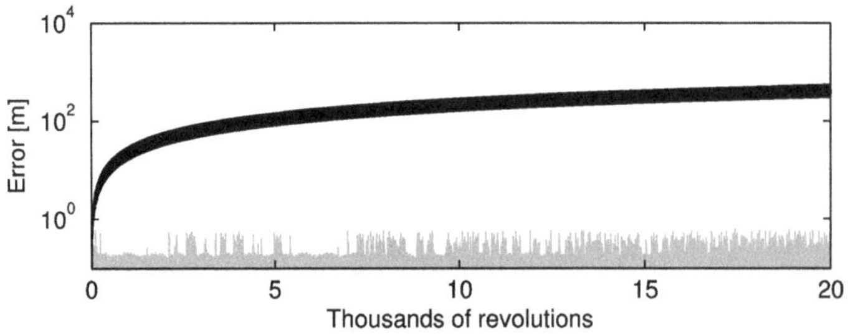

Fig. 2.6: Propagation error along a Keplerian orbit in Cartesian coordinates (black) and using a set of elements (gray).

whereas the system of equations (2.22–2.27) abides by

$$\frac{d\boldsymbol{\alpha}}{dt} = 0 \tag{2.29}$$

Equation (2.29) is integrated trivially and the values of the elements will not be affected by numerical errors. On the contrary, the right-hand side of Eq. (2.28) is nonzero and the solution may suffer from truncation errors. The reduction of the computational time is also a direct consequence of using Eq. (2.29) instead of Eq. (2.28). Figure 2.6 shows this phenomenon with a trivial example. A Keplerian orbit is integrated using Cowell's method (with the Störmer–Cowell integrator of order nine) and using a set of elements. The error in the integration of the Cartesian equations of motion grows in time. Conversely, the only source of error in the propagation using elements is the numerical resolution of Kepler's equation, and the round-off error in the transformation of the elements to the state vector.

When perturbations are present the right-hand side of Eqs. (2.22–2.27) will no longer be zero. However, it will scale with the magnitude of the perturbation. In practical scenarios the magnitude of the perturbations will typically be smaller than the Keplerian term. Thus, the derivatives of the elements will be small in magnitude compared to the derivatives of the coordinates. The evolution of the elements will be smoother, because they evolve with the time scale of the perturbation. All these factors are highly beneficial from a numerical perspective.

The classical elements have been the preferred choice for centuries due to their simplicity and intuitive interpretation.[8] The intrinsic singularities of the classical elements (for $e = 0$ and for $i = 0$) can be readily observed in Eqs. (2.22–2.27) and mo-

---

8 As a historical comment we recall that Johannes Kepler was the one who parameterized the orbits using the eccentricity and the axes. "*Ergo ellipsis est Planetæ iter*". With this sentence he stated that the planets follow elliptic orbits around the Sun. It is found in chapter 58 of his *Astronomia Nova*, published in 1609. His reasoning was based on observations of the orbit of Mars. It was not until 1775 that Leonhard Euler published his theory about the rotation of solids, entitled "Du mouvement d'un corps solide quelconque lorsq'il tourne autour d'un axe mobile" and presented in *Memoires de l'academie*

tivated researchers to propose alternative sets of elements. Broucke and Cefola (1972) suggested the use of a new set of elements, called the *equinoctial* elements, that avoid the referenced singularities of the classical elements. In the original notation proposed by the authors they form the set $\{a, h, k, \lambda_0, p, q\}$, and read

$$h = e\sin(\omega + \Omega)\,, \quad k = e\cos(\omega + \Omega)\,, \quad \lambda_0 = M_0 + \omega + \Omega\,,$$
$$p = \sin\Omega\tan\frac{i}{2}\,, \quad q = \cos\Omega\tan\frac{i}{2}$$

A slightly improved version from the implementation point of view was proposed by Walker et al. (1985).

Like Euler, Lagrange, Laplace, D'Alembert, Clairaut and many other great names, Charles-Eugène Delaunay devoted himself to improving the theory of lunar motion. In the very first chapter of his book (Delaunay, 1860, pp. 1–13) he suggested the use of a new set of elements for parameterizing orbital motion, $\{\ell, g, h, L, G, H\}$ (in the original order and notation). They relate to the classical orbital elements by means of

$$\ell = M\,, \qquad g = \omega\,, \qquad h = \Omega$$
$$L = \sqrt{\mu a}\,, \qquad G = L\sqrt{1 - e^2}\,, \qquad H = G\cos i$$

Their evolution in terms of the disturbing potential is also given explicitly by Delaunay (1860, p. 13, eq. 9). The canonicity of these elements in Kepler's problem, proved by the simplified form of the resulting Lagrange and Poisson brackets, makes them appealing for analytic studies. The interested reader is referred to works such as those by Brouwer (1959), Deprit (1981), Alfriend and Coffey (1984), or Gurfil and Lara (2014) for examples of their use. The Delaunay equations become singular for small eccentricities and inclinations, and the Poincaré elements $\{\Lambda, \xi, p, \lambda, \eta, q\}$ are often used instead, being (Arnold et al., 2007, p. 260):

$$\Lambda = L\,, \qquad \lambda = \ell + g + h$$
$$\xi = +\sqrt{2(L - G)}\cos(g + h)\,, \qquad \eta = -\sqrt{2(L - G)}\sin(g + h)$$
$$p = +\sqrt{2(G - \Theta)}\cos h\,, \qquad q = -\sqrt{2(G - \Theta)}\sin h$$

At only 21 years of age, Bengt Strömgren published a special perturbation method based on a vectorial set of elements. He parameterized the orbit in terms of the instantaneous angular velocity associated to the rotation of the orbit ($\mathrm{d}\Psi/\mathrm{d}t$ in his original notation), the eccentricity, the mean motion, and the mean longitude at epoch (Strömgren, 1929). The angular momentum vector and the Laplace–Runge–Lenz vector, constant in magnitude and direction in the Keplerian case, were exploited by Milankovitch (1939). In the work by Roy and Moran (1973) not only a more compact for-

des sciences de Berlin, vol. 16, pp. 176–227. He introduced what were later called the Euler angles. Gauss (born just two years after) already made extensive use of the angles $\{i, \omega, \Omega\}$ in most of his work, and they also appear in the work of Laplace (1799, chap. 4).

mulation of Milankovitch's method was presented, but they also derived a modified method for dealing with rectilinear and quasirectilinear orbits. In the introduction they discussed the connection with Hansen's lunar theory and the parameterization by Herrick (1953). The latter introduced the semimajor axis and semiminor axis vectors as the elements defining the orbit, and the mean anomaly behaved as an independent variable. Recently, Rosengren and Scheeres (2014) exploited the properties of Milankovitch's vectorial elements in Hamiltonian formalism. As an example of application they solved in closed form the secular motion of a particle accelerated by the solar radiation pressure.

Laplace noted the potential of accounting for the radial motion and the rotation of the radial direction separately (see for example the derivations leading to the system of equations labeled (H) in his "Traité de Mécanique Céleste", p. 149). This technique was developed in all its splendor by Deprit (1975), who recovered the concept of Hansen ideal frames (Sect. 4.1), and by Ferrándiz (1988), who introduced the concept of the *projective decomposition* of the dynamics when developing his dimension-rising canonical transformation. Deprit's elements and later those proposed by Vitins (1978) used quaternions for modeling the dynamics of the orbital plane. These concepts were later unified by Peláez et al. (2007), who published a new special perturbation method that they called Dromo. Chapter 4 is devoted to this particular formulation.

Time elements

From a numerical point of view, the main advantage of element-based parameterizations of orbital motion is that the propagation in the Keplerian case is trivial. Moreover, in the presence of weak perturbations the problem is still well behaved since the derivatives are small in magnitude and evolve in the time scales of the perturbations.

When the physical time t is replaced by a fictitious time s the Sundman transformation needs to be integrated numerically in order to compute the physical time. In its simplest form the equation to be integrated is

$$\frac{dt}{ds} = r$$

The right hand-side does not vanish (except for collision trajectories): the derivative is not necessarily small, it evolves in the time scale of orbital motion, and suffers from truncation errors even in the Keplerian case.

In order for the time transformation to benefit from the properties of element-like variables, a *time element* is introduced. Stiefel and Scheifele (1971, §18) presented possibly the first dedicated analysis of the theory of time elements. They opened their presentation explaining how the mean anomaly can be considered an element, and somehow behaves as a simplified time element. They then defined a time element τ attached to the set of natural elements in KS space.

Given a generic time element τ, if the perturbations vanish the evolution of the time element will reduce to

$$\frac{d\tau}{ds} = \gamma$$

where γ is constant. If $\gamma \neq 0$ then τ is a *linear* time element, whereas $\gamma = 0$ yields a *constant* time element. The physical time is recovered thanks to

$$t = \tau + f_t(s, \mathbf{r}, \mathbf{r}')$$

Here $f_t(s, \mathbf{r}, \mathbf{r}')$ denotes a generic nonlinear function. For example, using the eccentric anomaly as the independent variable and the time of periapsis passage as the time element ($\tau \equiv t_p$), it is

$$t = \tau + \frac{1}{n}(E - e \sin E)$$

The introduction of a time element was a technique that Baumgarte (1976a) explored as part of his efforts toward stabilizing the equations of motion. Nacozy (1981) considered different definitions of the time element based on the eccentric, the true, and the intermediate anomalies. He supported his analysis with numerical examples showing the improvements in the numerical performance when propagating an orbit perturbed by the Earth's J_2. Numerical explorations by Fukushima (2005) yielded two important conclusions about the use of time elements (in the context of the KS regularization): first, the error growth rate is reduced from quadratic to linear; second, it removes the periodic components of the error in unperturbed problems. A detailed review on time elements can be found in the work by Baù and Bombardelli (2014), where they also attached a constant and a linear time element to the set of Dromo elements.

2.6 Canonical transformations

Let $\mathbf{q} = [q_1, q_2, q_3]^\top$ and $\mathbf{p} = [p_1, p_2, p_3]^\top$ be the set of generalized coordinates and momenta. Given the Hamiltonian $\mathcal{H}$ and the canonical forces (Q_1, Q_2, Q_3) and (P_1, P_2, P_3), the canonical equations of motion read

$$\frac{dq_j}{dt} = +\frac{\partial \mathcal{H}}{\partial p_j} - Q_j \tag{2.30}$$

$$\frac{dp_j}{dt} = -\frac{\partial \mathcal{H}}{\partial q_j} + P_j \tag{2.31}$$

with $j = 1, 2, 3$. The canonical force P_j relates to the nonconservative perturbations and Q_j is introduced to preserve the symmetry of the equations. In its most general form the Hamiltonian can be written

$$\mathcal{H} = \mathcal{H}(t; q_1, q_2, q_3, p_1, p_2, p_3)$$

The time evolution of the Hamiltonian along a solution in phase space is governed by

$$\frac{\mathrm{d}\mathcal{H}}{\mathrm{d}t} = \frac{\partial \mathcal{H}}{\partial t} + \sum_j \left(\frac{\partial \mathcal{H}}{\partial q_j} \frac{\mathrm{d}q_j}{\mathrm{d}t} + \frac{\partial \mathcal{H}}{\partial p_j} \frac{\mathrm{d}p_j}{\mathrm{d}t} \right)$$

Introducing Eqs. (2.30–2.31) in the convective term yields

$$\frac{\mathrm{d}\mathcal{H}}{\mathrm{d}t} = \frac{\partial \mathcal{H}}{\partial t} - \sum_j \left(\frac{\partial \mathcal{H}}{\partial q_j} Q_j - \frac{\partial \mathcal{H}}{\partial p_j} P_j \right) \tag{2.32}$$

This is the law of energy. If there are no canonical forces acting on the system ($Q_j = P_j = 0$) then this equation reduces to the conservation law:

$$\frac{\mathrm{d}\mathcal{H}}{\mathrm{d}t} = \frac{\partial \mathcal{H}}{\partial t}$$

This expression proves that the Hamiltonian of a conservative system is constant if it does not depend explicitly on time.

Consider now a generic Sundman transformation

$$\frac{\mathrm{d}t}{\mathrm{d}s} = g(\mathbf{q}, \mathbf{p})$$

that introduces a fictitious time s. The physical time t becomes a dependent variable. We shall attach the time to the coordinates $\mathbf{q}$ by writing $q_0 = t$, so the extended vector results in $\mathbf{q} = [q_0, q_1, q_2, q_3]^\top$. In order to preserve the symmetry of the system the conjugate momentum p_0 is attached to $\mathbf{p}$. Following Stiefel and Scheifele (1971, chap. VIII), this construction yields the homogeneous Hamiltonian

$$\mathcal{H}_h = \mathcal{H} + p_0$$

Setting $j = 0$ in Eq. (2.30) leads to

$$\frac{\mathrm{d}q_0}{\mathrm{d}t} = \frac{\partial \mathcal{H}_h}{\partial p_0} - Q_0 = 1 - Q_0$$

The fact that $q_0 \equiv t$ shows that it must be $Q_0 = 0$. Similarly, Eq. (2.31) renders

$$\frac{\mathrm{d}p_0}{\mathrm{d}t} = -\frac{\partial \mathcal{H}_h}{\partial q_0} + P_0 = -\frac{\partial \mathcal{H}}{\partial t}\frac{\partial t}{\partial q_0} + P_0 = -\frac{\partial \mathcal{H}}{\partial t} + P_0 \tag{2.33}$$

If the canonical force P_0 is defined as

$$P_0 = -\sum_{j=1}^{3} \left(\frac{\partial \mathcal{H}}{\partial q_j} Q_j - \frac{\partial \mathcal{H}}{\partial p_j} P_j \right)$$

then Eq. (2.33) can be identified with Eq. (2.32) by making

$$p_0 \equiv -\mathcal{H}_0$$

and the canonical equation for p_0 is the law of energy.

In what remains of the section the canonical forces Q_j and P_j are neglected (Keplerian case). Using the coordinates (t, x, y, z) and momenta $(-\mathcal{H}_0, \dot{x}, \dot{y}, \dot{z})$, the homogeneous Hamiltonian in the Keplerian case reduces to

$$\mathcal{H}_h = \frac{1}{2}\left(p_1^2 + p_2^2 + p_3^2\right) - \frac{\mu}{\sqrt{q_1^2 + q_2^2 + q_3^2}} + p_0 \tag{2.34}$$

The canonical equations readily lead to the equations of motion and the conservation of the energy,

$$\frac{\mathrm{d}q_0}{\mathrm{d}t} = +\frac{\partial \mathcal{H}_h}{\partial p_0} = 1\,, \qquad \frac{\mathrm{d}q_j}{\mathrm{d}t} = +\frac{\partial \mathcal{H}_h}{\partial p_j} = p_j$$
$$\frac{\mathrm{d}p_0}{\mathrm{d}t} = -\frac{\partial \mathcal{H}_h}{\partial q_0} = 0\,, \qquad \frac{\mathrm{d}p_j}{\mathrm{d}t} = -\frac{\partial \mathcal{H}_h}{\partial q_j} = -\frac{\mu}{r^3}\,q_j$$

with $j = 1, 2, 3$.

Let $\mathcal{G}$ be the time transformation

$$\mathcal{G} : (t; q_0, \ldots, q_3, p_0, \ldots, p_3) \mapsto (s; q_0, \ldots, q_3, p_0, \ldots, p_3)$$

The fictitious time s is the new independent variable, with $\mathrm{d}t = g(r)\,\mathrm{d}s$. The coordinates and momenta are left unchanged for now. Considering the transformed Hamiltonian

$$\tilde{\mathcal{H}}_h = g(r)\,\mathcal{H}_h(q_0, \ldots, q_3, p_0, \ldots, p_3)$$

the canonical equations take the form

$$\frac{\mathrm{d}q_0}{\mathrm{d}s} = +\frac{\partial \tilde{\mathcal{H}}_h}{\partial p_0} = g(r)\,, \qquad \frac{\mathrm{d}q_j}{\mathrm{d}s} = +\frac{\partial \tilde{\mathcal{H}}_h}{\partial p_j} = g(r)p_j$$
$$\frac{\mathrm{d}p_0}{\mathrm{d}s} = -\frac{\partial \tilde{\mathcal{H}}_h}{\partial q_0} = 0\,, \qquad \frac{\mathrm{d}p_j}{\mathrm{d}s} = -\frac{\partial \tilde{\mathcal{H}}_h}{\partial q_j} = -\frac{\mu}{r^3}\,g(r)q_j - \frac{\partial g}{\partial q_j}\mathcal{H}_h$$

Given the form of $\mathrm{d}q_j/\mathrm{d}s$, it makes sense to introduce the change of variables

$$\tilde{p}_j = g(r)p_j \qquad \text{and} \qquad \tilde{q}_j = q_j$$

together with $\tilde{p}_0 = p_0$, $\tilde{q}_0 = q_0$. The new and the old momenta relate by means of

$$\frac{\mathrm{d}\tilde{p}_j}{\mathrm{d}s} = \frac{\mathrm{d}g}{\mathrm{d}s}p_j + g(r)\frac{\mathrm{d}p_j}{\mathrm{d}s} = \frac{1}{g(r)}\frac{\mathrm{d}g}{\mathrm{d}s}\tilde{p}_j + g(r)\frac{\mathrm{d}p_j}{\mathrm{d}s}$$

The equations of motion then reduce to

$$\frac{\mathrm{d}\tilde{q}_0}{\mathrm{d}s} = g(r), \qquad \frac{\mathrm{d}\tilde{q}_j}{\mathrm{d}s} = \tilde{p}_j$$
$$\frac{\mathrm{d}\tilde{p}_0}{\mathrm{d}s} = 0, \qquad \frac{\mathrm{d}\tilde{p}_j}{\mathrm{d}s} = -\frac{\mu}{r^3}g^2(r)\tilde{q}_j + \frac{1}{g(r)}\frac{\mathrm{d}g}{\mathrm{d}s}\tilde{p}_j - g(r)\frac{\partial g}{\partial \tilde{q}_j}\mathcal{H}_h$$

Identifying $\tilde{p}_j = r'_j$, $g(r) \equiv r$ and $\tilde{p}_0 \equiv -\mathcal{E}_k$, these equations become Eq. (2.11):

$$\frac{\mathrm{d}\tilde{p}_j}{\mathrm{d}s} \equiv \frac{\mathrm{d}^2\mathbf{r}}{\mathrm{d}s^2} = -\frac{\mu}{r}\mathbf{r} + \frac{r'}{r}\mathbf{r}' - \mathbf{r}\left[\frac{v^2}{2} - \frac{\mu}{r} - \mathcal{E}_k\right] = \frac{r'}{r}\mathbf{r}' - \mathbf{r}\left[\frac{(\mathbf{r}' \cdot \mathbf{r}')}{2r^2} - \mathcal{E}_k\right]$$

Interestingly, the control term appears naturally thanks to working with an homogeneous Hamiltonian.

Note, however, that under this last transformation

$$\mathscr{T} : (\mathbf{q}, \mathbf{p}) \mapsto (\tilde{\mathbf{q}}, \tilde{\mathbf{p}})$$

the Hamiltonian becomes

$$\mathcal{H}_h^* = \frac{1}{2g(r)}(\tilde{p}_1^2 + \tilde{p}_2^2 + \tilde{p}_3^2) - \frac{\mu g(r)}{\sqrt{\tilde{q}_1^2 + \tilde{q}_2^2 + \tilde{q}_3^2}} - g(r)\mathcal{H}_0$$

If we compute the partial derivatives with respect to $\tilde{p}_j$ we obtain

$$\frac{\partial \mathcal{H}_h^*}{\partial \tilde{p}_j} = \frac{\tilde{p}_j}{g(r)} \neq \frac{\mathrm{d}\tilde{q}_j}{\mathrm{d}s}$$

This means that the transformation $\mathscr{T}$ is not a canonical transformation (unless we consider the trivial case $g(r) = 1$), because the partial derivatives of the transformed Hamiltonian with respect to coordinates and momenta no longer yield the equations of motion.

The canonical nature (or noncanonical, in this case) of the transformation can be checked using a different method. First, we introduce the extended state vector $\mathbf{x} = [\mathbf{q}, \mathbf{p}]^\top$ so the canonical equations read

$$\frac{\mathrm{d}\mathbf{x}}{\mathrm{d}s} = \mathbf{J}\frac{\partial}{\partial \mathbf{x}}\tilde{\mathcal{H}}_h(\mathbf{x}) \tag{2.35}$$

It is written in terms of the matrix

$$\mathbf{J} = \begin{bmatrix} \mathbf{0} & \mathbf{I} \\ -\mathbf{I} & \mathbf{0} \end{bmatrix}$$

for which $\mathbf{J}^{-1} = -\mathbf{J}$. Now let $\tilde{\mathbf{x}} = [\tilde{\mathbf{q}}, \tilde{\mathbf{p}}]^\top$ define a generic set of transformed variables. If the transformation $(\mathbf{q}, \mathbf{p}) \mapsto (\tilde{\mathbf{q}}, \tilde{\mathbf{p}})$ is canonical, we should be able to write

$$\frac{\mathrm{d}\tilde{\mathbf{x}}}{\mathrm{d}s} = \mathbf{J}\frac{\partial}{\partial \tilde{\mathbf{x}}}\tilde{\mathcal{H}}_h(\tilde{\mathbf{x}}) \tag{2.36}$$

Using the chain rule we can refer the previous equation to the original variables:

$$\frac{\mathrm{d}\mathbf{x}}{\mathrm{d}s} = \mathbf{A}\frac{\mathrm{d}\tilde{\mathbf{x}}}{\mathrm{d}s}, \qquad \text{with} \qquad \mathbf{A} = \frac{\partial \mathbf{x}}{\partial \tilde{\mathbf{x}}}$$

Matrix $\mathbf{A}$ contains the partial derivatives of the old variables $(\mathbf{q}, \mathbf{p})$ with respect to the transformed variables $(\tilde{\mathbf{q}}, \tilde{\mathbf{p}})$. Introducing Eq. (2.36) yields

$$\frac{\mathrm{d}\mathbf{x}}{\mathrm{d}s} = \mathbf{A}\mathbf{J}\frac{\partial \tilde{\mathcal{H}}_h}{\partial \tilde{\mathbf{x}}} = \mathbf{A}\mathbf{J}\mathbf{A}^\top\frac{\partial \tilde{\mathcal{H}}_h}{\partial \mathbf{x}}$$

Equating this expression to Eq. (2.35) yields the condition for the transformation to be canonical, which is equivalent to **A** being symplectic:

$$\mathbf{A}\,\mathbf{J}\,\mathbf{A}^{\top} = \mathbf{A}^{\top}\,\mathbf{J}\,\mathbf{A} = \mathbf{J} \tag{2.37}$$

Going back to our example, we have $\tilde{\mathbf{q}} \equiv \mathbf{q}$ and $\tilde{\mathbf{p}} = g(r)\mathbf{p}$, so matrix **A** results in

$$\mathbf{A} = \frac{1}{g(r)}\begin{bmatrix} g(r)\mathbf{I} & \mathbf{0} \\ -\mathbf{v}\otimes\nabla_{\mathbf{r}}g(r) & \mathbf{I} \end{bmatrix}$$

The form of the product

$$\mathbf{A}^{\top}\mathbf{J}\,\mathbf{A} = \frac{1}{g(r)}\begin{bmatrix} \nabla_{\mathbf{r}}g(r)\otimes\mathbf{v} - \mathbf{v}\otimes\nabla_{\mathbf{r}}g(r) & \mathbf{I} \\ -\mathbf{I} & \mathbf{0} \end{bmatrix}$$

shows that matrix **A** will only be symplectic for $g(r) = 1$. Consequently, transformations of this type are not canonical. If **A** is symplectic then its inverse

$$\mathbf{A}^{-1} = -\mathbf{J}\,\mathbf{A}^{\top}\mathbf{J}$$

is symplectic too. Thus, the canonicity of a transformation can be verified both with the Jacobian of the new variables with respect to the old ones and with the Jacobian of the old variables with respect to the new ones.

The condition in Eq. (2.37) can be stated in terms of the *Lagrange brackets*. This operator applies to any pair of variables (a, b) and is referred to a set of coordinates and momenta. It reads

$$[a,\, b] = \sum_j \left(\frac{\partial p_j}{\partial a}\frac{\partial q_j}{\partial b} - \frac{\partial p_j}{\partial b}\frac{\partial q_j}{\partial a} \right) = \frac{\partial \mathbf{p}}{\partial a}\cdot\frac{\partial \mathbf{q}}{\partial b} - \frac{\partial \mathbf{p}}{\partial b}\cdot\frac{\partial \mathbf{q}}{\partial a}$$

The Lagrange brackets satisfy

$$[a,\, b] = -[b,\, a]$$

Following the reasoning that yielded Eq. (2.37), the transformation $\mathscr{T}$ is canonical if it leaves the Lagrange brackets unchanged, meaning that

$$[a,\, b]_{\mathbf{p},\mathbf{q}} = [a,\, b]_{\tilde{\mathbf{p}},\tilde{\mathbf{q}}}$$

A necessary and sufficient condition for the $\mathscr{T}$-invariance of the Lagrange brackets is

$$[\tilde{q}_k,\, \tilde{q}_\ell] = 0\,, \quad [\tilde{p}_k,\, \tilde{q}_\ell] = \delta_{k\ell}\,, \quad [\tilde{p}_k,\, \tilde{p}_\ell] = 0 \tag{2.38}$$

Introducing the *Poisson brackets*

$$\{a,\, b\} = \sum_j \left(\frac{\partial a}{\partial q_j}\frac{\partial b}{\partial p_j} - \frac{\partial a}{\partial p_j}\frac{\partial b}{\partial q_j} \right) = \frac{\partial a}{\partial \mathbf{q}}\cdot\frac{\partial b}{\partial \mathbf{p}} - \frac{\partial a}{\partial \mathbf{p}}\cdot\frac{\partial b}{\partial \mathbf{q}}$$

and exploiting the properties of symplectic maps the condition in Eq. (2.38) turns out to be equivalent to

$$\{\tilde{q}_k, \tilde{q}_\ell\} = 0\,, \quad \{\tilde{p}_k, \tilde{q}_\ell\} = \delta_{k\ell}\,, \quad \{\tilde{p}_k, \tilde{p}_\ell\} = 0$$

When applied to orbital motion, the Lagrange brackets typically relate the Cartesian state with a given set of elements, resulting in

$$[\text{œ}_i, \text{œ}_j] = \frac{\partial \mathbf{r}}{\partial \text{œ}_i} \cdot \frac{\partial \mathbf{v}}{\partial \text{œ}_j} - \frac{\partial \mathbf{v}}{\partial \text{œ}_i} \cdot \frac{\partial \mathbf{r}}{\partial \text{œ}_j}$$

Taking the time derivative of Lagrange's brackets shows that

$$\begin{aligned}\frac{\partial}{\partial t}[\text{œ}_i, \text{œ}_j] = {} & \frac{\partial}{\partial \text{œ}_i}\left(\frac{\partial \mathbf{r}}{\partial t}\right) \cdot \frac{\partial \mathbf{v}}{\partial \text{œ}_j} + \frac{\partial \mathbf{r}}{\partial \text{œ}_i} \cdot \frac{\partial}{\partial \text{œ}_j}\left(\frac{\partial \mathbf{v}}{\partial t}\right) - \frac{\partial}{\partial \text{œ}_i}\left(\frac{\partial \mathbf{v}}{\partial t}\right) \cdot \frac{\partial \mathbf{r}}{\partial \text{œ}_j} \\ & - \frac{\partial \mathbf{v}}{\partial \text{œ}_i} \cdot \frac{\partial}{\partial \text{œ}_j}\left(\frac{\partial \mathbf{r}}{\partial t}\right)\end{aligned}$$

Since $\partial \mathbf{v}/\partial t = -\partial \mathcal{V}/\partial \mathbf{r}$ and $\partial \mathbf{r}/\partial t = \mathbf{v}$, the derivative simplifies to

$$\begin{aligned}\frac{\partial}{\partial t}[\text{œ}_i, \text{œ}_j] &= \frac{\partial}{\partial \text{œ}_i}\left(\frac{\partial \mathcal{V}}{\partial \mathbf{r}}\right) \cdot \frac{\partial \mathbf{r}}{\partial \text{œ}_j} - \frac{\partial}{\partial \text{œ}_j}\left(\frac{\partial \mathcal{V}}{\partial \mathbf{r}}\right) \cdot \frac{\partial \mathbf{r}}{\partial \text{œ}_i} \\ &= \frac{\partial^2 \mathcal{V}}{\partial \text{œ}_i \partial \text{œ}_j} - \frac{\partial^2 \mathcal{V}}{\partial \text{œ}_j \partial \text{œ}_i} = 0\end{aligned}$$

That is, Lagrange's brackets do not depend on time. This is a classical result discussed in various references like in Brouwer and Clemence (1961, p. 278) or Battin (1999, p. 478). However, when introducing the fictitious time and referring the Lagrange and Poisson brackets to the fictitious velocity, these brackets will not necessarily be fictitious-time invariant.

2.7 Gauge freedom in celestial mechanics

The elements are the constants of integration in Kepler's problem. Considering a particular set of elements **œ**, the position and velocity vectors will be defined by nonlinear functions of the form

$$\mathbf{r}(t) = \mathbf{f}(t; \text{œ}_1, \ldots, \text{œ}_n) \qquad \text{and} \qquad \mathbf{v}(t) = \mathbf{g}(t; \text{œ}_1, \ldots, \text{œ}_n)$$

The state vector is the solution to the unperturbed two-body problem

$$\frac{\mathrm{d}^2\mathbf{r}}{\mathrm{d}t^2} = -\frac{\mu}{r^3}\mathbf{r} \equiv \mathbf{F}$$

Here **F** denotes the acceleration on the particle.

The motion under an external perturbation $\Delta\mathbf{F}$ is governed by the equation

$$\frac{d^2\mathbf{r}}{dt^2} = \mathbf{F} + \Delta\mathbf{F} \tag{2.39}$$

The solution to the perturbed motion can be obtained through the *variation of parameters* technique. The method was originally developed by Euler in his works about the precession of the equinoxes (Euler, 1749) and further developed by Laplace in a series of papers entitled "Théorie des variations séculaires des élémens des Planetes",[9] in which he studied the evolution of the orbital elements of the planets. In the presence of perturbations the orbital elements are no longer constant, but functions of time. Thus, the solution follows from

$$\begin{aligned} \mathbf{r}(t) &= \mathbf{f}(t; \alpha e_1(t), \ldots, \alpha e_n(t)) \\ \mathbf{v}(t) &= \mathbf{g}(t; \alpha e_1(t), \ldots, \alpha e_n(t)) \end{aligned}$$

The time derivative of the position vector yields

$$\frac{d\mathbf{r}}{dt} = \frac{\partial\mathbf{f}}{\partial t} + \sum \frac{\partial\mathbf{f}}{\partial \alpha e_i}\frac{d\alpha e_i}{dt} = \mathbf{g} + \mathbf{\Phi} \tag{2.40}$$

Note that it is the full derivative of the radius vector. It is defined in terms of $\mathbf{g} = \partial\mathbf{f}/\partial t$ and the convective terms, grouped under

$$\mathbf{\Phi} = \sum_j \frac{\partial\mathbf{f}}{\partial \alpha e_j}\frac{d\alpha e_j}{dt} \tag{2.41}$$

In order for the velocity of the perturbed problem, $d\mathbf{r}/dt$, to have the same form as the velocity of the unperturbed problem, $\mathbf{g}$, Lagrange imposed the constraint

$$\mathbf{\Phi} \equiv \sum_j \frac{\partial\mathbf{f}}{\partial \alpha e_j}\frac{d\alpha e_j}{dt} = 0$$

This ensures that the computed elements define the *osculating* orbit at every point. But this constraint can be relaxed in practice, as exploited in a series of papers by Efroimsky (2002, 2005). A discussion about this constraint can also be found in Brouwer and Clemence (1961, p. 274).

Let us transform the system (2.39) to a first-order system and assume that the right-hand side is uniformly Lipschitz continuous, which is a fair assumption provided that typically the perturbing terms are continuously differentiable. The Picard–Lindelöf theorem shows that the solution $(\mathbf{r}, \mathbf{v})$ is unique. This proves that $\mathbf{r}$ and $\mathbf{v}$ are invariant to the selection of the function $\mathbf{\Phi}$. Different forms of $\mathbf{\Phi}$ render different sets of

9 Published in *Nouveaux Mémoires de l'Académie Royale des Sciences et Belles-lettres (Berlin)* in three parts: the first one appeared in 1781 (pp. 199–276), the second in 1782 (pp. 169–292), and the third in 1783 (pp. 161–190).

nonosculating elements, but they will define the same state vector under the corresponding transformation. Motivated by this freedom of choice, Efroimsky (2002) suggested that the Lagrange constraint is rather arbitrary and it can be relaxed making $\boldsymbol{\Phi} \neq 0$. He referred to $\boldsymbol{\Phi}$ as a gauge function, representing a reparameterization of the solution $(\mathbf{r}, \mathbf{v})$. The set of all possible definitions of the gauge function forms the gauge group, a symmetry group denoted $\mathcal{G}$.

The gauge-generalized equations of motion

The invariance of $\mathbf{r}(t)$ under the action of the symmetry group $\mathcal{G}$ yields relations between different sets of elements emanating from alternative definitions of the gauge. In particular, denoting the osculating elements by $\hat{\boldsymbol{\oe}}_j$, it follows that

$$\mathbf{f}(t; \hat{\boldsymbol{\oe}}) = \mathbf{f}(t; \boldsymbol{\oe}) \tag{2.42}$$

Similarly, the velocity-matching condition renders

$$\mathbf{g}(t; \hat{\boldsymbol{\oe}}) = \mathbf{g}(t; \boldsymbol{\oe}) + \boldsymbol{\Phi}(t; \boldsymbol{\oe}) \tag{2.43}$$

These two systems of nonlinear algebraic equations can be solved for the nonosculating elements given the osculating ones, and vice-versa.

Differentiating Eq. (2.40) once more with respect to time renders

$$\frac{\mathrm{d}^2\mathbf{r}}{\mathrm{d}t^2} = \frac{\partial \mathbf{g}}{\partial t} + \sum \frac{\partial \mathbf{g}}{\partial \oe_j}\frac{\mathrm{d}\oe_j}{\mathrm{d}t} + \frac{\mathrm{d}\boldsymbol{\Phi}}{\mathrm{d}t}$$

where $\partial \mathbf{g}/\partial t \equiv \mathbf{F}$. Equating this result to Eq. (2.39) leads to

$$\sum \frac{\partial \mathbf{g}}{\partial \oe_j}\frac{\mathrm{d}\oe_j}{\mathrm{d}t} + \frac{\mathrm{d}\boldsymbol{\Phi}}{\mathrm{d}t} = \Delta\mathbf{F} \tag{2.44}$$

If Eq. (2.41) is multiplied by $\partial \mathbf{g}/\partial \oe_n$ and Eq. (2.44) is multiplied by $\partial \mathbf{f}/\partial \oe_n$, subtracting the resulting equations provides:

$$\frac{\partial \mathbf{f}}{\partial \oe_n}\cdot\frac{\mathrm{d}\boldsymbol{\Phi}}{\mathrm{d}t} + \frac{\partial \mathbf{f}}{\partial \oe_n}\cdot\sum_j \frac{\partial \mathbf{g}}{\partial \oe_j}\frac{\mathrm{d}\oe_j}{\mathrm{d}t} - \frac{\partial \mathbf{g}}{\partial \oe_n}\cdot\sum_j \frac{\partial \mathbf{f}}{\partial \oe_j}\frac{\mathrm{d}\oe_j}{\mathrm{d}t} = \frac{\partial \mathbf{f}}{\partial \oe_n}\cdot\Delta\mathbf{F} - \frac{\partial \mathbf{g}}{\partial \oe_n}\cdot\boldsymbol{\Phi}$$

The left-hand side simplifies to

$$\sum_j \left[\oe_n, \oe_j\right]\frac{\mathrm{d}\oe_j}{\mathrm{d}t} + \frac{\partial \mathbf{f}}{\partial \oe_n}\cdot\frac{\mathrm{d}\boldsymbol{\Phi}}{\mathrm{d}t} = \frac{\partial \mathbf{f}}{\partial \oe_n}\cdot\Delta\mathbf{F} - \frac{\partial \mathbf{g}}{\partial \oe_n}\cdot\boldsymbol{\Phi}$$

where $\left[\oe_n, \oe_j\right]$ are the Lagrange brackets using the position and velocity vectors as canonical variables in phase space ($\mathbf{q} \equiv \mathbf{f}$ and $\mathbf{p} \equiv \mathbf{g}$),

$$\left[\oe_n, \oe_j\right] = \frac{\partial \mathbf{f}}{\partial \oe_n}\cdot\frac{\partial \mathbf{g}}{\partial \oe_j} - \frac{\partial \mathbf{f}}{\partial \oe_j}\cdot\frac{\partial \mathbf{g}}{\partial \oe_n} \tag{2.45}$$

This result yields the gauge-generalized Lagrange planetary equations

$$\sum_j [\text{œ}_n, \text{œ}_j] \frac{\mathrm{d}\text{œ}_j}{\mathrm{d}t} = \left(\Delta \mathbf{F} - \frac{\mathrm{d}\boldsymbol{\Phi}}{\mathrm{d}t}\right) \cdot \frac{\partial \mathbf{f}}{\partial \text{œ}_n} - \frac{\partial \mathbf{g}}{\partial \text{œ}_n} \cdot \boldsymbol{\Phi} \tag{2.46}$$

Making $\boldsymbol{\Phi} = 0$ yields the classical form of the planetary equations. Note that this expression involves the full derivative of the gauge function, which reads

$$\frac{\mathrm{d}\boldsymbol{\Phi}}{\mathrm{d}t} = \frac{\partial \boldsymbol{\Phi}}{\partial t} + \sum_j \frac{\partial \boldsymbol{\Phi}}{\partial \text{œ}_j} \frac{\mathrm{d}\text{œ}_j}{\mathrm{d}t}$$

Therefore, if the gauge depends on the elements Eq. (2.46) should be resorted to

$$\sum_j \left([\text{œ}_n, \text{œ}_j] + \frac{\partial \boldsymbol{\Phi}}{\partial \text{œ}_j} \cdot \frac{\partial \mathbf{f}}{\partial \text{œ}_n} \right) \frac{\mathrm{d}\text{œ}_j}{\mathrm{d}t} = \left(\Delta \mathbf{F} - \frac{\partial \boldsymbol{\Phi}}{\partial t}\right) \cdot \frac{\partial \mathbf{f}}{\partial \text{œ}_n} - \frac{\partial \mathbf{g}}{\partial \text{œ}_n} \cdot \boldsymbol{\Phi} \tag{2.47}$$

The choice of a nontrivial gauge might have theoretical and practical advantages. Efroimsky and Goldreich (2003, 2004) showed that the gauge can be defined in order to preserve the canonicity of the formulation in the presence of velocity-dependent perturbations. Gurfil and Klein (2006) found the optimal form of the gauge that minimizes the integration error of a Runge–Kutta scheme. Later, Gurfil (2007) exploited the gauge freedom in the problem of spacecraft relative motion in order to find new families of frozen orbits. More details can be found in the book by Kopeikin et al. (2011, §1.5.3).

2.8 Conclusions

Regularization was born from the necessity of avoiding the singularities in the three-body problem. At first, regularized theories were mathematical contrivances that helped researchers to analyze the dynamics of the system. The formulations were based on different principles, from Hamiltonian mechanics to topological transformations.

During the design and operation of a space mission, the dynamical regime is far from the singularity and regularization might not seem useful. But it was soon realized that the transformations that worked so well in the three-body problem exhibited important advantages in the two-body problem too. Different techniques aim for an improved numerical scheme. For example, the introduction of special variables can potentially alleviate the negative effects that the Lyapunov instability of orbital motion has on the numerical integration of the Cartesian equations of motion. By replacing the time by an alternative independent variable one achieves an analytic step-size adaption that enhances the performance of the numerical integrator. Similarly, embedding integrals of motion in the equations behaves as a control term that slows down the error growth.

The use of orbital elements to model the dynamics is advantageous in many cases. Since weakly perturbed problems are close to Keplerian orbits, in such cases the evolution of the elements will be smoother than the evolution of the coordinates. From a numerical perspective, the right-hand side of the equations to be integrated will be smaller, meaning that the problem is less sensitive to truncation errors. Recent discoveries prove that the elements need not be referred to the osculating orbit (see, for example, the work by Dosopoulou and Kalogera, 2016, which derives a physical model for mass-transferring binary systems using nonosculating elements). The freedom of choice in the dynamical modeling has great potential for analytic and numerical studies.

In the following chapters the foundations of regularization are recovered and applied to different problems in astrodynamics. The techniques described in the present chapter will be exploited for finding new analytic solutions to some problems, and for deriving new formulations to describe particular phenomena. Chapter 3 focuses on the Kustaanheimo–Stiefel transformation as an example of a regularized formulation with a Sundman transformation of order one, and Chap. 4 studies the Dromo formulation, with a transformation of order two.

3 The Kustaanheimo–Stiefel space and the Hopf fibration

In the 1960s, Eduard Stiefel started to organize the Oberwolfach Meetings on Celestial Mechanics, in an attempt to draw the interest of mathematicians into this subject. The first of those meetings took place in 1964 and Paul Kustaanheimo presented his work on describing Keplerian motion using spinors. His work on spinors combined with Stiefel's experience in topology and numerical calculus gave birth to the celebrated extension of the Levi-Civita transformation to the three-dimensional case, known as the Kustaanheimo–Stiefel transformation or KS for short (Kustaanheimo and Stiefel, 1965). This extension had eluded researchers since Levi-Civita (1920) presented his original regularization of the planar problem and Hurwitz (1898) and Adams (1960) proved that transformations of this type only exist in spaces of dimension $n = 1, 2, 4, 8$.

Heinz Hopf discovered a particular transformation from the unit three-sphere $\mathbb{S}^3$ (a three-dimensional manifold embedded in $\mathbb{R}^4$) onto the unit two-sphere $\mathbb{S}^2$ (a two-dimensional manifold embedded in $\mathbb{R}^3$) so that the preimage of each point in three-dimensional space turns out to be a circle on $\mathbb{S}^3$, called a fiber (Hopf, 1931). All points in this fiber transform into the same point in three-dimensional space. Such transformation is referred to as the Hopf fibration. In fact, the Kustaanheimo–Stiefel transformation can be understood as a particular Hopf map (Stiefel and Scheifele, 1971, §44). Hopf was Stiefel's doctoral advisor and influenced other areas of his research, including the Hopf–Stiefel functions and the Stiefel manifolds. Davtyan et al. (1987) developed the generalization of KS transformation to the case $\mathbb{R}^8 \to \mathbb{R}^5$ in order to transform the problem of the five-dimensional hydrogen atom into an eight-dimensional oscillator. They successfully rewrote the Hamiltonian of the hydrogen atom as the Hamiltonian of an eight-dimensional oscillator. Deprit et al. (1994) published an exhaustive treaty on the transformations underlying KS regularization. They focused on the topic of linearization, connecting with prior work from Lagrange.

The KS transformation provides a robust regularization scheme for dealing with close approaches or even impact trajectories. Close encounters are one of the major challenges in N-body simulations. The first extensions of the KS transformation to N-body problems are due to Peters (1968), Aarseth (1971), and Bettis and Szebehely (1971). We refer to the work of Szebehely and Bettis (1971) for a review of the methods developed in those years. Aarseth and Zare (1974) focused on the three-body problem and their method was later generalized by Heggie (1974), who reformulated the Hamiltonian for dealing with an arbitrary number of particles. The shortcoming of Heggie's method is that it fails to reproduce collisions of more than two particles. Mikkola (1985) discovered a technique for avoiding this singularity by rewriting the Sundman time transformation in terms of the Lagrangian of the system. This method integrates $4N(N - 1) + 1$ equations of motion, so its use is recommended for few-body prob-

https://doi.org/10.1515/9783110559125-003

lems. The formulations based on the KS transformation have been improved throughout the years (Mikkola and Aarseth, 1998; Mikkola and Merritt, 2008), especially since the development of the chain regularization techniques (Mikkola and Aarseth, 1989, 1993). The introduction of relativistic corrections in the models has occupied different authors: Kupi et al. (2006) modified the KS regularization for two-body close encounters in N-body simulations by introducing post-Newtonian effects; Funato et al. (1996) published a reformulation of the KS transformation focused on time-symmetric algorithms. Aarseth (1999, 2003) presented several reviews of the evolution and keystones in the development of N-body simulations. The Levi-Civita variables have recently been recovered by Lega et al. (2011) to detect resonant close encounters in the three-body problem, and used by Astakhov and Farrelly (2004) in combination with an extension of the phase space to analyze the elliptic restricted three-body problem.

The topological implications and numerical performance of the KS transformation have motivated many studies on the subject. Stiefel was a committed teacher and supervised over 60 doctoral students. Some of them made significant contributions to regularization and celestial mechanics in general, names like Claude Burdet, Jiri Kriz, Gerhard Scheifele, Michael Vitins, and Jörg Waldvogel. Velte (1978) explored the representation of the KS transformation in the language of quaternions, a task that also occupied Vivarelli (1983) and Waldvogel (2006a,b). Different representations of KS regularization have been published by Vivarelli (1986, 1994), including a representation in hypercomplex algebra (Vivarelli, 1985). Hypercomplex numbers will be recovered in Chap. 5 to formulate a propagator based on the geometry of Minkowski space-time. Deprit et al. (1994) re-derived the KS transformation by doubling, using quaternions and octonions. They concluded by deriving the Burdet–Ferrándiz transformation from the foundations of KS regularization (Ferrándiz, 1988). Saha (2009) has recently reformulated the problem by combining quaternions with the Hamiltonian formalism. ElBialy (2007) approached the KS transformation from an alternative perspective, focusing on the connection with the Hopf map and the topological structure of the transformation. Moser (1970) determined the topological structure of Kepler's problem by regularizing the energy manifold. In a series of papers, Fukushima (2003, 2004, 2005) analyzed different numerical aspects of regularization, seeking scaling factors that guarantee the conservation of the integrals of motion and time elements for improving the stability of the time transformation. Recently, Roa et al. (2016c) studied a topological phenomenon in KS space related to the transition to chaos. By analyzing the structure supporting Hopf's fibration they showed that numerical errors related to the exponential divergence of the orbits may break down the KS transformation, as fibers will no longer be transformed to points in three-dimensions, and used this phenomenon to estimate numerical errors.

The present chapter approaches KS formalism focusing more on its topological nature. The first part of this chapter explains the need for an additional dimension when regularizing the three-dimensional problem. It is combined with historical notes about the relevance of this problem in mathematics. The equations of the KS transformation can be found in Sect. 3.2, with comments emphasizing the connection with the Hopf fibration. The inverse transformation is defined explicitly in terms of the Hopf map. Classical concepts of stability are translated to KS space using the Hopf fibration. These results serve as a foundation for introducing the concept of manifold of solutions and topological stability, which will lead to an alternative Lyapunov indicator (Sect. 3.4). Numerical examples of N-body systems are presented in Sect. 3.5 to show the consequences of topological instabilities. Finally, in Sect. 3.6 we derive the gauge-generalized equations for the elements attached to KS space by Stiefel and Scheifele (1971), and give them a geometrical interpretation. Section 3.7 serves as an appendix discussing geometrical and algebraic aspects of the KS transformation.

3.1 The need for an extra dimension: Fibrations of hyperspheres

The Levi-Civita variables, introduced in Sect. 2.4.1, yield a global regularization of the planar problem. Levi-Civita himself admitted that the extension of his transformation to the three-dimensional problem was not straightforward (in fact, he never achieved it):

> "The problem in space has long resisted my efforts, as I tried to approach it by similar coordinate changes. The usual canonical transformations relating to elliptical motion do not regularize either." Levi-Civita (1920, p. 2, in French)

Where is the difficulty coming from? Why did Kustaanheimo and Stiefel define a transformation in four dimensions, and why is it not possible to stay in three dimensions?

One possible answer comes from a classical problem in number theory, later extended to more profound topological results. The problem relates to quadratic forms of the type

$$\left(x_1^2 + \ldots + x_r^2\right)\left(y_1^2 + \ldots + y_s^2\right) = z_1^2 + \ldots + z_n^2 \tag{3.1}$$

where the terms z_k are bilinear functions of x_i and y_j. This identity is said to be of dimension (r, s, n). The problem consists in finding the conditions for which Eq. (3.1) holds in arbitrary dimensions. It is uncertain how he arrived at this result, but in the third century Diophantus of Alexandria stated (in words) that the identity $(2, 2, 2)$ holds for

$$z_1 = x_1 y_1 + x_2 y_2\,, \qquad z_2 = x_1 y_2 - x_2 y_1 \tag{3.2}$$

The reader might observe that taking two complex numbers $\mathfrak{p} = x_1 - \mathrm{i}\,x_2$ and $\mathfrak{q} = y_1 + \mathrm{i}\,y_2$ this identity reduces to

$$\mathfrak{z} = \mathfrak{p}\mathfrak{q}$$

with $\mathfrak{z} = z_1 + \mathrm{i}\,z_2$. Under this formalism, Eq. (3.1) is none other than the relation between the moduli of two complex numbers:

$$|\mathfrak{p}|^2 |\mathfrak{q}|^2 = |\mathfrak{z}|^2$$

Here $|\mathfrak{p}|^2 = \mathfrak{p}\mathfrak{p}^\dagger$ with $\mathfrak{p}^\dagger$ the conjugate of $\mathfrak{p}$. Writing $\mathfrak{p} = \mathfrak{q} = \xi + \mathrm{i}\,\eta$ and $\mathfrak{z} = x + \mathrm{i}\,y$ one obtains the fundamental relation of the Levi-Civita transformation presented in Eq. (2.13), namely

$$x^2 + y^2 = (\xi^2 + \eta^2)^2$$

Equation (3.2) becomes the Levi-Civita transformation itself, $x = \xi^2 - \eta^2$ and $y = 2\xi\eta$.

On May 4, 1748, Leonhard Euler wrote a letter to Christian Goldbach in which he noted that if $p = x_1^2 + x_2^2 + x_3^2 + x_4^2$ and $q = y_1^2 + y_2^2 + y_3^2 + y_4^2$, then $pq = z_1^2 + z_2^2 + z_3^2 + z_4^2$, with

$$\begin{aligned} z_1 &= x_1y_1 + x_2y_2 + x_3y_3 + x_4y_4 \\ z_2 &= x_1y_2 - x_2y_1 - x_3y_4 + x_4y_3 \\ z_3 &= x_1y_3 + x_2y_4 - x_3y_1 - x_4y_2 \\ z_4 &= x_1y_4 - x_2y_3 + x_3y_2 - x_4y_1 \end{aligned}$$

That is, he proved that Eq. (3.1) holds also for dimension $(4, 4, 4)$. Just like the case $(2, 2, 2)$ relates to multiplication in $\mathbb{C}$, Euler's polynomial identity is equivalent to the quaternion product (although defined almost a century before Hamilton described quaternions in 1843). Indeed, taking two quaternions $\mathfrak{p}, \mathfrak{q} \in \mathbb{H}$ defined as $\mathfrak{p} = x_1 - \mathrm{i}x_2 - \mathrm{j}x_3 - \mathrm{k}x_4$ and $\mathfrak{q} = y_1 + \mathrm{i}y_2 + \mathrm{j}y_3 + \mathrm{k}y_4$, Euler's result reduces to

$$\mathfrak{z} = \mathfrak{p}\mathfrak{q}$$

where again $|\mathfrak{p}|^2|\mathfrak{q}|^2 = |\mathfrak{z}|^2$ and $\mathfrak{z} \in \mathbb{H}$.

It took 150 years after Euler's discovery until Adolf Hurwitz proved that this kind of construction with dimension (n, n, n) exists only for $n = 1, 2, 4, 8$ (Hurwitz, 1898). This is an extremely beautiful result, since these are precisely the dimensions of the four only normed algebras: real numbers, $\mathbb{R}$, complex numbers, $\mathbb{C}$, quaternions, $\mathbb{H}$, and octonions, $\mathbb{O}$.[1]

Provided that there are no normed algebras of dimension $n = 3$, Kustaanheimo and Stiefel (1965) decided to move to the four-dimensional case, and then sought a transformation to three dimensions. They had the right tools since Heinz Hopf, who was Stiefel's doctoral advisor, had already presented important results regarding the

1 Dickson (1919) tackled the problem described in Eq. (3.1) and referred to it as the *eight square theorem*. He was possibly the first to note that the resulting formulas are indeed the norm of complex numbers, quaternions, and octonions (Cayley numbers). This leads to the Cayley–Dickson construction, which generates the higher-dimensional algebras by doubling the lower-dimensional ones. They allow one to define additional algebras, like sedenions with $n = 16$, but these are no longer normed algebras: the algebra $\mathbb{C}$ is commutative and associative under multiplication, $\mathbb{H}$ is associative but not commutative, and $\mathbb{O}$ is only alternative. Beyond octonions these properties are lost. Mathematicians like Ferrándiz (1988) and Deprit et al. (1994) made extensive use of these observations when deriving regularized theories. The latter connected the so-called *binary trees* with the standard basis in hypercomplex systems related to quaternions, octonions, and sedenions.

transformation of spheres from four dimensions to three dimensions (Hopf, 1931). Consider the four-dimensional sphere $\mathcal{S}^3$,

$$x_1^2 + x_2^2 + x_3^2 + x_4^2 = 1$$

and identify $\mathbb{R}^4$ with $\mathbb{C}^2$, i.e., $(x_1, x_2, x_3, x_4) \leftrightarrow (\mathfrak{z}_1, \mathfrak{z}_2) = (x_1 + \mathrm{i}x_2, x_3 + \mathrm{i}x_4)$. Thus, $\mathcal{S}^3$ admits an alternative definition:

$$|\mathfrak{z}_1|^2 + |\mathfrak{z}_2|^2 = 1$$

Hopf introduced the mapping $\mathscr{F} : \mathbb{C}^2 \mapsto \mathbb{C} \times \mathbb{R}$

$$\mathscr{F}(\mathfrak{z}_1, \mathfrak{z}_2) = (2\mathfrak{z}_1\mathfrak{z}_2{}^{\dagger}, |\mathfrak{z}_1|^2 - |\mathfrak{z}_2|^2) = (y_1 + \mathrm{i}y_2, y_3) \tag{3.3}$$

that is called the *Hopf fibration* or the *Hopf map*. The first component of its image is a complex number, whereas the second component is a real number. Thus, the Hopf fibration maps four-dimensional spheres to three-dimensional spheres, $\mathcal{S}^3 \to \mathcal{S}^2$. In addition, if we define a different point on a great circle of $\mathcal{S}^3$ in terms of $(\mathfrak{w}_1, \mathfrak{w}_2) = (\mathfrak{b}\,\mathfrak{z}_1, \mathfrak{b}\,\mathfrak{z}_2)$ and with $|\mathfrak{b}| = 1$ defining the great circle, it follows that

$$\mathscr{F}(\mathfrak{w}_1, \mathfrak{w}_2) = \mathscr{F}(\mathfrak{b}\,\mathfrak{z}_1, \mathfrak{b}\,\mathfrak{z}_2) = |\mathfrak{b}|\,\mathscr{F}(\mathfrak{z}_1, \mathfrak{z}_2) = \mathscr{F}(\mathfrak{z}_1, \mathfrak{z}_2) \tag{3.4}$$

This means that every point on the great circle $|\mathfrak{b}| = 1$ of $\mathcal{S}^3$ transforms into the same point on $\mathcal{S}^2$. The Hopf fibration transforms circles, $\mathcal{S}^1$, on the four-dimensional sphere to points on the three-dimensional sphere, defining the fiber bundle $\mathcal{S}^1 \hookrightarrow \mathcal{S}^3 \to \mathcal{S}^2$. The KS transformation is, in fact, a particular case of the Hopf map.

For completeness we note that a direct consequence of the fundamental theorem by Adams (1960) is that there are four admissible fiber bundles,

$$\mathcal{S}^p \hookrightarrow \mathcal{S}^n \to \mathcal{S}^{n-p}$$

which connect the spheres with (p, n) pairs $(0, 1)$, $(1, 3)$, $(3, 7)$, and $(7, 15)$. Their representation complies with $\mathbb{R}$, $\mathbb{C}$, $\mathbb{H}$, and $\mathbb{O}$, respectively.

3.2 The KS transformation as a Hopf map

Kustaanheimo and Stiefel (1965) introduced the four-dimensional vector $\mathbf{u} = [u_1, u_2, u_3, u_4]^\top$, defined in the parametric space $\mathbb{U}^4$, in order to regularize the motion in three-dimensional Euclidean space $\mathbb{E}^3$. The parametric space is identified with $\mathbb{C}^2$ thanks to considering the pair of complex numbers $(\mathfrak{v}_1, \mathfrak{v}_2) = (u_1 + \mathrm{i}u_4, u_2 - \mathrm{i}u_3)$. Introducing this pair in the Hopf fibration, Eq. (3.3) provides $(y + \mathrm{i}z, x)$, with:

$$\begin{aligned} x &= u_1^2 - u_2^2 - u_3^2 + u_4^2 \\ y &= 2(u_1u_2 - u_3u_4) \\ z &= 2(u_1u_3 + u_2u_4) \end{aligned} \tag{3.5}$$

By identifying (x, y, z) with the components of the position vector $\mathbf{r}$ in an inertial frame, Kustaanheimo and Stiefel defined the KS transformation $\mathscr{K} : \mathbb{U}^4 \mapsto \mathbb{E}^3$. It can be written in a compact matrix form just like the Levi-Civita transformation,

$$\mathbf{x} = \mathscr{K}(\mathbf{u}) = \mathsf{L}(\mathbf{u})\,\mathbf{u} \tag{3.6}$$

simply noting that $\mathbf{x} = [x, y, z, 0]^\top$ is the trivial extension to $\mathbb{R}^4$ of the position vector. The KS matrix $\mathsf{L}(\mathbf{u})$ reads:

$$\mathsf{L}(\mathbf{u}) = \begin{bmatrix} u_1 & -u_2 & -u_3 & u_4 \\ u_2 & u_1 & -u_4 & -u_3 \\ u_3 & u_4 & u_1 & u_2 \\ u_4 & -u_3 & u_2 & -u_1 \end{bmatrix}$$

Alternative representations of the KS transformation have been provided by Vivarelli (1983) and Waldvogel (2006b), in the language of quaternions, spinors, and hypercomplex numbers. The properties of the KS matrix are similar to those of the Levi-Civita matrix. To emphasize the connection with Hopf's results we refer the reader to Hopf (1931, eq. 1).

The KS matrix is r-orthogonal, i.e.,

$$\mathsf{L}^{-1}(\mathbf{u}) = \frac{1}{r}\,\mathsf{L}^\top(\mathbf{u}) \tag{3.7}$$

Every point $\mathbf{u}$ is KS-mapped to one single point in Cartesian space $\mathbb{E}^3$. The radial distance r relates to the KS variables by means of

$$r = u_1^2 + u_2^2 + u_3^2 + u_4^2 = \|\mathbf{u}\|^2 \tag{3.8}$$

The KS transformation maps fibers on the three-sphere of radius $\sqrt{r}$ in $\mathbb{U}^4$ to points on the two-sphere of radius r in $\mathbb{E}^3$.

Regularizing the equations of orbital motion by means of the KS transformation requires the time transformation due to Sundman:

$$\mathrm{d}t = r\,\mathrm{d}s \tag{3.9}$$

which was widely discussed in the previous chapter.

3.2.1 Defining the fibers

Hopf (1931) proved that the transformation from the three-sphere to the two-sphere maps circles to single points, defining the structure $\mathbb{S}^1 \hookrightarrow \mathbb{S}^3 \to \mathbb{S}^2$. A direct consequence of the result shown in Eq. (3.4) is the fact that Eq. (3.6) is invariant under the gauge transformation $\mathscr{R} : \mathbf{u} \mapsto \mathbf{w}$,

$$\mathbf{x} = \mathsf{L}(\mathbf{u})\,\mathbf{u} = \mathsf{L}(\mathbf{w})\,\mathbf{w} \tag{3.10}$$

Vector $\mathbf{w} = [w_1, w_2, w_3, w_4]^\top$ takes the form:

$$\mathbf{w} = \mathscr{R}(\phi; \mathbf{u}) = \mathsf{R}(\phi)\,\mathbf{u} \tag{3.11}$$

where $\mathsf{R}(\phi)$ is the matrix

$$\mathsf{R}(\phi) = \begin{bmatrix} \cos\phi & 0 & 0 & -\sin\phi \\ 0 & \cos\phi & \sin\phi & 0 \\ 0 & -\sin\phi & \cos\phi & 0 \\ \sin\phi & 0 & 0 & \cos\phi \end{bmatrix} \tag{3.12}$$

This matrix is orthogonal, and also

$$\mathsf{R}^\top(\phi) = \mathsf{R}(-\phi)$$

Being $\mathsf{R}(\phi)$ orthogonal, Eq. (3.11) can be inverted to provide

$$\mathbf{u} = \mathscr{R}^{-1}(\phi; \mathbf{w}) = \mathsf{R}(-\phi)\,\mathbf{w} \tag{3.13}$$

The transformation $\mathscr{R}$ preserves the radius r, i.e.,

$$r = \mathbf{u} \cdot \mathbf{u} = \mathbf{w} \cdot \mathbf{w}$$

Since the radius is invariant to the selection of the point in the fiber it follows that the physical time, defined by Eq. (3.9), is $\mathscr{R}$-invariant as well.[2] The identity in Eq. (3.10) and the r-orthogonality of matrix L furnish a useful relation:

$$\mathbf{w} = \mathsf{L}^{-1}(\mathbf{w})\,\mathsf{L}(\mathbf{u})\,\mathbf{u} = \mathsf{R}(\phi)\,\mathbf{u} \implies \mathsf{L}^\top(\mathbf{w})\,\mathsf{L}(\mathbf{u}) = r\mathsf{R}(\phi)$$

The angular variable ϕ parameterizes the Hopf fibration in four-dimensional space. In fact, Eq. (3.11) defines explicitly the fiber $\mathcal{F}$: changing the value of ϕ defines different points in $\mathbb{U}^4$ that are KS transformed to the same point in $\mathbb{E}^3$. This yields the definition of fiber as the subset of all points in four-dimensional space that are mapped into the same point in $\mathbb{E}^3$ by means of the KS transformation,

$$\mathcal{F} = \{\mathbf{w}(\phi) \in \mathbb{U}^4 \,|\, \mathbf{x} = \mathscr{K}(\mathbf{w}),\ \forall\phi \in [0, 2\pi)\}$$

A different fiber transforms into a different point. Consequently, two fibers cannot intersect because the intersection point will then be transformed into the same point in $\mathbb{E}^3$ despite belonging to two different fibers (Stiefel and Scheifele, 1971, p. 271). The stereographic projection of the fibers onto $\mathbb{E}^3$ reveals that two fibers in KS space are connected by a Hopf link, as sketched in Fig. 3.1.

2 Alternative forms of the time transformation can be found in the literature, generalized as $\mathrm{d}t/\mathrm{d}s = g(\mathbf{x}, \dot{\mathbf{x}})$. We refer to Sect. 2.2 for a survey of transformations involving different powers of the radial distance, the potential, the Lagrangian, or combinations of the relative separations for the case of N-body problems. The vectors $\mathbf{x}$ and $\dot{\mathbf{x}}$ are $\mathscr{R}$-invariant, so the uniqueness of the physical time is also guaranteed for more general transformations.

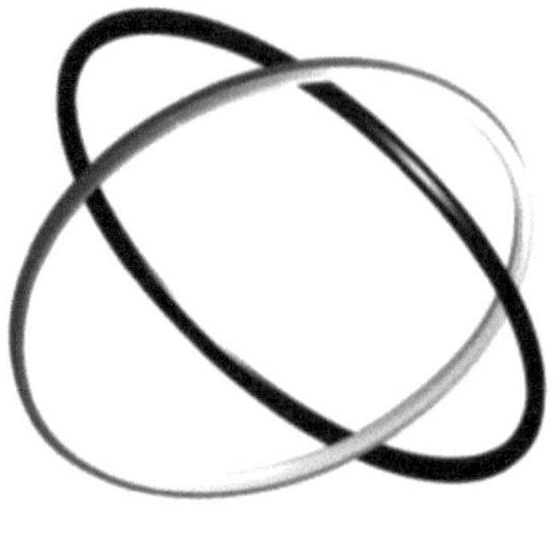

Fig. 3.1: Hopf link connecting two different fibers in KS space. The Hopf fibration is visualized by means of the stereographic projection to $\mathbb{E}^3$.

3.2.2 The velocity and the bilinear relation

Let $\mathbf{u}, \mathbf{w} \in \mathbb{U}^4$. The KS matrix satisfies the property

$$\mathsf{L}(\mathbf{u})\,\mathbf{w} = \mathsf{L}(\mathbf{w})\,\mathbf{u} \iff \ell(\mathbf{u}, \mathbf{w}) = 0$$

where $\ell(\mathbf{u}, \mathbf{w})$ denotes the bilinear relation

$$\ell(\mathbf{u}, \mathbf{w}) = u_1 w_4 - u_2 w_3 + u_3 w_2 - u_4 w_1$$

In addition, if $\ell(\mathbf{u}, \mathbf{w}) = 0$ then

$$(\mathbf{u} \cdot \mathbf{u})\,\mathsf{L}(\mathbf{w})\,\mathbf{w} - 2(\mathbf{u} \cdot \mathbf{w})\,\mathsf{L}(\mathbf{u})\,\mathbf{w} + (\mathbf{w} \cdot \mathbf{w})\,\mathsf{L}(\mathbf{u})\,\mathbf{u} = 0 \tag{3.14}$$

This property was referred to as the "reciprocity of KS-operators" by Deprit et al. (1994). It is easy to verify that $\ell(\mathbf{u}, \mathbf{u}) = 0$ is responsible for the fourth component of $\mathsf{L}(\mathbf{u})\,\mathbf{u}$ being null, and the result is the extension to $\mathbb{U}^4$ of a three-dimensional vector. The fourth component of $\mathsf{L}(\mathbf{u})\,\mathbf{w}$ is null if, and only if, $\ell(\mathbf{u}, \mathbf{w}) = 0$. The operator $\mathsf{L}(\mathbf{u})$ is linear, meaning that

$$\mathsf{L}(k\mathbf{u}) = k\mathsf{L}(\mathbf{u})\,, \qquad \mathsf{L}(\mathbf{u} + \mathbf{w}) = \mathsf{L}(\mathbf{u}) + \mathsf{L}(\mathbf{w})$$

Differentiating Eq. (3.6) with respect to fictitious time and taking into account the time transformation from Eq. (3.9) yields

$$\dot{\mathbf{x}} = \frac{2}{r}\,\mathsf{L}(\mathbf{u})\,\mathbf{u}' \tag{3.15}$$

where $\dot{\mathbf{x}} = [v_x, v_y, v_z, 0]^\top$ is the velocity vector extended to $\mathbb{R}^4$. Note that the fourth component is zero, which means

$$\ell(\mathbf{u}, \mathbf{u}') = 0$$

Moreover, Stiefel and Scheifele (1971, p. 29) proved that $\ell(\mathbf{u}, \mathbf{u}') = 0$ is a first integral of orbital motion. Provided that the KS transformation is $\mathscr{R}$-invariant, it follows that the bilinear relation holds for all points in a given fiber,

$$\ell(\mathbf{w}, \mathbf{w}') = \ell(\mathbf{u}, \mathbf{u}') = 0 \tag{3.16}$$

Let $\mathbf{t}(\phi) \in \mathbb{U}^4$ denote the vector that is tangent to a fiber $\mathcal{F}$ at $\mathbf{w}(\phi)$. The direction of $\mathbf{t}$ can be obtained by differentiating Eq. (3.11) with respect to ϕ. It reads

$$\mathbf{t} = \mathsf{R}^*(\phi)\,\mathbf{u}$$

where $\mathsf{R}^*(\phi) = \mathrm{d}\mathsf{R}(\phi)/\mathrm{d}\phi$ is obtained by differentiating Eq. (3.12). Taking as an example $\phi = 0$ yields the components of the tangent vector $\mathbf{t}$,

$$\mathbf{t} = [-u_4, u_3, -u_2, u_1]^\top \tag{3.17}$$

This unveils a geometrical interpretation of the bilinear relation $\ell(\mathbf{u}, \mathbf{v}) = 0$: it can be understood as an orthogonality condition (Stiefel and Scheifele, 1971, §43), since

$$\ell(\mathbf{u}, \mathbf{v}) = 0 \iff \mathbf{v} \cdot \mathbf{t} = 0$$

Two vectors $\mathbf{u}$ and $\mathbf{v}$ satisfy the bilinear relation $\ell(\mathbf{u}, \mathbf{v}) = 0$ if $\mathbf{v}$ is orthogonal to the fiber through $\mathbf{u}$. Provided that $\ell(\mathbf{u}, \mathbf{u}') = 0$ holds naturally and it is an integral of motion it follows that the velocity in KS space, $\mathbf{u}'$, is always orthogonal to the fiber at $\mathbf{u}$. The fiber bundle $\mathbb{S}^1 \hookrightarrow \mathbb{S}^3 \to \mathbb{S}^2$ shows that the fibers constituting the three-sphere are circles, corresponding to points on the two-sphere. Indeed, the tangent vector $\mathbf{t}(\phi)$ is always perpendicular to the position vector $\mathbf{w}(\phi)$,

$$\mathbf{w} \cdot \mathbf{t} = [\mathsf{R}(\phi)\,\mathbf{u}] \cdot [\mathsf{R}^*(\phi)\,\mathbf{u}] = \mathbf{u} \cdot \{\mathsf{R}^\top(\phi)\,[\mathsf{R}^*(\phi)\mathbf{u}]\} = 0$$

no matter the value of ϕ. Section 3.7 is devoted to the definition of orthogonal bases and a cross product in $\mathbb{U}^4$.

3.2.3 The two-body problem

Equation (2.5) governs the evolution of the two-body problem in terms of the fictitious time s. Using the notation introduced in the present chapter it takes the form

$$\mathbf{x}'' - \frac{r'}{r}\mathbf{x}' + \frac{\mu}{r}\mathbf{x} = r^2\mathbf{a}_p$$

Vector $\mathbf{a}_p$ denotes the perturbing acceleration with a null fourth component in order to extend it to four dimensions. Differentiating Eq. (3.10) with respect to fictitious time renders

$$r\mathsf{L}(\mathbf{u}')\,\mathbf{u}' + \mathsf{L}(\mathbf{u})\left(r\mathbf{u}'' - r'\mathbf{u}' + \frac{\mu}{2}\,\mathbf{u}\right) = \frac{r^3}{2}\mathbf{a}_p \tag{3.18}$$

The fictitious radial velocity r' transforms into

$$r' = \frac{\mathbf{r} \cdot \mathbf{r}'}{r} = 2(\mathbf{u} \cdot \mathbf{u}')$$

Identifying $\mathbf{w} \equiv \mathbf{u}'$, Eq. (3.14) provides the relation

$$r\mathsf{L}(\mathbf{u}')\,\mathbf{u}' = \mathsf{L}(\mathbf{u})\left[2(\mathbf{u} \cdot \mathbf{u}')\,\mathbf{u}' - (\mathbf{u}' \cdot \mathbf{u}')\,\mathbf{u}\right]$$

Introducing this expression in Eq. (3.18) and premultiplying the result by $\mathbf{L}^{-1}(\mathbf{u})$ yields

$$\mathbf{u}'' + \frac{1}{r}\left[\frac{\mu}{2} - (\mathbf{u}' \cdot \mathbf{u}')\right]\mathbf{u} = \frac{r}{2}\mathbf{L}^\top(\mathbf{u})\,\mathbf{a}_p$$

The term in brackets turns out to be equivalent to the Keplerian energy of the system, $\mathcal{E}_k$:

$$\omega^2 = \frac{1}{r}\left[\frac{\mu}{2} - (\mathbf{u}' \cdot \mathbf{u}')\right] = \frac{1}{2}\left(\frac{\mu}{r} - \frac{v^2}{2}\right) = -\frac{\mathcal{E}_k}{2}$$

In sum, the perturbed two-body problem reduces to

$$\mathbf{u}'' + \omega^2\mathbf{u} = \frac{(\mathbf{u} \cdot \mathbf{u})}{2}\mathbf{L}^\top(\mathbf{u})\,\mathbf{a}_p \tag{3.19}$$

under the KS regularization.

3.2.4 The inverse mapping

The inverse KS transformation $\mathscr{K}^{-1}: \mathbf{x} \mapsto \mathbf{u}$ maps points to fibers. Let us introduce the auxiliary vector $\mathbf{v} = [v_1, v_2, v_3, v_4]^\top$, defined by the mapping

$$\begin{aligned} v_1 &= R\sin\theta \\ v_2 &= \frac{1}{2R}(y\sin\theta - z\cos\theta) \\ v_3 &= \frac{1}{2R}(y\cos\theta + z\sin\theta) \\ v_4 &= -R\cos\theta \end{aligned} \tag{3.20}$$

Here $R^2 = (r + |x|)/2$. The angle θ is different from ϕ: the points on the fiber are parameterized by θ, which is measured with respect to a certain axis; given two points $\mathbf{u}$ and $\mathbf{w}$ obtained by setting $\theta = \theta_1$ and θ_2 in Eq. (3.20), respectively, they relate by virtue of Eq. (3.11). This equation then provides the relation:

$$\theta_2 - \theta_1 = \phi$$

meaning that the variable ϕ denotes the angular *separation* between points along the same fiber. The value of θ depends on the position of the reference axis, whereas ϕ is independent from the selection of the origin of angles.

The point $\mathbf{u}$ is finally defined as

$$\begin{aligned} \mathbf{u} &= [v_1, v_2, v_3, v_4]^\top \quad \text{if } x \geq 0 \\ \mathbf{u} &= [v_2, v_1, v_4, v_3]^\top \quad \text{if } x < 0 \end{aligned} \tag{3.21}$$

Two alternative expressions are considered for avoiding potential singularities. They differ in the selection of the axes in KS space. From this result any point $\mathbf{w}_0$ in the initial fiber $\mathcal{F}_0$ can be obtained attending to

$$\begin{aligned} \mathbf{w}_0(\phi) &= \mathbf{R}(\phi)\,\mathbf{u}_0 \qquad \text{if } x_0 \geq 0 \\ \mathbf{w}_0(\phi) &= \mathbf{R}(-\phi)\,\mathbf{u}_0 \qquad \text{if } x_0 < 0 \end{aligned} \tag{3.22}$$

so that $\mathbf{x}_0 = \mathsf{L}(\mathbf{w}_0)\mathbf{w}_0$. The sign criterion complies with the different definitions of axes in KS space. See Sect. 3.7 for a discussion of the orthogonal frames attached to the fiber. The linear form of the inverse transformation can be found in Sect. D.3.

The velocity in $\mathbb{U}^4$ is obtained by inverting Eq. (3.15), taking into account the orthogonality relation in Eq. (3.7):

$$\mathbf{u}' = \frac{1}{2}\mathsf{L}^\top(\mathbf{u})\,\dot{\mathbf{x}}$$

The geometry of the inverse KS transformation can be studied from Fig. 3.2. On the left, there is a three-dimensional sphere of radius r which contains points in $\mathbb{E}^3$. We can select a collection of points on the surface of this sphere, defining the dark semicircumference. Applying the inverse KS transformation to this set of points transforms the semicircumference in a manifold in four dimensions, which can be represented in three dimensions by taking its stereographic projection. The result is the dark surface on the right of the figure. If we are only interested in how single points $\mathbf{r}_i$ are transformed to KS space, we will take only one point from the sphere on the left and it will define only one fiber on the right, represented in white and called $\mathcal{F}_j$. The KS transformation will map all points on this fiber to $\mathbf{r}_i$.

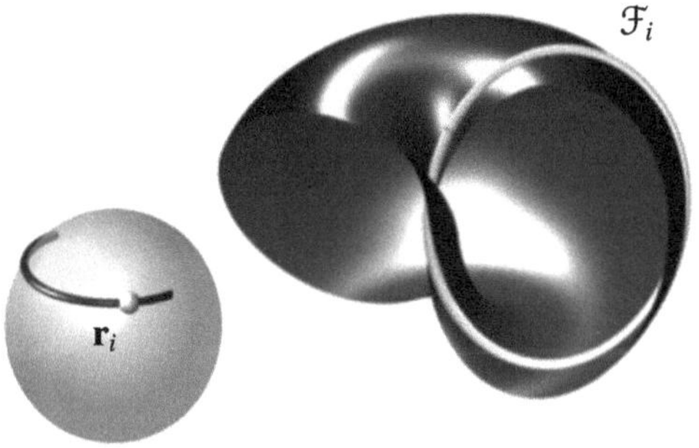

Fig. 3.2: Stereographic projection to $\mathbb{E}^3$ of the Hopf fibration corresponding to a set of initial positions on the three-dimensional sphere of radius r. The black semitorus consists of all the fibers that transform into the semi-circumference on the two-sphere on the bottom-left corner. One single fiber $\mathcal{F}_i$ is plotted in white, corresponding to $\mathbf{r}_i$.

3.3 Stability in KS space

The classical concepts of stability from Lyapunov and Poincaré can be translated to KS language by considering the topology of the transformation. First, we introduce an important new theorem regarding the geometry of the fibers (Roa et al., 2016c). From this theorem the concept of the fundamental manifold arises naturally.

The stability concepts presented here are not based on numerical analyses; previous studies concerning the stability of KS transformation (Baumgarte, 1972a, 1976b; Arakida and Fukushima, 2000) focused on the behavior of the *numerical* procedure. We aim for a series of definitions that capture the *physical* behavior, which should be independent from the formulation of the dynamics.

3.3.1 A central theorem

Two fibers can never intersect, as discussed when formally defining a fiber. It is now possible to advance on this statement and to formulate a fundamental property of the KS transformation:

Theorem 2. *(Roa et al., 2016c) The angular separation between two trajectories emanating from $\mathcal{F}_0$, measured along every fiber, is constant. That is*

$$\mathbf{w}_0 = \mathbf{R}(\phi_0)\,\mathbf{u}_0 \implies \mathbf{w}(s) = \mathbf{R}(\phi_0)\,\mathbf{u}(s)$$

for any value of ϕ_0 and the fictitious time s. This is an intrinsic property of KS space and does not depend on the dynamics of the system.

Proof. Consider two trajectories in KS space, $\mathbf{u} = \mathbf{u}(s)$ and $\mathbf{w} = \mathbf{w}(s)$, departing from the same fiber $\mathcal{F}_0$. They relate by means of Eq. (3.11). In the most general case the angle ϕ can be described by a function $\phi = \phi(s)$ and initially it is $\phi(0) = \phi_0$. The trajectories evolve according to

$$\mathbf{w}(s) = \mathscr{R}(\phi; \mathbf{u}(s)) = \mathbf{R}(\phi)\mathbf{u}(s) \tag{3.23}$$

Differentiating this equation with respect to fictitious time yields

$$\mathbf{w}'(s) = \mathbf{R}'(\phi)\,\mathbf{u}(s) + \mathbf{R}(\phi)\,\mathbf{u}'(s) \tag{3.24}$$

Equation (3.16) proved that the bilinear relation holds for any trajectory in KS space, meaning that $\ell(\mathbf{w}, \mathbf{w}') = \ell(\mathbf{u}, \mathbf{u}') = 0$. This renders:

$$\ell(\mathbf{w}, \mathbf{w}') = \ell(\mathbf{R}(\phi)\mathbf{u}, \mathbf{R}'(\phi)\,\mathbf{u} + \mathbf{R}(\phi)\,\mathbf{u}') = 0$$

after substituting Eqs. (3.23) and (3.24). Expanding the bilinear relation in the previous expression shows that

$$\ell(\mathbf{R}(\phi)\mathbf{u}, \mathbf{R}'(\phi)\,\mathbf{u} + \mathbf{R}(\phi)\,\mathbf{u}') = r\,\frac{\mathrm{d}\phi}{\mathrm{d}s} + \ell(\mathbf{u}, \mathbf{u}') = 0 \tag{3.25}$$

Assuming that $r > 0$ and considering that $\ell(\mathbf{u}, \mathbf{u}') = 0$ one gets

$$\frac{\mathrm{d}\phi}{\mathrm{d}s} = 0 \implies \phi(s) = \phi_0$$

so the angular separation along every fiber remains constant. We emphasize that no assumptions about the dynamics have been made. □

A direct consequence of this result is the relation between the velocities along the trajectories $\mathbf{u}(s)$ and $\mathbf{w}(s)$:

$$\mathbf{w}'(s) = \mathscr{R}(\phi; \mathbf{u}'(s))$$

3.3.2 The fundamental manifold Γ

A trajectory in Cartesian space, understood as a continuum of points in $\mathbb{E}^3$, is represented by a continuum of fibers in $\mathbb{U}^4$. Each fiber is KS transformed to a point of the trajectory. The fibers form the *fundamental manifold*, Γ.

Equation (3.22) defines the initial fiber $\mathcal{F}_0$, which yields a whole family of solutions parameterized by the angular variable ϕ. Every trajectory $\mathbf{w}(s)$ is confined to the fundamental manifold. Thanks to Thm. 2 the manifold Γ can be constructed following a simple procedure: first, a reference trajectory $\mathbf{u}(s)$ is propagated from any point in $\mathcal{F}_0$; then, mapping the transformation $\mathscr{R}$ over it renders a fiber $\mathcal{F}_i$ for each point $\mathbf{u}(s_i)$ of the trajectory. The set $\cup_i \mathcal{F}_i$ defines Γ. Recall that

$$\bigcap_i \mathcal{F}_i = \emptyset$$

The fact that all trajectories emanating from $\mathcal{F}_0$ are confined to Γ is what makes an arbitrary choice of θ in Eq. (3.20) possible. Figure 3.3 depicts the construction of the fundamental manifold Γ.

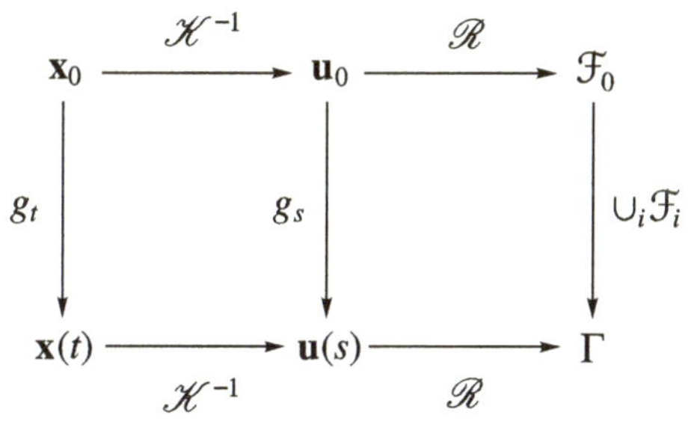

Fig. 3.3: Construction of the fundamental manifold. The mapping $g_t\colon \mathbf{x}_0 \mapsto \mathbf{x}(t)$ denotes the integration of the trajectory from t_0 to t. Similarly, g_s refers to the propagation using the fictitious time.

3.3.3 Fixed points, limit cycles, and attractors

Points in $\mathbb{E}^3$ transform into fibers in $\mathbb{U}^4$. Thus, a fixed point in Cartesian space, $\mathbf{x}_0$, translates into a fixed fiber in KS space, $\mathcal{F}_0$. Asymptotically stable fixed fibers (to be defined formally in the next section) attract the fundamental manifold of solutions, $\Gamma \to \mathcal{F}_0$. Asymptotic instability is equivalent to the previous case under a time reversal.

Limit cycles are transformed to fundamental manifolds, referred to as limit fundamental manifolds Γ_0. A fundamental manifold Γ originating in the basin of attraction of a limit fundamental manifold will converge to it after sufficient time. For $\Gamma \to \Gamma_0$ convergence means that each fiber in Γ approaches the corresponding fiber in Γ_0. Correspondence between fibers is governed by the time synchronism of the solutions.[3]

In a more general sense, attractors in $\mathbb{U}^4$ are invariant sets of the flow. The point-to-fiber correspondence connects attractors in $\mathbb{E}^3$ with attractors in KS space. The basin of attraction of an attractive set $Y_u \subset \mathbb{U}^4$ is built from its definition in three dimensions. Let $X \subset \mathbb{E}^3$ be the basin of attraction of Y. It can be transformed to KS space, $X \to X_u$, thanks to

$$X_u = (\mathscr{R} \circ \mathscr{K}^{-1})(X) = \mathscr{R}(\mathscr{K}^{-1}(X))$$

This construction transforms arbitrary sets in $\mathbb{E}^3$ to $\mathbb{U}^4$. The inverse KS transformation constitutes a dimension raising mapping, so in general $\dim(X_u) = \dim(X) + 1$.

3.3.4 Relative dynamics and synchronism

The theories about the local stability of dynamical systems are based on the relative dynamics between nearby trajectories. The concepts of stability formalize how the separation between two (initially close) trajectories evolves in time. But the concept of *time evolution* requires a further discussion because of having introduced an alternative time variable via the Sundman transformation.

Keplerian motion is known to be Lyapunov unstable; see Chap. 2 for details. Small differences in the semimajor axes of two orbits result in a separation that grows in time because of having different periods. However, Kepler's problem transforms into a harmonic oscillator by means of the KS transformation, with the fictitious time being equivalent to the eccentric anomaly. The resulting system is stable: for fixed values of the eccentric anomaly the separation between points in each orbit will be small, because of the structural (or Poincaré) stability of the motion. These considerations are critical for the numerical integration of the equations of motion. But in this chapter we seek a theory of stability in $\mathbb{U}^4$ expressed in the language of the physical time t, because of its physical and practical interest. The conclusions about the stability of the system will be equivalent to those obtained in Cartesian space.

The spectrum of the linearized and normalized form of Kepler's problem written in Cartesian coordinates exhibits one eigenvalue with a positive real part, $\lambda = (2\mu/r^3)^{1/2}$ – see Eq. (2.2) –. Lyapunov's theory of linear stability states that the system is unstable.

3 A dedicated discussion about the synchronism of the solutions can be found in Chapter 7. An interesting and closely related concept is the *isochronous correspondence* defined by Szebehely (1967, p. 233).

Under the action of the KS transformation Kepler's problem transforms into

$$\frac{d^2\mathbf{u}}{ds^2} = -\frac{h}{2}\mathbf{u} \tag{3.26}$$

where h is minus the Keplerian energy. Although the linear analysis is not decisive in this case, selecting a candidate Lyapunov function $V(\mathbf{u}, \mathbf{u}') = h(\mathbf{u} \cdot \mathbf{u})/4 + (\mathbf{u}' \cdot \mathbf{u}')/2$ the stability of the system is proved. In order to represent the Lyapunov instability of the motion with respect to time t the Sundman transformation needs to be considered. Given two circular orbits of radii r_1 and r_2, the time delay between both solutions is $\Delta t = t_2 - t_1 = (r_2 - r_1)s$. The time delay grows with fictitious time and small values of $r_2 - r_1$ do not guarantee that Δt remains small.

This phenomenon relates to the synchronism of the solutions. Solutions to the system defined in Eq. (3.26) are stable if they are synchronized in fictitious time, but unstable if they are synchronized in physical time. We adopt this latter form of synchronism for physical coherence. Chapter 7 discusses the differences between the two approaches when introducing the theory of asynchronous relative motion.

3.3.5 Stability of the fundamental manifold

3.3.5.1 Lyapunov stability

A trajectory $\mathbf{r}(t)$ in $\mathbb{E}^3$ is said to be Lyapunov stable if, for every small $\varepsilon > 0$, there is a value $\delta > 0$ such that for any other solution $\mathbf{r}^*(t)$ satisfying $\|\mathbf{r}(t_0) - \mathbf{r}^*(t_0)\| < \delta$ it is $\|\mathbf{r}(t) - \mathbf{r}^*(t)\| < \varepsilon$, with $t > t_0$. In KS language trajectory translates into fundamental manifold. In order to extend the definition of Lyapunov stability accordingly, an adequate metric d to measure the distance between manifolds is required.

Let Γ_1 and Γ_2 be two (distinct) fundamental manifolds. The fibers in Γ_1 can never intersect the fibers in Γ_2. But both manifolds may share certain fibers, corresponding to the points of intersection between the two resulting trajectories in Cartesian space. The distance between the manifolds at $t \equiv t(s_1) = t(s_2)$ is the distance between the corresponding fibers. Setting θ to a reference value θ_{ref} in Eq. (3.20) so that $\theta_1 = \theta_2 \equiv \theta_{\text{ref}}$, we introduce the metric:

$$d(t; \Gamma_1, \Gamma_2) = \frac{1}{2\pi} \int_0^{2\pi} \|\mathbf{w}_1(s_1; \phi) - \mathbf{w}_2(s_2; \phi)\| \, d\phi \tag{3.27}$$

with $d(t; \Gamma_1, \Gamma_2) \equiv d(\mathcal{F}_1, \mathcal{F}_2)$. It is measured by computing the distance between points in Γ_1 and Γ_2 with the same value of ϕ, and then integrating over the entire fiber. It is defined for given values of physical time, and not fictitious time. The reason is that the goal of this section is to define a theory of stability such that the fundamental manifold inherits the stability properties of the trajectory in Cartesian space. This theory is based on the physics of the system, not affected by a reformulation of the equations of motion.

Consider a fundamental manifold Γ, referred to a nominal trajectory $\mathbf{r}(t)$, and a second manifold Γ^* corresponding to a perturbed trajectory $\mathbf{r}^*(t)$. If the nominal trajectory is Lyapunov stable, then for every $\varepsilon_u > 0$ there is a number $\delta_u > 0$ such that

$$d(t_0; \Gamma, \Gamma^*) < \delta_u \implies d(t; \Gamma, \Gamma^*) < \varepsilon_u$$

If the initial separation between the manifolds is small it will remain small according to the metric defined in Eq. (3.27).

The nominal solution $\mathbf{r}(t)$ is said to be asymptotically stable if $\|\mathbf{r}(t) - \mathbf{r}^*(t)\| \to 0$ for $t \to \infty$. Similarly, the fundamental manifold Γ will be asymptotically stable if $d(t; \Gamma, \Gamma^*) \to 0$ for sufficiently large times. The opposite behavior $d(t; \Gamma, \Gamma^*) \to \infty$ corresponds to an asymptotically unstable fundamental manifold. It behaves as if it were asymptotically stable if the time is reversed.

3.3.5.2 Poincaré maps and orbital stability

The notion of Poincaré (or orbital) stability is particularly relevant when analyzing the fundamental manifold due to its geometrical implications. Kepler's problem is unstable in the sense of Lyapunov but it is orbitally stable: disregarding the time evolution of the particles within their respective orbits, the separation between the orbits remains constant.

The definition of the Poincaré map in $\mathbb{E}^3$ involves a two-dimensional section Σ that is transversal to the flow. Denoting by $\mathbf{p}_1, \mathbf{p}_2, \ldots$ the successive intersections of a periodic orbit with Σ, the Poincaré map $\mathscr{P}$ renders

$$\mathscr{P}(\mathbf{p}_n) = \mathbf{p}_{n+1}$$

The generalization of the Poincaré section to KS space $\mathscr{K} : \Sigma \to \Sigma_u$ results in a subspace embedded in $\mathbb{U}^4$. In Sect. 3.2.2 we showed that the trajectories intersect the fibers at right angles, provided that the velocity $\mathbf{u}'$ is orthogonal to the vector tangent to the fiber. Thus, every fiber defines a section that is transversal to the flow. The transversality condition for Σ translates into the section containing the fiber at $\mathbf{u}$.

The Poincaré section Σ_u can be constructed by combining the set of fibers that are KS transformed to points in Σ. Let $\mathbf{n} = [n_x, n_y, n_z]^\top$ be the unit vector normal to Σ in $\mathbb{E}^3$, projected onto an inertial frame. The Poincaré section takes the form

$$\Sigma \equiv n_x(x - x_0) + n_y(y - y_0) + n_z(z - z_0) = 0 \tag{3.28}$$

where (x_0, y_0, z_0) are the coordinates of the first intersection point. Equation (3.28) can be written in parametric form as $\Sigma(x(\eta, \xi), y(\eta, \xi), z(\eta, \xi))$, with η and ξ two free parameters. The extended Poincaré section Σ_u is obtained by transforming points on Σ to KS space and then mapping the fibration $\mathscr{R}$:

$$\Sigma_u = (\mathscr{R} \circ \mathscr{K}^{-1})(\Sigma)$$

The choice of the Poincaré section Σ is not unique, and therefore the construction of Σ_u is not unique either. The resulting Poincaré section Σ_u is a subspace of dimension three embedded in $\mathbb{U}^4$. Indeed, the transformation $(\mathscr{R} \circ \mathscr{K}^{-1})(\Sigma)$ provides:

$$\Sigma \mapsto \Sigma_u(u_1(\eta, \xi, \phi), u_2(\eta, \xi, \phi), u_3(\eta, \xi, \phi), u_4(\eta, \xi, \phi))$$

meaning that points in Σ_u are fixed by three parameters, (η, ξ, ϕ). The dimension is raised by $(\mathscr{R} \circ \mathscr{K}^{-1})$.

The intersection between a given fundamental manifold and the Poincaré section Σ_u results in a fiber,

$$\Gamma \cap \Sigma_u = \mathfrak{F}$$

Successive intersections can be denoted $\mathfrak{F}_1, \mathfrak{F}_2, \ldots$ The Poincaré map in $\mathbb{U}^4$, $\mathscr{P}: \Sigma_u \to \Sigma_u$, is

$$\mathscr{P}(\mathfrak{F}_n) = \mathfrak{F}_{n+1}$$

Every point in a fiber intersects Σ_u simultaneously. Because of the $\mathscr{R}$-invariance of the Sundman transformation the time period between crossings is the same for every trajectory connecting $\mathfrak{F}_n$ and $\mathfrak{F}_{n+1}$.

Let Γ denote a fundamental manifold representing a nominal periodic orbit, and let Γ^* be a perturbed solution. They differ in the conditions at the first Σ-crossing, $\mathfrak{F}_1$ and $\mathfrak{F}_1^*$, respectively. The manifold Γ is said to be Poincaré (or orbitally) stable if

$$d\left(\mathfrak{F}_1^*, \mathfrak{F}_1\right) < \delta_u \implies d\left(\mathscr{P}^n(\mathfrak{F}_1^*), \mathfrak{F}_1\right) < \varepsilon_u$$

If the separation between the fibers at the first crossing is small, the separation will remain small after n crossings.

3.4 Order and chaos

In the previous section we generalized the key concepts of dynamical stability to KS space. The approach we followed aims for a theory that captures the physical properties of the system, instead of focusing on its purely numerical conditioning. The next step is the analysis of chaos in $\mathbb{U}^4$.

Chaotic systems are extremely sensitive to numerical errors due to the strong divergence of the integral flow. This is specially important in the vicinity of singularities, and it is precisely here where KS regularization exhibits all its potential. This section focuses on characterizing the exponential divergence of trajectories in $\mathbb{U}^4$ due to highly unstable dynamics.

By definition the fundamental manifold is mapped to a trajectory in $\mathbb{E}^3$. The equations of motion in $\mathbb{U}^4$ are no more than a reformulation of a dynamical system originally written in $\mathbb{E}^3$. For sufficiently smooth perturbations the Picard–Lindelöf theorem ensures the uniqueness of the solution. Thus, the corresponding fundamental manifold is also unique and its KS transform defines only one trajectory. This means that any trajectory in the fundamental manifold is mapped to the same exact trajectory in $\mathbb{E}^3$, no matter the position within the initial fiber. An observer in three-dimensional space, unaware of the extra degree of freedom introduced by the gauge $\mathscr{R}$, will always perceive the same trajectory no matter the values of ϕ.

3.4.1 The $\mathscr{K}$-separation

In order to integrate the equations of motion numerically in $\mathbb{U}^4$ the initial values of $\mathbf{u}_0$ and $\mathbf{u}_0'$ need to be fixed. This means choosing a point in the fiber $\mathcal{F}_0$. Since all the points in $\mathcal{F}_0$ are KS transformed to the same exact state vector in $\mathbb{E}^3$, the selection of the point is typically arbitrary. But for an observer in $\mathbb{U}^4$ different values of ϕ yield different initial conditions, and therefore the initial value problem to be integrated may behave differently. Ideally[4] all trajectories emanating from $\mathcal{F}_0$ remain in the same fundamental manifold, which is unique. However, numerical errors leading to the exponential divergence of the trajectories can cause the trajectories to depart from the fundamental manifold. In other words, after sufficient time two trajectories originating from the same fiber $\mathcal{F}_0$, $\mathbf{w}_0 = \mathscr{R}(\phi; \mathbf{u}_0)$, will no longer define the same fiber $\mathcal{F}(s)$, $\mathbf{w}(s) \neq \mathscr{R}(\phi; \mathbf{u}(s))$. In this case Thm. 2 will be violated. Multiple fundamental manifolds will appear, obtained by mapping the transformation $\mathscr{R}$ over each of the trajectories. The observer in $\mathbb{E}^3$ will see a collection of trajectories that depart from the same exact state vector and they separate in time, as if the problem had a random component. This behavior can only be understood in four dimensions.

These topological phenomena yield a natural way of measuring the error growth in KS space without the need of a precise solution. Let $\mathbf{u}(s)$ be a reference trajectory in $\mathbb{U}^4$, and let $\mathbf{w}(s)$ be a second trajectory defined by $\mathbf{w}_0 = \mathscr{R}(\phi; \mathbf{u}_0)$. It is possible to build the fundamental manifold Γ from the solution $\mathbf{u}(s)$. The second solution is expected to be $\mathbf{w}^*(s) = \mathscr{R}(\phi; \mathbf{u}(s))$ by virtue of Thm. 2. When numerical errors are present $\mathbf{w}(s)$ and its expected value $\mathbf{w}^*(s)$ (the projection of the fundamental manifold) may not coincide. Note that $\mathbf{w}(s) = \mathbf{w}^*(s)$ ensures the uniqueness of the solution, but says nothing about its accuracy. The separation between $\mathbf{w}(s)$ and its projection on Γ is an indicator of the breakdown of the topological structure supporting the KS transformation, meaning that the solutions can no longer be trusted.

Motivated by this discussion we introduce the concept of the $\mathscr{K}$-separation, $d_{\mathscr{K}}$

$$d_{\mathscr{K}}(s) = \|\mathbf{w}(s) - \mathbf{w}^*(s)\| = \|\mathbf{w}(s) - \mathscr{R}(\phi; \mathbf{u}^*(s))\| \tag{3.29}$$

4 Because of the limited precision of floating-point arithmetic, even the fact that all points generated with Eq. (3.22) and varying ϕ will be KS-transformed to the same exact point in $\mathbb{E}^3$ should be questioned. The loss of accuracy in the computation of the initial conditions in $\mathbb{U}^4$ will eventually introduce errors of random nature. As a result, Eq. (3.22) provides points that are not exactly in the true fiber. Although the separation is small (of the order of the round-off error) and negligible in most applications, it may have an impact on the numerical integration of chaotic systems.

defined as the Euclidean distance between an integrated trajectory and its projection on the manifold of solutions. Monitoring the growth of the $\mathscr{K}$-separation is a way of quantifying the error growth of the integration. In the context of N-body simulations, Quinlan and Tremaine (1992) discussed how the separation between nearby trajectories evolves: the divergence is exponential in the linear regime when the separation is small, but the growth rate is reduced when the separation is large. At this point the separation might be comparable to the interparticle distance. The $\mathscr{K}$-separation will grow exponentially at first (for $d_{\mathscr{K}} \ll 1$) until it is no longer small ($d_{\mathscr{K}} \sim \mathcal{O}(1)$), and then its growth slows down. Locating the transition point is equivalent to finding the time scale $t_{\rm cr}$ in which the solution in KS space can no longer be trusted: for $t < t_{\rm cr}$ the topological structure of $\mathbb{U}^4$ is preserved, but for $t > t_{\rm cr}$ the uniqueness of the manifold of solutions Γ is not guaranteed.

For $t < t_{\rm cr}$ the $\mathscr{R}$-invariance of the Sundman transformation holds. The time for all the points in a fiber coincides. Thus, $t_{\rm cr}$ and $s_{\rm cr}$ are interchangeable: at $t < t_{\rm cr}$ it is also $s < s_{\rm cr}$. The behavior of the solutions can be equally analyzed in terms of the physical or the fictitious time.

In practice, the $\mathscr{K}$-separation is evaluated as follows:

1. Choosing a reference θ in Eqs. (3.20) and (3.21), for example $\theta = 0$, integrate $\mathbf{u}^*(s)$.
2. Propagate a second trajectory $\mathbf{w}(s)$ generated from Eq. (3.22) with $\phi \neq 0$.
3. Build the expected trajectory $\mathbf{w}^*(s)$ by mapping $\mathscr{R}(\phi)$ over $\mathbf{u}^*(s)$, i.e., $\mathbf{w}^* = \mathsf{R}(\phi)\mathbf{u}^*$. The $\mathscr{K}$-separation is the Euclidean distance between $\mathbf{w}(s)$ and $\mathbf{w}^*(s)$.

3.4.2 Topological stability

The uniqueness of Γ can be understood as *topological stability*. KS space is said to be topologically stable if all the trajectories emanating from the same fiber define a unique manifold of solutions, and therefore they are all KS-transformed to the same trajectory in $\mathbb{E}^3$. For an observer in $\mathbb{E}^3$ a topologically unstable system seems nondeterministic, with solutions departing from the same initial conditions but separating in time with no apparent reason.

A system is topologically stable in the interval $t < t_{\rm cr}$. The trajectories diverge exponentially,

$$d_{\mathscr{K}}(t)/d_{\mathscr{K}}(0) \sim e^{\gamma_t t} \qquad \text{or} \qquad d_{\mathscr{K}}(s)/d_{\mathscr{K}}(0) \sim e^{\gamma_s s}$$

Here γ is equivalent to a Lyapunov exponent. For $t > t_{\rm cr}$ this equation no longer models the growth of the $\mathscr{K}$-separation and the system is topologically unstable. Simulations over the transition time $t_{\rm cr}$ integrated in $\mathbb{U}^4$ can no longer be trusted. Depending on the integrator, the integration tolerance, the floating-point arithmetic, the compiler, etc. the values of $t_{\rm cr}$ for a given problem might change. Thus, topological stability is a property of a certain propagation, which requires all the previous factors to be defined.

The validity of the solution for an integration over the critical time t_{cr} is not guaranteed. When $t_{cr} < t_{esc}$ (with t_{esc} denoting the escape time) not even the value of t_{esc} can be estimated accurately. In such a case solutions initialized at different points in the fiber may yield different escape times.

The method presented in this section provides an estimate of the interval in which the propagation is topologically stable. The exponent γ depends on the integration scheme and the dynamics, but it is not strongly affected by the integration tolerance. An estimate of the value of γ provides an estimate of the critical time for a given integration tolerance ε. Assuming $d_{\mathscr{K}}(t_{cr}) \sim 1$:

$$t_{cr} \sim -\frac{1}{\gamma_t} \log \varepsilon \tag{3.30}$$

Conversely, if the simulation needs to be carried out up to a given t_f, the required integration tolerance is approximately

$$\varepsilon \sim e^{-\gamma_t t_f}$$

This simple criterion proves useful for tuning and evaluating the numerical integration. In the following examples of application the values of γ_t are estimated by finding the slope of the exponential growth of the $\mathscr{K}$-separation in logarithmic scale. Although more rigorous algorithms could be developed, this approximation provides a good estimate of transition time between regimes.

3.5 Topological stability in *N*-body problems

Two examples of N-body problems are analyzed in this section. The first example is the Pythagorean three-body problem. The second example is a nonplanar configuration of the four-body problem. This problem simulates the dynamics of two field stars interacting with a stellar binary. The experiments are designed to show the practical aspects of the new concept of stability introduced in this chapter: the topological stability of KS space.

The problems are integrated using the regularization of the N-body problem based on the KS transformation proposed by Mikkola (1985) as a reformulation of the method by Heggie (1974). The initialization of the method is modified so that different points on the initial fiber $\mathcal{F}_0$ can be chosen. This means generalizing the relative coordinates $\mathbf{u}_{ij}$ to $\mathbf{w}_{ij}$ by means of the transformation $\mathscr{R}: \mathbf{u} \mapsto \mathbf{w}$. The trajectories depart from the same initial conditions in $\mathbb{E}^3$. We inherit the normalization proposed in the referenced paper, so that the gravitational constant is equal to one.

Heggie–Mikkola's method is implemented in Fortran. The Hamiltonian nature of the N-body problem has motivated the use of symplectic integrators. These algorithms do not suffer from the long-term accumulation of the error in the conservation of the energy (Wisdom and Holman, 1991; Sanz-Serna, 1992). Symplectic integrators rely on a fixed-step integration, which hinders the integration of close encounters. Although Duncan et al. (1998) developed SYMBA, a multistep and truly symplectic integrator that adjusts the step size according to the perturbations, and Chambers (1999) invented a hybrid symplectic integrator that deals with close encounters using a conventional integrator, conventional integrators with variable step size are preferred for numerical explorations of binary encounters because handling the integration step size is critical (Mikkola, 1983; Hut and Bahcall, 1983; Bacon et al., 1996; Mikkola and Tanikawa, 1999). As Mikkola recommends, we will integrate the problem with the Bulirsch and Stoer (1966) extrapolation scheme (Hairer et al., 1991, §II.9). See also the work by Murison (1989) for an analysis of the performance of this integrator in the three-body problem. The total $\mathscr{K}$-separation is computed by combining the $\mathscr{K}$-separations for the relative dynamics of each pair of bodies. Writing $\mathbf{u}_{ij} \equiv \mathbf{u}_\ell$ it is

$$d_{\mathscr{K}} = \sqrt{\sum_{\ell} d^2_{\mathscr{K},\ell}}$$

where $d_{\mathscr{K},\ell}$ is the $\mathscr{K}$-separation computed for $\mathbf{u}_\ell$.

3.5.1 The Pythagorean three-body problem

Originally developed by Burrau (1913), the Pythagorean problem consists in three particles of masses $m_1 = 3$, $m_2 = 4$, and $m_3 = 5$. The particles will be denoted (1), (2), and (3). At $t = 0$ the bodies are at rest and lying on the vertices of a Pythagorean right triangle of sides 3, 4, and 5. The initial conditions are summarized in Table 3.1. This problem has been solved and discussed in detail by Szebehely and Peters (1967), so the solution to the problem is known.

The solution is displayed in Fig. 3.4. Initially the bodies approach the origin and after a number of close-approaches body (1) is ejected along a trajectory in the first quadrant, whereas (2) and (3) form a binary that escapes in the opposite direction. The escape occurs at approximately $t_{esc} \sim 60$. The solution shown in the figure is obtained by setting the integration tolerance to $\varepsilon = 10^{-13}$.

The problem is first integrated from a set of initial conditions obtained with $\theta = 0$ in Eq. (3.20). Then, a second trajectory initialized with $\theta \equiv \phi = 120$ deg is integrated

Tab. 3.1: Initial configuration of the Pythagorean problem. "Id" refers to the identification index of each body.

Id	x	y	z	v_x	v_y	v_z
(1)	1.0000	3.0000	0.0000	0.0000	0.0000	0.0000
(2)	−2.0000	−1.0000	0.0000	0.0000	0.0000	0.0000
(3)	1.0000	−1.0000	0.0000	0.0000	0.0000	0.0000

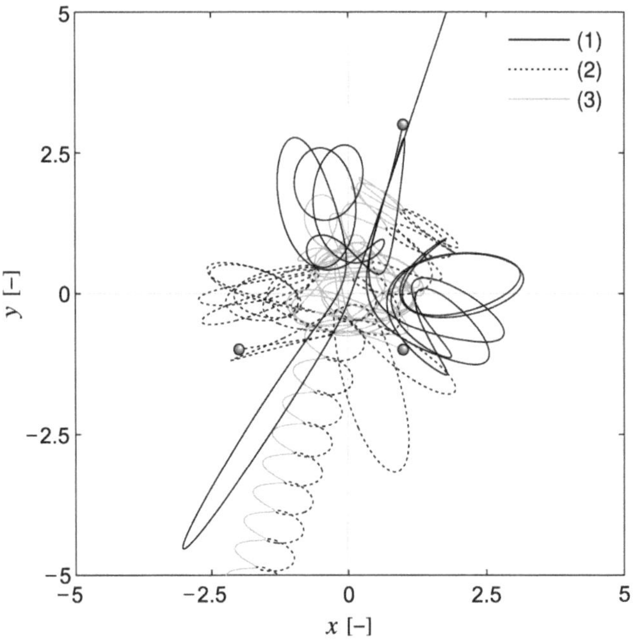

Fig. 3.4: Solution to the Pythagorean three-body problem. The thick dots represent the initial configuration of the system.

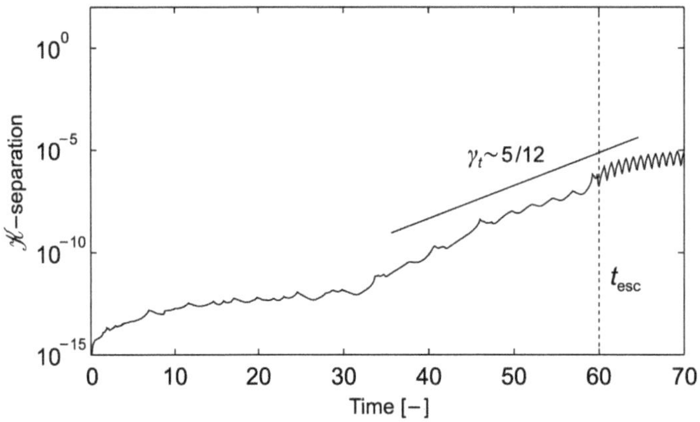

Fig. 3.5: $\mathscr{K}$-separation for the Pythagorean problem computed from a reference trajectory with $\theta = 0$ and $\phi = 120$ deg.

and their $\mathscr{K}$-separation is shown in Fig. 3.5. After a transient the separation grows exponentially with $\gamma_t \sim 5/12$, and no transitions are observed until the escape time ($t_{cr} > t_{esc}$). As discussed in the previous section this is equivalent to saying that the $\mathscr{K}$-separation remains small, and consequently the integration in $\mathbb{U}^4$ is topologically stable. The transformed solution in $\mathbb{E}^3$ will be unique no matter the initial position in the fiber $\mathcal{F}_0$.

3.5.2 Field stars interacting with a stellar binary

This second example analyzes the gravitational interaction of a binary system (1,2), of masses $m_1 = m_2 = 5$, with two incoming field stars (3) and (4) of masses $m_3 = m_4 = 3$. The initial conditions, presented in Table 3.2, have been selected so that both field stars reach the binary simultaneously.

The manifold of solutions is constructed from a reference solution with $\theta = 90$ deg. A second solution with $\theta = 120$ deg (or $\phi = 30$ deg) is propagated and the corresponding $\mathcal{K}$-separation is plotted in Fig. 3.6. The are two different regimes in the growth of the $\mathcal{K}$-separation: the first part corresponds to the linear regime where the $\mathcal{K}$-separation is small, whereas in the second part the separation is no longer small. Both regimes are separated by $t_{cr} \sim 42$, when solutions in $\mathbb{E}^3$ no longer coincide. This result is in good agreement with the value predicted by Eq. (3.30), which is $t_{cr} \sim 40$. Since for $t = t_{cr}$ the bodies have not yet escaped and the integration continues, the solution is topologically unstable. The escape time associated to the reference solution, $t_{esc} \sim 75$, might not be representative because it corresponds to the interval $t > t_{cr}$.

A direct consequence of the topological instability of the integration is the fact that solutions departing from the initial fiber $\mathcal{F}_0$ no longer represent the same solution in $\mathbb{E}^3$. Figure 3.7 shows two solutions that emanate from different points of the initial fiber. Ideally they should coincide exactly; but because the integration is topologically unstable for $t > t_{cr}$ the difference between both solutions becomes appreciable and the accuracy of the integration over t_{cr} cannot be guaranteed.

The topological instability is not directly related to the conservation of the energy. Although for $t > t_{cr}$ the integration becomes topologically unstable, Fig. 3.8 shows that the energy is conserved down to the integration tolerance until t_{esc}, well beyond t_{cr}. This is a good example of the fact that the preservation of the integrals of motion is a necessary but not sufficient condition for concluding that a certain integration is correct.

Tab. 3.2: Dimensionless initial conditions for the binary system, (1,2), and the field stars, (3) and (4). "Id" refers to the identification index of each star.

Id	x	y	z	v_x	v_y	v_z
(1)	0.6245	0.6207	0.0000	−0.7873	0.0200	−0.0100
(2)	0.6245	−0.6207	0.0000	0.7873	0.0200	0.0100
(3)	3.0000	3.0000	3.0000	−0.3000	−0.3000	−0.3000
(4)	−5.0817	−3.0000	−3.0000	0.3000	0.2333	0.3000

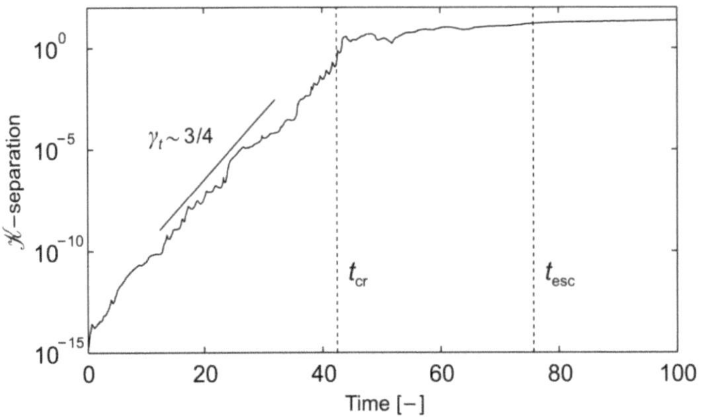

Fig. 3.6: $\mathscr{K}$-separation for the four-body problem computed from a reference trajectory with θ = 90 deg and a second trajectory with ϕ = 30 deg.

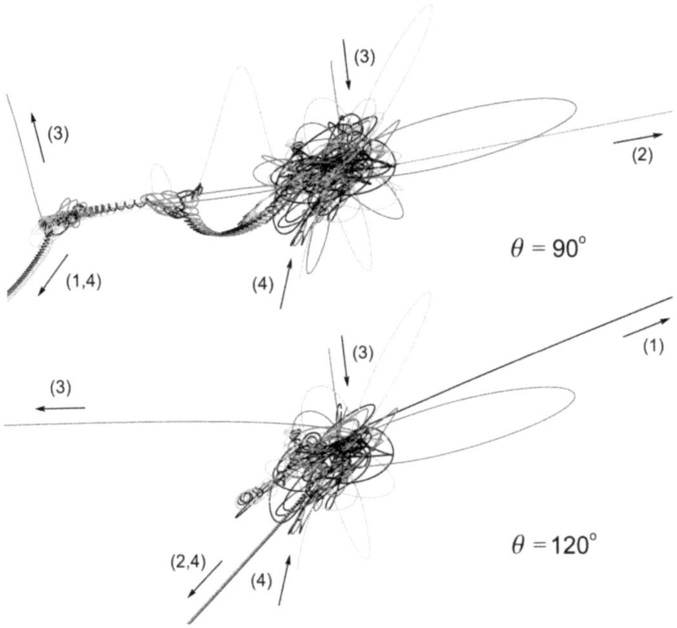

Fig. 3.7: Two solutions to the four-body problem departing from the same fiber $\mathcal{F}_0$: the top figure corresponds to θ = 90 deg, and the bottom figure has been generated with θ = 120 deg.

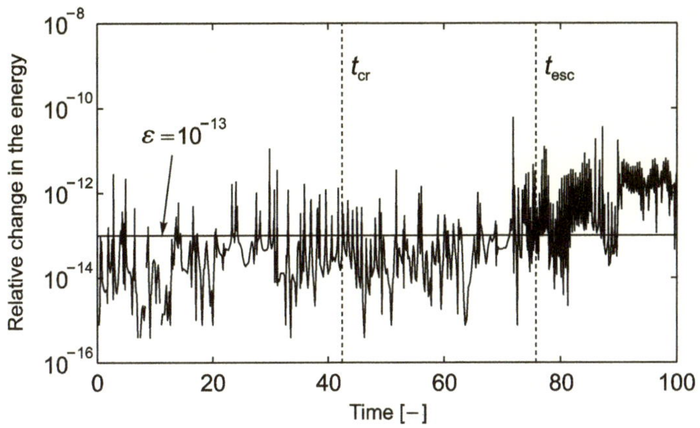

Fig. 3.8: Relative change in the energy referred to its initial value, $(\mathcal{E}(t) - \mathcal{E}_0)/\mathcal{E}_0$.

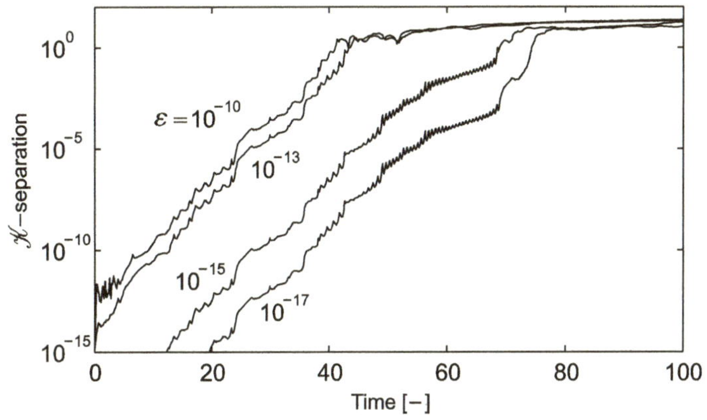

Fig. 3.9: $\mathcal{K}$-separation for the four-body problem for different integration tolerances. The solutions for $\varepsilon = 10^{-15}$ and 10^{-17} are computed in quadruple precision floating-point arithmetic.

The evolution of the $\mathcal{K}$-separation depends on the integration scheme and the tolerance. In order to analyze this dependency Fig. 3.9 shows the results of integrating the problem with four different tolerances and of changing from double to quadruple precision floating-point arithmetic. It is observed that refining the integration tolerance might extend the interval of topological stability. However, the dynamics of the system remain chaotic and the solutions will eventually diverge for sufficiently long times.

3.6 Gauge-generalized elements in KS space

Stiefel and Scheifele (1971, §19) attached a set of vectorial elements to KS space, with important advantages from the numerical integration perspective. In this section we provide a new perspective on the geometry of the elements and present, for the first time, the gauge-generalized elements in KS space.

The perturbation can be decomposed in

$$\mathbf{a}_p = -\nabla \mathcal{V} + \mathbf{p}$$

where $\mathcal{V}$ is a disturbing potential and $\mathbf{p}$ refers to the perturbations that do not derive from a potential. With this, Eq. (3.19) transforms into

$$\frac{\mathrm{d}^2\mathbf{u}}{\mathrm{d}s^2} + \omega^2\mathbf{u} = -\frac{\mathcal{V}}{2}\mathbf{u} - \frac{r}{4}\left[\frac{\partial \mathcal{V}}{\partial \mathbf{u}} - 2\mathsf{L}^\top(\mathbf{u})\,\mathbf{p}\right] \tag{3.31}$$

under the KS transformation. The angular frequency ω relates to the total energy $\mathcal{E}$ by means of

$$\omega^2 = -\frac{\mathcal{E}}{2} = \frac{1}{2}\left(\frac{\mu}{r} - \frac{v^2}{2} - \mathcal{V}\right) = -\frac{1}{2}(\mathcal{E}_k + \mathcal{V})$$

It differs from the Keplerian energy $\mathcal{E}_k$ by the disturbing potential $\mathcal{V}$. The relation in Eq. (3.14) combined with Eq. (3.15) renders a useful expression:

$$\mathsf{L}(\mathbf{u}')\,\mathbf{u}' = \frac{(\mathbf{x}\cdot\dot{\mathbf{x}})}{2}\,\dot{\mathbf{x}} - \frac{v^2}{4}\,\mathbf{x}$$

between the velocity and the derivative of $\mathbf{u}$.

3.6.1 Natural sets of elements: Geometrical interpretation

Differentiating Kepler's equation in the unperturbed case yields the time evolution of the eccentric anomaly. The result can be inverted to provide an alternative form of Sundman's transformation:

$$\frac{\mathrm{d}t}{\mathrm{d}E} = \frac{r}{2\omega_k} \tag{3.32}$$

where ω_k is the natural frequency referred to the Keplerian energy, $\omega_k = (-\mathcal{E}_k/2)^{1/2}$. The eccentric anomaly replaces the fictitious time as the independent variable, $E = 2\omega_k s$. Derivatives with respect to E, denoted by $\square^*$, relate to the derivatives with respect to s by means of

$$\frac{\mathrm{d}}{\mathrm{d}E} = \frac{1}{2\omega_k}\frac{\mathrm{d}}{\mathrm{d}s}$$

The velocity, given in Eq. (3.15), transforms into

$$\dot{\mathbf{x}} = \frac{4\omega_k}{r}\,\mathsf{L}(\mathbf{u})\,\mathbf{u}^* \tag{3.33}$$

Introducing the eccentric anomaly as the new independent variable and neglecting perturbations, the right-hand side of Eq. (3.31) vanishes, and the solution is simply

$$\mathbf{u}(E) = \mathbf{u}_0 \cos\frac{E}{2} + 2\mathbf{u}_0^* \sin\frac{E}{2} \tag{3.34}$$

The initial conditions $\mathbf{u}_0 = \mathbf{u}(0)$ and $\mathbf{u}_0^* = \mathbf{u}^*(0)$ are defined in terms of the position and velocity at $E = 0$. The state at periapsis, written in the perifocal frame $\mathcal{P} = \{\mathbf{i}_\mathcal{P}, \mathbf{j}_\mathcal{P}, \mathbf{k}_\mathcal{P}\}$, reads

$$\mathbf{r}_0 = r_p\,\mathbf{i}_\mathcal{P} \qquad \text{and} \qquad \mathbf{v}_0 = \sqrt{\mu(1+e)/r_p}\,\mathbf{j}_\mathcal{P}$$

with r_p the radius at periapsis. The radius at apoapsis will be denoted r_a.

Let $\hat{\mathbf{a}}, \hat{\mathbf{b}} \in \mathbb{U}^4$ be the representation of vectors $\mathbf{i}_\mathcal{P}$ and $\mathbf{j}_\mathcal{P}$ in KS space. The initial conditions $\mathbf{r}_0$ and $\mathbf{v}_0$ yield $\mathbf{u}_0 = \sqrt{r_p}\,\hat{\mathbf{a}}$ and $\mathbf{u}_0^* = \sqrt{r_a/4}\,\hat{\mathbf{b}}$. These expressions transform Eq. (3.34) into

$$\mathbf{u}(E) = \sqrt{r_p}\,\hat{\mathbf{a}} \cos\frac{E}{2} + \sqrt{r_a}\,\hat{\mathbf{b}} \sin\frac{E}{2} \tag{3.35}$$

This equation and its derivative can be inverted to provide the definition of $\hat{\mathbf{a}}$ and $\hat{\mathbf{b}}$ at periapsis,

$$\hat{\mathbf{a}} = \frac{\mathbf{u}_0}{\sqrt{r_p}}, \qquad \text{and} \qquad \hat{\mathbf{b}} = \frac{2\mathbf{u}_0^*}{\sqrt{r_a}} \tag{3.36}$$

The vectors $\hat{\mathbf{a}}$ and $\hat{\mathbf{b}}$ satisfy the bilinear relation

$$\ell(\hat{\mathbf{a}}, \hat{\mathbf{b}}) = \frac{2}{a\sqrt{1-e^2}}\,\ell(\mathbf{u}_0, \mathbf{u}_0^*) = 0$$

Equation (3.35) shows that the motion in the parametric space $\mathbb{U}^4$ is confined to the plane spanned by vectors $\hat{\mathbf{a}}, \hat{\mathbf{b}}$. This plane is in fact of Levi-Civita type,[5] or an $\mathscr{L}$-plane, provided that the relation $\ell(\hat{\mathbf{a}}, \hat{\mathbf{b}}) = 0$ holds. The $\mathscr{L}$-plane attached to $\hat{\mathbf{a}}$ and $\hat{\mathbf{b}}$ is the representation of the orbital plane in $\mathbb{U}^4$. Vector $\hat{\mathbf{a}}$ determines the direction of the minor axis, and $\hat{\mathbf{b}}$ corresponds to the major axis.

Figure 3.10 depicts the geometry of the problem in the parametric space $\mathbb{U}^4$. Equation (3.35) is in fact the equation of an ellipse centered at the origin. It is worth emphasizing that vector $\mathbf{u}$ is not referred to the focus of the ellipse, but to its center. The minor and major axes correspond to the square roots of the radii at periapsis and apoapsis, respectively. The position in the ellipse is given by the half-angle $E/2$. The ellipse is contained in the $\mathscr{L}$-plane defined by the vector elements $(\hat{\mathbf{a}}, \hat{\mathbf{b}})$.

5 Stiefel and Scheifele (1971, §43) defined an $\mathscr{L}$-plane as the two-dimensional plane in $\mathbb{U}^4$ passing through the origin and spanned by two vectors $\hat{\mathbf{a}}, \hat{\mathbf{b}} \in \mathbb{U}^4$ satisfying the bilinear relation $\ell(\hat{\mathbf{a}}, \hat{\mathbf{b}}) = 0$. This definition can be extended to state that any pair of vectors in the $\mathscr{L}$-plane satisfy the bilinear relation. The transformation to planes in the Cartesian space is a Levi-Civita mapping, which doubles the angles and squares the distances.

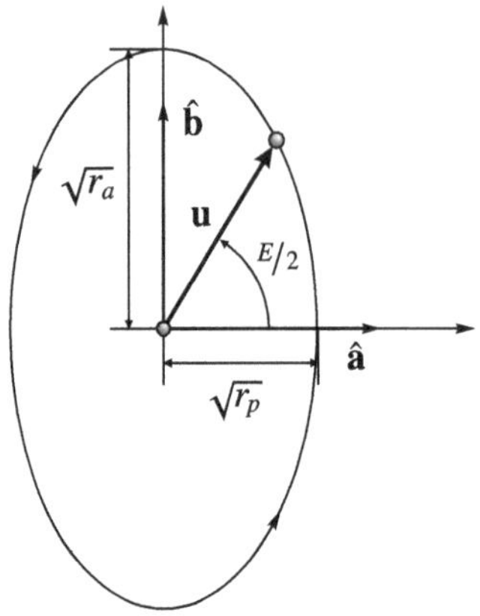

Fig. 3.10: Elliptic orbit in the $\mathscr{L}$-plane spanned by $(\hat{\mathbf{a}}, \hat{\mathbf{b}})$ in $\mathbb{U}^4$.

By KS-transforming the expression defining $\mathbf{u}_0$ it follows that

$$\mathbf{x}_0 = \mathsf{L}(\mathbf{u}_0)\,\mathbf{u}_0 = r_p\,\mathsf{L}(\hat{\mathbf{a}})\,\hat{\mathbf{a}} \implies \mathbf{i}_{\mathcal{P}} = \mathsf{L}(\hat{\mathbf{a}})\,\hat{\mathbf{a}}$$

We have relaxed the notation by not distinguishing the unit vector $\mathbf{i}_{\mathcal{P}}$ from its extension to $\mathbb{U}^4$, because the fourth component vanishes. Similarly, vector $\mathbf{u}_0^*$ transforms into

$$\dot{\mathbf{x}}_0 = \frac{4\omega_k}{r_p}\mathsf{L}(\mathbf{u}_0)\,\mathbf{u}_0^* = \sqrt{\frac{\mu(1+e)}{r_p}}\,\left[\mathsf{L}(\hat{\mathbf{a}})\,\hat{\mathbf{b}}\right] \implies \mathbf{j}_{\mathcal{P}} = \mathsf{L}(\hat{\mathbf{a}})\,\hat{\mathbf{b}}$$

Provided that the bilinear relation $\ell(\hat{\mathbf{a}}, \hat{\mathbf{b}}) = 0$ holds, it is $\mathsf{L}(\hat{\mathbf{a}})\,\hat{\mathbf{b}} = \mathsf{L}(\hat{\mathbf{b}})\,\hat{\mathbf{a}}$. Note that vector $\hat{\mathbf{b}}$ can be written $\hat{\mathbf{b}} = 2r_a^{-1/2}\,\mathbf{u}_0^*$, and therefore

$$\mathsf{L}(\hat{\mathbf{b}})\,\hat{\mathbf{b}} = \frac{4}{r_a}\mathsf{L}\left(\mathbf{u}_0^*\right)\,\mathbf{u}_0^* = -\frac{\mathbf{x}_0}{r_p}$$

This result yields a useful relation

$$\mathsf{L}(\hat{\mathbf{b}})\,\hat{\mathbf{b}} = -\mathsf{L}(\hat{\mathbf{a}})\,\hat{\mathbf{a}}$$

Finally, it is worth proving the orthogonality of $\hat{\mathbf{a}}$ and $\hat{\mathbf{b}}$. It is shown by the inner product[6]

$$\hat{\mathbf{a}}\cdot\hat{\mathbf{b}} = \left[\mathsf{L}^\top(\hat{\mathbf{a}})\,\mathbf{i}_{\mathcal{P}}\right]\cdot\left[\mathsf{L}^\top(\hat{\mathbf{a}})\,\mathbf{j}_{\mathcal{P}}\right] = \mathbf{i}_{\mathcal{P}}\cdot\left[\mathsf{L}(\hat{\mathbf{a}})\,\mathsf{L}^\top(\hat{\mathbf{a}})\,\mathbf{j}_{\mathcal{P}}\right] = \mathbf{i}_{\mathcal{P}}\cdot\mathbf{j}_{\mathcal{P}} = 0$$

Since $\hat{\mathbf{a}}$ and $\hat{\mathbf{b}}$ are unit vectors the linear operators $\mathsf{L}(\hat{\mathbf{a}})$ and $\mathsf{L}(\hat{\mathbf{b}})$ are orthogonal – see Eq. (3.7). Equipping the KS space with a cross product as defined in Sect. 3.7.2 one can

6 Recall the general property of the inner product involving vectors $\mathbf{x}, \mathbf{y} \in \mathbb{R}^n$ and a matrix $\mathbf{A} \in \mathbb{R}^{n\times n}$:

$$(\mathbf{A}\,\mathbf{x})\cdot\mathbf{y} = \mathbf{x}\cdot(\mathbf{A}^\top\mathbf{y}).$$

prove that $\mathbf{k}_{\mathcal{P}} = \mathbf{i}_{\mathcal{P}} \times \mathbf{j}_{\mathcal{P}}$ is given by the first three components of $\hat{\mathbf{a}} \times \hat{\mathbf{b}}$. Selecting, for example the definition from Eq. (3.57) given by Vivarelli (1987), the vectors in the triple $\{\hat{\mathbf{a}}, \hat{\mathbf{b}}, \hat{\mathbf{a}} \times \hat{\mathbf{b}}\}$ are mutually orthogonal.

Equation (3.36) defines vectors $\hat{\mathbf{a}}$ and $\hat{\mathbf{b}}$ from the initial conditions $\mathbf{u}_0$ and $\mathbf{u}_0^*$. Since the initial conditions are given at periapsis, vectors $\hat{\mathbf{a}}$ and $\hat{\mathbf{b}}$ are tied to the axes of the ellipse. If the orbit is circular, Eq. (3.36) still holds and can be used to initialize these two vectors. They will no longer correspond to the axes of the ellipse, but simply to the direction of $\mathbf{u}_0$ and $\mathbf{u}_0^*$.

Consider now the more general problem in which $\mathbf{p} \neq 0$ and $\mathcal{V} \neq 0$. The eccentric anomaly E is replaced by the generalized eccentric anomaly, φ. The new independent variable is given a dynamical definition: it evolves with an angular velocity of the same form as the time rate of change of the Keplerian eccentric anomaly – Eq. (3.32) –:

$$\frac{\mathrm{d}t}{\mathrm{d}\varphi} = \frac{r}{2\omega} \tag{3.37}$$

The Keplerian natural frequency ω_k is replaced by the frequency related to the total energy $\mathcal{E}$

$$\omega = \sqrt{\frac{-\mathcal{E}}{2}} = \sqrt{\frac{\mu - r\mathcal{V}}{2(r + 4\,\|\mathbf{u}^*\|^2)}} \tag{3.38}$$

At departure the generalized eccentric anomaly coincides with the eccentric anomaly. Both angles in the parametric space $\mathbb{U}^4$ relate by means of

$$\varphi = E + \sigma \tag{3.39}$$

The angle $\sigma/2$ is the angle traversed by the osculating perifocal frame, defined by $\hat{\mathbf{a}}$ and $\hat{\mathbf{b}}$, due to the perturbations. If no perturbations are considered ($\mathcal{V} = 0$ and $\mathbf{p} = 0$) the generalized eccentric anomaly coincides with the eccentric anomaly. The following expression relates to their evolution

$$\frac{\mathrm{d}\sigma}{\mathrm{d}t} = \frac{\mathrm{d}\varphi}{\mathrm{d}t} - \frac{\mathrm{d}E}{\mathrm{d}t}$$

where $\mathrm{d}E/\mathrm{d}t$ is the full derivative of the osculating eccentric anomaly. It can be obtained by differentiating Kepler's equation with respect to time, taking into account the convective terms from the rate of change of the elements.

The generalized eccentric anomaly is measured from a particular reference frame. This frame, referred to as the *intermediate frame* $\mathcal{A}$, is the one that rotates with respect to the perifocal frame $\mathcal{P}$ with angular velocity

$$\boldsymbol{\omega}_{\mathcal{A}\mathcal{P}} = \frac{1}{2}\frac{\mathrm{d}\sigma}{\mathrm{d}t}(\hat{\mathbf{a}} \times \hat{\mathbf{b}})$$

The intermediate frame is defined dynamically. At departure it coincides with the perifocal frame, so that $\mathbf{i}_{\mathcal{A}}$ has the direction of $\mathbf{e}_0$. Vectors $(\hat{\mathbf{a}}, \hat{\mathbf{b}})$ define the perifocal frame in $\mathbb{U}^4$, which differs from the intermediate frame in a rotation of magnitude $\sigma/2$. In what remains of the chapter the generalized eccentric anomaly is used as the independent variable. For the sake of simplicity we shall use the symbol $\square^*$ for derivatives with respect to φ.

Taking the generalized eccentric anomaly as the new independent variable the equations of motion become

$$\frac{\mathrm{d}^2\mathbf{u}}{\mathrm{d}\varphi^2} + \frac{\mathbf{u}}{4} = \mathbf{Q} \tag{3.40}$$

Here $\mathbf{Q}$ defines the forcing terms:

$$\mathbf{Q} = -\frac{1}{16\omega^2}\left\{2\mathcal{V}\mathbf{u} + r\left[\frac{\partial \mathcal{V}}{\partial \mathbf{u}} - 2\mathsf{L}^{\mathsf{T}}(\mathbf{u})\,\mathbf{p}\right]\right\} - \frac{1}{\omega}\frac{\mathrm{d}\omega}{\mathrm{d}\varphi}\frac{\mathrm{d}\mathbf{u}}{\mathrm{d}\varphi}$$

The evolution of the natural frequency ω is governed by the differential equation:

$$\frac{\mathrm{d}\omega}{\mathrm{d}\varphi} = -\frac{r}{8\omega^2}\frac{\partial \mathcal{V}}{\partial t} - \frac{1}{2\omega}[\mathbf{u}^* \cdot \mathsf{L}^{\mathsf{T}}(\mathbf{u})\mathbf{p}] \tag{3.41}$$

The first term vanishes when the disturbing potential does not depend explicitly on time.

We seek a solution of the form

$$\mathbf{u}(\varphi) = \mathbf{a}(\varphi)\cos\frac{\varphi}{2} + \mathbf{b}(\varphi)\sin\frac{\varphi}{2} \tag{3.42}$$

where vectors $\mathbf{a}, \mathbf{b} \in \mathbb{U}^4$ are the elements of the unperturbed motion defined by Stiefel and Scheifele (1971, §19). Introducing Eq. (3.39) and equating to Eq. (3.35) yields the linear transformation between the elements $(\hat{\mathbf{a}}, \hat{\mathbf{b}})$ and $(\mathbf{a}, \mathbf{b})$:

$$\begin{aligned}\mathbf{a} &= \sqrt{r_p}\,\hat{\mathbf{a}}\cos\frac{\sigma}{2} - \sqrt{r_a}\,\hat{\mathbf{b}}\sin\frac{\sigma}{2}\\ \mathbf{b} &= \sqrt{r_p}\,\hat{\mathbf{a}}\sin\frac{\sigma}{2} + \sqrt{r_a}\,\hat{\mathbf{b}}\cos\frac{\sigma}{2}\end{aligned}$$

Recall that in the unperturbed problem it is $\sigma = 0$. Thus, $\mathbf{a}$ and $\mathbf{b}$ will be equal to $\hat{\mathbf{a}}$ and $\hat{\mathbf{b}}$ multiplied by the square root of the radius at periapsis and apoapsis, respectively.

The determinant of the resulting (linear) transformation is $a\sqrt{1-e^2}$. This result shows that the transformation is not a rotation, in general. Only when the orbit is circular will vectors $(\mathbf{a}, \mathbf{b})$ be obtained from $(\hat{\mathbf{a}}, \hat{\mathbf{b}})$ by a rotation of amplitude $\sigma/2$ and a dilation of magnitude $\sqrt{a}$. This can also be understood geometrically; let γ_a denote the angle between $\mathbf{a}$ and $\hat{\mathbf{a}}$, and γ_b the angle between $\mathbf{b}$ and $\hat{\mathbf{a}}$. They can be solved from the previous equations:

$$\tan\gamma_a = -\sqrt{\frac{1+e}{1-e}}\tan\frac{\sigma}{2}, \quad \text{and} \quad \tan\gamma_b = +\sqrt{\frac{1+e}{1-e}}\cot\frac{\sigma}{2}$$

to prove that $\gamma_a \neq -\sigma/2$ and $\gamma_b \neq -(\pi+\sigma)/2$ except for $e = 0$.

In general, vectors $\mathbf{a}$ and $\mathbf{b}$ will not be orthogonal. The inner product

$$\mathbf{a}\cdot\mathbf{b} = \frac{r_p - r_a}{2}\sin\sigma$$

shows that only if $\sigma = 0$ or $r_p = r_a$ will the vectors $(\mathbf{a}, \mathbf{b})$ form a set of orthogonal elements. The former occurs when no perturbations are considered. The latter requires the eccentricity to be null during the entire motion. This will only occur under very specific conditions, for example an initially circular orbit subject only to perturbations that are normal to the orbital plane.

The norm of **a** and **b** is given by

$$\mathbf{a} \cdot \mathbf{a} = a(1 - e \cos \sigma) \qquad \text{and} \qquad \mathbf{b} \cdot \mathbf{b} = a(1 + e \cos \sigma)$$

showing that they are not unit vectors, in general. The geometry of the orbital problem in $\mathbb{U}^4$ is sketched in Fig. 3.11. The origin of the generalized eccentric anomaly is the intermediate frame $\mathcal{A}$. When mapped to the parametric space $\mathbb{U}^4$ frames $\mathcal{P}$ and $\mathcal{A}$ relate by a rotation of magnitude $\sigma/2$. The osculating eccentric anomaly is measured with respect to the perifocal frame.

The state of the particle in $\mathbb{U}^4$ has been parameterized in terms of the vector elements **a** and **b**. The total energy or, more precisely, the natural frequency ω, is the ninth element. When no perturbations act on the system these variables are constant and their numerical integration is trivial. Note, however, that the time transformation (3.37) needs to be integrated even in the pure Keplerian problem. For the sake of improving the numerical performance of the method the physical time is decomposed in (Stiefel and Scheifele, 1971, p. 92)

$$t = \tau - \frac{1}{\omega}(\mathbf{u} \cdot \mathbf{u}^*) \tag{3.43}$$

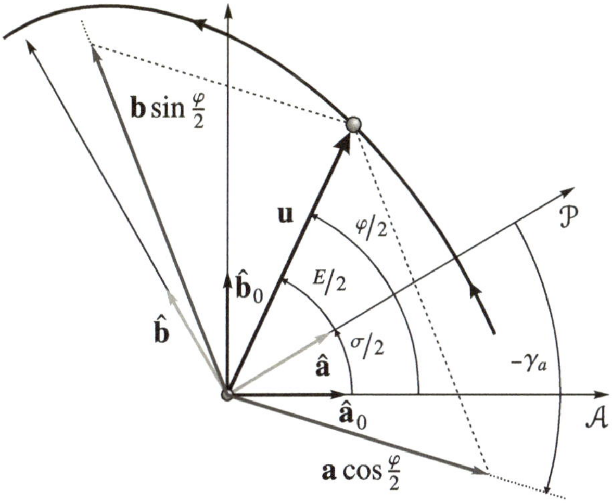

Fig. 3.11: Geometrical interpretation of the elements **a** and **b** in the $\mathscr{L}$-plane.

where τ is a time element, governed by the perturbations. Differentiating this expression yields the equation of evolution of the time element:

$$\frac{d\tau}{d\varphi} = \frac{1}{\omega^3}(\mu - 2r\mathcal{V}) - \frac{r}{16\omega^3}\left\{\mathbf{u}\cdot\left[\frac{\partial\mathcal{V}}{\partial\mathbf{u}} - 2\mathbf{L}^\top(\mathbf{u})\,\mathbf{p}\right]\right\} - \frac{2}{\omega^3}\frac{d\omega}{d\varphi}(\mathbf{u}\cdot\mathbf{u}^*) \tag{3.44}$$

When the disturbing terms vanish the time element grows linearly in time.

Incorporating the time element the problem is completely defined by a set of ten orbital elements, œ_j (with $j = 1, \ldots, 10$). They are the natural frequency ω, the time element τ, and the components of the four-vectors **a** and **b**.

3.6.2 Gauge freedom in KS space

In the following lines we extend the concept of the gauge freedom to the natural elements in KS space. As discussed in the previous section, in the absence of perturbations Eq. (3.40) reduces to a harmonic oscillator that admits the trivial solution

$$\mathbf{u}(\varphi) = \mathbf{a}\cos\frac{\varphi}{2} + \mathbf{b}\sin\frac{\varphi}{2} \equiv \mathbf{f}(\varphi;\mathbf{a},\mathbf{b}) \tag{3.45}$$

with **a** and **b** constant. In addition, the velocity in $\mathbb{U}^4$ takes the form

$$\mathbf{u}^*(\varphi) = -\frac{\mathbf{a}}{2}\sin\frac{\varphi}{2} + \frac{\mathbf{b}}{2}\cos\frac{\varphi}{2} \equiv \mathbf{g}(\varphi;\mathbf{a},\mathbf{b})$$

The symmetry of these equations yields a simple form of the Lagrange and Poisson brackets. The only nonzero terms are

$$[a_i,\, b_i] = -[b_i,\, a_i] = 1/2$$
$$\{a_i,\, b_i\} = -\{b_i,\, a_i\} = 2$$

in which a_i and b_i are the components of **a** and **b**, and $i = 1, 2, 3, 4$. Following the procedure presented in Sect. 2.7, we define the gauge as

$$\boldsymbol{\Phi} = \frac{\partial\mathbf{f}}{\partial\mathbf{a}}\circ\frac{d\mathbf{a}}{d\varphi} + \frac{\partial\mathbf{f}}{\partial\mathbf{b}}\circ\frac{d\mathbf{b}}{d\varphi} = \frac{d\mathbf{a}}{d\varphi}\cos\frac{\varphi}{2} + \frac{d\mathbf{b}}{d\varphi}\sin\frac{\varphi}{2} \tag{3.46}$$

Its full derivative takes the form

$$\frac{d\boldsymbol{\Phi}}{d\varphi} = \frac{\partial\boldsymbol{\Phi}}{\partial\varphi} + \frac{\partial\boldsymbol{\Phi}}{\partial\mathbf{a}}\circ\frac{d\mathbf{a}}{d\varphi} + \frac{\partial\boldsymbol{\Phi}}{\partial\mathbf{b}}\circ\frac{d\mathbf{b}}{d\varphi} \tag{3.47}$$

Equation (2.46) translates to KS language as

$$\sum_j [\text{œ}_n,\, \text{œ}_j]\frac{d\text{œ}_j}{d\varphi} = -\left(\mathbf{Q} + \frac{d\boldsymbol{\Phi}}{d\varphi}\right)\cdot\frac{\partial\mathbf{f}}{\partial\text{œ}_n} - \frac{\partial\mathbf{g}}{\partial\text{œ}_n}\cdot\boldsymbol{\Phi}$$

with

$$\mathbf{Q} = \frac{1}{16\omega^2}\left\{2\mathcal{V}\mathbf{u} + r\left[\frac{\partial\mathcal{V}}{\partial\mathbf{u}} - 2\mathbf{L}^\top(\mathbf{u})\,\mathbf{p}\right]\right\} + \frac{1}{\omega}\frac{d\omega}{d\varphi}\frac{d\mathbf{u}}{d\varphi}$$

It renders the gauge-generalized form of the equation given by Stiefel and Scheifele (1971, p. 89, eq. 19,52):

$$\frac{\mathrm{d}\mathbf{a}}{\mathrm{d}\varphi} = +2\left(\mathbf{Q} + \frac{\mathrm{d}\boldsymbol{\Phi}}{\mathrm{d}\varphi}\right)\sin\frac{\varphi}{2} + \boldsymbol{\Phi}\cos\frac{\varphi}{2} \tag{3.48}$$

$$\frac{\mathrm{d}\mathbf{b}}{\mathrm{d}\varphi} = -2\left(\mathbf{Q} + \frac{\mathrm{d}\boldsymbol{\Phi}}{\mathrm{d}\varphi}\right)\cos\frac{\varphi}{2} + \boldsymbol{\Phi}\sin\frac{\varphi}{2} \tag{3.49}$$

Equation (3.41) defined the evolution of the osculating total energy. Because of the gauge invariance of this variable this equation still holds and defines the evolution of the energy as:

$$\frac{\mathrm{d}\omega}{\mathrm{d}\varphi} = -\frac{r}{8\omega^2}\frac{\partial \mathcal{V}}{\partial t} - \frac{1}{2\omega}[\mathbf{u}^* \cdot \mathbf{L}^\mathrm{T}(\mathbf{u})\mathbf{p}] \tag{3.50}$$

The decomposition of the time in a secular part and a time element is $\mathcal{G}$-independent so Eq. (3.44) still holds.

Equations (3.48–3.49) include the full derivative of the gauge. But this term involves the derivatives of the elements, as shown by Eq. (3.47). It is then convenient to rearrange these equations so that the derivatives of the corresponding elements are all grouped in the left-hand side:

$$\mathbf{M}(\varphi)\frac{\mathrm{d}\mathbf{a}}{\mathrm{d}\varphi} = +2\left(\mathbf{Q} + \frac{\partial\boldsymbol{\Phi}}{\partial\varphi}\right)\sin\frac{\varphi}{2} + \left(\mathbf{I}\cos\frac{\varphi}{2} + 2\frac{\partial\boldsymbol{\Phi}}{\partial\mathbf{b}}\right)\boldsymbol{\Phi}$$

$$\mathbf{M}(\varphi)\frac{\mathrm{d}\mathbf{b}}{\mathrm{d}\varphi} = -2\left(\mathbf{Q} + \frac{\partial\boldsymbol{\Phi}}{\partial\varphi}\right)\cos\frac{\varphi}{2} + \left(\mathbf{I}\sin\frac{\varphi}{2} - 2\frac{\partial\boldsymbol{\Phi}}{\partial\mathbf{a}}\right)\boldsymbol{\Phi}$$

This is achieved by introducing the matrix

$$\mathbf{M}(\varphi) = \mathbf{I} - 2\frac{\partial\boldsymbol{\Phi}}{\partial\mathbf{a}}\sin\frac{\varphi}{2} + 2\frac{\partial\boldsymbol{\Phi}}{\partial\mathbf{b}}\cos\frac{\varphi}{2}$$

with $\mathbf{I}$ the identity matrix of dimension four.

In sum, the gauge-generalized equations of motion in KS space are:

$$\frac{\mathrm{d}\tau}{\mathrm{d}\varphi} = \frac{1}{\omega^3}(\mu - 2r\mathcal{V}) - \frac{r}{16\omega^3}\left\{\mathbf{u}\cdot\left[\frac{\partial\mathcal{V}}{\partial\mathbf{u}} - 2\mathbf{L}^\mathrm{T}(\mathbf{u})\,\mathbf{p}\right]\right\} - \frac{2}{\omega^3}\frac{\mathrm{d}\omega}{\mathrm{d}\varphi}(\mathbf{u}\cdot\mathbf{u}^*) \tag{3.51}$$

$$\frac{\mathrm{d}\omega}{\mathrm{d}\varphi} = -\frac{r}{8\omega^2}\frac{\partial\mathcal{V}}{\partial t} - \frac{1}{2\omega}[\mathbf{u}^* \cdot \mathbf{L}^\mathrm{T}(\mathbf{u})\mathbf{p}] \tag{3.52}$$

$$\frac{\mathrm{d}\mathbf{a}}{\mathrm{d}\varphi} = \mathbf{M}^{-1}\left[+2\left(\mathbf{Q} + \frac{\partial\boldsymbol{\Phi}}{\partial\varphi}\right)\sin\frac{\varphi}{2} + \left(\mathbf{I}\cos\frac{\varphi}{2} + 2\frac{\partial\boldsymbol{\Phi}}{\partial\mathbf{b}}\right)\boldsymbol{\Phi}\right] \tag{3.53}$$

$$\frac{\mathrm{d}\mathbf{b}}{\mathrm{d}\varphi} = \mathbf{M}^{-1}\left[-2\left(\mathbf{Q} + \frac{\partial\boldsymbol{\Phi}}{\partial\varphi}\right)\cos\frac{\varphi}{2} + \left(\mathbf{I}\sin\frac{\varphi}{2} - 2\frac{\partial\boldsymbol{\Phi}}{\partial\mathbf{a}}\right)\boldsymbol{\Phi}\right] \tag{3.54}$$

The equations are completed with the initial conditions:

$$\varphi_0 = E_0: \quad \tau = 0 \quad \omega = \sqrt{\frac{\mu - r_0\mathcal{V}_0}{2\,(r_0 + \|\mathbf{u}_0'\|^2)}}, \quad \mathbf{a} = \mathbf{a}_0 \quad \mathbf{b} = \mathbf{b}_0$$

The nonosculating elements $\mathbf{a}_0$ and $\mathbf{b}_0$ are solved from the osculating elements $\hat{\mathbf{a}}_0$ and $\hat{\mathbf{b}}_0$ given the equations

$$\mathbf{f}(\varphi_0; \tilde{\mathbf{a}}_0, \tilde{\mathbf{b}}_0) = \mathbf{f}(\varphi_0; \mathbf{a}_0, \mathbf{b}_0) \tag{3.55}$$

$$\mathbf{g}(\varphi_0; \tilde{\mathbf{a}}_0, \tilde{\mathbf{b}}_0) = \mathbf{g}(\varphi_0; \mathbf{a}_0, \mathbf{b}_0) + \boldsymbol{\Phi}(\varphi_0; \mathbf{a}_0, \mathbf{b}_0) \tag{3.56}$$

Recall that $\mathbf{f} = \mathbf{a}\cos(\varphi/2) + \mathbf{b}\sin(\varphi/2)$ and $\mathbf{g} = \partial\mathbf{f}/\partial\varphi$. The choice of the gauge must make the matrix $\mathsf{M}(\varphi)$ regular.

3.7 Orthogonal bases

3.7.1 Basis attached to the fiber

In Section 3.2.2 it is shown that fibers are circles in $\mathbb{U}^4$. Let $\mathbf{u}$ and $\mathbf{w} = \mathscr{R}(\phi; \mathbf{u})$ be two vectors attached to a fiber $\mathfrak{F}$. They span a plane containing the fiber. This plane is not a plane of the Levi-Civita type because $\ell(\mathbf{u}, \mathbf{w}) \neq 0$. Since trajectories intersect fibers at right angles this subspace is transversal to the flow. An orthogonal basis can be attached to the resulting plane, allowing projections on the transversal subspace. Although arbitrary orthonormal bases can be constructed via the Gram–Schmidt procedure (Nayfeh and Balachandran, 2004, pp. 529–530), the basis described in this section appears naturally in the formulation.

Associated to every vector $\mathbf{u}$ there is a KS matrix $\mathsf{L}(\mathbf{u})$. The columns of the matrix define a vector basis $\mathcal{B} = \{\mathbf{u}^1, \mathbf{u}^2, \mathbf{u}^3, \mathbf{u}^4\}$, with $\mathbf{u}^1 \equiv \mathbf{u}$. The basis $\mathcal{B}$ is orthogonal,

$$\mathbf{u}^i \cdot \mathbf{u}^j = r\delta_{ij}$$

Here δ_{ij} denotes Kronecker's delta. Assuming $x \geq 0$ every point in the fiber generated by $\mathbf{w} = \mathscr{R}(\phi; \mathbf{u})$ lies in the plane spanned by $\mathbf{u}^1$ and $\mathbf{u}^4$, i.e., $\mathbf{w} \cdot \mathbf{u}^2 = \mathbf{w} \cdot \mathbf{u}^3 = 0$ for all ϕ. Conversely, for $x < 0$ the fiber is confined to the $\mathbf{u}^2\mathbf{u}^3$ plane. The basis $\mathcal{B}$ is an orthogonal basis attached to the fiber at $\mathbf{u}$. In addition,

$$\mathbf{u}^4 \cdot \mathbf{u}' = \ell(\mathbf{u}, \mathbf{u}') = 0$$

meaning that $\mathbf{u}'$ is perpendicular to $\mathbf{u}^4$. In fact, $\mathbf{u}^4 = -\mathbf{t}$, as shown by Eq. (3.17).

The vectors arising from the products $\mathsf{L}(\mathbf{u})\mathbf{u}^i$, $i = 1, 2, 3$ have a vanishing fourth component. They are equivalent to vectors in $\mathbb{E}^3$. However, the fourth component of the product $\mathsf{L}(\mathbf{u})\mathbf{u}^4$ is not zero. The three vectors obtained by these transformations correspond to the position vector $\mathbf{r}$, and a pair of orthogonal vectors spanning the plane tangential to the two-sphere at $\mathbf{r}$ in $\mathbb{E}^3$. These vectors are the columns of the associated Cailey matrix.

3.7.2 Cross product

Stiefel and Scheifele (1971, pp. 277–281) sought a definition of cross product in the parametric space $\mathbb{U}^4$ when discussing the orthogonality conditions of vectors and Levi-Civita planes. Although the cross product of two vectors in $\mathbb{R}^3$ is intuitive, its generalization to higher dimensions is not straightforward. Independent proofs from different authors (see for example Brown and Gray, 1967) show that the cross product of two vectors only exists in dimensions 1, 3, 7; for n dimensions the cross product involves $n-1$ vectors. Stiefel and Scheifele (1971) defined the product $\mathbf{p} = \mathbf{u} \times \mathbf{v}$ as

$$\mathbf{p} = \mathsf{L}(\mathbf{u})\, \mathbf{v}^4$$

where $\mathbf{v}^4 = [v_4, -v_3, v_2, -v_1]^\top$ is the fourth column of $\mathsf{L}(\mathbf{v})$. The properties of this construction motivated the authors to call (p_1, p_2, p_3) the cross product of $\mathbf{u} \times \mathbf{v}$, with $p_4 = \mathbf{u} \cdot \mathbf{v}$. In the following lines we analyze in more detail this construction and connect with alternative definitions provided by Vivarelli (1987) and Deprit et al. (1994).

Let $\{\mathbf{e}_1, \mathbf{e}_2, \dots, \mathbf{e}_n\}$ be an orthogonal basis in $\mathbb{R}^n$. The Grassmann exterior product gives rise to the bivectors $\mathbf{e}_i \wedge \mathbf{e}_j$, trivectors $\mathbf{e}_i \wedge \mathbf{e}_j \wedge \mathbf{e}_k$, and successive blades of grade $m \leq n$ (Flanders, 1989, §II). They constitute the subspaces $\bigwedge^m \mathbb{R}^n$ of the exterior algebra:

$$\bigwedge \mathbb{R}^n = \mathbb{R} \oplus \mathbb{R}^n \oplus \bigwedge^2 \mathbb{R}^n \oplus \dots \oplus \bigwedge^n \mathbb{R}^n$$

noting that $\bigwedge^0 \mathbb{R}^n = \mathbb{R}$ and $\bigwedge^1 \mathbb{R}^n = \mathbb{R}^n$. Without being exhaustive we simply recall that such exterior algebra is associative with unity, satisfying $\mathbf{e}_i \wedge \mathbf{e}_j = -\mathbf{e}_j \wedge \mathbf{e}_i$ and $\mathbf{e}_i \wedge \mathbf{e}_i = 0$. The exterior product of two parallel vectors vanishes. We shall write $\mathbf{e}_{ij\dots k} = \mathbf{e}_i \wedge \mathbf{e}_j \wedge \dots \wedge \mathbf{e}_k$ for brevity.

In Section 3.7.1 an orthogonal basis attached to $\mathbf{u}$ was defined, where two of its vectors are KS-transformed to vectors spanning the plane tangent to the two-sphere in $\mathbb{E}^3$. Identifying $\mathbf{u}^i = \sqrt{r}\,\mathbf{e}_i$, the exterior product of vectors $\{\mathbf{u}^1, \mathbf{u}^2, \mathbf{u}^3, \mathbf{u}^4\}$ generates the oriented hypervolume

$$\mathbf{u}^1 \wedge \mathbf{u}^2 \wedge \mathbf{u}^3 \wedge \mathbf{u}^4 = -r^2 \mathbf{e}_{1234}$$

provided that $\det(\mathsf{L}(\mathbf{u})) = -r^2$. In three dimensions the exterior product is equivalent to the cross product, given $\mathbf{e}_1 \times \mathbf{e}_2 = \mathbf{e}_3$, $\mathbf{e}_1 \times \mathbf{e}_3 = -\mathbf{e}_2$, and $\mathbf{e}_2 \times \mathbf{e}_3 = \mathbf{e}_1$. Applying the cross product to the first three elements of $\mathcal{B}$ provides

$$\mathbf{u}^1 \times \mathbf{u}^2 \times \mathbf{u}^3 = r\mathbf{u}^4$$

This result confirms that $\mathcal{B}$ is, indeed, an orthogonal basis.

Vivarelli (1987) and Deprit et al. (1994) worked in the more general Clifford algebra $\mathfrak{Cl}_3$. Introducing the Clifford product of two vectors

$$\mathbf{ab} = \mathbf{a} \cdot \mathbf{b} + \mathbf{a} \wedge \mathbf{b}$$

the exterior algebra over $\mathbb{R}^3$ can be identified with the Clifford algebra $\mathfrak{Cl}_3$: bivectors and trivectors become $\mathbf{e}_i \wedge \mathbf{e}_j \rightarrow \mathbf{e}_i\mathbf{e}_j$ and $\mathbf{e}_1 \wedge \mathbf{e}_2 \wedge \mathbf{e}_3 \rightarrow \mathbf{e}_1\mathbf{e}_2\mathbf{e}_3$. Note that $\mathbf{ba} = \mathbf{a}\cdot\mathbf{b} - \mathbf{a}\wedge\mathbf{b} \neq \mathbf{ab}$, so the Clifford product is not commutative. Thus, the even subalgebra $\mathfrak{Cl}_3^+ = \mathbb{R} \oplus \bigwedge^2 \mathbb{R}^3$, isomorphic to the quaternion algebra $\mathbb{H}$, is not commutative either.

The algebra $\mathbb{H} \simeq \mathfrak{Cl}_3^+$ is endowed with the multiplication rules $\mathbf{e}_1^2 = \mathbf{e}_2^2 = \mathbf{e}_3^2 = -1$, and the product of vectors is anticommutative, $\mathbf{e}_i\mathbf{e}_j = -\mathbf{e}_j\mathbf{e}_i$. Identifying these bivectors with the quaternion basis elements

$$\mathbf{e}_2\mathbf{e}_3 = \mathrm{i}\,, \quad \mathbf{e}_3\mathbf{e}_1 = \mathrm{j}\,, \quad \mathbf{e}_1\mathbf{e}_2 = \mathrm{k}$$

the product of two quaternions $\mathfrak{u}$ and $\mathfrak{v}$ is established. Vivarelli (1987) rewrote Stiefel and Scheifele's form of the cross product in terms of the quaternion product

$$\mathfrak{u} \times \mathfrak{v} = \frac{1}{2}(\mathfrak{u}\mathrm{k}\mathfrak{v}_* - \mathfrak{v}\mathrm{k}\mathfrak{u}_*) \tag{3.57}$$

where $_*$ denotes the involution $\mathfrak{u}_* = u_1 + u_2\mathrm{i} + u_3\mathrm{j} - u_4\mathrm{k}$. Disregarding the arrangement of the components, Deprit et al. (1994) defined the cross product of two quaternions as

$$\mathfrak{u} \times \mathfrak{v} = \frac{1}{2}(\mathfrak{v}\mathfrak{u}^\dagger - \mathfrak{u}\mathfrak{v}^\dagger)$$

Here $\square^\dagger$ denotes the quaternion conjugate. The difference between these quaternionic definitions and the original one from Stiefel and Scheifele is the fact that $\mathfrak{u}\times\mathfrak{v}$ is a pure quaternion, i.e., $\mathfrak{R}(\mathfrak{u}\times\mathfrak{v}) = 0$, whereas the fourth component of Stiefel and Scheifele's product $\mathbf{u} \times \mathbf{v}$ is $\mathbf{u} \cdot \mathbf{v}$.

3.8 Conclusions

The topology of the KS transformation has important consequences for the stability and accuracy of the solutions in KS space. There are two key aspects to consider when studying the stability of the motion. First, the presence of a fictitious time that replaces the physical time as the independent variable. Second, the dimension-raising nature of the Hopf fibration.

Classical theories of stability are based on the separation between nearby trajectories. Having introduced a fictitious time, the question of how to synchronize the trajectories arises. The numerical stabilization of the equations of motion by KS regularization relates to solutions synchronized in fictitious time. But a theory of stability synchronized in physical time allows the translation of concepts such as attractive sets, Lyapunov stability, Poincaré maps, etc. to KS language.

The additional dimension provides a degree of freedom to the solution in parametric space. In general, the free parameter can be fixed arbitrarily, with little or no impact on the resulting trajectory in Cartesian space. However, as strong perturbations destabilize the system, different values of the free parameter may result in completely different solutions in time. This phenomenon is caused by numerical errors and destroys the topological structure of KS transformation: points in a fiber are no longer transformed into one single point. By monitoring the topological stability of the integration it is possible to estimate an indicator similar to the Lyapunov time.

The geometry of the orbital motion in KS space attains a symmetric form. The natural vectorial elements introduced by Stiefel and Scheifele (1971, §19) can be given a geometrical interpretation, thanks to having characterized orthogonal bases in KS space. From this construction we derived the gauge-generalized equations of orbital motion in KS space.

4 The Dromo formulation

Defining an orbit in space using orbital elements can be decomposed into two steps: finding the orientation of the orbital plane with respect to a certain reference, and introducing a conic section that represents the instantaneous orbit. This approach stems naturally from the classical orbital elements, for example; (i, Ω, ω) are the Euler angles that fix the orbital plane, (a, e) define the size and shape of the corresponding conic section, and M determines where in the orbit the particle is. The natural decomposition of the motion has been recursively exploited in the literature. Some sets of elements relying on this construction have been introduced in Sect. 2.5, like the ones by Deprit (1975), Palacios and Calvo (1996), Broucke and Cefola (1972), Roy and Moran (1973), Lara (2016), Peláez et al. (2007), etc. As an example to show how these methods work, the present chapter studies the latter formulation, called *Dromo*. This formulation is chosen because it features most of the classical regularization techniques including Sundman transformations, the use of elements, embedding integrals of motion, and introducing time elements. Chapter 3 focused on a formulation with a Sundman transformation of order one, and the present chapter deals with a transformation of order two.

The Dromo elements are the components of the eccentricity vector in an ideal frame, the inverse of the angular momentum, and the four components of a quaternion that defines the orbital plane with respect to an inertial reference. Replacing the traditional Euler angles by the Euler parameters avoids potential singularities and transforms trigonometric relations into polynomial equations, at the cost of increasing the dimension of the system. Possibly the first application of this technique to perturbed orbital motion was due to Broucke et al. (1971) and Deprit (1975). Dromo features a second-order Sundman transformation that introduces the *ideal anomaly*, which behaves as the independent variable. Later studies defined a time element to improve the integration of the time transformation (Baù et al., 2014; Baù and Bombardelli, 2014). The formulation has been progressively improved and the latest version is that by Urrutxua et al. (2015b). A complete and very instructive review can be found in Urrutxua (2015, Chap. 4). In a series of works, Baù et al. (2015) presented elegant modifications of the method based on a different time transformation and the introduction of a perturbing potential. Roa and Peláez (2015d) recently derived a reformulation of Dromo that is specifically conceived for hyperbolic orbits. They formulated the evolution of the eccentricity vector in Minkowski space-time to take advantage of the hyperbolic nature of this metric. Chapter 5 is devoted to this particular formulation.

Dromo not only presents an efficient integration scheme, but it has also proven useful for many analytical studies. Bombardelli et al. (2011) published an asymptotic solution to the dynamics of a spacecraft perturbed by a constant tangential acceleration, while Gonzalo and Bombardelli (2014) and Urrutxua et al. (2015a) focused on the radial case. Chapter 9 includes a dedicated review of methods and techniques aimed

https://doi.org/10.1515/9783110559125-004

at solving the problem of orbital motion under a continuous acceleration. Bombardelli (2014) arrived at an analytical description of collision avoidance maneuvers, a study that resulted in an interesting strategy for finding optimal avoidance strategies (Bombardelli and Hernando-Ayuso, 2015). Amato and Bombardelli (2014) and Amato et al. (2016) studied the potential of Dromo for propagating resonant trajectories of Potentially Hazardous Asteroids focusing on the error growth after successive close encounters.

Dromo uses a gyroscopic description of the dynamics of the orbital plane. It is based on the evolution of the angular momentum vector following the typical approach for modeling rigid body dynamics. In fact, Urrutxua and Peláez (2014) took advantage of this connection to derive a roto-translational propagator based on Dromo. Formulations that decompose the dynamics in the motion of the orbital plane plus the motion inside the orbital plane typically become singular when the angular momentum vanishes, because the orbital plane is not defined.

In this chapter, we first introduce the concept of Hansen ideal reference frames. The Dromo formulation is built upon this mathematical contrivance in Sect. 4.2. The numerical advantages of Dromo are discussed in Sect. 4.3 over a few examples. Next, Sect. 4.4 discusses the noncanonical nature of the Dromo formulation. In Sect. 4.5, we present the gauge-generalized equations of Dromo following the procedure described in Sect. 2.7, together with the Lagrange and Poisson brackets related to the Dromo elements. This chapter also characterizes the singularities of formulations of this type (Sect. 4.6).

4.1 Hansen ideal frames

Let $\mathcal{I} = \{\mathbf{i}_\mathcal{I}, \mathbf{j}_\mathcal{I}, \mathbf{k}_\mathcal{I}\}$ be the inertial frame, and $\mathcal{A} = \{\mathbf{i}_\mathcal{A}, \mathbf{j}_\mathcal{A}, \mathbf{k}_\mathcal{A}\}$ an arbitrary rotating reference frame. The velocity of a particle, defined by the time derivative of the radius vector, can be computed from the perspective of the inertial or the rotating reference frame. The absolute and relative velocities relate through

$$\left.\frac{\mathrm{d}\mathbf{r}}{\mathrm{d}t}\right|_\mathcal{I} = \left.\frac{\mathrm{d}\mathbf{r}}{\mathrm{d}t}\right|_\mathcal{A} + \boldsymbol{\omega}_{\mathcal{A}\mathcal{I}} \times \mathbf{r}$$

where $\boldsymbol{\omega}_{\mathcal{A}\mathcal{I}}$ is the angular velocity vector (or the Darboux vector) of frame $\mathcal{A}$ with respect to the inertial space. The absolute velocity decomposes in the velocity relative to the moving reference plus the evolution of the rotating reference itself. If the moving frame $\mathcal{A}$ is attached to the orbital plane, then the angular velocity $\boldsymbol{\omega}_{\mathcal{A}\mathcal{I}}$ governs the dynamics of this plane. Thus, in the unperturbed problem it is $\boldsymbol{\omega}_{\mathcal{A}\mathcal{I}} = 0$ (the orbital plane is fixed) meaning that

$$\left.\frac{\mathrm{d}\mathbf{r}}{\mathrm{d}t}\right|_\mathcal{I} = \left.\frac{\mathrm{d}\mathbf{r}}{\mathrm{d}t}\right|_\mathcal{A} \tag{4.1}$$

That is, the time derivatives of $\mathbf{r}$ in frames $\mathcal{I}$ and $\mathcal{A}$ coincide. This phenomenon motivated Peter A. Hansen to propose the concept of *ideal coordinates*, back in 1857. The original definition, taken from Hansen (1857, p. 66), reads

> "I call ideal coordinates of a planet, comet, or satellite those having the property that not only them but also their derivatives with respect to time have the same form in the perturbed and unperturbed motion."

Although Hansen did not work with vectors, in modern applications the concept of Hansen ideal coordinates has been extended to define Hansen *ideal frames*. Let us recover the rotating frame $\mathcal{A}$, for which $\boldsymbol{\omega}_{\mathcal{A}\mathcal{I}} = 0$ when there are no perturbations. In the Keplerian case Eq. (4.1) holds because the orbital plane is fixed. Frame $\mathcal{A}$ is said to be Hansen ideal if the velocity relative to $\mathcal{A}$ in the perturbed problem (with $\boldsymbol{\omega}_{\mathcal{A}\mathcal{I}} \neq 0$) is still given by Eq. (4.1). This yields the condition

$$\left.\frac{\mathrm{d}\mathbf{r}}{\mathrm{d}t}\right|_{\mathcal{I}} = \left.\frac{\mathrm{d}\mathbf{r}}{\mathrm{d}t}\right|_{\mathcal{A}} + \boldsymbol{\omega}_{\mathcal{A}\mathcal{I}} \times \mathbf{r} = \left.\frac{\mathrm{d}\mathbf{r}}{\mathrm{d}t}\right|_{\mathcal{A}} \implies \boldsymbol{\omega}_{\mathcal{A}\mathcal{I}} \times \mathbf{r} = 0$$

The angular velocity vector $\boldsymbol{\omega}_{\mathcal{A}\mathcal{I}}$ must be parallel to the radial direction for frame $\mathcal{A}$ to be ideal, a phenomenon that was already noted by Hansen (1857, p. 67). Thus, the angular velocity of a Hansen ideal frame follows the direction of the radius vector. More details and some new results regarding Hansen ideal frames were recently presented by Jochim (2012). An illuminating result is the connection between the condition for a frame to be Hansen ideal and the Lagrange constraint stemming from the variation of the parameters method; both require the solution to take the same form in the perturbed and unperturbed problems, and thus the Lagrange constraint satisfies the condition for Hansen ideal frames.

The main advantage of introducing ideal frames is that it allows a separate treatment of the in-plane motion and the motion of the orbital plane itself. Herget (1962) explored this concept when trying to eliminate the singularity related to the definition of the apses when $e \to 0$. Broucke et al. (1971) combined Hansen's results with replacing the Euler angles by the Euler parameters, an approach that was later refined by Deprit (1975) when he published a special perturbation method based on ideal elements. In a sequel, he characterized ideal frames from an algebraic perspective (Deprit, 1976).

4.2 Dromo

Under a normalization such that $\mu = 1$, the governing equations of motion for the two-body problem reduce to

$$\frac{\mathrm{d}^2\mathbf{r}}{\mathrm{d}t^2} + \frac{\mathbf{r}}{r^3} = \mathbf{a}_p\,, \qquad r = \|\mathbf{r}\|$$

The initial value of the semimajor axis is taken as the unit of length, meaning that the mean motion reduces to unity. Let $\mathcal{L}$ be the orbital frame, with $\mathcal{L} = \{\mathbf{i}, \mathbf{j}, \mathbf{k}\}$. Such a basis is defined by the expressions:

$$\mathbf{i} = \frac{\mathbf{r}}{r}, \qquad \mathbf{k} = \frac{\mathbf{h}}{h}, \qquad \mathbf{j} = \mathbf{k} \times \mathbf{i} \tag{4.2}$$

where $\mathbf{h} \in \mathbb{R}^3$ is the angular momentum vector of the particle.

The dynamics of the orbital plane are modeled as if it were a rigid body. A reference frame, equivalent to a body-fixed frame, is attached to it and the motion of this frame describes the evolution of the orbital plane. The orbital plane can be defined geometrically by infinitely many frames sharing their z-axis and rotating with different angular velocities. This yields different definitions of the perifocal frames $\mathcal{P} = \{\mathbf{i}_{\mathcal{P}}, \mathbf{j}_{\mathcal{P}}, \mathbf{k}_{\mathcal{P}}\}$: for example, the one defined by the osculating values of the eccentricity and angular momentum vectors is the osculating perifocal frame. However, we shall look for an alternative definition of the perifocal frame: a definition that makes the perifocal frame ideal in Hansen's sense. Describing the motion of the orbital plane with a Hansen ideal frame will yield the dynamical decomposition discussed in the previous section. This frame, referred to as the perifocal *departure* frame (a term coined by Deprit, 1975) is given a dynamical definition derived in the following lines.

The evolution of the orbital frame is governed by the differential equations

$$\left.\frac{\mathrm{d}\mathbf{i}}{\mathrm{d}t}\right|_{\mathcal{I}} = \boldsymbol{\omega}_{\mathcal{L}\mathcal{I}} \times \mathbf{i} = \frac{\mathrm{d}}{\mathrm{d}t}\left(\frac{\mathbf{r}}{r}\right) = \frac{1}{r}\left(\mathbf{v} - \frac{\mathrm{d}r}{\mathrm{d}t}\mathbf{i}\right)$$
$$\left.\frac{\mathrm{d}\mathbf{k}}{\mathrm{d}t}\right|_{\mathcal{I}} = \boldsymbol{\omega}_{\mathcal{L}\mathcal{I}} \times \mathbf{k} = \frac{\mathrm{d}}{\mathrm{d}t}\left(\frac{\mathbf{h}}{h}\right) = \frac{1}{h}\left(\frac{\mathrm{d}\mathbf{h}}{\mathrm{d}t} - \frac{\mathrm{d}h}{\mathrm{d}t}\mathbf{k}\right)$$

Let $\boldsymbol{\omega}_{\mathcal{L}\mathcal{I}} = \omega_x\,\mathbf{i} + \omega_y\,\mathbf{j} + \omega_z\,\mathbf{k}$ be the angular velocity of the orbital frame with respect to the inertial frame. Projecting the previous equations onto $\mathcal{L}$ yields:

$$\omega_x = -\frac{1}{h}\left(\frac{\mathrm{d}\mathbf{h}}{\mathrm{d}t}\cdot\mathbf{j}\right) = -\frac{1}{h}(\mathbf{r}\times\mathbf{a}_p)\cdot\mathbf{j} = \frac{r}{h}(\mathbf{a}_p\cdot\mathbf{k})$$
$$\omega_y = 0$$
$$\omega_z = \frac{1}{r}(\mathbf{v}\cdot\mathbf{j}) = -\frac{1}{r}\,\mathbf{v}\cdot(\mathbf{i}\times\mathbf{k}) = \frac{1}{r^2}(\mathbf{k}\cdot\mathbf{h}) = \frac{h}{r^2}$$

meaning that

$$\boldsymbol{\omega}_{\mathcal{L}\mathcal{I}} = \frac{h}{r^2}\mathbf{k} + \frac{r}{h}(\mathbf{a}_p\cdot\mathbf{k})\,\mathbf{i} = \boldsymbol{\omega}_{\mathcal{L}\mathcal{P}} + \boldsymbol{\omega}_{\mathcal{P}\mathcal{I}} \tag{4.3}$$

The orbital frame is defined by the rotation of an intermediate frame along the radial direction $\mathbf{i}$, plus a rotation along the $\mathbf{k}$ direction. Motivated by this result, Peláez et al. (2007) defined the perifocal departure frame as the one that rotates with an angular velocity

$$\boldsymbol{\omega}_{\mathcal{P}\mathcal{I}} = \frac{r}{h}(\mathbf{a}_p\cdot\mathbf{k})\,\mathbf{i} \tag{4.4}$$

It is $\boldsymbol{\omega}_{\mathcal{P}\mathcal{I}} \parallel \mathbf{r}$ and therefore $\mathcal{P}$ is a Hansel ideal frame. The orientation of this frame at departure can be fixed as required, but the obvious choice is to make it coincide with the osculating perifocal frame at the departure epoch:

$$\mathbf{i}_{\mathcal{P}} = \frac{\mathbf{e}_0}{e_0}, \qquad \mathbf{k}_{\mathcal{P}} = \frac{\mathbf{h}_0}{h_0}, \qquad \mathbf{j}_{\mathcal{P}} = \mathbf{k}_{\mathcal{P}} \times \mathbf{i}_{\mathcal{P}}$$

The relative motion between the orbital and the perifocal departure frame abides by

$$\boldsymbol{\omega}_{\mathcal{L}\mathcal{P}} = \frac{h}{r^2}\mathbf{k}$$

The independent variable in Dromo is the *ideal anomaly*, σ, defined in terms of the time transformation

$$\frac{\mathrm{d}t}{\mathrm{d}\sigma} = \frac{r^2}{h} \tag{4.5}$$

The ideal anomaly decomposes in

$$\sigma = \beta + \vartheta$$

Here β is the angle between the osculating eccentricity vector and $\mathbf{e}_0$, and ϑ is the true anomaly. Figure 4.1 shows the geometry of the different reference frames that have been introduced. The relation between the ideal anomaly and the true anomaly can be observed in this figure.

4.2.1 The eccentricity vector

The velocity of the particle is given by

$$\mathbf{v} = \frac{\mathrm{d}\mathbf{r}}{\mathrm{d}t} = \frac{\mathrm{d}r}{\mathrm{d}t}\mathbf{i} + \frac{h}{r}\mathbf{j}$$

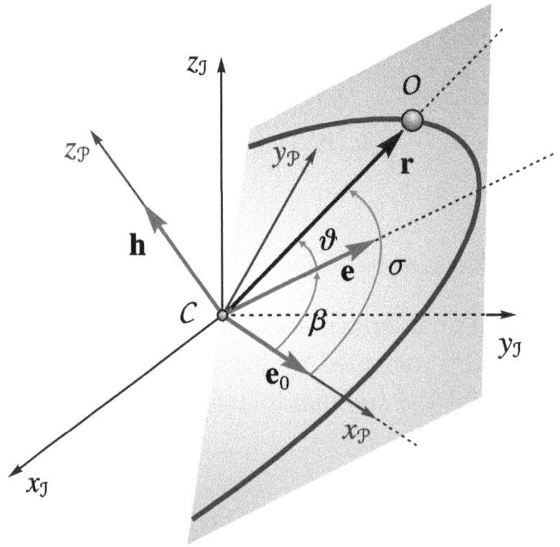

Fig. 4.1: Geometrical definition of the problem.

Thus, the eccentricity vector $\mathbf{e}$ admits the following definition

$$\mathbf{e} = \frac{\mathbf{v} \times \mathbf{h}}{\mu} - \frac{\mathbf{r}}{r} = \left(\frac{h^2}{r} - 1\right)\mathbf{i} - h\frac{\mathrm{d}r}{\mathrm{d}t}\mathbf{j} \tag{4.6}$$

This expression is projected on the departure perifocal frame $\mathcal{P}$ to provide:

$$\begin{aligned}\mathbf{e} &= \left[\left(\frac{h^2}{r} - 1\right)\cos\sigma + h\frac{\mathrm{d}r}{\mathrm{d}t}\sin\sigma\right]\mathbf{i}_{\mathcal{P}} + \left[\left(\frac{h^2}{r} - 1\right)\sin\sigma - h\frac{\mathrm{d}r}{\mathrm{d}t}\cos\sigma\right]\mathbf{j}_{\mathcal{P}} \\ &= \zeta_1\,\mathbf{i}_{\mathcal{P}} + \zeta_2\,\mathbf{j}_{\mathcal{P}}\end{aligned}$$

That is, the pair (ζ_1, ζ_2) defines the components of the eccentricity vector in the departure perifocal frame $\mathcal{P}$. Recall that the angle β describes the evolution of the osculating eccentricity vector, i.e., $\tan\beta = \zeta_2/\zeta_1$. The projections ζ_1 and ζ_2 admit two alternative definitions:

$$\zeta_1 = \left(\frac{h^2}{r} - 1\right)\cos\sigma + h\frac{\mathrm{d}r}{\mathrm{d}t}\sin\sigma \equiv e\,\cos\beta$$

$$\zeta_2 = \left(\frac{h^2}{r} - 1\right)\sin\sigma - h\frac{\mathrm{d}r}{\mathrm{d}t}\cos\sigma \equiv e\,\sin\beta$$

Introducing the inverse of the angular momentum,

$$\zeta_3 = \frac{1}{h}$$

the triple $(\zeta_1, \zeta_2, \zeta_3)$ defines the first three Dromo elements. In the unperturbed case it is $\beta = 0$, and therefore $\zeta_1 = e$, $\zeta_2 = 0$, and $\zeta_3 = 1/h_0$ for all t.

4.2.2 The orbital plane

Let $\mathfrak{r}_{\mathcal{I}}, \mathfrak{r}_{\mathcal{P}} \in \mathbb{H}$ be two pure quaternions associated to the components of vector $\mathbf{r}$ in the inertial and in the departure perifocal reference frames, $\mathcal{I}$ and $\mathcal{P}$, respectively. Both quaternions relate through the rotation action

$$\mathfrak{r}_{\mathcal{I}} = \mathfrak{n}\,\mathfrak{r}_{\mathcal{P}}\,\mathfrak{n}^{\dagger}, \qquad \mathfrak{n} \in \mathbb{H} \tag{4.7}$$

Quaternion $\mathfrak{n}^{\dagger}$ is the conjugate of $\mathfrak{n}$. See Appendix A for details on the algebra of quaternions.

The unit quaternion $\mathfrak{n} = \eta_4 + \mathrm{i}\,\eta_1 + \mathrm{j}\,\eta_2 + \mathrm{k}\,\eta_3$ relates to the Euler angles by means of successive applications of Euler's theorem:

$$\begin{aligned}\eta_1 &= \sin\frac{i}{2}\cos\frac{\Omega - \tilde{\omega}}{2}, \qquad \eta_2 = \sin\frac{i}{2}\sin\frac{\Omega - \tilde{\omega}}{2} \\ \eta_3 &= \cos\frac{i}{2}\sin\frac{\Omega + \tilde{\omega}}{2}, \qquad \eta_4 = \cos\frac{i}{2}\cos\frac{\Omega + \tilde{\omega}}{2}\end{aligned} \tag{4.8}$$

with i the osculating inclination, Ω the osculating right ascension of the ascending node, and $\tilde{\omega}$ the departure argument of the pericenter. It differs from the osculating argument of the pericenter by the angle $\beta = \arctan \zeta_2/\zeta_1$, although at $t = 0$ they both coincide, $\omega_0 \equiv \tilde{\omega}_0$. The Euler parameters $(\eta_1, \eta_2, \eta_3, \eta_4)$ are the last four Dromo elements.

4.2.3 Evolution equations

The Dromo elements form the set $(\zeta_1, \zeta_2, \zeta_3, \eta_1, \eta_2, \eta_3, \eta_4)$. In this section we will derive the equations governing the evolution of the elements.

The radial distance in Dromo variables reads

$$r = \frac{h^2}{1 + e\cos\vartheta} = \frac{1}{\zeta_3^2\,(1 + e\cos\vartheta)} = \frac{1}{\zeta_3^2\, s}$$

with $s = 1 + e\cos\vartheta = 1 + \zeta_1 \cos\sigma + \zeta_2 \sin\sigma$. The evolution of the elements with σ is integrated together with the Sundman transformation (4.5), in order to propagate the physical time. The time behaves as a dependent variable. This differential equation is rewritten in Dromo elements to provide:

$$\frac{\mathrm{d}t}{\mathrm{d}\sigma} = \frac{1}{\zeta_3^3 s^2} \tag{4.9}$$

The equation of the hodograph is obtained from the cross product of $\mathbf{k}$ and the eccentricity vector, given by Eq. (4.6),

$$\mathbf{v} = \zeta_3(\mathbf{j} + \mathbf{k} \times \mathbf{e})$$

Considering that $\mathbf{e} = \zeta_1\, \mathbf{i}_{\mathcal{P}} + \zeta_2\, \mathbf{j}_{\mathcal{P}}$, the hodograph equation transforms into

$$\zeta_1\, \mathbf{j}_{\mathcal{P}} - \zeta_2\, \mathbf{i}_{\mathcal{P}} = -\mathbf{j} + \frac{\mathbf{v}}{\zeta_3}$$

This equation is differentiated with respect to ideal anomaly σ and projected along $\mathbf{i}_{\mathcal{P}}$ and $\mathbf{j}_{\mathcal{P}}$ to provide

$$\begin{aligned}\frac{\mathrm{d}\zeta_1}{\mathrm{d}\sigma} &= s\,\sin\sigma\,(\tilde{\mathbf{a}}_p \cdot \mathbf{i}) + \left[\zeta_1 + (1+s)\cos\sigma\right](\tilde{\mathbf{a}}_p \cdot \mathbf{j}) \\ \frac{\mathrm{d}\zeta_2}{\mathrm{d}\sigma} &= -s\,\cos\sigma(\tilde{\mathbf{a}}_p \cdot \mathbf{i}) + \left[\zeta_2 + (1+s)\sin\sigma\right](\tilde{\mathbf{a}}_p \cdot \mathbf{j})\end{aligned}$$

where the term $\tilde{\mathbf{a}}_p$ stands for $\tilde{\mathbf{a}}_p = \mathbf{a}_p / \left(\zeta_3^4 s^3\right)$. It is the normalized perturbing acceleration.

The angular momentum vector changes according to the expression

$$\frac{\mathrm{d}\mathbf{h}}{\mathrm{d}t} = \mathbf{r} \times \mathbf{a}_p = r(\mathbf{i} \times \mathbf{a}_p)$$

where it is $\mathbf{h} = h\,\mathbf{k}$. Considering that

$$\frac{\mathrm{d}\mathbf{h}}{\mathrm{d}t} = \frac{\mathrm{d}h}{\mathrm{d}t}\mathbf{k} + h\frac{\mathrm{d}\mathbf{k}}{\mathrm{d}t} = \frac{\mathrm{d}h}{\mathrm{d}t}\mathbf{k} + h(\boldsymbol{\omega}_{\mathcal{L}\mathcal{I}} \times \mathbf{k}) = \frac{\mathrm{d}h}{\mathrm{d}t}\mathbf{k} - r(\mathbf{a}_p \cdot \mathbf{k})\,\mathbf{j}$$

the evolution equation for $\mathbf{h}$ is projected along $\mathbf{k}$ to provide

$$\frac{\mathrm{d}h}{\mathrm{d}t} = r(\mathbf{a}_p \cdot \mathbf{j}) \implies \frac{\mathrm{d}h}{\mathrm{d}\sigma} = \frac{r^3}{h}(\mathbf{a}_p \cdot \mathbf{j})$$

The derivative of ζ_3 results in

$$\frac{\mathrm{d}\zeta_3}{\mathrm{d}\sigma} = -\zeta_3\,(\tilde{\mathbf{a}}_p \cdot \mathbf{j})$$

Note that the angular momentum of the particle is only affected by the projection of the perturbing acceleration in the along-track direction.

Let $\mathfrak{w}_{\mathcal{P}\mathcal{I}}$ be the pure quaternion defined by the projections of vector $\boldsymbol{\omega}_{\mathcal{P}\mathcal{I}}/2$ in frame $\mathcal{P}$. This quaternion relates to quaternion $\mathfrak{n}$ by means of

$$2\mathfrak{w}_{\mathcal{P}\mathcal{I}} = 2\mathfrak{n}^{\dagger}\,\frac{\mathrm{d}\mathfrak{n}}{\mathrm{d}t}$$

This equation is inverted to provide the time evolution of quaternion $\mathfrak{n}$,

$$\frac{\mathrm{d}\mathfrak{n}}{\mathrm{d}t} = \frac{1}{2}\,\mathfrak{n}(2\mathfrak{w}_{\mathcal{P}\mathcal{I}})$$

Expanding the quaternion product and taking into account the definition of $\boldsymbol{\omega}_{\mathcal{P}\mathcal{I}}$ given in Eq. (4.4) provides the evolution equations for the components of $\mathfrak{n}$. The full system of eight ordinary differential equations becomes

$$\frac{\mathrm{d}t}{\mathrm{d}\sigma} = \frac{1}{\zeta_3^3 s^2} \tag{4.10}$$

$$\frac{\mathrm{d}\zeta_1}{\mathrm{d}\sigma} = s\,\sin\sigma\,(\tilde{\mathbf{a}}_p \cdot \mathbf{i}) + \left[\zeta_1 + (1+s)\cos\sigma\right](\tilde{\mathbf{a}}_p \cdot \mathbf{j}) \tag{4.11}$$

$$\frac{\mathrm{d}\zeta_2}{\mathrm{d}\sigma} = -s\,\cos\sigma\,(\tilde{\mathbf{a}}_p \cdot \mathbf{i}) + \left[\zeta_2 + (1+s)\sin\sigma\right](\tilde{\mathbf{a}}_p \cdot \mathbf{j}) \tag{4.12}$$

$$\frac{\mathrm{d}\zeta_3}{\mathrm{d}\sigma} = -\zeta_3\,(\tilde{\mathbf{a}}_p \cdot \mathbf{j}) \tag{4.13}$$

$$\frac{\mathrm{d}\eta_1}{\mathrm{d}\sigma} = \frac{(\tilde{\mathbf{a}}_p \cdot \mathbf{k})}{2}(\ \eta_4\,\cos\sigma - \eta_3\,\sin\sigma) \tag{4.14}$$

$$\frac{\mathrm{d}\eta_2}{\mathrm{d}\sigma} = \frac{(\tilde{\mathbf{a}}_p \cdot \mathbf{k})}{2}(\ \eta_3\,\cos\sigma + \eta_4\,\sin\sigma) \tag{4.15}$$

$$\frac{\mathrm{d}\eta_3}{\mathrm{d}\sigma} = \frac{(\tilde{\mathbf{a}}_p \cdot \mathbf{k})}{2}(-\eta_2\,\cos\sigma + \eta_1\,\sin\sigma) \tag{4.16}$$

$$\frac{\mathrm{d}\eta_4}{\mathrm{d}\sigma} = \frac{(\tilde{\mathbf{a}}_p \cdot \mathbf{k})}{2}(-\eta_1\,\cos\sigma - \eta_2\,\sin\sigma) \tag{4.17}$$

Equations (4.10–4.17) need to be integrated from the initial conditions

$$\sigma = \sigma_0: \quad t = t_0\,, \quad \zeta_1 = e_0\,, \quad \zeta_2 = 0\,, \quad \zeta_3 = 1/h_0\,, \quad \mathfrak{n} = \mathfrak{n}_0$$

Quaternion $\mathfrak{n}_0$ is initialized following Eq. (4.8). Recall that $\tilde{\mathbf{a}}_p = \mathbf{a}_p/\left(\zeta_3^4 s^3\right)$.

The position and velocity vectors written in the orbital frame $\mathcal{L}$ read

$$\mathbf{r} = \frac{1}{\zeta_3^2 s}\mathbf{i} \qquad \text{and} \qquad \mathbf{v} = \zeta_3(u\,\mathbf{i} + s\,\mathbf{j})$$

with

$$s = 1 + \zeta_1 \cos\sigma + \zeta_2 \sin\sigma\,, \qquad u = \zeta_1 \sin\sigma - \zeta_2 \cos\sigma$$

The unit vectors **i** and **j** can be referred to the departure perifocal frame thanks to

$$\mathbf{i} = \cos\sigma\,\mathbf{i}_{\mathcal{P}} + \sin\sigma\,\mathbf{j}_{\mathcal{P}}\,, \qquad \mathbf{j} = -\sin\sigma\,\mathbf{i}_{\mathcal{P}} + \cos\sigma\,\mathbf{j}_{\mathcal{P}}$$

Finally, the transformation from the components in the perifocal departure frame to the inertial frame is given by the quaternionic product in Eq. (4.7), which can be written in terms of the rotation matrix

$$\mathbf{R}(\mathfrak{n}) = \begin{bmatrix} 1 - 2\left(\eta_2^2 + \eta_3^2\right) & 2(\eta_1\eta_2 - \eta_3\eta_4) & 2(\eta_1\eta_3 + \eta_2\eta_4) \\ 2(\eta_1\eta_2 + \eta_3\eta_4) & 1 - 2\left(\eta_1^2 + \eta_3^2\right) & 2(\eta_2\eta_3 - \eta_1\eta_4) \\ 2(\eta_1\eta_3 - \eta_2\eta_4) & 2(\eta_1\eta_4 + \eta_2\eta_3) & 1 - 2\left(\eta_1^2 + \eta_2^2\right) \end{bmatrix}$$

The position and velocity vectors in the inertial frame then read

$$\mathbf{r}_{\mathcal{I}} = \frac{1}{\zeta_3^2 s}\mathbf{R}(\mathfrak{n}) \begin{bmatrix} \cos\sigma \\ \sin\sigma \\ 0 \end{bmatrix}, \qquad \mathbf{v}_{\mathcal{I}} = \zeta_3 \mathbf{R}(\mathfrak{n}) \begin{bmatrix} u\cos\sigma - s\sin\sigma \\ u\sin\sigma + s\cos\sigma \\ 0 \end{bmatrix}$$

4.3 Improved performance

The motivation behind the development of Dromo was to improve the numerical performance of the propagation compared to the straightforward integration in Cartesian coordinates (specially for highly elliptical or weakly perturbed orbits), as well as to achieve the regularization of the equations of motion. For completeness, this section presents two examples of application to evaluate the nominal performance of Dromo compared to Cowell's method. Other examples can be found in, for example, the work of Urrutxua (2015, Chap. 8). When comparing the numerical performance one should always keep in mind that the results depend strongly on the problem under consideration, and should avoid making claims about the overall superiority of a method for generic applications.

The first example (Problem 1) is the test problem 2b proposed by Stiefel and Scheifele (1971, p. 122). It is a highly elliptical orbit perturbed by the gravitational attraction of the Moon and the Earth's J_2. The orbit of the Moon is assumed circular and Keplerian. The orbit of the particle is integrated for about 50 revolutions, and the osculating elements at departure are defined in Table 4.1.

The performances of Dromo and Cowell are compared in Fig. 4.2. The figure shows the relation between the final error in position and the number of function calls. An efficient formulation will require less function calls for the same level of accuracy, thanks to providing an adequate discretization of the orbit. In this example, the orbits are integrated using a variable step-size RKF5(4) scheme. The performance curves are generated by changing the integration tolerance from $\varepsilon = 10^{-6}$ to 10^{-14} and the error is measured with respect to an accurate solution integrated in quadruple precision floating-point arithmetic with different formulations, making sure that the first 16 digits coincide. Dromo reduces the required number of function calls by almost an order of magnitude when compared to the integration in Cartesian coordinates.

Tab. 4.1: Initial conditions of the test problems.

	Units	Problem 1	Problem 2
MJD_0 (ET)	–	N/A	51, 544.0000
t_f	days	288.1277	20.0000
a	km	136, 000.4185	10, 000.0000
e	–	0.9500	0.3000
i	deg	30.0000	50.0000
ω	deg	−90.0000	180.0000
Ω	deg	0.0000	150.0000
M_0	deg	0.0000	162.3605

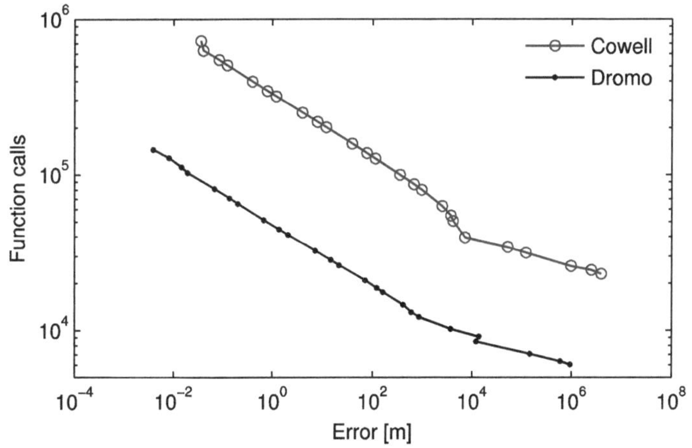

Fig. 4.2: Performance of the integration of Problem 1.

The second example (Problem 2) is a realistic orbit around the Earth, with the initial conditions defined in Table 4.1 (given in the ICRF/J2000 reference system with the reference plane defined by the Earth mean equator and equinox at epoch). The Earth's gravity field is modeled with a 40×40 grid using the harmonics from the GGM03S solution. The orientation of the Earth is given by the IAU 2006/2000A, CIO based standard (using X, Y series). In order to model the effects of the atmospheric drag, the area-to-mass ratio of the spacecraft is assigned the value $A/m = 0.0075\ \mathrm{m^2/kg}$, and the drag coefficient is $c_D = 2.0$. Space weather data with the three-hour values of the atmospheric parameters, as well as the Earth Orientation Parameters, are taken from Celestrak. The atmospheric density is given by the MSISE90 model. The acceleration due to the solar radiation pressure is computed assuming a reflectivity coefficient $c_R = 1.2$, a nominal solar flux of $\Phi = 1367\ \mathrm{W/m^2}$ with periodic seasonal variations, and a conic shadow model. The perturbations from the Sun, the Moon, and all the planets in the Solar System are constructed from the DE430 ephemeris. The perturbation model is realistic, and the forces are small compared to the Keplerian term.

The orbit is integrated with LSODAR, a variable order, variable step size and multistep Adams-based method implemented in the library ODEPACK (Hindmarsh, 1983). In addition, Cowell's method is also integrated with an eighth-order Störmer–Cowell scheme. Figure 4.3 displays the performance of each integration. Although the Störmer–Cowell integration is more efficient than Dromo for low-accuracy integrations, its accuracy does not match the one provided by LSODAR. With this integrator Dromo proves more efficient than Cowell for a given level of accuracy. In addition, the minimum error reached by Dromo is two orders of magnitude smaller than the minimum error reached by the integration in Cartesian variables. The improvements in performance come from the combination of two factors. First, a more efficient step-size control thanks to the second-order Sundman transformation. This is particularly

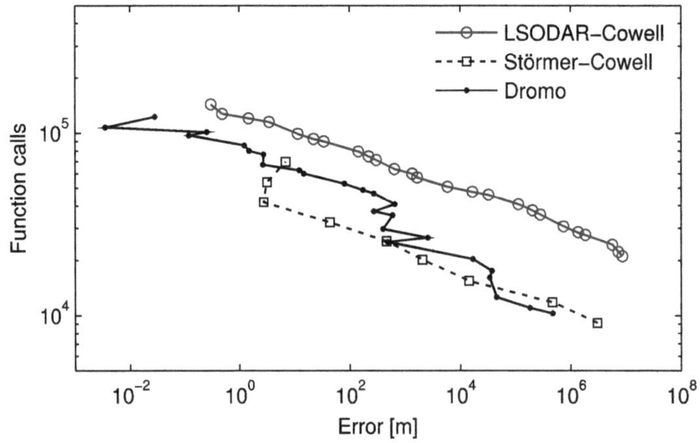

Fig. 4.3: Performance of the integration of Problem 2.

relevant in Problem 1, due to the high eccentricity of the orbit. Second, a smoother evolution of the right-hand side of the differential equations to be integrated because of using elements in lieu of coordinates. The advantage of this technique has been widely discussed in Sect. 2.5, and exhibits all its potential in Problem 2, where perturbations are small.

4.4 Variational equations and the noncanonicity of Dromo

The present section is devoted to deriving the variational equations of Dromo, and presenting the Lagrange and Poisson brackets. The variational equations and the Lagrange brackets will be recovered in the next section for deriving the gauge-generalized Dromo equations, and in Chap. 8 for solving the relative dynamics in Dromo variables. With these results at hand, some comments on the noncanonicity of Dromo will be included.

In Sect. 2.6 we proved that the canonical equations based on a homogeneous Hamiltonian can be easily referred to a fictitious time. The physical time becomes a dependent variable associated to an additional coordinate, $q_0 \equiv t$. The corresponding canonical equation referred to the ideal anomaly renders

$$\frac{\mathrm{d}q_0}{\mathrm{d}\sigma} = \frac{\partial \tilde{\mathcal{H}}_h}{\partial p_0} = \frac{r^2}{h}$$

which turns out to be the Sundman transformation. Following the procedure in Sect. 2.6 Dromo can be understood as a transformation

$$\mathscr{D}\colon (\sigma; q_0, \ldots, q_3, p_0, \ldots, p_3) \mapsto (\sigma; \bar{q}_0, \ldots, \bar{q}_3, \bar{p}_0, \ldots, \bar{p}_3)$$

where $\bar{q}_0 = q_0$, $\bar{q}_i = \zeta_i$ (with $i = 1, 2, 3$), $\bar{p}_0 = p_0$, and $\bar{p}_i$ are three of the components of quaternion $\mathfrak{n}$. The fourth component is not independent from the others; in order to ensure that the Jacobian of the transformation is regular the constraint:

$$\eta_1^2 + \eta_2^2 + \eta_3^2 + \eta_4^2 = 1 \implies \eta_1\,\delta\eta_1 + \eta_2\,\delta\eta_2 + \eta_3\,\delta\eta_3 + \eta_4\,\delta\eta_4 = 0$$

needs to be satisfied. A similar discussion can also be found in the work of Broucke et al. (1971) or Gurfil (2005a), among others.

The conditions for canonicity depend on the form of the Jacobian of $\mathscr{D}$, denoted **A**. This matrix is built from the partial derivatives of the position and velocity vectors, $\mathbf{r} = \mathbf{f}(\sigma; \boldsymbol{\oe})$ and $\mathbf{v} = \mathbf{g}(\sigma; \boldsymbol{\oe})$:

$$\frac{\partial \mathbf{f}}{\partial \zeta_1} = -\frac{\cos\sigma}{\zeta_3^2 s^2}\mathbf{i}, \qquad \frac{\partial \mathbf{f}}{\partial \zeta_2} = -\frac{\sin\sigma}{\zeta_3^2 s^2}\mathbf{i}, \qquad \frac{\partial \mathbf{f}}{\partial \zeta_3} = -\frac{2}{\zeta_3^3 s}\mathbf{i}$$

$$\frac{\partial \mathbf{f}}{\partial \eta_1} = +\frac{2}{\zeta_3^2 s \eta_4}\left[N_{1324}^{+}\,\mathbf{j} + (N_{1144}^{+}\sin\sigma + N_{3412}^{-}\cos\sigma)\mathbf{k}_{\mathcal{P}}\right]$$

$$\begin{aligned}
\frac{\partial \mathbf{f}}{\partial \eta_2} &= -\frac{2}{\zeta_3^2 s \eta_4}\left[N^-_{1423}\,\mathbf{j} + (N^+_{2244}\cos\sigma - N^+_{1234}\sin\sigma)\mathbf{k}_{\mathcal{P}}\right]\\
\frac{\partial \mathbf{f}}{\partial \eta_3} &= +\frac{2}{\zeta_3^2 s \eta_4}\left[N^+_{3344}\,\mathbf{j} - (N^+_{1423}\cos\sigma - N^-_{1324}\sin\sigma)\mathbf{k}_{\mathcal{P}}\right]
\end{aligned}$$

and

$$\begin{aligned}
\frac{\partial \mathbf{g}}{\partial \zeta_1} &= +\zeta_3\,\mathbf{j}_{\mathcal{P}}\,, \qquad \frac{\partial \mathbf{g}}{\partial \zeta_2} = -\zeta_3\,\mathbf{i}_{\mathcal{P}}\\
\frac{\partial \mathbf{g}}{\partial \zeta_3} &= -(\zeta_2 + \sin\sigma)\,\mathbf{i}_{\mathcal{P}} + (\zeta_1 + \cos\sigma)\,\mathbf{j}_{\mathcal{P}}\\
\frac{\partial \mathbf{g}}{\partial \eta_1} &= -\frac{2\zeta_3}{\eta_4}\{N^+_{1324}[(\zeta_1 + \cos\sigma)\mathbf{i}_{\mathcal{P}} + (\zeta_2 + \sin\sigma)\mathbf{j}_{\mathcal{P}}]\\
&\quad - [N^-_{1234}(\zeta_2 + \sin\sigma) + N^+_{1144}(\zeta_1 + \cos\sigma)]\mathbf{k}_{\mathcal{P}}\}\\
\frac{\partial \mathbf{g}}{\partial \eta_2} &= +\frac{2\zeta_3}{\eta_4}\{N^-_{1423}[(\zeta_1 + \cos\sigma)\mathbf{i}_{\mathcal{P}} + (\zeta_2 + \sin\sigma)\mathbf{j}_{\mathcal{P}}]\\
&\quad + [N^+_{2244}(\zeta_2 + \sin\sigma) + N^+_{1234}(\zeta_1 + \cos\sigma)]\mathbf{k}_{\mathcal{P}}\}\\
\frac{\partial \mathbf{g}}{\partial \eta_3} &= -\frac{2\zeta_3}{\eta_4}\{N^+_{3344}[(\zeta_1 + \cos\sigma)\mathbf{i}_{\mathcal{P}} + (\zeta_2 + \sin\sigma)\mathbf{j}_{\mathcal{P}}]\\
&\quad - [N^+_{1423}(\zeta_2 + \sin\sigma) + N^-_{1324}(\zeta_1 + \cos\sigma)]\mathbf{k}_{\mathcal{P}}\}
\end{aligned}$$

They are written in terms of the coefficient $N^{\pm}_{ijk\ell} = \eta_i\eta_j \pm \eta_k\eta_\ell$. In this case, η_4 has been chosen as the dependent component.

Having identified $[q_1, q_2, q_3]^\top \equiv \mathbf{r}$, $[p_1, p_2, p_3]^\top \equiv \mathbf{v}$, $\bar{q}_i = \zeta_i$ and $\bar{p}_i = \eta_i$ (with $i = 1, 2, 3$), the nonzero Lagrange brackets read

$$\begin{aligned}
[\zeta_1, \zeta_2] &= \frac{1}{\zeta_3 s^2} & [\zeta_1, \zeta_3] &= \frac{2s\sin\sigma - u\cos\sigma}{\zeta_3^2 s^2}\\
[\zeta_2, \zeta_3] &= -\frac{\zeta_1 + (1+s)\cos\sigma}{\zeta_3^2 s^2} & [\zeta_3, \eta_1] &= \frac{2}{\eta_4\zeta_3^2}N^+_{1324}\\
[\zeta_3, \eta_2] &= \frac{2}{\eta_4\zeta_3^2}N^-_{2314}\,, & [\zeta_3, \eta_3] &= \frac{2}{\eta_4\zeta_3^2}N^+_{3344}\\
[\eta_1, \eta_2] &= \frac{\eta_4}{\eta_1}[\eta_1, \eta_3] = \frac{\eta_4}{\eta_2}[\eta_2, \eta_3] & &= \frac{4}{\zeta_3}
\end{aligned}$$

For completeness, the Poisson brackets read:

$$\begin{aligned}
\{\zeta_1, \eta_1\} &= -\frac{\eta_2}{\eta_1}\{\zeta_1, \eta_2\} = \frac{\eta_2}{\eta_4}\{\zeta_1, \eta_3\} &&= -\frac{\zeta_3\eta_2}{2}[\zeta_1 + (1+s)\cos\sigma]\\
\{\zeta_2, \eta_1\} &= -\frac{\eta_2}{\eta_1}\{\zeta_2, \eta_2\} = \frac{\eta_2}{\eta_4}\{\zeta_2, \eta_3\} &&= \frac{\zeta_3\eta_2}{2}(u\cos\sigma - 2s\sin\sigma)
\end{aligned}$$

$$
\begin{aligned}
\{\zeta_3, \eta_1\} &= -\frac{\eta_2}{\eta_1}\{\zeta_3, \eta_2\} = \frac{\eta_2}{\eta_4}\{\zeta_3, \eta_3\} = \frac{\zeta_3^2\eta_2}{2} \\
\{\zeta_1, \zeta_2\} &= \zeta_3 s^2, \qquad \{\eta_2, \eta_3\} = \frac{\zeta_3}{4}N^+_{1324} \\
\{\eta_1, \eta_2\} &= \frac{\zeta_3}{4}N^+_{3344}, \qquad \{\eta_1, \eta_3\} = \frac{\zeta_3}{4}N^-_{1423}
\end{aligned}
$$

The condition (2.38) is violated meaning that Dromo is not canonical. This is equivalent to noting that the Jacobian matrix is not symplectic.

4.5 Gauge-generalized Dromo formulation

Equations (4.10–4.17) can be generalized by introducing a nontrivial gauge, following the procedure presented in Sect. 2.7. Thanks to having defined the Lagrange brackets explicitly, Eq. (2.46) leads to the general form of the gauge-generalized Dromo equations when considering the time transformation from Eq. (4.9):

$$
\sum_j [\text{œ}_n, \text{œ}_j]\frac{\mathrm{d}\text{œ}_j}{\mathrm{d}\sigma} = \left(\frac{\Delta\mathbf{F}}{\zeta_3^3 s^2} - \frac{\mathrm{d}\boldsymbol{\Phi}}{\mathrm{d}\sigma}\right)\cdot\frac{\partial\mathbf{f}}{\partial\text{œ}_n} - \frac{\partial\mathbf{g}}{\partial\text{œ}_n}\cdot\boldsymbol{\Phi}
$$

In this case, the gauge function $\boldsymbol{\Phi}$ relates to the elements by means of

$$
\boldsymbol{\Phi} = \sum_j \frac{\partial\mathbf{f}}{\partial\text{œ}_j}\frac{\mathrm{d}\text{œ}_j}{\mathrm{d}\sigma}
$$

Let $\tilde{\boldsymbol{\Phi}} = \boldsymbol{\Phi}/(\zeta_3 s)$ be the adjusted gauge function. Denoting $(\tilde{a}_{p,x}, \tilde{a}_{p,y}, \tilde{a}_{p,z})$ and $(\tilde{\Phi}_x, \tilde{\Phi}_y, \tilde{\Phi}_z)$ the components of $\tilde{\mathbf{a}}_p$ and $\tilde{\boldsymbol{\Phi}}$ in the orbital frame $\mathcal{L}$, the gauge-generalized version of Eqs. (4.11–4.17) is

$$
\begin{aligned}
\frac{\mathrm{d}\zeta_1}{\mathrm{d}\sigma} &= +s\sin\sigma\left(\tilde{a}_{p,x} - \tilde{\Phi}'_x\right) + [\zeta_1 + (1+s)\cos\sigma]\left(\tilde{a}_{p,y} - \tilde{\Phi}'_y\right) \\
&\quad + (\zeta_1 + \cos\sigma)\tilde{\Phi}_x + \left(2\zeta_2 + \ell^2\sin\sigma/s\right)\tilde{\Phi}_y \\
\frac{\mathrm{d}\zeta_2}{\mathrm{d}\sigma} &= -s\cos\sigma\left(\tilde{a}_{p,x} - \tilde{\Phi}'_x\right) + [\zeta_2 + (1+s)\sin\sigma]\left(\tilde{a}_{p,y} - \tilde{\Phi}'_y\right) \\
&\quad + (\zeta_2 + \sin\sigma)\tilde{\Phi}_x - \left(2\zeta_1 + \ell^2\cos\sigma/s\right)\tilde{\Phi}_y \\
\frac{\mathrm{d}\zeta_3}{\mathrm{d}\sigma} &= -\frac{\zeta_3}{s}\left[s(\tilde{a}_{p,y} - \tilde{\Phi}'_y) + s\tilde{\Phi}_x - u\tilde{\Phi}_y\right] \\
\frac{\mathrm{d}\eta_1}{\mathrm{d}\sigma} &= \frac{1}{2}(+\eta_4\cos\sigma - \eta_3\sin\sigma)\left(\tilde{a}_{p,z} - \tilde{\Phi}'_z\right) \\
&\quad + \frac{1}{2s}\left\{s\eta_2\tilde{\Phi}_y + [\eta_3(\zeta_1 + \cos\sigma) + \eta_4(\zeta_2 + \sin\sigma)]\tilde{\Phi}_z\right\} \\
\frac{\mathrm{d}\eta_2}{\mathrm{d}\sigma} &= \frac{1}{2}(+\eta_3\cos\sigma + \eta_4\sin\sigma)\left(\tilde{a}_{p,z} - \tilde{\Phi}'_z\right) \\
&\quad - \frac{1}{2s}\left\{s\eta_1\tilde{\Phi}_y + [\eta_4(\zeta_1 + \cos\sigma) - \eta_3(\zeta_2 + \sin\sigma)]\tilde{\Phi}_z\right\}
\end{aligned}
$$

$$\frac{d\eta_3}{d\sigma} = \frac{1}{2}(-\eta_2 \cos\sigma + \eta_1 \sin\sigma)\left(\tilde{a}_{p,z} - \tilde{\Phi}'_z\right)$$
$$+ \frac{1}{2s}\left\{s\eta_4\,\tilde{\Phi}_y - [\eta_1(\zeta_1 + \cos\sigma) + \eta_2(\zeta_2 + \sin\sigma)]\tilde{\Phi}_z\right\}$$
$$\frac{d\eta_4}{d\sigma} = \frac{1}{2}(-\eta_1 \cos\sigma - \eta_2 \sin\sigma)\left(\tilde{a}_{p,z} - \tilde{\Phi}'_z\right)$$
$$- \frac{1}{2s}\left\{s\eta_3\,\tilde{\Phi}_y - [\eta_2(\zeta_1 + \cos\sigma) - \eta_1(\zeta_2 + \sin\sigma)]\tilde{\Phi}_z\right\}$$

In these equations $\tilde{\Phi}'$ denotes the full derivative of the gauge. It is important to note that Eq. (4.10) needs to be integrated together with this system of equations. The fact that the time transformation depends on ζ_3 apart from r is responsible for the physical time not being invariant to the selection of the gauge anymore. Thus, the time transformation should be defined as

$$\frac{dt}{d\sigma} = \hat{\zeta}_3 r^2 = \frac{1}{\hat{\zeta}_3^3 s^2}$$

in terms of the *osculating* value of ζ_3. The transformation between the osculating and the nonosculating elements is given in Eqs. (2.42–2.42).

The referenced equation involves the full derivative of the gauge function, given by

$$\frac{d\mathbf{\Phi}}{d\sigma} = \frac{\partial\mathbf{\Phi}}{\partial\sigma} + \sum_j \frac{\partial\mathbf{\Phi}}{\partial œ_j}\frac{dœ_j}{d\sigma}$$

When the convective terms do not vanish the evolution of the Dromo elements is given explicitly by

$$\sum_j \left([œ_n, œ_j] - \frac{\partial\mathbf{\Phi}}{\partial œ_j} \cdot \frac{\partial\mathbf{f}}{\partial œ_n} \right) \frac{dœ_j}{d\sigma} = \left(\frac{\Delta\mathbf{F}}{\zeta_3^3 s^2} - \frac{\partial\mathbf{\Phi}}{\partial\sigma} \right) \cdot \frac{\partial\mathbf{f}}{\partial œ_n} - \frac{\partial\mathbf{g}}{\partial œ_n} \cdot \mathbf{\Phi}$$

The terms multiplying the derivatives of the elements in the left-hand side of this equation form a matrix that needs to be inverted in order to define the evolution equations. The final form of the equations depends on the selection of the gauge function.

4.6 Singularities

There are two specific situations that make the integration of the Dromo equations (and any other method relying on the gyroscopic decomposition) problematic:

Asymptotes: Equations (4.10–4.17) depend explicitly on the perturbing term $\tilde{\mathbf{a}}_p = \mathbf{a}_p/(\zeta_3^4 s^3)$ and become singular for $(\zeta_3^4 s^3) \to 0$. The term $s = 1 + e\cos\vartheta$ vanishes in two cases: (i) in the case of a parabolic orbit ($e = 1$), when the particle is sufficiently far from the attractive center, as $\vartheta \to \pi$. (ii) In the case of a hyperbolic orbit ($e > 1$), when the particle approaches the asymptote of the osculating hyperbola $\vartheta \to \arccos(-1/e)$. In addition, if $(\zeta_3^3 s^2)$ vanishes Eq. (4.10) becomes singular.

Even for Keplerian orbits with constant nonzero angular momentum s vanishes when approaching infinity along an asymptote.

Angular momentum: If the position and velocity vectors are parallel, $\mathbf{r} \times \mathbf{v} = 0$, the angular momentum vanishes and the departure perifocal frame $\mathcal{P}$ is undetermined. This problem may appear at departure if $\mathbf{r}_0 \parallel \mathbf{v}_0$, or during the integration process due to external perturbations. Strong perturbations in the along-track direction have an important effect on the angular momentum of the particle, and may make it zero.

Both situations are intimately related to flyby trajectories. The first problem is a geometrical issue that appears only in hyperbolic and parabolic orbits. The second problem relates to the magnitude and the effective direction of the perturbing term. In practice, both phenomena are usually coupled.

The angular momentum appears recursively in the derivation of the equations of Dromo. The orbital plane is defined by means of the angular momentum vector. The angular velocity of the perifocal departure frame $\mathcal{P}$, given in Eq. (4.4), becomes singular when the angular momentum becomes zero. Indeed, if the position and velocity vectors are parallel they do not define any plane, but a straight line. The direction of this line is given by the radius vector of the particle, which follows a rectilinear orbit along this line. The trajectory of the particle can be considered a degenerated conic section.

The description of the orbital plane adopted in Dromo is based on the dynamics of the rigid-body. Relying on the angular momentum of the particle a gyroscopic description of the orbital plane is provided. The quaternion $\mathfrak{n}$ is introduced to overcome the singularities associated to the usual Euler angles. But the gyroscopic approach is not valid when the equivalent rigid-body is no longer spinning, i.e., when the angular momentum vanishes.

Equation (4.13) shows the dependency of the angular momentum with the perturbing terms. The angular momentum is affected by the component of the perturbing term along the direction of $\mathbf{j}$. Numerical instabilities might be encountered when this particular projection of the perturbing acceleration grows. This could be the case when the particle approaches a massive planet (such as Jupiter) along an incoming trajectory that leads to a deep flyby (Roa et al., 2015b).

4.7 Conclusions

The power of Hansen ideal frames resides in the effective decoupling between the in-plane dynamics and the motion of the orbital plane itself. Once the perturbation is projected in the orbital frame the normal component only affects the dynamics of the orbital plane. Dromo exploits all the potential of this decomposition by introducing a quaternion to model the evolution of the orbital plane. In addition, the dynamical decomposition yields an intuitive interpretation of the effects of each component of the perturbing acceleration.

Dromo combines different regularization and stabilization techniques (Chap. 2). The use of the Sundman transformation enhances the step-size regulation. Moreover, the degree two of the transformation increases the benefits of the time transformation: the new independent variable is equivalent to the true anomaly, which captures in more detail the dynamics around periapsis. The integration of elements instead of coordinates yields a smoother integration scheme, specially when facing weak perturbations.

Formulations relying on modeling the evolution of the orbital plane might encounter a singularity when the angular momentum vanishes, because the orbital plane is no longer determined. The singularity, which is not common in practical scenarios, relates to rectilinear trajectories or hyperbolic motions very close to the asymptotes.

5 Dedicated formulation: Propagating hyperbolic orbits

The numerical integration of perturbed hyperbolic orbits is challenging due to the rapid changes in the relative velocity during pericenter passage, which may lead to considerable uncertainties (Hajduková Jr, 2008). A proper formulation should also exhibit numerical stability during the integration. Cowell's method provides a simple integration scheme for arbitrary orbits. Because of the Lyapunov instability of orbital motion, numerical errors in Cowell's method grow rapidly (Chapter 2). The instability of the motion is governed by the eigenvalues shown in Eq. (2.2): the exponential behavior due to the eigenvalue $(2\mu/r^3)^{1/2}$ is responsible for the strong divergence of the integral flow during close flybys. This chapter will show how regularization techniques are useful for defining dedicated formulations conceived for specific purposes, like improving the stability of the propagation of hyperbolic orbits.

Hyperbolic orbits are closely related to the dynamics and evolution mechanisms of bodies and particles leaving the solar system. Comets are a relevant example. Long-period comets ejected from the Oort cloud due to the effect of passing stars typically enter the inner solar system with quasiparabolic orbits (Hills, 1981; Królikowska, 2001). Gravitational and nongravitational effects may cause the comets to leave the solar system following hyperbolic orbits (Królikowska, 2006). Special attention has been paid to the numerical propagation of such orbits (Yabushita, 1979; Thomas and Morbidelli, 1996), since they can be integrated backward in time in order to analyze the origin and evolution of comets (Everhart, 1976; Ipatov, 1999). Another example is the role of hyperbolic orbits in the dynamics of dust particles, in particular the flux of β-meteoroids incoming from the inner solar system following hyperbolic orbits and detected by the Ulysses mission (Wehry and Mann, 1999).

In asteroid-planet close approaches, the relative orbit of the asteroid with respect to the planet is typically hyperbolic. Such phenomena are important because impacts may occur. Nowadays, one major concern is evaluating the collision risk of the Near-Earth Asteroids (Milani and Knežević, 1990; Muinonen et al., 2001). The dynamics of the encounter are complex and orbital resonances are particularly critical. Planetary perturbations may lead to orbital resonances that produce high risk close encounters (Kozai, 1985; Ferraz-Mello et al., 1998). Milani et al. (1989) explored the behavior of resonant asteroids through massive numerical simulations. Valsecchi et al. (2003) developed an analytical theory for modeling resonant returns. If sufficient observational data is available, the asteroid mass and gravitational distribution may be modeled when considering its dynamics (Yeomans et al., 1997; Miller et al., 2002). Roa and Handmer (2015) studied the likely fate of asteroid fragments after disruption in a lunar distant-retrograde orbit using various campaigns of numerical simulations, in which flybys around the Moon have a strong scattering effect. An accurate description of hyperbolic orbits increases the scientific value of the results.

https://doi.org/10.1515/9783110559125-005

Many exploration missions involve hyperbolic orbits. On their way to interstellar space, both Voyager 1 and 2 spacecraft followed gravity-assist swingby trajectories about Jupiter and Saturn. Voyager 2 also visited Uranus and Neptune, and it is currently traveling through the heliosheath toward interstellar space (see Kohlhase and Penzo, 1977, for the original design, and Fig. 1.1 for its orbit). The Cassini-Huygens mission is another example of gravity-assist trajectories; to reach Saturn, the spacecraft performed three planetary flybys of Venus, the Earth, and Jupiter (Burton et al., 2001, and Fig. 1.2). During the mission remarkable advances in several fields transpired, from detecting interstellar dust (Altobelli et al., 2003) to general relativity experiments (Bertotti et al., 2003), including characterization of Titan's atmosphere (Waite Jr et al., 2005) and Jupiter observations (Porco et al., 2003). The Europa mission, the next one in NASA's Flagship Program, will perform 45 flybys of Europa looking for suitable conditions for life. The unprecedented scientific results expected from the mission require extremely accurate propagations for modeling the dynamics of the probe.

In Chapter 4 we introduced the concept of Hansen ideal frames and discussed its potential for numerical propagation. The works by Deprit (1975), Palacios et al. (1992), and Peláez et al. (2007) exploited this idea to derive a set of regular elements. The unperturbed radial motion in hyperbolic orbits is well known in terms of hyperbolic trigonometry (Battin, 1999, p. 167):

$$r = a(1 - e \cosh H)$$

For coherence with the radial solution the motion of the eccentricity vector will be described using hyperbolic angles and rotations, and not the traditional Euclidean definitions. We will formulate the dynamics using the mathematical constructions underlying Minkowski space-time. In the field of special relativity, Hermann Minkowski (1923) created a four-dimensional description of time and space to simplify the formulation of Einstein's theory. The geometry of the two-dimensional Minkowski space-time (namely the Minkowski plane $\mathbb{R}^2_1$) naturally introduces hyperbolic trigonometry (Saloom and Tari, 2012). Some authors refer to this plane as the Lorentz plane (Sobczyk, 1995).

Hypercomplex numbers provide useful geometrical representations. The components of a two-dimensional hypercomplex number can be understood as the coordinates of a point on a certain plane. Depending on the metrics, hypercomplex numbers define different geometries. For instance, the modulus of a complex number is equivalent to the Euclidean distance from the origin to the point its components define. Similarly, the modulus of a hyperbolic number leads to the definition of distance on the Minkowski plane (Boccaletti et al., 2008, chap. 4). Complex numbers and hyperbolic numbers are special cases of hypercomplex numbers. Hyperbolic numbers are also known as dual numbers (Kantor et al., 1989) or perplex numbers: the latter was introduced by Fjelstad (1986), who described the hyperbolic constructions in special relativity through them. Ulrych (2005) extended hyperbolic numbers to quantum physics and Motter and Rosa (1998) stated their formal inclusion in Clifford algebras. Cariñena et al. (1991) explored Minkowskian metrics to represent orbits in the configuration and in the velocity spaces. Multicomplex numbers, a special subset of hypercomplex numbers, generalized complex numbers to higher dimensions (Price, 1991). Lantoine et al. (2012) applied these numbers to the computation of high-order derivatives.

Rotations and metrics in the Minkowski plane $\mathbb{R}^2_1$ are described using hyperbolic numbers. The goal of this chapter is to provide a more accurate and stable description of such orbits, in which the propagation error is not affected by periapsis passage. The set of eight elements, according to the definition of *element* by Stiefel and Scheifele (1971, chap. 3), derives from the theory of hypercomplex numbers. The first element is a time element. The second and third elements are the components of the eccentricity vector on $\mathbb{R}^2_1$. The fourth element is the semimajor axis. The components of the eccentricity vector are referred to a certain reference frame, attached to the orbital plane. The motion of the orbital plane is uniquely determined by the motion of this frame. It is defined on the inertial reference by means of a quaternion. The components of the quaternion are the last four elements. Quaternions are treated as particular instances of hypercomplex numbers.

This chapter is organized as follows. Section 5.1 defines the orbital problem, and the evolution of the eccentricity vector on the Minkowski plane is determined. This section closes with the definition of the bijection between the components of the eccentricity vector on the Euclidean and the Minkowski plane. The variation of parameters technique is applied in Sect. 5.3, where the evolution equations for the orbit geometry are obtained. Section 5.4 describes the motion of the orbital plane. The time element is defined in Sect. 5.5. A brief summary of the equations is presented. Finally, Sect. 5.6 analyzes the performance of the method through a set of numerical experiments. Hypercomplex numbers are described in Appendix A. General definitions and properties of the associated algebras are stated.

5.1 Orbital motion

Let $\mathbf{r} \in \mathbb{R}^3$ denote the position vector of a particle in an inertial reference frame $\mathcal{I}$. The perturbed motion in a central gravity field is governed by the equation:

$$\frac{\mathrm{d}^2\mathbf{r}}{\mathrm{d}t^2} = -\frac{\mathbf{r}}{r^3} + \mathbf{a}_p \tag{5.1}$$

with $r = ||\mathbf{r}||$ and a normalized gravitational parameter $\mu = 1$.

The orbital frame $\mathcal{L} = \{\mathbf{i}, \mathbf{j}, \mathbf{k}\}$ rotates with the particle and it is given the usual definition:

$$\mathbf{i} = \frac{\mathbf{r}}{r}, \quad \mathbf{k} = \frac{\mathbf{h}}{h}, \quad \mathbf{j} = \mathbf{k} \times \mathbf{i}$$

The angular momentum vector is $\mathbf{h}$, and $h = ||\mathbf{h}||$. The departure perifocal frame, $\mathcal{P}$, is described by the basis $\{\mathbf{i}_\mathcal{P}, \mathbf{j}_\mathcal{P}, \mathbf{k}_\mathcal{P}\}$,

$$\mathbf{i}_\mathcal{P} = \frac{\mathbf{e}_0}{e_0}, \quad \mathbf{k}_\mathcal{P} = \frac{\mathbf{h}_0}{h_0}, \quad \mathbf{j}_\mathcal{P} = \mathbf{k}_\mathcal{P} \times \mathbf{i}_\mathcal{P}$$

just like in Chap. 4.

Let $\mathcal{A}$ denote an intermediate frame, with $\mathcal{A} = \{\mathbf{i}_\mathcal{A}, \mathbf{j}_\mathcal{A}, \mathbf{k}_\mathcal{A}\}$. The $x_\mathcal{A}y_\mathcal{A}$-plane is contained in the orbital plane, i.e., $\mathbf{k}_\mathcal{A} \parallel \mathbf{k}_\mathcal{P}$. At $t = t_0$ this frame coincides with frame $\mathcal{P}$ but there exists a relative angular velocity between both frames, which is normal to the orbital plane. There is a degree of freedom in the definition of frame $\mathcal{A}$: the angle between the $x_\mathcal{A}$-axis and the $x_\mathcal{P}$-axis. The reason for introducing the intermediate frame is that the eccentricity vector can be written

$$\mathbf{e} = \rho_1 \mathbf{i}_\mathcal{A} + \rho_2 \mathbf{j}_\mathcal{A}$$

with $(\rho_1, \rho_2) \in \mathbb{R}\times\mathbb{R}$ to be defined as required. Once the components of the eccentricity vector are defined (Sect. 5.1.1), an explicit expression for the relative velocity $\mathcal{A}/\mathcal{P}$ will be provided (Sect. 5.4). Initially, it is $\rho_1(0) = e_0$ and $\rho_2(0) = 0$. Figure 5.1 depicts the described reference frames.

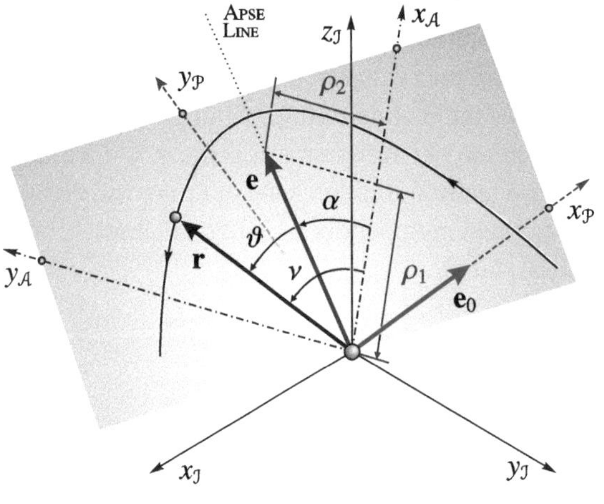

Fig. 5.1: Schematic view of the reference frames to be used. The true anomaly is denoted by ϑ, and ν defines the angle between versor $\mathbf{i}_\mathcal{A}$ and the radius vector, that is $\nu = \alpha + \vartheta$.

5.1.1 The eccentricity vector

The evolution of the eccentricity vector is defined in the intermediate reference $\mathcal{A}$ in terms of

$$\mathbf{e} = \rho_1 \mathbf{i}_\mathcal{A} + \rho_2 \mathbf{j}_\mathcal{A}$$

Let $\mathfrak{z}_e \in \mathbb{C}$ be the representation of **e** on the Gauss–Argand plane. This number is written

$$\mathfrak{z}_e = \rho_1 + \mathrm{i}\rho_2$$

with $\mathrm{i} = \sqrt{-1}$ the imaginary unit.

The Euclidean norm of the eccentricity vector is given by

$$||\mathbf{e}||^2_{\mathbb{E}^2} = \mathfrak{z}_e\mathfrak{z}_e^\dagger = \rho_1^2 + \rho_2^2 \equiv |\mathfrak{z}_e|^2_{\mathbb{C}} \tag{5.2}$$

where $\mathfrak{z}_e^\dagger$ is the complex conjugate of $\mathfrak{z}_e$.

Alternatively, an equivalent representation of **e** can be found on the Minkowski plane, namely

$$\mathfrak{w}_e = \lambda_1 + \mathrm{j}\lambda_2\,, \qquad \mathfrak{w}_e \in \mathbb{D},\ (\lambda_1, \lambda_2) \in \mathbb{R}\times\mathbb{R}$$

Here $\mathbb{D}$ denotes the field of hyperbolic numbers,

$$\mathbb{D} = \{x + \mathrm{j}y \mid (x, y) \in \mathbb{R}\times\mathbb{R},\ \mathrm{j}^2 = +1\}$$

and j is the hyperbolic imaginary unit. Note that $\mathrm{j} = \sqrt{+1} \notin \mathbb{R}$. The formal definition of hyperbolic numbers and its connection with Minkowskian geometry can be found in Appendix A.

The norm is defined by the metrics in $\mathbb{R}^2_1$,

$$||\mathbf{e}||^2_{\mathbb{R}^2_1} = \mathfrak{w}_e\mathfrak{w}_e^\dagger = \lambda_1^2 - \lambda_2^2 \equiv |\mathfrak{w}_e|^2_{\mathbb{D}} \tag{5.3}$$

The norm of the eccentricity vector is the osculating eccentricity, e. The different definitions of norm in the Euclidean and the Minkowskian vector space provide two different expressions for the osculating eccentricity e, given in Eqs. (5.2) and (5.3). But the eccentricity of the orbit has a clear physical meaning and it is possible to assume that both definitions of norm yield the same physical quantity (Roa and Peláez, 2015d,e). Thus, equating Eqs. (5.2) and (5.3) establishes a relation between the Euclidean and the Minkowskian components of **e**:

$$e^2 = \rho_1^2 + \rho_2^2 = \lambda_1^2 - \lambda_2^2$$

The evolution of the eccentricity vector is defined in terms of a circular angle α on the Gauss–Argand plane, and in terms of a hyperbolic angle γ on the Minkowski plane. This is easily inferred from the polar form of $\mathfrak{z}_e$ and $\mathfrak{w}_e$,

$$\mathfrak{z}_e = e\,\mathrm{e}^{\mathrm{i}\alpha} = e(\cos\alpha + \mathrm{i}\sin\alpha)$$
$$\mathfrak{w}_e = e\,\mathrm{e}^{\mathrm{j}\gamma} = e(\cosh\gamma + \mathrm{j}\sinh\gamma)$$

That is, vector **e** is determined by the circular angle α or, equivalently, by the hyperbolic angle γ, together with the osculating eccentricity e. The numbers $\mathfrak{z}_e$ and $\mathfrak{w}_e$ satisfy

$$\mathfrak{z}_e = e(\cos\alpha + \mathrm{i}\sin\alpha) = \rho_1 + \mathrm{i}\rho_2$$
$$\mathfrak{w}_e = e(\cosh\gamma + \mathrm{j}\sinh\gamma) = \lambda_1 + \mathrm{j}\lambda_2$$

Given the previous expressions it is natural to write

$$\rho_1 = e\cos\alpha\,, \qquad \lambda_1 = e\cosh\gamma$$
$$\rho_2 = e\sin\alpha\,, \qquad \lambda_2 = e\sinh\gamma$$

Provided that $\mathfrak{z}_e$ and $\mathfrak{w}_e$ represent the same physical entity – the eccentricity vector $\mathbf{e}$ – there exists a bijective map between the Gauss–Argand plane and the Minkowski plane, $(\rho_1, \rho_2) \leftrightarrow (\lambda_1, \lambda_2)$. This bijection can be derived geometrically when analyzing the components of $\mathbf{e}$ on both planes. Figure 5.2 shows the geometrical construction of the problem when superposing both planes. When α changes the osculating eccentricity vector defines a circumference of radius e, $\mathcal{c}^e$, on the Gauss–Argand plane. In such a case, it defines a rectangular (or equilateral) hyperbola on the Minkowski plane with the semimajor axis equal to the osculating eccentricity, $\mathcal{h}^e$. Hyperbolic rotations are described in Appendix A.

The geometry shown in Fig. 5.2 leads to the relation:

$$\alpha = \mathfrak{gd}\,(\gamma)$$

where $\mathfrak{gd}\,(\gamma)$ refers to the Gudermannian function (Battin, 1999, p. 168). The explicit expression of the Gudermannian is

$$\mathfrak{gd}\,(\gamma) = \int_0^\gamma \operatorname{sech} t\, \mathrm{d}t = \arctan(\sinh\gamma) = 2\arctan \mathrm{e}^\gamma - \frac{\pi}{2} \tag{5.4}$$

This function satisfies

$$\cos\mathfrak{gd}\,(\gamma) = \operatorname{sech}\gamma\,, \qquad \sin\mathfrak{gd}\,(\gamma) = \tanh\gamma$$

From these properties of the Gudermannian the mapping $(\lambda_1, \lambda_2) \mapsto (\rho_1, \rho_2)$ is established,

$$\rho_1 = \lambda_1\left(1 - \frac{\lambda_2^2}{\lambda_1^2}\right), \qquad \rho_2 = \frac{\lambda_2}{\lambda_1}\sqrt{\lambda_1^2 - \lambda_2^2} \tag{5.5}$$

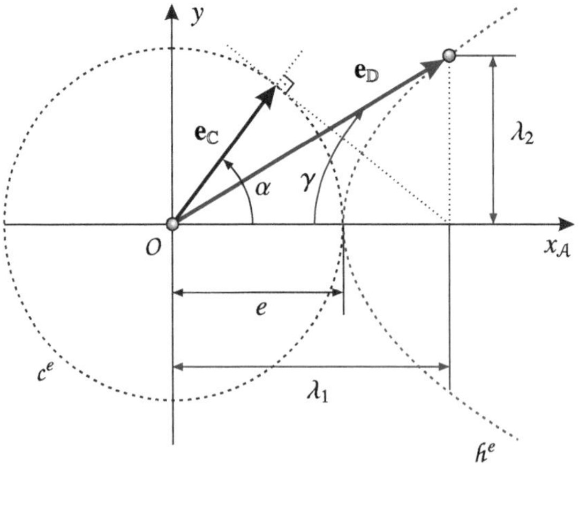

Fig. 5.2: Geometrical representation of the eccentricity vector superposing the Gauss–Argand plane and the Minkowski plane.

and the inverse mapping $(\rho_1, \rho_2) \mapsto (\lambda_1, \lambda_2)$ reads

$$\lambda_1 = \rho_1 \left(1 + \frac{\rho_2^2}{\rho_1^2}\right), \qquad \lambda_2 = \frac{\rho_2}{\rho_1}\sqrt{\rho_1^2 + \rho_2^2}$$

5.1.2 The hyperbolic anomaly

The hyperbolic anomaly H is the hyperbolic angle between the osculating apse line and the direction from the center of the hyperbola, $\mathcal{O}$, to the particle, $\mathcal{P}$. It is the hyperbolic equivalent to the true anomaly, ϑ (the circular angle between the osculating apse line and the radius vector). The direction of the apse line is given either by the circular angle α or by the hyperbolic angle γ. The axes on the Minkowski plane are oriented as frame $\mathcal{A}$, and centered at $\mathcal{O}$.

In the perifocal reference frame, the eccentricity vector points toward the positive abscissa. The orbit corresponds to the left branch of the hyperbola. Let $\mathfrak{v}_e \in \mathbb{D}$ be a hyperbolic number that fixes the direction of the apse line with respect to the center of the hyperbola, $\mathcal{O}$, in reference $\mathcal{A}$. On $\hbar_+^1$ it is $\mathfrak{v}_e^+ = \mathrm{e}^{\mathrm{j}\gamma}$, whereas on $\hbar_-^1$ it is $\mathfrak{v}_e^- = -\mathrm{e}^{\mathrm{j}\gamma}$, for compliance with the sign criterion. Hyperbolic rotations will then be defined on $\hbar^1$. Recall that this hyperbola is different from the osculating orbit.

Let $\mathfrak{v}_r^- \in \mathbb{D}$ be a hyperbolic number that provides the direction from $\mathcal{O}$ to the particle, with $\mathfrak{v}_r^- \subset \hbar_-^1$. From the definition of the hyperbolic anomaly it follows that

$$\mathfrak{v}_r^- = \mathfrak{v}_e^- \, \mathrm{e}^{-\mathrm{j}H} = -\mathrm{e}^{\mathrm{j}\gamma}\,\mathrm{e}^{-\mathrm{j}H} = -\mathrm{e}^{-\mathrm{j}(H-\gamma)} = -\mathrm{e}^{-\mathrm{j}u} \tag{5.6}$$

That is, the direction to the particle can be directly referred to the $x_{\mathcal{A}}$-axis using the hyperbolic angle $u = H - \gamma$. This angle becomes the new independent variable. Figure 5.3 shows the geometry of the problem.

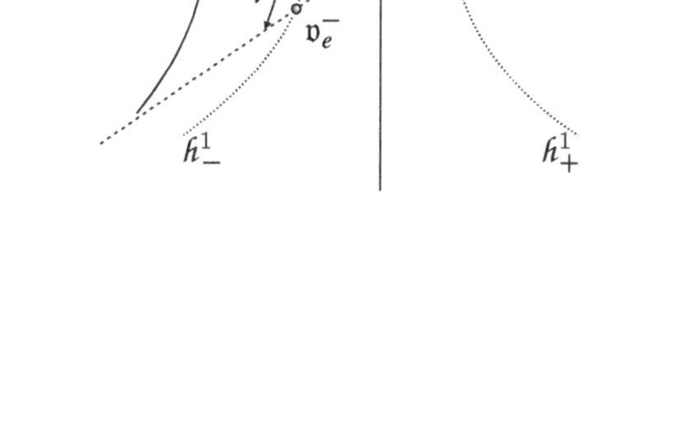

Fig. 5.3: Geometrical interpretation of the hyperbolic anomaly and the independent variable u on the Minkowski plane. The focus of the hyperbola is denoted by $\mathcal{F}$.

The hyperbolic anomaly satisfies the relations:

$$\cosh H = \cosh y \cosh u + \sinh y \sinh u \tag{5.7}$$

$$\sinh H = \sinh y \cosh u + \cosh y \sinh u \tag{5.8}$$

If there are no perturbations it is $y = 0$ and $H \equiv u$. The true anomaly can be obtained from the hyperbolic anomaly,

$$\cos \vartheta = \frac{e - \cosh H}{e \cosh H - 1}, \qquad \cosh H = \frac{e + \cos \vartheta}{e \cos \vartheta + 1}$$

$$\sin \vartheta = \frac{\ell \sinh H}{e \cosh H - 1}, \qquad \sinh H = \frac{\ell \sin \vartheta}{e \cos \vartheta + 1}$$

with $\ell = \sqrt{e^2 - 1}$. Initially it is $y = 0$ and $u_0 = H_0$, so the apse line coincides with the $x_{\mathcal{A}}$-axis.

5.2 Hyperbolic rotations and the Lorentz group

The eccentricity vector can be written $\mathbf{e} = e\,\mathbf{u}_e$, in which $\mathbf{u}_e$ follows the direction of the line of apses. According to Eq. (5.6) the eccentricity vector can be projected in $\mathcal{A}$ by means of the linear mapping

$$\mathbf{R}(y) = \begin{bmatrix} \cosh y & \sinh y \\ \sinh y & \cosh y \end{bmatrix}$$

with $\det(\mathbf{R}) = 1$.

Let $\mathbf{T}$ denote the metric tensor attached to the coordinate space. In Euclidean space $\mathbb{E}^2$ the signature of $\mathbf{T}$ is $\langle +, + \rangle$. Although the determinant of $\mathbf{R}(y)$ is unity, in $\mathbb{E}^2$ the operator $\mathbf{R}(y)$ is not orthogonal and does not define a rotation. Indeed, considering that $\mathbf{R}(y) = \mathbf{R}^\top(y)$ it is easy to verify that the orthogonality condition

$$\mathbf{R}(y)\,\mathbf{T}\,\mathbf{R}^\top(y) = \mathbf{T}$$

is not satisfied, because

$$\mathbf{R}(y)\,\mathbf{T}\,\mathbf{R}^\top(y) \equiv \mathbf{R}(y)\,\mathbf{R}(y) = \begin{bmatrix} 2\cosh^2 y - 1 & 2\cosh y \sinh y \\ 2\cosh y \sinh y & 2\cosh^2 y - 1 \end{bmatrix}$$

Since the operator is not orthogonal it does not represent a rotation in $\mathbb{E}^2$. But if one migrates to the two-dimensional Minkowski space-time $\mathbb{M}$ the metric tensor is endowed with a signature $\langle +, - \rangle$ (see for example Penrose, 2004, chap. 18). The modification of the signature is equivalent to introducing a time-like component, and in this case the condition

$$\mathbf{R}(y)\,\mathbf{T}\,\mathbf{R}^\top(y) = \mathbf{T}$$

holds. The operator $\mathbf{R}(\gamma)$ is orthogonal in $\mathbb{M}$ and $\det \mathbf{R}(\gamma) = 1$, so it can be seen as a rotation action.

This construction is simply a subgroup of the more general Lorentz group $O(3, 1)$. If one identifies $\beta = v/c$ and the Lorentz factor $g = 1/\sqrt{1-\beta^2}$ with $g = \cosh\gamma$ and $-\beta g = \sinh\gamma$, the two-dimensional Lorentz boost

$$\begin{bmatrix} g & -\beta g \\ -\beta g & g \end{bmatrix}$$

reduces to matrix $\mathbf{R}$. The Lorentz group leaves the space-time origin invariant.

5.3 Variation of parameters

The solution to the generic perturbed problem is found using the variation of parameters procedure. First, the solution $r_k = r_k(t)$ to the Keplerian problem is obtained in terms of different constant parameters. Seeking a solution of the same form, the perturbed problem is solved considering that the parameters are no longer constant but functions of u.

5.3.1 The new independent variable

The new independent variable u is related to the physical time through the Sundman transformation

$$\frac{\mathrm{d}t}{\mathrm{d}u} = r\sqrt{a} \tag{5.9}$$

and defines a fictitious time. Baù et al. (2014,2015) explored a similar transformation for the elliptic case. In the following the semimajor axis is assumed to be positive, $a \equiv |a| > 0$, as is the energy, $\mathcal{E} > 0$. The radial velocity is

$$\frac{\mathrm{d}r}{\mathrm{d}t} = \frac{1}{r\sqrt{a}}\frac{\mathrm{d}r}{\mathrm{d}u}$$

Derivation with respect to physical time yields

$$\frac{\mathrm{d}^2r}{\mathrm{d}t^2} = \frac{1}{r^2a}\frac{\mathrm{d}^2r}{\mathrm{d}u^2} - \frac{1}{r^3a}\left(\frac{\mathrm{d}r}{\mathrm{d}u}\right)^2 - \frac{1}{2a^2r^2}\frac{\mathrm{d}r}{\mathrm{d}u}\frac{\mathrm{d}a}{\mathrm{d}u}$$

This expression relates the radial acceleration to the fictitious radial acceleration. In particular,

$$\frac{\mathrm{d}^2r}{\mathrm{d}u^2} = r^2a\frac{\mathrm{d}^2r}{\mathrm{d}t^2} + \frac{1}{r}\left(\frac{\mathrm{d}r}{\mathrm{d}u}\right)^2 + \frac{1}{2a}\frac{\mathrm{d}r}{\mathrm{d}u}\frac{\mathrm{d}a}{\mathrm{d}u} \tag{5.10}$$

Equation (5.10) governs the radial motion, $r(u)$. To complete this equation, explicit expressions for the terms in the right-hand side are required, namely the radial acceleration, the fictitious radial velocity, and the evolution of the semimajor axis. The three following propositions derive these expressions.

Proposition 1. *(Roa and Peláez, 2015d,e) The radial acceleration is given by*

$$\frac{\mathrm{d}^2 r}{\mathrm{d}t^2} = -\frac{1}{r^2} + \frac{h^2}{r^3} + (\mathbf{a}_p \cdot \mathbf{i}) \tag{5.11}$$

where h is the angular momentum of the particle.

Proof. Let $\mathbf{a} \in \mathbb{R}^3$ denote the acceleration vector. Equation (5.1) is rewritten as

$$\mathbf{a}_{\mathcal{L}\mathcal{I}} = -\frac{1}{r^2}\mathbf{i} + \mathbf{a}_p$$

This expression provides the acceleration of the particle relative to the inertial reference $\mathcal{I}$. The in-plane motion is governed by the acceleration $\mathbf{a}_{\mathcal{L}\mathcal{P}}$, considering the inertia terms

$$\mathbf{a}_{\mathcal{L}\mathcal{P}} = -\frac{1}{r^2}\mathbf{i} + \mathbf{a}_p - \mathbf{a}_{\mathcal{P}\mathcal{I}} - 2\,\boldsymbol{\omega}_{\mathcal{P}\mathcal{I}} \times \mathbf{v}_{\mathcal{L}\mathcal{P}}$$

The angular velocities that appear in the problem are

$$\boldsymbol{\omega}_{\mathcal{L}\mathcal{P}} = \frac{h}{r^2}\mathbf{k}\,, \qquad \boldsymbol{\omega}_{\mathcal{P}\mathcal{I}} = \frac{r}{h}(\mathbf{a}_p \cdot \mathbf{k})\,\mathbf{i} \tag{5.12}$$

The definition of these angular velocities by Peláez et al. (2007) can be found in Eq. (4.3). The acceleration $\mathbf{a}_{\mathcal{L}\mathcal{P}}$ is expanded,

$$\mathbf{a}_{\mathcal{L}\mathcal{P}} = -\frac{1}{r^2}\mathbf{i} + \left[\mathsf{I} - (\mathbf{k} \otimes \mathbf{k})\right] \circ \mathbf{a}_p \tag{5.13}$$

where I is the second-order unit tensor, represented by the identity matrix, $(\mathbf{k} \otimes \mathbf{k})$ denotes the dyadic product, and $(\circ)$ refers to the contracted product. The tensorial term yields

$$\left[\mathsf{I} - (\mathbf{k} \otimes \mathbf{k})\right] \circ \mathbf{a}_p = \mathbf{a}_p - (\mathbf{a}_p \cdot \mathbf{k})\,\mathbf{k} = (\mathbf{a}_p \cdot \mathbf{i})\,\mathbf{i} + (\mathbf{a}_p \cdot \mathbf{j})\,\mathbf{j} \equiv \mathbf{a}_{\parallel,p}$$

Since $\mathbf{k}$ is normal to the orbital plane $\mathbf{a}_{\parallel,p}$ defines the in-plane perturbing acceleration.

The in-plane acceleration can alternatively be determined through its kinematic definition in polar coordinates. That is,

$$\mathbf{a}_{\mathcal{L}\mathcal{P}} = \left(\frac{\mathrm{d}^2 r}{\mathrm{d}t^2} - \frac{h^2}{r^3}\right)\mathbf{i} + \frac{1}{r}\frac{\mathrm{d}h}{\mathrm{d}t}\mathbf{j} \tag{5.14}$$

By equating Eqs. (5.13) and (5.14) the following relation is obtained:

$$-\frac{1}{r^2}\mathbf{i} + \mathbf{a}_{\parallel,p} = \left(\frac{\mathrm{d}^2 r}{\mathrm{d}t^2} - \frac{h^2}{r^3}\right)\mathbf{i} + \frac{1}{r}\frac{\mathrm{d}h}{\mathrm{d}t}\mathbf{j}$$

If projected along the direction of $\mathbf{i}$, this expression leads to the radial acceleration,

$$\frac{\mathrm{d}^2 r}{\mathrm{d}t^2} = -\frac{1}{r^2} + \frac{h^2}{r^3} + (\mathbf{a}_p \cdot \mathbf{i})$$

affected by the perturbing term $(\mathbf{a}_p \cdot \mathbf{i}) = (\mathbf{a}_{\parallel,p} \cdot \mathbf{i})$. □

Proposition 2. *The evolution of the semimajor axis in terms of the fictitious time u abides by*

$$\frac{\mathrm{d}a}{\mathrm{d}u} = -2a^2 \left[\frac{\mathrm{d}r}{\mathrm{d}u}(\mathbf{a}_p \cdot \mathbf{i}) + h\sqrt{a}(\mathbf{a}_p \cdot \mathbf{j}) \right] \tag{5.15}$$

Proof. The Keplerian energy of the system is of the form:

$$\mathcal{E} = \frac{v^2}{2} - \frac{1}{r} = \frac{1}{2a}, \qquad \text{with } a > 0, \quad \mathcal{E} > 0$$

The time evolution of the energy is obtained in terms of the power performed by the perturbing forces. That is,

$$\frac{\mathrm{d}\mathcal{E}}{\mathrm{d}t} = \mathbf{a}_p \cdot \mathbf{v} \tag{5.16}$$

This equation is a particular case of a more general result: consider a generic constant in Kepler's problem, κ. Its time evolution due to external perturbations $\mathbf{a}_p$ is

$$\frac{\mathrm{d}\kappa}{\mathrm{d}t} = \frac{\partial \kappa}{\partial \mathbf{v}} \cdot \mathbf{a}_p$$

Since the partial derivative of the energy with respect to the velocity vector is the velocity vector itself, for $\kappa = \mathcal{E}$ the previous equation reduces to Eq. (5.16).

The velocity is projected in the orbital frame as

$$\mathbf{v} = \frac{\mathrm{d}r}{\mathrm{d}t}\mathbf{i} + \frac{h}{r}\mathbf{j} \tag{5.17}$$

and the power becomes

$$\frac{\mathrm{d}\mathcal{E}}{\mathrm{d}t} = \frac{\mathrm{d}r}{\mathrm{d}t}(\mathbf{a}_p \cdot \mathbf{i}) + \frac{h}{r}(\mathbf{a}_p \cdot \mathbf{j})$$

This equation provides a simple expression for the time evolution of the energy.

The time evolution of the semimajor axis is obtained from

$$\frac{\mathrm{d}a}{\mathrm{d}t} = -2a^2 \frac{\mathrm{d}\mathcal{E}}{\mathrm{d}t}$$

and, considering the aforementioned definition of the power, this becomes

$$\frac{\mathrm{d}a}{\mathrm{d}t} = -2a^2 \left[\frac{\mathrm{d}r}{\mathrm{d}t}(\mathbf{a}_p \cdot \mathbf{i}) + \frac{h}{r}(\mathbf{a}_p \cdot \mathbf{j}) \right] \tag{5.18}$$

Equation (5.18) is transformed into the derivative with respect to u by applying the transformation given in Eq. (5.9),

$$\frac{\mathrm{d}a}{\mathrm{d}u} = -2a^2 \left[\frac{\mathrm{d}r}{\mathrm{d}u}(\mathbf{a}_p \cdot \mathbf{i}) + h\sqrt{a}(\mathbf{a}_p \cdot \mathbf{j}) \right]$$

□

Proposition 3. *The fictitious radial velocity squared is*

$$\left(\frac{\mathrm{d}r}{\mathrm{d}u}\right)^2 = r^2 - h^2 a + 2ra$$

Proof. Considering Eq. (5.17) the definition given to the energy can be expanded,

$$2\mathcal{E} = \left(\frac{\mathrm{d}r}{\mathrm{d}t}\right)^2 + \frac{h^2}{r^2} - \frac{2}{r} = \frac{1}{a}$$

The term $(1/a)$ reads

$$\frac{1}{a} = \left(\frac{\mathrm{d}r}{\mathrm{d}t}\right)^2 + \frac{h^2}{r^2} - \frac{2}{r} \tag{5.19}$$

From Eq. (5.19) and the definition of the fictitious time it follows that

$$\left(\frac{\mathrm{d}r}{\mathrm{d}u}\right)^2 = \left(\frac{\mathrm{d}r}{\mathrm{d}t}\right)^2 \left(\frac{\mathrm{d}t}{\mathrm{d}u}\right)^2 = r^2 - h^2 a + 2ra$$

□

Considering Props. 1 and 3, Eq. (5.10) becomes

$$\frac{\mathrm{d}^2 r}{\mathrm{d}u^2} - a - r = r^2 a(\mathbf{a}_p \cdot \mathbf{i}) + \frac{1}{2a}\frac{\mathrm{d}r}{\mathrm{d}u}\frac{\mathrm{d}a}{\mathrm{d}u} \tag{5.20}$$

The last term is intentionally left unchanged.

5.3.1.1 Solution to the unperturbed case

In the absence of perturbations ($\mathbf{a}_p = 0$) the semimajor axis remains constant and Eq. (5.20) reduces to

$$\frac{\mathrm{d}^2 r}{\mathrm{d}u^2} - a - r = 0$$

Integration of this equation yields

$$r(u) = a(B_1 \cosh u + B_2 \sinh u - 1) \tag{5.21}$$

where $a > 0$ is constant, and B_1 and B_2 are two integration constants to be determined. In Keplerian hyperbolic orbits the radius vector is a well known function of the form:

$$r = a(e \cosh H - 1)\,. \tag{5.22}$$

Equating Eqs. (5.21) and (5.22) leads to

$$B_1 = e\,, \qquad B_2 = 0\,, \qquad u \equiv H$$

The fictitious radial velocity of the particle is simply

$$\frac{\mathrm{d}r}{\mathrm{d}u} = a(B_1 \sinh u + B_2 \cosh u) \tag{5.23}$$

which reduces to

$$\frac{\mathrm{d}r}{\mathrm{d}u} = ae \sinh H$$

in the unperturbed case.

5.3.1.2 Extension to the perturbed case

The solution to the general perturbed problem is constructed through the variation of parameters procedure. The equations developed in this section define the evolution of the triple $\{\lambda_1, \lambda_2, \lambda_3\}$ with respect to fictitious time u, which is the new independent variable. From Eq. (5.21) a more general solution is suggested,

$$r(u) = \lambda_3(u)[\lambda_1(u)\cosh u + \lambda_2(u)\sinh u - 1] \tag{5.24}$$

where λ_1, λ_2, and $\lambda_3 \equiv a$ are no longer constants because of the perturbing forces. This expression can be obtained directly from Eq. (5.22),

$$r(u) = a(u)[e(u)\cosh H(u) - 1]$$

and introducing Eq. (5.7):

$$\begin{aligned} r(u) &= a(u)[e(u)(\cosh \gamma \cosh u + \sinh \gamma \sinh u) - 1] \\ &= a(u)[\lambda_1(u)\cosh u + \lambda_2(u)\sinh u - 1] \end{aligned}$$

In sum, the solution $r = r(u)$ is written

$$r(u) = \lambda_3(u)[\lambda_1(u)\cosh u + \lambda_2(u)\sinh u - 1] = \lambda_3(u)\hat{r}(u) \tag{5.25}$$

with $\hat{r}(u) = \lambda_1(u)\cosh u + \lambda_2(u)\sinh u - 1$.

To compute the fictitious radial velocity Eq. (5.25) is derived to provide:

$$\frac{\mathrm{d}r}{\mathrm{d}u} = \frac{\mathrm{d}\lambda_3}{\mathrm{d}u}\hat{r} + \lambda_3\left(\frac{\mathrm{d}\lambda_1}{\mathrm{d}u}\cosh u + \frac{\mathrm{d}\lambda_2}{\mathrm{d}u}\sinh u\right) + \lambda_3(\lambda_1 \sinh u + \lambda_2 \cosh u)$$

We assume a null gauge,

$$\Phi \equiv \frac{\mathrm{d}\lambda_3}{\mathrm{d}u}\hat{r} + \lambda_3\left(\frac{\mathrm{d}\lambda_1}{\mathrm{d}u}\cosh u + \frac{\mathrm{d}\lambda_2}{\mathrm{d}u}\sinh u\right) = 0 \tag{5.26}$$

Imposing this condition guarantees that the fictitious velocity on the Keplerian orbit defined by the instantaneous values of $\{\lambda_1, \lambda_2, \lambda_3\}$ coincides with that of the true orbit. Hence, such a Keplerian orbit is in fact the osculating orbit. The derivative of r reduces to

$$\frac{\mathrm{d}r}{\mathrm{d}u} = \lambda_3(\lambda_1 \sinh u + \lambda_2 \cosh u) \tag{5.27}$$

5.3.2 Evolution equations

The second derivative of Eq. (5.27) yields

$$\frac{\mathrm{d}^2 r}{\mathrm{d}u^2} = \frac{\mathrm{d}\lambda_3}{\mathrm{d}u}\hat{q} + \lambda_3\left(\frac{\mathrm{d}\lambda_1}{\mathrm{d}u}\sinh u + \frac{\mathrm{d}\lambda_2}{\mathrm{d}u}\cosh u\right) + \lambda_3(\hat{r} + 1)$$

with $\hat{q} = \lambda_1 \sinh u + \lambda_2 \cosh u$. Using this result, Eq. (5.20) becomes

$$\frac{d\lambda_1}{du} \sinh u + \frac{d\lambda_2}{du} \cosh u = \lambda_3^2 \hat{r}^2 (\mathbf{a}_p \cdot \mathbf{i}) - \frac{\hat{q}}{2\lambda_3} \frac{d\lambda_3}{du} \tag{5.28}$$

The derivative of the semimajor axis, defined in Eq. (5.15), is completed when considering the form of $r = r(u)$,

$$\frac{d\lambda_3}{du} = -2\lambda_3^3 [\hat{q}(\mathbf{a}_p \cdot \mathbf{i}) + \ell(\mathbf{a}_p \cdot \mathbf{j})]$$

where

$$\ell = \sqrt{\lambda_1^2 - \lambda_2^2 - 1} = \sqrt{e^2 - 1}$$

Equations (5.26) and (5.28) form a system of equations from which the derivatives of the variables λ_1 and λ_2 can be solved. They provide the governing equations, Eqs. (5.29–5.31),

$$\frac{d\lambda_1}{du} = \lambda_3^2 \left\{ (\ell^2 \sinh u + 2\lambda_2 \hat{r})(\mathbf{a}_p \cdot \mathbf{i}) + \ell[\lambda_1 + (\hat{r} - 1) \cosh u](\mathbf{a}_p \cdot \mathbf{j}) \right\} \tag{5.29}$$

$$\frac{d\lambda_2}{du} = \lambda_3^2 \left\{ (-\ell^2 \cosh u + 2\lambda_1 \hat{r})(\mathbf{a}_p \cdot \mathbf{i}) + \ell[\lambda_2 - (\hat{r} - 1) \sinh u](\mathbf{a}_p \cdot \mathbf{j}) \right\} \tag{5.30}$$

$$\frac{d\lambda_3}{du} = -2\lambda_3^3 \left[\hat{q}(\mathbf{a}_p \cdot \mathbf{i}) + \ell(\mathbf{a}_p \cdot \mathbf{j}) \right] \tag{5.31}$$

These equations are equivalent to Gauss' form of the Lagrange planetary equations, in terms of the new set of variables.

The bijection between the Gauss–Argand plane and the Minkowski plane, defined in Eq. (5.5), relates the variables (ρ_1, ρ_2) and (λ_1, λ_2). These expressions may be differentiated to provide

$$\frac{d\rho_1}{du} = \left(2 - \frac{e^2}{\lambda_1^2}\right) \frac{d\lambda_1}{du} - 2\frac{\lambda_2}{\lambda_1} \frac{d\lambda_2}{du}$$

$$\frac{d\rho_2}{du} = \frac{1}{e} \left[\frac{\lambda_2^3}{\lambda_1^2} \frac{d\lambda_1}{du} + \frac{1}{\lambda_1} \left(\lambda_1 - 2\lambda_2^2\right) \frac{d\lambda_2}{du} \right]$$

Replacing the value of the derivatives of (λ_1, λ_2), given in Eqs. (5.29–5.31), leads to the evolution equations for variables ρ_1 and ρ_2:

$$\frac{d\rho_1}{du} = \frac{\lambda_3^2}{\lambda_1^2} \Big\{ [\ell^2 e^2 \sinh u + 2\lambda_2(\ell^2 - \hat{r})](\mathbf{a}_p \cdot \mathbf{i}) + \ell[(\ell^2 + 1)(1 - \hat{r}) \cosh u + \lambda_1(2\hat{r}^2 + \ell^2 - 1)](\mathbf{a}_p \cdot \mathbf{j}) \Big\} \tag{5.32}$$

$$\frac{d\rho_2}{du} = \frac{\lambda_3^2}{e\lambda_1^2} \Big\{ - [\ell^2 e^2 \lambda_1 \cosh u + (1 - \hat{r})\ell^4 + \ell^2(1 - 3\hat{r} - (1 + \hat{r})\lambda_1^2) - 2\hat{r}](\mathbf{a}_p \cdot \mathbf{i}) + \ell[e^2(1 - \hat{r})\hat{q} + \lambda_1\lambda_2(\hat{r}^2 + \ell^2)](\mathbf{a}_p \cdot \mathbf{j}) \Big\} \tag{5.33}$$

Equations (5.29–5.30) provide the evolution of the eccentricity vector on the Minkowski plane, whereas Eqs. (5.32–5.33) provide the evolution of the eccentricity vector on the Cartesian plane with $e > 1$.

5.4 Orbital plane dynamics

The orbital frame $\mathcal{L}$ rotates around the normal to the orbital plane with angular velocity $\boldsymbol{\omega}_{\mathcal{L}\mathcal{A}}$. The motion of frame $\mathcal{L}$ referred to the inertial reference is composed of the in-plane motion and the motion of the orbital plane itself. That is,

$$\boldsymbol{\omega}_{\mathcal{L}\mathcal{I}} = \boldsymbol{\omega}_{\mathcal{L}\mathcal{A}} + \boldsymbol{\omega}_{\mathcal{A}\mathcal{I}}$$

The term $\boldsymbol{\omega}_{\mathcal{L}\mathcal{A}}$ refers to the in-plane motion, whereas $\boldsymbol{\omega}_{\mathcal{A}\mathcal{I}}$ accounts for the motion of the orbital plane with respect to the inertial reference $\mathcal{I}$. The angular velocity of the orbital plane is then given by

$$\boldsymbol{\omega}_{\mathcal{A}\mathcal{I}} = \boldsymbol{\omega}_{\mathcal{L}\mathcal{I}} - \boldsymbol{\omega}_{\mathcal{L}\mathcal{A}} \tag{5.34}$$

The second term corresponds to the time evolution of angle ν,

$$\boldsymbol{\omega}_{\mathcal{L}\mathcal{A}} = \frac{\mathrm{d}\nu}{\mathrm{d}t}\mathbf{k}$$

Recall that Eq. (5.12) provided explicit expressions for the dynamics of the departure perifocal frame $\mathcal{P}$, namely $\boldsymbol{\omega}_{\mathcal{L}\mathcal{P}}$ and $\boldsymbol{\omega}_{\mathcal{P}\mathcal{I}}$:

$$\boldsymbol{\omega}_{\mathcal{L}\mathcal{I}} = \boldsymbol{\omega}_{\mathcal{L}\mathcal{P}} + \boldsymbol{\omega}_{\mathcal{P}\mathcal{I}}$$

In sum, Eq. (5.34) can be written

$$\boldsymbol{\omega}_{\mathcal{A}\mathcal{I}} = \boldsymbol{\omega}_{\mathcal{L}\mathcal{P}} + \boldsymbol{\omega}_{\mathcal{P}\mathcal{I}} - \boldsymbol{\omega}_{\mathcal{L}\mathcal{A}}$$

which results in

$$\boldsymbol{\omega}_{\mathcal{A}\mathcal{I}} = \frac{r}{h}(\mathbf{a}_p \cdot \mathbf{k})\,\mathbf{i} + \left(\frac{h}{r^2} - \frac{\mathrm{d}\nu}{\mathrm{d}t}\right)\mathbf{k} = \omega_1\mathbf{i} + \omega_3\mathbf{k} \tag{5.35}$$

The pair (ω_1, ω_3) defines the components of $\boldsymbol{\omega}_{\mathcal{A}\mathcal{I}}$ in frame $\mathcal{L}$. Projected onto frame $\mathcal{A}$ the angular velocity reads

$$\boldsymbol{\omega}_{\mathcal{A}\mathcal{I}} = p_a\mathbf{i}_{\mathcal{A}} + q_a\mathbf{j}_{\mathcal{A}} + r_a\mathbf{k}$$

The angular velocity referred to the fictitious time becomes

$$\boldsymbol{\Omega}_{\mathcal{A}\mathcal{I}} = \hat{p}_a\mathbf{i}_{\mathcal{A}} + \hat{q}_a\mathbf{j}_{\mathcal{A}} + \hat{r}_a\mathbf{k}\,, \qquad \text{with} \quad \boldsymbol{\Omega}_{\mathcal{A}\mathcal{I}} = \boldsymbol{\omega}_{\mathcal{A}\mathcal{I}}\frac{\mathrm{d}t}{\mathrm{d}u}$$

Let $\mathfrak{r}_{\mathcal{A}}, \mathfrak{r}_{\mathcal{I}}$ be two pure quaternions associated to the components of $\mathbf{r}$ in frames $\mathcal{A}$ and $\mathcal{I}$, respectively. The connection of quaternions with the set of four-dimensional hypercomplex numbers $\mathbb{H}^4$ is discussed in Appendix A. Let $\mathfrak{q} \in \mathbb{H}^4$ be the unit quaternion that defines the position of frame $\mathcal{A}$ on the inertial frame. Quaternion $\mathfrak{q}$ is written in terms of its components as $\mathfrak{q} = \chi_4 + \mathfrak{i}\chi_1 + \mathfrak{j}\chi_2 + \mathfrak{k}\chi_3$. It defines the rotation $\mathfrak{r}_{\mathcal{A}} \mapsto \mathfrak{r}_{\mathcal{I}}$,

$$\mathfrak{r}_{\mathcal{I}} = \mathfrak{q}\, \mathfrak{r}_{\mathcal{A}}\, \mathfrak{q}^{\dagger}$$

The derivative of quaternion $\mathfrak{q}$ with respect to fictitious time u is given by

$$\frac{d\mathfrak{q}}{du} = \frac{1}{2}\mathfrak{q}(2\mathfrak{m}_{\mathcal{A}\mathcal{I}})$$

with $2\mathfrak{m}_{\mathcal{A}\mathcal{I}} = 0 + \mathfrak{i}\,\hat{p}_a + \mathfrak{j}\,\hat{q}_a + \mathfrak{k}\,\hat{r}_a$. Computing the quaternion product leads to the equations:

$$\frac{d\chi_1}{du} = \frac{1}{2}(\ \chi_4\hat{p}_a - \chi_3\hat{q}_a + \chi_2\hat{r}_a) \tag{5.36}$$

$$\frac{d\chi_2}{du} = \frac{1}{2}(\ \chi_3\hat{p}_a + \chi_4\hat{q}_a - \chi_1\hat{r}_a) \tag{5.37}$$

$$\frac{d\chi_3}{du} = \frac{1}{2}(-\chi_2\hat{p}_a + \chi_1\hat{q}_a + \chi_4\hat{r}_a) \tag{5.38}$$

$$\frac{d\chi_4}{du} = \frac{1}{2}(-\chi_1\hat{p}_a - \chi_2\hat{q}_a - \chi_3\hat{r}_a) \tag{5.39}$$

The components $(\hat{p}_a, \hat{q}_a, \hat{r}_a)$ of the fictitious angular velocity in frame $\mathcal{A}$ are obtained from

$$\hat{p}_a = \hat{\omega}_1 \cos\nu\,, \qquad \hat{q}_a = \hat{\omega}_1 \sin\nu\,, \qquad \hat{r}_a = \hat{\omega}_3$$

with $\mathbf{\Omega}_{\mathcal{A}\mathcal{I}} = \hat{\omega}_1\mathbf{i} + \hat{\omega}_3\mathbf{k}$. The first component $\hat{\omega}_1$ of the angular velocity is obtained directly from Eq. (5.35), when referred to the fictitious time,

$$\hat{\omega}_1 = \lambda_3^2 \frac{\hat{r}^2}{\ell}(\mathbf{a}_p \cdot \mathbf{k})$$

The second component, $\hat{\omega}_3$, is

$$\hat{\omega}_3 = \frac{\lambda_3^2}{e^2\lambda_1}\Big\{\big[\ell\lambda_1(\hat{r} - \ell^2) + e(\ell^2(1-\hat{r}) - 2\hat{r})\big](\mathbf{a}_p \cdot \mathbf{i}) + \hat{q}\big[\lambda_1(\ell^2 + \hat{r}) + \ell e(\hat{r} - 1)\big](\mathbf{a}_p \cdot \mathbf{j})\Big\} \tag{5.40}$$

5.5 Time element

Equations (5.29–5.31) provide the geometry of the orbit, and Eqs. (5.36–5.39) define the motion of the orbital plane. The right-hand side of these equations vanishes when there are no perturbations. However, the evolution of the time given by Eq. (5.9) does not vanish in the unperturbed case, since $\lambda_3^{3/2}\hat{r}$ is not zero. As discussed in Sect. 2.5, introducing time elements typically improves the numerical performance of the integration. The time element is obtained as the correction to the unperturbed time due to perturbations. That is,

$$t = t_{\rm np} + t_{\rm te} \tag{5.41}$$

where $t_{\rm np}$ refers to the term that does not include perturbations, and $t_{\rm te}$ is the time element, satisfying $t_{\rm te} = 0$ if $\mathbf{a}_p = 0$. The term $t_{\rm np}$ abides by

$$t_{\rm np} - t_{\rm np0} = a^{3/2}(e \sinh H - H)$$

with $t_{\rm np0} = -a_0^{3/2}(e_0 \sinh H_0 - H_0)$. Introducing the new set of variables the previous equation becomes

$$t_{\rm np} - t_{\rm np0} = \lambda_3^{3/2}(\hat{q} - u)$$

Deriving the definition given in Eq. (5.41) provides the evolution of the time element,

$$\frac{dt_{\rm te}}{du} = \frac{dt}{du} - \frac{dt_{\rm np}}{du}$$

resulting in

$$\frac{dt_{\rm te}}{du} = \lambda_3^{7/2}[(\hat{r}^2 + 2(2\hat{r} - \ell^2) - 3\hat{q}u)(\mathbf{a}_p \cdot \mathbf{i}) + \ell(2\hat{q} - 3u)(\mathbf{a}_p \cdot \mathbf{j})] \tag{5.42}$$

Summary

The final system of equations is:

$$\frac{dt_{\rm te}}{du} = \lambda_3^{7/2}\left\{[\hat{r}^2 + 2(2\hat{r} - \ell^2) - 3\hat{q}u](\mathbf{a}_p \cdot \mathbf{i}) + \ell(2\hat{q} - 3u)(\mathbf{a}_p \cdot \mathbf{j})\right\} \tag{5.43}$$

$$\frac{d\lambda_1}{du} = \lambda_3^2\left\{(\ \ell^2 \sinh u + 2\lambda_2\hat{r})(\mathbf{a}_p \cdot \mathbf{i}) + \ell[\lambda_1 + (\hat{r} - 1)\cosh u](\mathbf{a}_p \cdot \mathbf{j})\right\} \tag{5.44}$$

$$\frac{d\lambda_2}{du} = \lambda_3^2\left\{(-\ell^2 \cosh u + 2\lambda_1\hat{r})(\mathbf{a}_p \cdot \mathbf{i}) + \ell[\lambda_2 - (\hat{r} - 1)\sinh u](\mathbf{a}_p \cdot \mathbf{j})\right\} \tag{5.45}$$

$$\frac{d\lambda_3}{du} = -2\lambda_3^3\left[\hat{q}(\mathbf{a}_p \cdot \mathbf{i}) + \ell(\mathbf{a}_p \cdot \mathbf{j})\right] \tag{5.46}$$

$$\frac{d\chi_1}{du} = \frac{1}{2}\left[\hat{\omega}_1(\ \chi_4 \cos v - \chi_3 \sin v) + \hat{\omega}_3\chi_2\right] \tag{5.47}$$

$$\frac{d\chi_2}{du} = \frac{1}{2}\left[\hat{\omega}_1(\ \chi_3 \cos v + \chi_4 \sin v) - \hat{\omega}_3\chi_1\right] \tag{5.48}$$

$$\frac{d\chi_3}{du} = \frac{1}{2}\left[\hat{\omega}_1(-\chi_2 \cos v + \chi_1 \sin v) + \hat{\omega}_3\chi_4\right] \tag{5.49}$$

$$\frac{d\chi_4}{du} = \frac{1}{2}\left[\hat{\omega}_1(-\chi_1 \cos v - \chi_2 \sin v) - \hat{\omega}_3\chi_3\right] \tag{5.50}$$

Equations (5.43–5.50) need to be integrated from the initial conditions:

$$t = t_0: \quad u_0 = H_0\,, \quad \lambda_1 = e_0\,, \quad \lambda_2 = 0\,, \quad \lambda_3 = a_0\,, \quad t_{\rm te} = 0$$

When the time element is not included, Eq. (5.43) is replaced by Eq. (5.9),

$$\frac{dt}{du} = \lambda_3^{3/2}\hat{r}$$

The reader willing to implement this algorithm is encouraged to follow the sequence:

1. Given the hyperbolic angle u and the pair (λ_1, λ_2) compute the auxiliary terms:

$$\hat{r} = \lambda_1 \cosh u + \lambda_2 \sinh u - 1$$
$$\hat{q} = \lambda_1 \sinh u + \lambda_2 \cosh u$$
$$e = \sqrt{\lambda_1^2 - \lambda_2^2} \quad \text{and} \quad \ell = \sqrt{\lambda_1^2 - \lambda_2^2 - 1}$$

2. The state vector is obtained from

$$\mathbf{r} = \lambda_3\, \hat{r}\, \mathbf{i}\,, \qquad \mathbf{v} = \frac{1}{\hat{r}\sqrt{\lambda_3}}(\hat{q}\,\mathbf{i} + \ell\,\mathbf{j})$$

   Equation (5.51) provides the transformation to the components in the inertial reference. The physical time reads

$$t = \lambda_3^{3/2}(\hat{q} - u) - a_0^{3/2}(e_0 \sinh H_0 - H_0) + t_{\text{te}}$$

   The perturbing term $\mathbf{a}_p$ can now be obtained.

3. Compute the angle α and the hyperbolic angle γ using Eq. (5.4),

$$\tanh\gamma = \frac{\lambda_2}{\lambda_1}\,, \qquad \alpha = \mathfrak{gd}\,(\gamma) = \arctan\sinh\gamma$$

4. Solve the hyperbolic anomaly, $H = u + \gamma$.
5. Obtain the true anomaly,

$$\vartheta = \operatorname{atan2}\,(\ell \sinh H, e - \cosh H)$$

   and the angle ν, which provides the evolution of frame $\mathcal{L}$ with respect to frame $\mathcal{A}$, $\nu = \alpha + \vartheta$.

6. Compute the components of the angular velocity in $\mathcal{A}$, $\hat{\omega}_1$ and $\hat{\omega}_3$:

$$\hat{\omega}_1 = \lambda_3^2\frac{\hat{r}^2}{\ell}(\mathbf{a}_p \cdot \mathbf{k})$$
$$\hat{\omega}_3 = \frac{\lambda_3^2}{e^2\lambda_1}\Big\{\big[\ell\lambda_1(\hat{r} - \ell^2) + e(\ell^2(1-\hat{r}) - 2\hat{r})\big](\mathbf{a}_p \cdot \mathbf{i}) + \hat{q}\big[\lambda_1(\ell^2 + \hat{r}) + \ell e(\hat{r} - 1)\big](\mathbf{a}_p \cdot \mathbf{j})\Big\}$$

7. Integrate Eqs. (5.43–5.50) to advance one step, and return to 1.

The Cartesian coordinates of the particle in the inertial reference are obtained from

$$\mathfrak{r}_{\mathcal{I}} = \mathfrak{s}\,\mathfrak{r}_{\mathcal{L}}\,\mathfrak{s}^\dagger\,, \quad \text{and} \quad \mathfrak{v}_{\mathcal{I}} = \mathfrak{s}\,\mathfrak{v}_{\mathcal{L}}\,\mathfrak{s}^\dagger \tag{5.51}$$

where quaternions $\mathfrak{r}_{\mathcal{L}}$ and $\mathfrak{v}_{\mathcal{L}}$ are $\mathfrak{r}_{\mathcal{L}} = 0 + \mathrm{i}\, x_{\mathcal{L}} + \mathrm{j}\, 0 + \mathrm{k}\, 0$ and $\mathfrak{v}_{\mathcal{L}} = 0 + \mathrm{i}\, \dot{x}_{\mathcal{L}} + \mathrm{j}\, \dot{y}_{\mathcal{L}} + \mathrm{k}\, 0$. The components of quaternion $\mathfrak{s} = s_4 + \mathrm{i}\, s_1 + \mathrm{j}\, s_2 + \mathrm{k}\, s_3$ are

$$s_1 = \chi_1 \cos\frac{\nu}{2} + \chi_2 \sin\frac{\nu}{2}\,, \quad s_2 = \chi_2 \cos\frac{\nu}{2} - \chi_1 \sin\frac{\nu}{2}$$

$$s_3 = \chi_3 \cos\frac{\nu}{2} + \chi_4 \sin\frac{\nu}{2}\,, \quad s_4 = \chi_4 \cos\frac{\nu}{2} - \chi_3 \sin\frac{\nu}{2}$$

The following expressions provide the orbital elements from the new set of elements:

$$\begin{aligned}
a &= \lambda_3 \\
e &= \left(\lambda_1^2 - \lambda_2^2\right)^{1/2} \\
i &= \operatorname{atan2}\left(\chi_1^2 + \chi_2^2,\, \chi_3^2 + \chi_4^2\right) \\
\omega &= \operatorname{atan2}\left(C_1^- \cos\alpha + C_2^+ \sin\alpha,\; C_2^+ \cos\alpha - C_1^- \sin\alpha\right) \\
\Omega &= \operatorname{atan2}\left(C_1^+,\, C_2^-\right)
\end{aligned}$$

with $C_1^\pm = (\chi_1\chi_3 \pm \chi_2\chi_4)$ and $C_2^\pm = (\chi_1\chi_4 \pm \chi_2\chi_3)$.

5.6 Numerical evaluation

The performance of the proposed formulation is evaluated in this section. The orbits of four hyperbolic comets (Table 5.1), and the geocentric flybys of NEAR, Cassini, and Rosetta (Table 5.3) are integrated using different formulations. They are compared in terms of accuracy and runtime. The Minkowskian formulation is compared against Cowell's method, the Kustaanheimo–Stiefel (KS) transformation, the Sperling–Burdet (SB) regularization, and the Dromo formulation. The simulations are conducted with the LSODAR integrator. Claims about the advantages of one method over others should always be understood in the context of specific examples and simulation setups, in order not to overstate the behavior of the formulation. This is particularly important when numerical experiments become more complex, because they are hardly reproducible.

In order to determine the accuracy of the propagation a reference solution is constructed. First, the orbits are integrated with Cowell's method in quadruple precision

Tab. 5.1: Orbits of the four selected hyperbolic comets.

Comet	q **[au]**	e **[–]**	i **[deg]**	Ω **[deg]**	ω **[deg]**	**Osculation**
C/1999 U4 Catalina-Skiff	4.92	1.0076	51.93	32.29	77.52	2002-Jan-03
C/2002 B3 Linear	6.06	1.0075	73.69	289.36	123.21	2002-Feb-11
C/2008 J4 McNaught	0.45	1.0279	87.37	289.69	92.18	2008-May-17
C/2011 K1 Schwartz-H.	3.38	1.0053	122.61	70.74	167.09	2011-Jun-03

floating-point arithmetic, setting the integrator tolerance to $\varepsilon_{\text{tol}} = 10^{-22}$. A dense output is generated providing the position of the bodies with high accuracy. The results are now truncated to the first 16 digits, so the solution is reduced to double precision. The numerical test cases will be integrated in double precision floating-point arithmetic and compared to these reference values that are considered the exact solution to the problem.

The error ϵ is defined as the mean value of the error in position across all the points in the dense output. Denoting by ϵ_j the error at a certain date j, the error in the solution is

$$\epsilon = \frac{1}{N}\sum_{j=1}^{N}\epsilon_j\,, \qquad \text{with } \epsilon_j = \left\|\mathbf{r}_j - \mathbf{r}_j^{\text{ref}}\right\| \tag{5.52}$$

Each point from the reference solution is denoted by $\mathbf{r}_j^{\text{ref}}$, and N is the total number of points in the dense output.

The runtime is the time from which the integration starts until that at which the final point is reached. It includes the time for computing the perturbations at each integration step. The runtime is measured as the mean runtime after 40 consecutive integrations. This procedure averages the effect of possible instabilities in the integration.

5.6.1 Hyperbolic comets

Table 5.1 summarizes the orbital parameters defining the orbits of four hyperbolic comets. The orbits of the comets are defined by means of the perihelion distance, q, eccentricity, e, inclination, i, right ascension of the ascending node, Ω, and argument of periapsis, ω. Angles are referred to the ICRF/J2000.0 reference frame, and the reference plane is the Earth Mean Equator and Equinox of reference epoch. Their heliocentric orbits are integrated for four years: two years before periapsis passage, and two years after.

The propagator includes three sources of perturbations: third-body perturbations, relativistic corrections, and the Sun's oblateness. The perturbations from the major planets, Pluto, and the Moon, as well as the four major asteroids (Ceres, Vesta, Pallas, and Hygiea) are computed from the DE431 ephemeris. Relativistic corrections account for the effects of the spherical central body, Lense–Thirring, oblateness, and rotational energy. Additional nongravitational accelerations are not considered in the model.

Figure 5.4 displays the machine runtime for different levels of accuracy in the solution. The error in position displayed in this figure corresponds to the definition given in Eq. (5.52). Each point is obtained by varying the integration tolerance down to the machine precision. The machine zero for the selected compiler is 2.2204×10^{-16}, so the tolerance spans from $\varepsilon_{\text{tol}} = 10^{-5}$ to $\varepsilon_{\text{tol}} = 10^{-15}$. The smallest tolerance corresponds to the last points to the left of the curves. In these particular examples, the Minkowskian

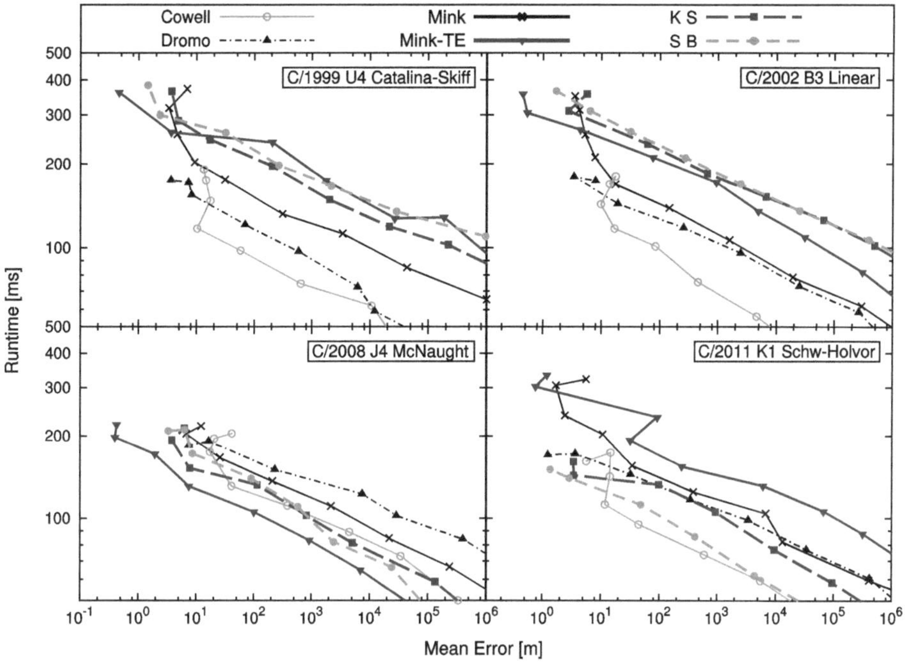

Fig. 5.4: Integration runtime vs. accuracy of the solution for the integration of the hyperbolic comets. The curves correspond to Cowell's formulation, Dromo, the Minkowskian solution without ("Mink") and with the time element ("Mink-TE"), the Kustaanheimo–Stiefel transformation ("KS"), and the Sperling–Burdet regularization ("SB"). The curves are parameterized in terms of the tolerance, ε_{tol}.

formulation with the time element (Mink-TE) is the most accurate. Improvements in accuracy beyond one order of magnitude with respect to Cowell's method are observed when integrating the orbits of the four comets. To achieve these results the integration of Mink-TE requires more time, because it involves more algebraic operations than the rest. But the excess in runtime depends on the case: for comets C/1999 U4 Catalina-Skiff, C/2002 B3 Linear, and C/2011 K1 Schwartz-Holvorcem the total runtime of Mink-TE is about twice the runtime for Cowell's method, which is the fastest. Mink-TE exhibits its best performance when propagating the orbit of C/2008 J4 McNaught: it is faster than KS and SB, and the accuracy is improved by about one order of magnitude even with respect to these formulations. The Minkowskian formulation without the time element does not exhibit clear improvements in performance nor accuracy.

The time evolution of the error is plotted in Fig. 5.5. This figure shows the error in position at date j, ϵ_j. The values correspond to the tolerance for which each formulation is more accurate. Mink-TE exhibits the most stable behavior, because the formulation is not affected by perihelion passage: while the error of the reference formulations grows around the perihelion (specially for comets C/1999 U4 Catalina-Skiff,

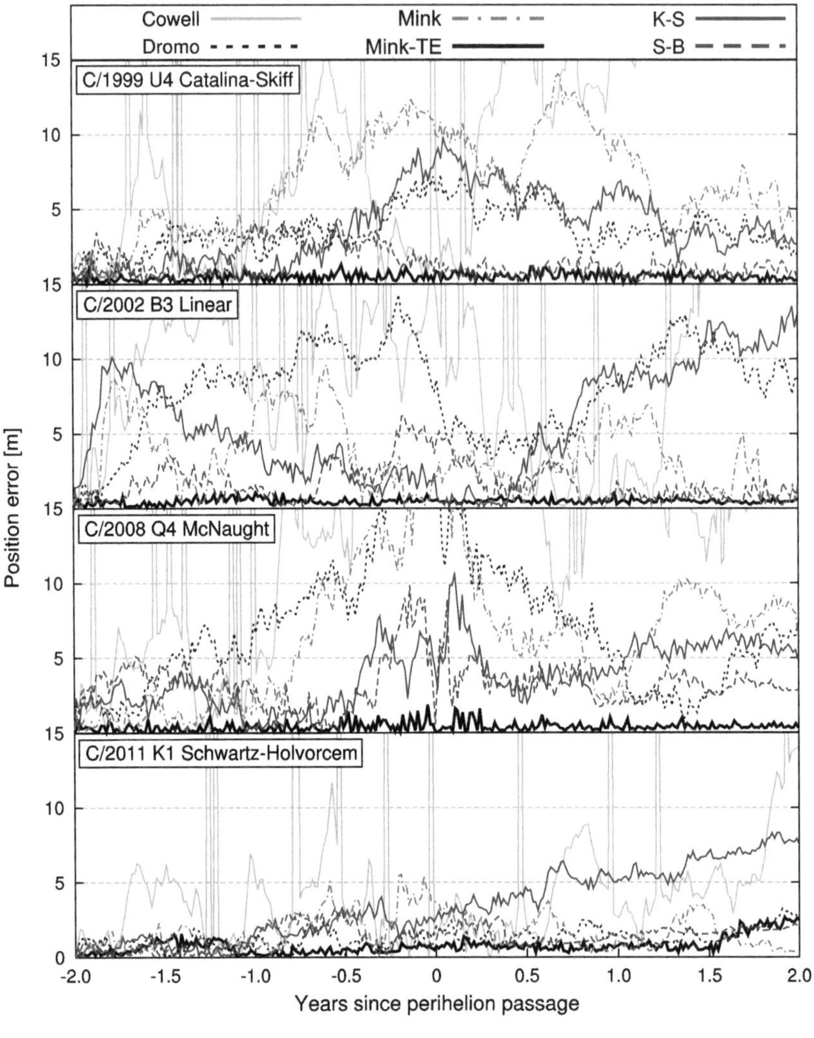

Fig. 5.5: Time evolution of the error in position at epoch, ϵ_j, for the four hyperbolic comets.

and C/2008 J4 McNaught) the error curve of Mink-TE remains flat, thanks to the natural representation of the geometry. For comet C/2011 K1 Schwartz-Holvorcem this formulation is still the most stable during the integration, although the error grows slightly toward the end. The SB is the second formulation in terms of accuracy and stability of the error, but error peaks about the perihelion of comet C/2008 J4 McNaught still appear. In this figure it is easily observed how the introduction of the time element improves the performance of the proposed formulation.

Table 5.2 shows the mean value and the standard deviation of the error distribution for the cases displayed in Fig. 5.5. These values correspond to the last point to the

Tab. 5.2: Mean value (μ_ϵ) and standard deviation (σ_ϵ) of the most accurate error profile from each formulation.

	C/1999 U4		C/2002 B3		C/2008 J4		C/2011 K1	
	μ_ϵ [m]	σ_ϵ [m]	μ_ϵ [m]	σ_ϵ [m]	μ_ϵ [m]	σ_ϵ [m]	μ_ϵ [m]	σ_ϵ [m]
Cowell	14.43	15.97	13.88	14.54	17.25	23.68	5.67	11.51
Dromo	3.61	1.48	7.92	3.09	7.49	5.06	1.24	1.40
Mink	6.82	3.39	3.50	2.45	6.73	8.28	1.69	2.07
Mink-TE	0.46	0.26	0.45	0.19	0.40	0.50	0.76	0.94
KS	3.69	2.49	5.64	3.73	3.82	4.35	3.44	4.18
SB	1.44	1.04	1.69	1.32	3.29	1.59	1.37	1.57

left (best accuracy) in Fig. 5.4. The numerical values in this table confirm that the error profile for Mink-TE is the most stable in the proposed test cases. Note that the mean and standard deviation of the error is below one meter only for Mink-TE. For the rest of the formulations the mean value and standard deviation of the error remains over one meter in all four cases.

5.6.2 Geocentric flybys

This section deals with the geocentric flybys of NEAR, Cassini, and Rosetta spacecraft. The integration starts approximately when the spacecraft enters the sphere of influence of the Earth, and stops when it leaves the sphere. This corresponds to a time span of 4 days for NEAR, 2 days for Cassini, and 6 days for Rosetta. Table 5.3 defines the reference orbits.

The propagator now includes different sources of perturbations: third-body perturbations, relativistic corrections, and the Earth gravity field. The perturbations from the major planets and the Moon, as well as the four major asteroids (Ceres, Vesta, Pallas, and Hygiea) are computed from the DE431 ephemeris. Relativistic corrections account for the effects of the spherical central body, Lense–Thirring, oblateness, and rotational energy. A 100 × 100 gravity field is implemented (based on the GGM03S model) in order to make the evaluation of the perturbations the most expensive part of the call to the function to be integrated. The Earth rotation model corresponds to the IAU 2006/2000A, CIO based (X-Y series) standard. The Earth Orientation Parameters

Tab. 5.3: Definition of the geocentric flybys of NEAR, Cassini and Rosetta.

S/C	a [km]	h_{per} [km]	e [–]	i [deg]	Ω [deg]	ω [deg]	Osculation
NEAR	8496	533	1.81	108.0	88.3	145.1	1998-Jan-23
Cassini	1555	1172	5.86	25.4	3.2	248.5	1999-Aug-18
Rosetta	26702	1954	1.31	144.9	170.2	143.1	2005-Mar-04

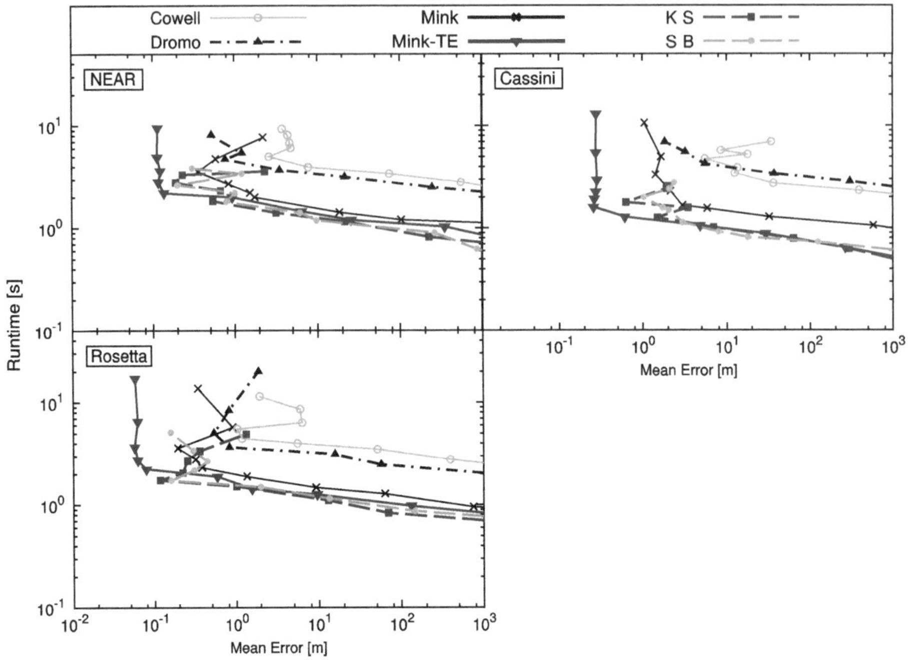

Fig. 5.6: Integration runtime vs. accuracy of the solution for the integration of the geocentric flybys of NEAR, Cassini, and Rosetta.

from Vallado and Kelso (2013) are considered. Additional nongravitational accelerations are not computed in the model.

Figure 5.6 displays the machine runtime for different levels of accuracy in the solution. The error in position displayed in this figure corresponds to the definition given in Eq. (5.52). It is observed that the Minkowskian formulation with the time element (Mink-TE) is the most accurate in the three test cases. The fact that perturbations are expensive to compute (in terms of runtime) gives preference to formulations that are integrated more efficiently, i.e., that require less function calls. The overhead related to a formulation being more complicated to evaluate becomes less important. As a consequence, and despite not being as simple as Cowell's method, the Minkowskian formulation turns out to be faster.

The time evolution of the error is plotted in Fig. 5.7. This figure shows the error in position at date j, ϵ_j. The values correspond to the smallest tolerance limited by the machine zero, $\varepsilon_{\text{tol}} = 10^{-15}$. The Mink-TE solution exhibits the most stable behavior, and it is also the most accurate. Note that this formulation is not affected significantly by the perigee passage, while the rest of the formulations show certain error peaks around perigee. In particular, the integration error of the orbits of NEAR and Rosetta show significant jumps at periapsis when propagated in KS variables. Table 5.4 shows the mean value and the standard deviation of the error distribution for the cases dis-

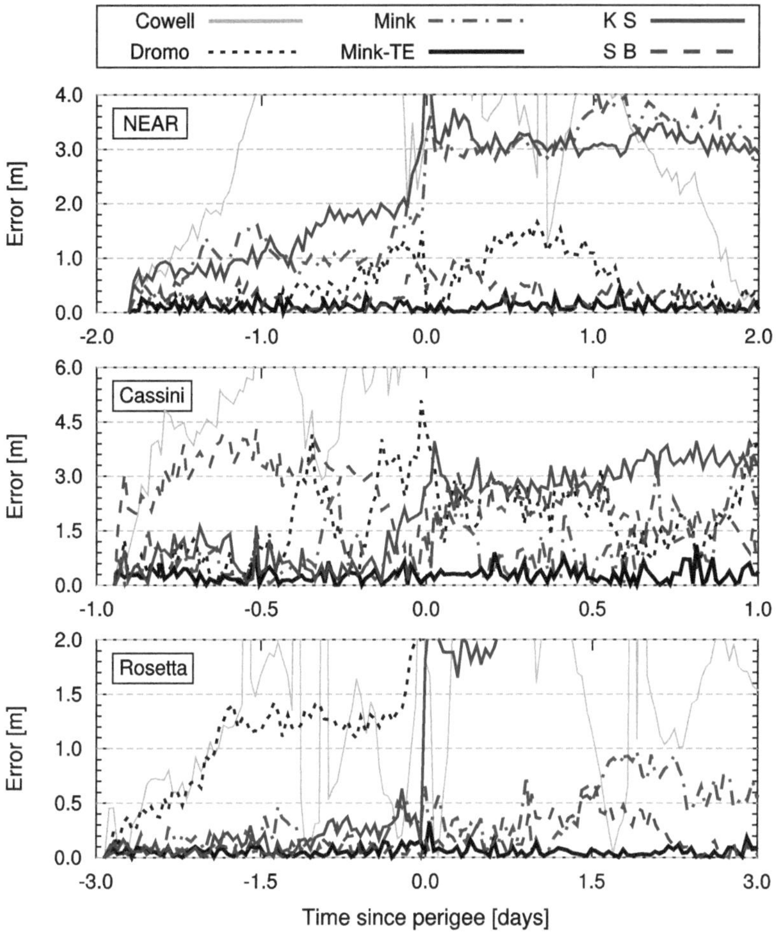

Fig. 5.7: Time evolution of the error in position at epoch, ϵ_j, for NEAR, Cassini, and Rosseta. The orbit is sampled every 40 min for NEAR, 20 min for Cassini and, 1 h for Rosetta.

Tab. 5.4: Mean value (μ_ϵ) and standard deviation (σ_ϵ) of the position error for the smallest admissible tolerance, $\varepsilon_{tol} = 10^{-15}$.

	NEAR		**Cassini**		**Rosetta**	
	μ_ϵ [m]	σ_ϵ [m]	μ_ϵ [m]	σ_ϵ [m]	μ_ϵ [m]	σ_ϵ [m]
Cowell	3.75	3.79	34.25	32.51	1.96	2.24
Dromo	0.52	0.46	1.84	1.16	1.90	0.96
Mink	2.21	1.19	1.05	0.91	0.34	0.27
Mink-TE	0.11	0.09	0.28	0.21	0.06	0.04
KS	2.32	1.04	2.00	1.25	1.34	1.16
SB	0.30	0.22	2.38	0.88	0.16	0.15

played in Fig. 5.7. These values are obtained with the smallest admissible tolerance, corresponding to the last point to the left in Fig. 5.6. These numerical values show that Mink-TE might lead to reductions in the mean error of about one or two orders of magnitude, depending on the reference formulation. These results suggest that using the Minkowskian propagation method in generic examples might improve the overall performance of the integration.

5.7 Conclusions

Describing hyperbolic orbits is intimately related to hyperbolic geometry. The formulation presented in this chapter fills the gap between hyperbolic problems in astrodynamics and the mathematical foundations of hyperbolic geometry. The advances in special relativity, in particular the algebraic and geometrical characterization of Minkowski space-time, provide a number of useful mathematical tools. The geometry in Minkowski space-time is successfully extended to celestial mechanics. The techniques presented in Chapter 2 can be adapted to a particular problem in order to obtain formulations that are tailored to that specific application.

More specifically, three main conclusions are drawn from the previous analysis: (i) Despite being more intuitive, Euclidean geometry is not the only possible choice for defining vectors and rotations. They can be defined equivalently in the two-dimensional Minkowski space-time. This geometrical consideration is motivated by the fact that the unperturbed motion of a particle along a hyperbolic orbit is typically parameterized using hyperbolic trigonometry. The evolution of the eccentricity vector on the Minkowski plane is formulated in terms of hyperbolic functions, that comply with the solution on the osculating orbit. This leads to an adapted version of Dromo, conceived for hyperbolic orbits. (ii) The bijection between the Minkowski plane and the Euclidean plane has been established rigorously. (iii) The resulting formulation improves the stability of the numerical integration. This has a direct impact on the accuracy of the solution, having reduced the propagation error in the proposed test cases.

The strongest point of this particular formulation is that the error evolution is not affected by periapsis passage, a typical downside of Cowell's method. In addition, it reaches levels of accuracy that cannot be achieved with the compared formulations. This extra level of accuracy may be important in some critical scenarios. The complexity of the equations might increase the computational time with respect to similar methods when perturbations are easy to evaluate, although the eventual increase in CPU time is quite affordable.

6 Evaluating the numerical performance

Many different formulations and numerical schemes for propagating the equations of orbital motion can be found in the literature. Almost always, the numerical experiments following the derivation of a new formulation show the apparent superiority of the new method with respect to the existing ones. Statements of this sort should be made cautiously, because the results are strongly tied to the particular problem that is being considered. In addition, there are many parameters that can be tuned during the experiments. Consequently, aiming for standardization and reproducibility, as well as realizing that generic claims about the overall superiority of a method might be too optimistic, are the pillars of a precise performance analysis. A rigorous approach should focus on analyzing the effect that different regularizing and stabilizing techniques have on a given formulation and how well they work together. This way, the user can understand when to expect improvements in the performance. The present chapter discusses a number of practical points related to setting up the infrastructure for numerical comparisons. The main tool used in this study is the software package PERFORM (Performance Evaluation of Regularized Formulations of Orbital Motion), a dedicated program for orbit propagation including different formulations and integrators. The ultimate goal is to have a way to compare the numerical performance of the formulations when running on the same platform, with the same force models, etc (Roa, 2016b, chap. 6).

Table 6.2 presents a list of the compared formulations and Appendix B contains a detailed explanation. This chapter makes no claims about the performance of specific formulations. It simply shows how tools like PERFORM shall be used to decide which combination of orbital formulations and numerical integrators yields the most efficient solution for a given problem. One should not expect one formulation to outperform the rest in every single scenario. This kind of analysis is important when thousands of similar propagations need to be carried out. Examples of relevant applications are trajectory optimization and planetary protection studies. When optimizing a trajectory, the solver will call the cost function many times. Evaluating the cost function and its gradient typically requires propagating the trajectory. Algorithms for local optimization will restrict the search to bounded neighborhoods around the nominal trajectory. Thus, small differences in performance between trajectories in the search space should be expected, and finding the most efficient strategy for integrating the nominal orbit will improve the overall exploration. Most interplanetary mission designs include planetary protection studies. They reduce to campaigns of Monte-Carlo simulations (or equivalent techniques) in which the nominal trajectory is perturbed slightly to find the probability of crashing with celestial bodies without a sufficiently thick atmosphere in which the spacecraft could disintegrate. One interesting example is the Cassini spacecraft, which is powered by a plutonium-filled radioisotope thermoelectric generator. The planetary protection teams had to make sure that Cassini would

https://doi.org/10.1515/9783110559125-006

not hit the Saturnian moons. Placing a radioactive heat source on an icy moon like Enceladus might have important consequences; if life were ever found in the moon, it would be unclear whether it was originated by Cassini's heat source, or by natural processes.

Section 6.1 explains the setup that will be used for numerical comparisons. Details on the orbital formulations, numerical integrators, and force models are provided. Section 6.2 shows how PERFORM can be used for evaluating the performance of different formulations when integrating a given problem. The concept of the sequential performance diagram (SPD) is introduced to compare the behavior of the formulations.

6.1 Implementation

The core of PERFORM is implemented in Fortran. The program can be compiled and run both in double and quadruple precision floating-point arithmetic.

6.1.1 Force models

Gravitational perturbations from the planets in the Solar System are computed according to the JPL DE series of ephemerides. Solar radiation pressure is computed accounting for seasonal variations of the solar flux and a conic shade model. The atmospheric density models implemented in PERFORM are the MSISE90 (Hedin, 1991), the Jacchia 70 and 77 (Jacchia, 1977), and a simple exponential model. Space weather data (including the three-hour variations) are retrieved from Celestrack. The gravity field of the Earth, the Moon, and Mars are modeled with the Grace Gravity Model (GGM03S), Grail's lunar model, and the MRO95A model from the Mars Reconnaissance Orbiter. The perturbations and gravitational harmonics are computed following the technique by Cunningham (1970). PERFORM has been integrated with the SOFA package provided by the IAU. The orientation of the Earth can be determined using different standards:

a) IAU 2000A, CIO based, using (X, Y) series.
b) IAU 2000B, CIO based, using (X, Y) series.
c) IAU 2000A, equinox based, using classical angles.
d) IAU 2006/2000A, CIO based, using classical angles.
e) IAU 2006/2000A, CIO based, using (X, Y) series.

The Earth Orientation Parameters (EOP) providing the corrections to the position of the pole and the length of day are retrieved from either the International Earth Rotation and Reference System Service (IERS) or the US Naval Observatory (USNO). The simplified model without nutation or precession by Archinal et al. (2011) is also available in PERFORM. Table 6.1 shows the differences between the different conventions and EOP. Given an equatorial vector defined in the GCRS, it is then transformed to the ITRS using two different models. Then, the angle between the resulting vectors provides the offset between the conventions. The first row of the table corresponds to the difference in the rotation defined by model a) and the rest of the models. The second row corresponds to the rotation difference between the same model but when using the IERS and the USNO EOP. Finally, the last row shows the total runtime for computing the GCRS to ITRS rotation matrix 5000 consecutive times. Using the B series over the A series reduces the runtime one order of magnitude. The difference with respect to the rotation using the A series is about 62.5 μas. The approximate model from Archinal et al. (2011) yields an angular difference of $237''$, but the total runtime is reduced by a factor of 500. The difference in using one source of EOP or the other is 140 mas in all five cases.

Relativistic corrections follow from a simplification of the Parameterized Post-Newtonian (PPN) model, including a term due to the gravitational and rotational energy of the central body, its oblateness, and the Lense–Thirring effect. For heliocentric propagations the contribution from the Sun J_2 is accounted for. Specific perturbations like the continuous thrust from the engines or the nongravitational actions upon comets (following Królikowska, 2006) have also been implemented.

Tab. 6.1: Performance of the different GCRS to ITRS transformations at JD(TT) = 2,457,066.50.

	Model a)	Model b)	Model c)	Model d)	Model e)	Approx.
Offset w.r.t. a) [mas]	–	0.0625	3.10E–6	0.0777	0.0776	2.37E5
IERS vs USNO [mas]	140.70	140.870	140.68	140.70	140.70	–
Runtime [s]	6.651	0.578	6.460	6.549	4.945	0.015

6.1.2 Formulations

An orbit propagator is defined by two major elements. First, the formulation chosen to model the dynamics of the system. Second, the integrator used to propagate numerically the corresponding equations of motion. Both elements are important when assessing the performance of a given integration.

Chapter 2 presented a plethora of methods and techniques for improving the numerical performance of the propagation, resulting in different formulations of the equations of orbital motion. The formulations implemented in PERFORM are listed in Table 6.2, and explained in Appendix B. The first property of a given formulation is the dimension of the system, i.e., the number of equations to be integrated (N_{eq}). Next, the number of elements N_{el} is important because it defines the number of equations whose right-hand side will be zero (or at least constant) during the propagation of

Tab. 6.2: List of propagation methods implemented in PERFORM (for a detailed description of each formulation see Appendix B).

	N_{eq}	N_{el}	dt/ds	Time Elem.	ID	Class
Cowell	6	0	–	–	COW	–
Stabilized Cowell	8	2	r	Linear	SCW	4
Dromo	8	7	r^2/h	–	DRO	2
Time-Dromo	8	7	–	–	TDR	3
Minkowskian + T.E.	8	8	$r/\sqrt{2\mathcal{E}}$	Const.	HDT	1
Minkowskian	8	7	$r/\sqrt{2\mathcal{E}}$	–	HDR	1
KS-a	9	0	r	–	KS_	4
KS-b	9	1	r	Linear	KST	4
KS-c	10	1	r	–	KSR	4
KS-d	10	2	r	Linear	KRT	4
Sperling-a	7	0	r	–	SB_	4
Sperling-b	7	1	r	Linear	SBT	4
Sperling-c	13	4	r	–	SBR	4
Sperling-d	13	5	r	Linear	SRT	4
Deprit	8	7	–	–	DEP	3
Palacios	8	5	r^2/h	–	PAL	2
Milankovitch	7	6	–	–	MIL	3
Stiefel–Scheifele	10	10	$r/\sqrt{-2\mathcal{E}}$	Linear	SSc	1
Equinoctial	6	5	–	–	EQU	3
Classical	6	5	–	–	CLA	3
Burdet elem. (BG14)	11	11	r	Const.	BCP	1
Unified State Model	7	3	–	–	USM	–
EDromo-0	8	7	$r/\sqrt{-2\mathcal{E}}$	–	ED0	1
EDromo-1	8	8	$r/\sqrt{-2\mathcal{E}}$	Linear	ED1	1
EDromo-2	8	8	$r/\sqrt{-2\mathcal{E}}$	Const.	ED2	1
BF-0H	10	1	r^2/h	–	B0H	2
BF-1H	10	1	r^2/h	Linear	B1H	2
BF-2H	10	2	r^2/h	Const.	B2H	2
BF-3H	11	1	r^2/h	–	B3H	2
BF-0C	10	1	r^2/h	–	B0C	2
BF-1C	10	1	r^2/h	Linear	B1C	2
BF-2C	10	2	r^2/h	Const.	B2C	2
BF-3C	11	1	r^2/h	–	B3C	2
Stiefel-1	10	9	r	–	ST0	1
Stiefel-2	10	10	r	Linear	ST1	1
Stiefel-3	10	10	r	Const.	ST2	1

Keplerian orbits, and will evolve slowly in the presence of weak perturbations. For $N_{eq} = N_{el}$ all the variables are elements, and all the benefits presented in Sect. 2.5 may be expected when integrating weakly perturbed problems. The table also shows the time transformation featured in the formulation. The time transformation is closely related to the use of time elements, which may be constant (vanishing derivative in the

Keplerian case) or linear (constant derivative in the Keplerian case). The ID assigned to each formulation is the internal identification used by PERFORM. The formulations are divided into four classes:

Class 1: element-based formulations (i.e., $N_{\mathrm{el}} \geq N_{\mathrm{eq}}/2$) including a Sundman transformation of order one.

Class 2: element-based formulations including a Sundman transformation of order two.

Class 3: element-based formulations with no time transformation (the physical time is the independent variable).

Class 4: formulations based on coordinates including a Sundman transformation of order one.

6.1.3 Numerical integration

A sophisticated numerical integrator can improve the efficiency of the propagation in Cartesian coordinates, just like regularized formulations reduce the error growth rate and improve the stability compared to Cowell's method. An adequate combination of formulation/integrator might maximize the potential of both elements. For this reason, formulations should be tested with numerical integration schemes of different nature. For details on the methods, we refer to the classic book by Hairer et al. (1991).

The baseline integrator is LSODE, the Livermore Solver for Ordinary Differential Equations implemented by Hindmarsh (1983). This algorithm is distributed together with many variants in the library ODEPACK. It is based on an Adams–Moulton multistep scheme, and switches automatically to implicit backward differentiation when the integrator detects a stiff problem. In addition, Runge–Kutta methods like RKF5(4), DVERK, RKF7(6), and RKF8(7) will be added to the comparison, and the DoPri853 scheme is incorporated for completeness (Dormand and Prince, 1980).

The second-order equations of motion can be integrated with a Störmer–Cowell integration scheme. The implementation in PERFORM follows the guidelines in Berry (2004). This is a second-order multistep method, which is particularly well suited for integrating the equations of motion in Cartesian coordinates.

6.1.4 Variational equations

For many scientific and operational applications, propagating the nominal state of the particle is not enough; information about the sensitivity of the trajectory is required, which means that the variational equations need to be solved too. The solution provides the partial derivatives of the trajectory with respect to the initial conditions and, in some cases, certain physical parameters. This section briefly introduces the propagation of the variational equations using regularized variables, and Sect. 8.6 will later analyze this topic in further detail.

Denoting ν the generic independent variable, and $\mathbf{y} = \mathbf{y}(\nu)$ the state vector, propagating the orbit means solving the initial value problem

$$\frac{\mathrm{d}\mathbf{y}}{\mathrm{d}\nu} = \mathbf{f}(\nu; \mathbf{y}), \qquad \text{with} \quad \mathbf{y}(\nu_0) = \mathbf{y}_0 \tag{6.1}$$

Given a nominal orbit $\mathbf{y} = \mathbf{y}(\nu)$, in many problems it is interesting to monitor the evolution of nearby solutions, defined initially as $\mathbf{y}'_0 = \mathbf{y}_0 + \delta\mathbf{y}_0$. If the initial separation $\delta\mathbf{y}_0$ is small, the relative dynamics can be described using a linear model,

$$\delta\mathbf{y}(\nu) = \Xi(\nu; \mathbf{y}_0)\, \delta\mathbf{y}_0$$

in which $\Xi(\nu; \mathbf{y}_0)$ represents the state-transition matrix. In practice this matrix is obtained numerically because it depends on the perturbations and the state vector at each integration step. Its evolution is governed by the initial value problem

$$\frac{\partial \Xi}{\partial \nu} = \left.\frac{\partial \mathbf{f}}{\partial \mathbf{y}}\right|_{\nu} \Xi(\nu; \mathbf{y}_0), \qquad \text{with} \quad \Xi(\nu_0; \mathbf{y}_0) = \mathbf{I}$$

These equations are integrated together with Eq. (6.1), and the components of the state-transition matrix are attached to the state vector. The Jacobian matrix $\partial\mathbf{f}/\partial\mathbf{y}|_{\nu}$ is defined by the partial derivatives of the right-hand side of Eq. (6.1) with respect to the state vector, keeping constant ν. It is computed numerically at each integration step using finite central differences.

6.2 Evaluating the performance

This section presents two examples designed to evaluate the performance of different formulations. The first example is the test problem already introduced in Table 4.1. A highly eccentric geocentric orbit is perturbed by the Moon and the oblateness of the Earth, as defined by Stiefel and Scheifele (1971, p. 122). The second example simulates a low-thrust transfer from LEO to GEO. The communications satellite ABS-3A (a Boeing 702SP spacecraft) is taken as a representative example.

For convenience, the initial conditions and the final state of the two problems are summarized in Table 6.3. The reference solution is constructed by integrating the problem in quadruple precision floating-point arithmetic with the finest tolerance using different formulations. Then, only the common digits are retained. Because the numerical experiments will be carried out in double precision, the solution can be considered exact as long as the first 16 digits coincide (the machine zero is 2.2204×10^{-16}).

Tab. 6.3: Initial conditions and last reference point of the numerical test cases.

	Units	Problem 1	Problem 2
MJD_0	–	N/A	57083
t_f	days	288.12768941	100.00000000
x_0	km	0.000000000000000E+00	1.115741121014817E+04
y_0	km	−5.888972700000000E+03	1.019260166000129E+03
z_0	km	−3.400000000000000E+03	−2.268170222538688E+03
$v_{x,0}$	km/s	1.069133800000000E+01	7.309103529900698E−01
$v_{y,0}$	km/s	0.000000000000000E+00	3.662867091587677E+00
$v_{z,0}$	km/s	0.000000000000000E+00	3.991045017757724E+00
x_f	km	−2.421905011593605E+04	−1.814081886073619E+04
y_f	km	−2.279621063730220E+05	1.145173099550784E+04
z_f	km	−1.297534424000825E+05	−5.027819073450440E+02
$v_{x,f}$	km/s	−3.072444684207348E−01	−1.256074877885349E+00
$v_{y,f}$	km/s	1.539502056878936E−01	−2.613579325742426E+00
$v_{z,f}$	km/s	7.809786648117930E−02	−3.413479931546313E+00

6.2.1 Problem 1

Problem 1 is integrated with all the formulations listed in Table 6.2 except for the Minkowskian propagator (Chap. 5), which is only valid for hyperbolic orbits. The orbit is propagated with two different integrators as representative examples: RKF5(4) and LSODAR. The performance of the integration is evaluated with the traditional plots that present the error in the propagation and the number of function calls (it is equivalent to the total runtime). The performance curves are built by changing the integration tolerance down to the machine precision. In order to simplify the visualization of the results, the data is presented in the form of sequential performance diagrams (SPD).

Figure 6.1 is the SPD corresponding to the integration with the RKF5(4) scheme. The central figure shows the performance curves of the most efficient formulation of each class, plus the solutions using Cowell's method and the unified state model, which do not fit in any of the four classes. The criterion for choosing the most efficient formulation is based on a trade-off between speed and accuracy. The second level of performance plots in the SPD corresponds to the performance of each formulation within a given class. The curves have been arranged so that the most efficient formulation (the one selected to be represented in the central figure), is plotted first. Certain formulations have different versions. In the second level of figures only the most efficient version of each formulation is presented. The performance of all versions of a given formulation can be analyzed in the third level of figures. The introduction of a Sundman transformation determines the performance of the methods: the central figure shows that using the time as the independent variable (Class 3 formulations, COW, and USM) renders performance curves with the same slope. Indeed, the use of a

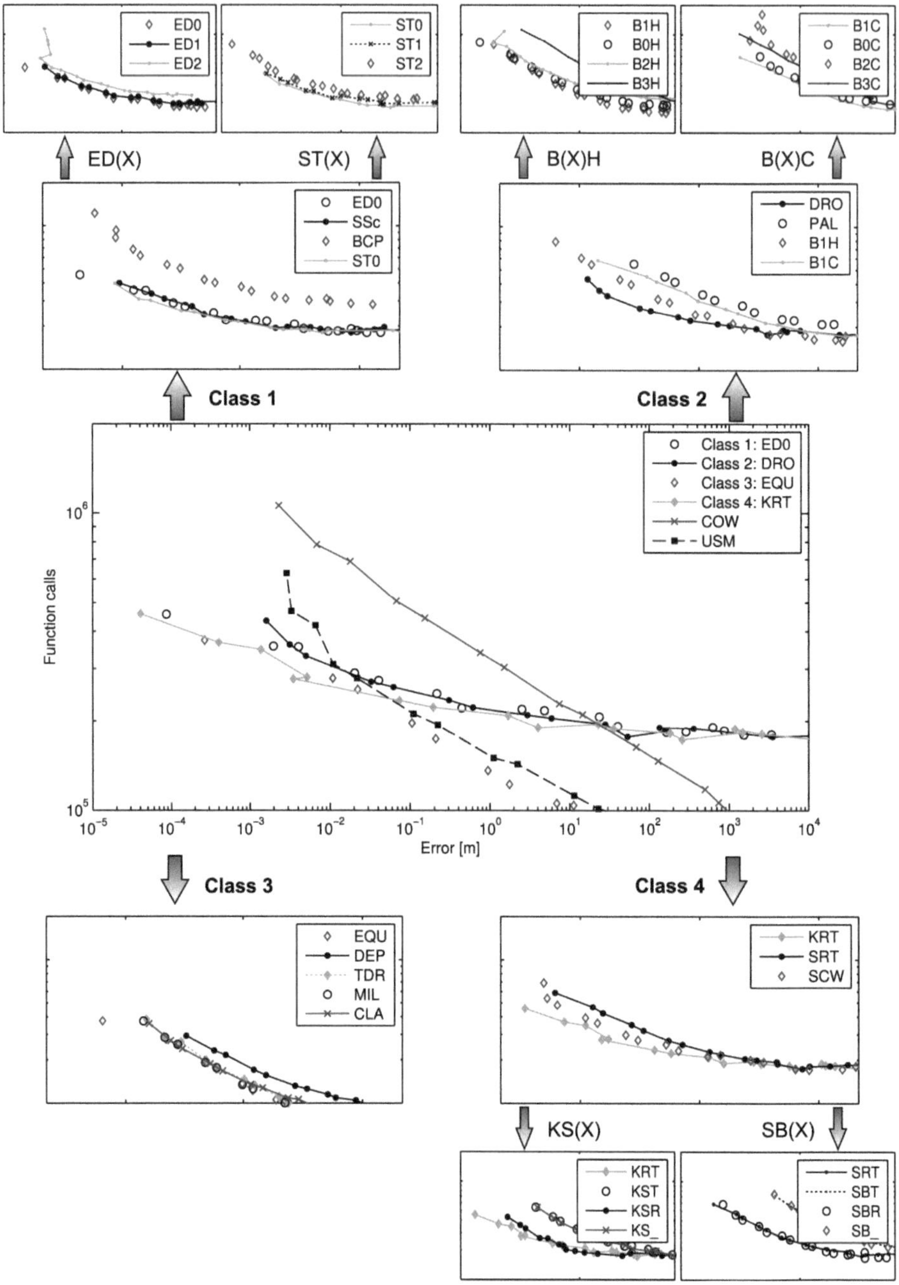

Fig. 6.1: Sequential performance diagram (SPD) for the Stiefel–Scheifele problem integrated with a RKF5(4) scheme.

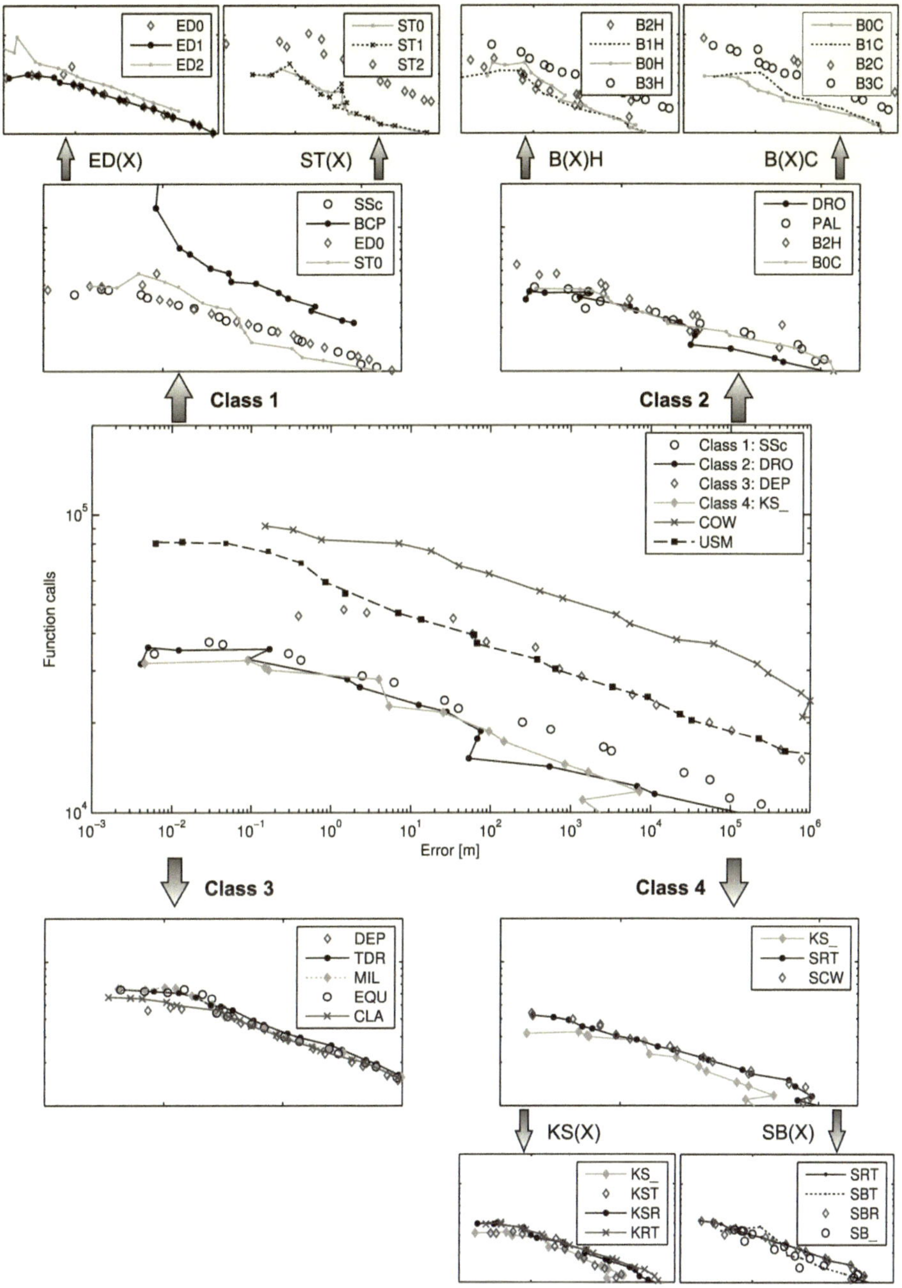

Fig. 6.2: SPD for the Stiefel–Scheifele problem integrated with LSODAR.

fictitious time behaves as an analytic step-size adaption (Sect. 2.2), and has a direct effect on the error-control routines. The KS transformation with redundant equations and a time element (KRT) exhibits the best compromise between maximum accuracy and speed. It is interesting to note that the SPD shows that the integration of the redundant equations (both for the KS and the SB formulations) has a significant effect on the performance, much more significant than introducing a time element.

The SPD for LSODAR is depicted in Fig. 6.2, which follows the same structure as Fig. 6.1. This particular integrator includes a large collection of features that make the algorithm more robust and efficient. For the same level of accuracy it is one order of magnitude faster than the RKF5(4). However, the minimum error that can be reached is about two orders of magnitude larger than the one obtained with the RKF5(4). The limit comes from the minimum admissible tolerance, which is constrained by the machine zero. Disregarding the discontinuities related to the changes in the order of the integrator, LSODAR reduces the differences in performance between formulations in the same family. This is because the integrator is highly flexible and adaptable. In particular, for the KS and SB families integrating redundant equations does not improve the overall performance. The SPD shows that Dromo (DRO) and the KS regularization (KS_) yield the best performance, followed closely by the use of the Stiefel–Scheifele elements (SSc). There is a clear separation between the methods involving a time transformation and those using the physical time as the independent variable (COW, USM, and Class 3 formulations).

In some scenarios, the integration of the equations in Cartesian coordinates is preferred; for instance, when high-order variational equations are propagated together with the state vector, because the dimension of the system grows quickly with the number of variables. Finding the most adequate integrator for a given application can have the same effect as using a regularized formulation instead of Cowell's method. In fact, many research groups focus their attention on the integrator part,

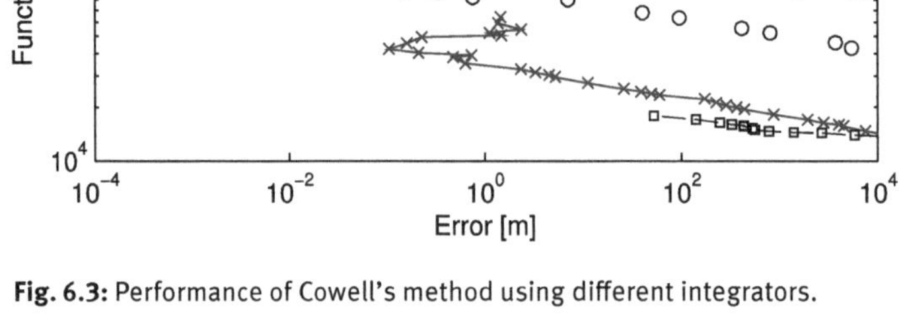

Fig. 6.3: Performance of Cowell's method using different integrators.

rather than on the formulation part.[1] Figure 6.3 depicts the performance of Cowell's method when using different integrators, including LSODAR, RKF5(4), RKF8(7), DoPri853, and Störmer–Cowell of orders 8 and 12. The performance curves presented in this figure reveal that significant differences should be expected depending on the choice of the integration scheme. In particular, Störmer–Cowell is the fastest, and the accuracy is comparable to that of LSODAR. Despite being slower, the methods of the Runge–Kutta family are more accurate in this particular application.

6.2.2 Problem 2

The spacecraft mass is 1954 kg and the area-to-mass ratio is assigned the value $A/m = 0.0767\,\mathrm{m}^2/\mathrm{kg}$. The drag coefficient and the reflectivity are $c_D = 2.0$ and $c_R = 1.2$, respectively. The atmospheric density is retrieved from the MSISE90 model. The Earth gravity field is modeled with a 10×10 expansion, and the orientation of the Earth is computed neglecting the precession and nutation effects. The perturbations from the Sun and the Moon are defined using the DE430 ephemeris. The spacecraft is provided with three ion engines, each generating 165 mN of thrust. The thrust is assumed constant and directed along the velocity vector. The changes in the mass of the spacecraft are neglected. The orbit is propagated for 100 days, which corresponds to roughly 870 revolutions of the spiral. The initial semimajor axis is 10,000 km, and the eccentricity at departure is 0.15.

The performance of the different formulations integrated with the RKF5(4) scheme can be studied in the SPD in Fig. 6.4. Because of the perturbations being small, the use of elements versus coordinates drives the classification of the methods in terms of their performance. Cowell's method is entirely based on coordinates, and exhibits the worst performance. Next, the unified state model (USM) and the KS formulation with redundant equations (KSR) show similar performances. These methods include both coordinates and elements. Finally, the Milankovitch elements (MIL), the Stiefel–Scheifele elements (SSc), and Dromo (DRO) show similar performances, as they are all based on elements. As the eccentricity is small, the difference between using a first-order, a second-order, or no time transformation at all is not important. In this particular example the performance curves of B1H, B2H, B1C, and B2C are out of scale: the time element in this version of the Burdet–Ferrándiz formulation is not well suited for numerical integration with a RKF5(4) scheme.

The SPD using LSODAR is shown in Fig. 6.5. As in Problem 1, the adaptability of the integrator makes the performance of formulations within the same class very similar. In addition, the overall speed of the propagations is improved with respect to the RKF5(4) integrator, with little or no penalties in the accuracy. In this particular case the

1 An example is DIVA (Krogh, 1974), an advanced integrator developed at the Jet Propulsion Laboratory that has been used ever since for generating ephemerides.

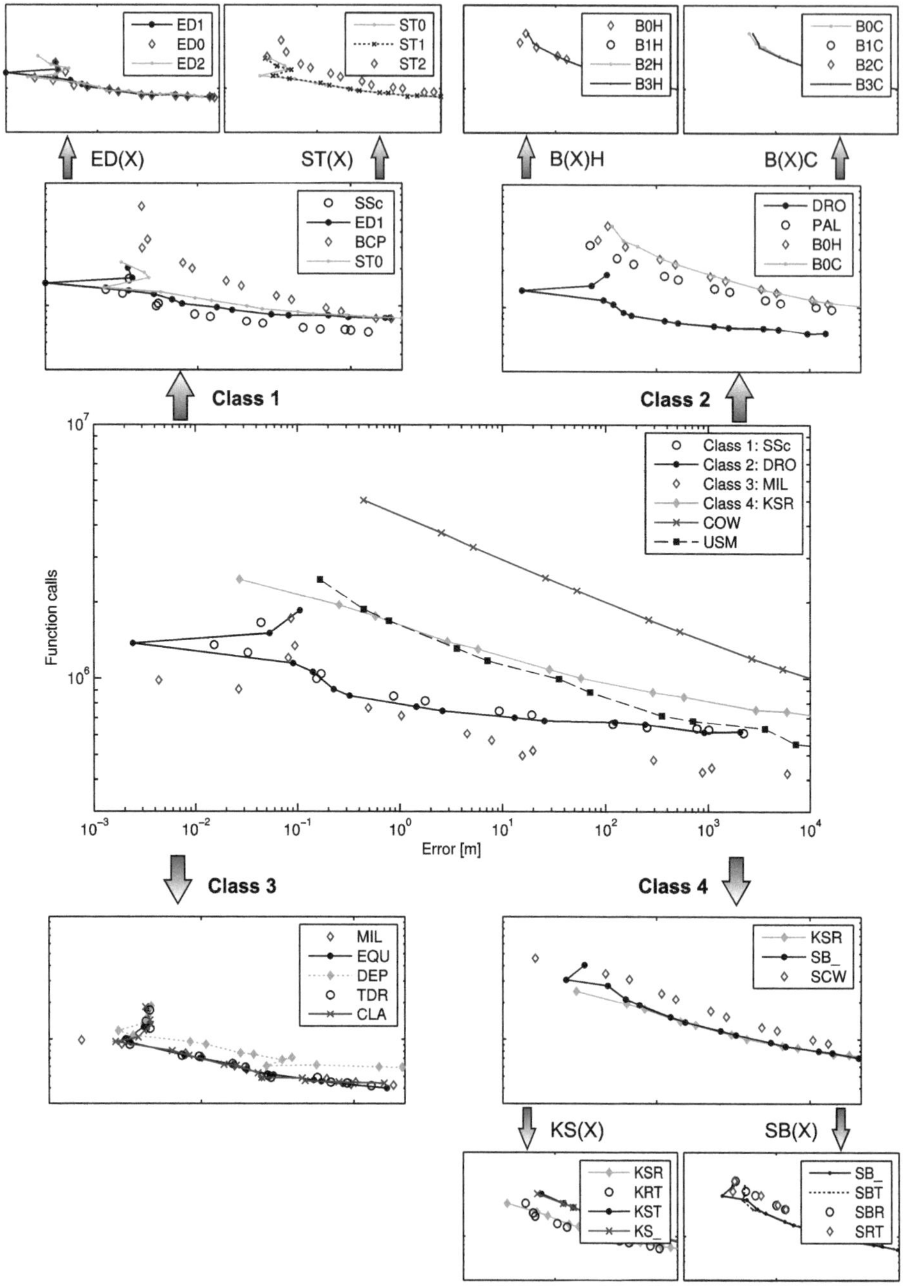

Fig. 6.4: SPD for Problem 2 integrated with RKF5(4).

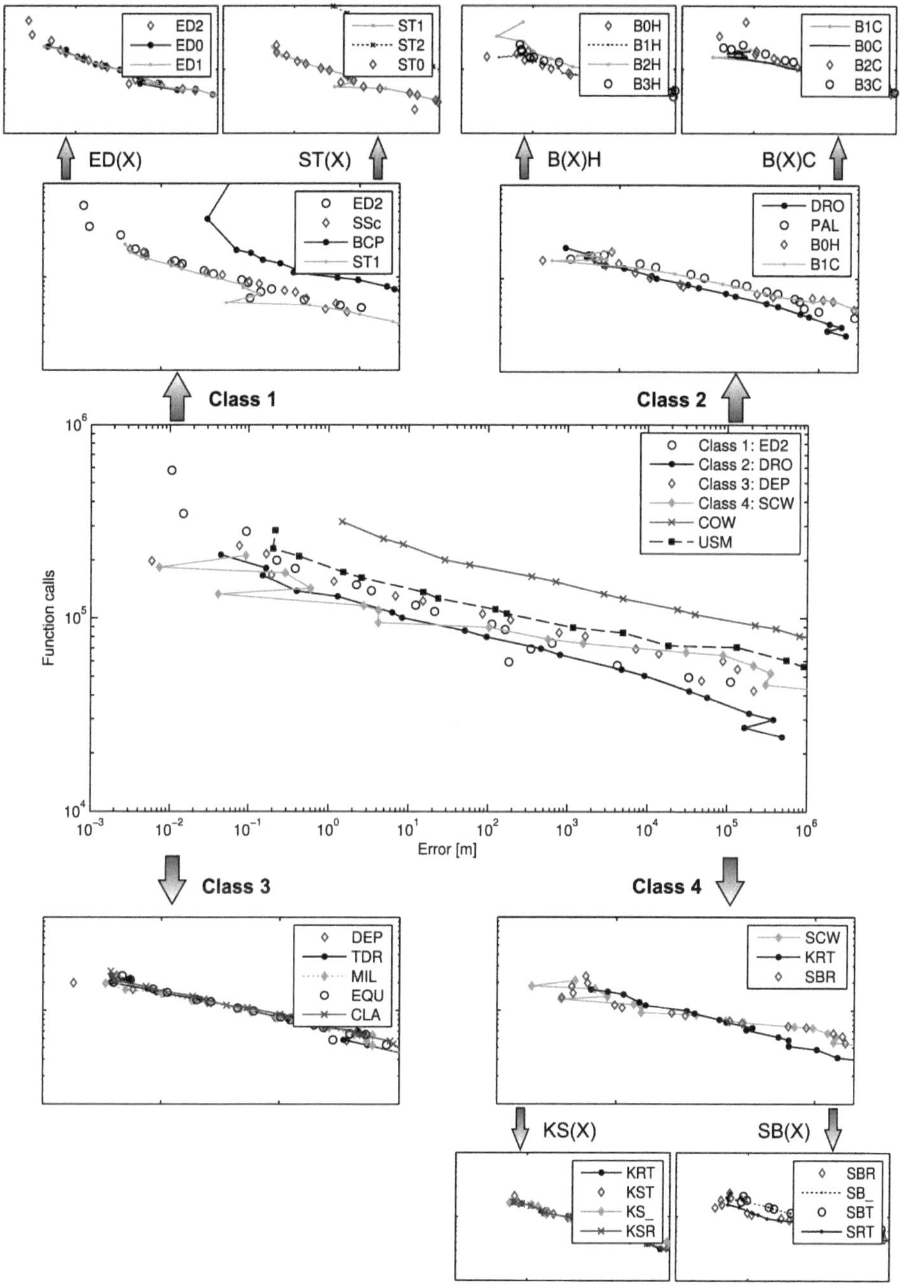

Fig. 6.5: SPD for Problem 2 integrated with LSODAR.

use of elements is as important as the time transformation. Indeed, the performance of Class 3 (elements with no time transformation) and Class 4 methods (coordinates with time transformation) are comparable, and similar to the performance of formulations using both elements and a fictitious time. Cowell's method is the least efficient in this example, and its performance improves significantly when embedding the energy and using a time element (SCW). The unified state model (USM) models the orbit with three elements, a technique that results in significant improvements in the computational efficiency.

6.3 Conclusions

There are many problems in astrodynamics and orbital mechanics that require the integration of millions of trajectories. Typically, the reach of such numerical explorations is limited by the computational resources. Before launching massive campaigns of simulations, it is worth finding which is the best combination of numerical integrator and orbital formulation to solve the problem. In this way, the runtime may be reduced by several orders of magnitude while retaining the same level of accuracy. Even if the computational time is not the driving factor, the right choice of the formulation and integrator may significantly improve the accuracy of the propagations. Because each formulation will perform differently depending on the problem to be integrated, it is important to understand the role that each regularizing and stabilizing technique plays, so the user can decide which formulation to use when facing a particular problem.

The goal of the numerical comparisons presented in this chapter was not to find the best formulation, but to show the effect that the various regularization techniques (introduced in Chap. 2) have on the propagation. Moreover, by grouping the formulations in classes with similar properties, the differences between formulations can be understood in terms of the techniques that they implement.

Part II: **Applications**

"And here there dawned on me the notion that we must admit, in some sense,
a fourth dimension of space for the purpose of calculating with triples...
An electric circuit seemed to close, and a spark flashed forth."

– Sir William Rowan Hamilton

7 The theory of asynchronous relative motion

Linear solutions to relative motion are preferred for many practical applications like, for example, their integration in navigation algorithms or their implementation in onboard computers. In their pioneering work, Clohessy and Wiltshire (1960) arrived at the exact solution to the linear equations that govern the circular rendezvous problem. These equations were originally posed by Laplace (1799, p. 152) and further developed by Hill (1878). They linearized the equations of motion assuming that the relative distance is small compared to the radius of the reference orbit. This hypothesis has been adopted in most of the subsequent studies. Tschauner and Hempel (1964, 1965) addressed the elliptic case and formulated the governing equations of motion. In deriving his theory of the primer vector, Lawden (1954) had published a set of equations of motion that are equivalent to those of the elliptic rendezvous. He provided a semianalytical solution to these equations in his book, which requires the numerical evaluation of an integral (Lawden, 1963, p. 85). Carter (1990, 1998) paid special attention to Lawden's integral and proposed different methods to simplify it. Lawden and Carter's work motivated Yamanaka and Ankersen (2002) to derive a state-transition matrix for the elliptic case that represents the exact solution to the linearized problem. De Vries (1963) formulated the same problem making extensive use of the elliptic elements. He provided an approximate series solution by means of an expansion in powers of the eccentricity. His solution is written in terms of the true anomaly.

Alfriend et al. (2000) and Broucke (2003) solved the linear problem through a geometrical construction. The relative orbit is determined by applying a set of small differences to the orbital elements defining the reference orbit. Alfriend et al. (2000) pioneered this technique, and referred to it as the *geometric method*. It allows one to solve the problem using different formulations and different sets of elements. The variational form of the solution helps to interpret the dynamics of the problem. Schaub (2004) presented a geometrical method based entirely on the use of differences in the classical elements of the leader's orbit, accounting for the effect of the Earth's oblateness. Given this solution, the secular drift due to differences on the semimajor axes can be easily analyzed. Gim and Alfriend (2003, 2005) posed the problem using the set of nonsingular equinoctial elements and published a state-transition matrix including the J_2 perturbation. Casotto (1993) conducted an interesting study of the transformations between differences in the initial state vector and differences in the orbital elements. D'Amico and Montenbruck (2006) introduced the concept of the differential eccentricity/inclination vector. By parameterizing the relative dynamics in terms of the differences in the projections of the eccentricity vector and a combination of the relative inclination with the relative ascending node, they arrived at a compact formulation. The versatility of the formulation was proven by accounting for the effects of both differential drag and J_2. We refer to the work by Lee et al. (2007) for a formal proof of the validity of the solutions obtained through the variational form of

https://doi.org/10.1515/9783110559125-007

the equations of motion. Melton (2000) found an alternative state-transition matrix by means of an expansion in powers of the eccentricity that is written with time as the independent variable. Sullivan et al. (2017) published an exhaustive survey of different formulations.

The solution to the linear equations of motion includes secular terms. Such terms may be canceled if the initial conditions satisfy the so-called energy-matching condition (Alfriend et al., 2009, chap. 4). This condition implies co-orbital motion of leader and follower. Special attention has been paid to how the initial conditions relate to the relative drift between the spacecraft (Gurfil, 2005b; Gurfil and Lara, 2013). In the general case, the linear solution depreciates as time advances and the separation between the spacecraft grows. For sufficiently large relative distances nonlinear effects are no longer negligible and affect the accuracy of the solution. Different approaches to accounting for nonlinear effects can be found in the literature. Gurfil and Kholshevnikov (2006) presented an elegant characterization of the manifolds of nonlinear relative motion. Gim and Alfriend (2003) proposed an ingenious interpretation of the problem to reduce the effects of nonlinearities. They introduced a set of curvilinear coordinates that relate naturally to the geometry of the problem, a technique that was later recovered by Bombardelli et al. (2015). The linear solution remains valid as long as the difference between the semimajor axes is small. Vallado and Alfano (2014) analyzed in detail the transformation between Cartesian and curvilinear coordinates, assessing their accuracy. Bombardelli et al. (2017) derived a compact approximate solution for relative motion between orbits with moderate eccentricity and inclination. Some authors have studied the second-order equations of the circular problem (London, 1963; Karlgaard and Lutze, 2003), deriving solutions through different methods. Vaddi et al. (2003) tackled the eccentric case relying on a number of asymptotic expansions. They provided the required corrections on the initial conditions that yield periodic solutions when considering higher order terms. Sengupta et al. (2007) developed a tensorial representation of the second-order state-transition matrix for the elliptic case, coupled with differential gravitational perturbations.

The exact solution to the unperturbed fully nonlinear problem can be obtained by subtracting the state vectors of the leader and follower spacecraft. Gurfil and Kasdin (2004) proposed a method for obtaining solutions of arbitrary orders, relying on an expansion in Fourier series. In a collection of papers, Condurache and Martinuşi (2007b,c, 2009) wrote the exact solution to the problem by referring the state of the follower spacecraft to the orbital elements defining the reference orbit. A more compact representation in quaternionic form has been presented, which is independent from any geometrical assumptions (Condurache and Martinuşi, 2010). This method was later recovered by Martinuşi and Gurfil (2011) to incorporate different terms of the gravity field of the central body. In a different context, Gurfil and Kasdin (2003) derived the set of epicyclic orbital elements. They arise from the canonical modeling of the problem of co-orbital relative motion and are suitable for accounting for the mutual gravitational attraction between bodies.

The relative state vector is computed at a given time. It represents the difference in position and velocity between two spacecraft, measured at that particular time. This renders the time-synchronous solution. Since regularization deals with time transformations, a new definition of time is introduced. And the question about the synchronism of relative motion arises. Should the relative state vector be synchronized in time, or in fictitious time? This question has been widely discussed in Chap. 2 when exploring the Lyapunov instability and the structural stability of orbital motion. Solutions synchronized in time suffer from Lyapunov instability, whereas solutions synchronized in fictitious time enjoy orbital stability. Thus, we will explore an alternative definition of relative motion, in which the states are computed for the same fictitious time. A time delay will appear, and it will be corrected a posteriori to recover the physical meaning of the solution. The ultimate goal is to improve the accuracy of the dynamical model.

In this chapter, we formulate the key concepts of the novel *theory of asynchronous relative motion*. The theory has several advantages. First, it provides the variational equations for any formulation relying on an independent variable different from time. It greatly simplifies the derivation of the partial derivatives of the orbital elements in the unperturbed case. In the literature there are very few studies about how the variational equations of regularized formulations should be propagated, and the present theory fills that gap. Second, it allows one to introduce nonlinear effects easily in the linear solution to relative motion, leading to important improvements in accuracy. The nonlinear correction is based on the dynamics of the problem and it is computed a posteriori. Third, the proposed theory is a physical concept: it can be applied to any of the existing linear solutions, and it admits the introduction of arbitrary perturbations. It is not restricted to the Keplerian case. The correction depends on the time delay in the asynchronous solution. A general expression for the time delay is provided in this chapter by means of the equinoctial orbital elements. Using this set of elements guarantees that the problem is free of singularities. As representative examples, the method will be applied to the Clohessy–Wiltshire (CW) solution in the circular case, and to the solution from Yamanaka–Ankersen's (YA) state-transition matrix in the elliptic case. The new form of these solutions provides more accurate results both in position and velocity.

First, the concept of synchronism is introduced and the nonlinear correction is derived from the main definitions. A simple example is included to show how the proposed method is applied. Section 7.3 generalizes the theory to any formulation. The explicit form of the time delay is given in terms of the equinoctial orbital elements. The connection with the solution provided by Gim and Alfriend (2005) is discussed. The circular case is addressed in Sect. 7.4 and the corrected form of the CW solution is presented. Special attention is paid to how the nonlinear correction relates to high-order solutions. Section 7.5 includes a number of numerical test cases, both perturbed and unperturbed, where the performance of the corrected solutions is compared with the original methods. The inverse transformations can be found in Appendix D.

It is worth emphasizing that the applications of this theory go beyond spacecraft relative motion. This chapter focuses on this particular topic, but the concepts presented here can be applied to any problem involving the partial derivatives of the states. This includes orbit determination, optimization, the search for periodic orbits, and many others.

7.1 Definition of the problem

The problem of relative motion reduces to describing the relative dynamics of a follower spacecraft, f, with respect to a leader spacecraft, ℓ. Let $\mathcal{L} = \{\mathbf{i}, \mathbf{j}, \mathbf{k}\}$ be the Euler–Hill reference frame, centered at the leader spacecraft and defined by the basis:

$$\mathbf{i} = \frac{\mathbf{r}_\ell}{r_\ell}, \qquad \mathbf{k} = \frac{\mathbf{h}_\ell}{h_\ell}, \qquad \mathbf{j} = \mathbf{k} \times \mathbf{i}$$

where $\mathbf{r}$ is the radius vector and $\mathbf{h}$ denotes the angular momentum. The subscript ℓ refers to the leader spacecraft, while the subscript f refers to the follower spacecraft. Frame $\mathcal{L}$ rotates about the inertial reference $\mathcal{I} = \{\mathbf{i}_\mathcal{I}, \mathbf{j}_\mathcal{I}, \mathbf{k}_\mathcal{I}\}$ with angular velocity:

$$\boldsymbol{\omega}_{\mathcal{L}\mathcal{I}} = \frac{h_\ell}{r_\ell^2}\mathbf{k}$$

The relative position is defined by

$$\delta\mathbf{r} = x\,\mathbf{i} + y\,\mathbf{j} + z\,\mathbf{k}$$

The relative velocity referred to the fixed reference is denoted by $\delta\mathbf{v}$, whereas the relative velocity from the perspective of the rotating frame is $\delta\dot{\mathbf{r}}$. They relate through the inertia terms

$$\delta\dot{\mathbf{r}} = \delta\mathbf{v} - \boldsymbol{\omega}_{\mathcal{L}\mathcal{I}} \times \delta\mathbf{r}$$

and their components in the rotating reference are $\delta\mathbf{v} = v_x\mathbf{i} + v_y\mathbf{j} + v_z\mathbf{k}$ and $\delta\dot{\mathbf{r}} = \dot{x}\,\mathbf{i} + \dot{y}\,\mathbf{j} + \dot{z}\,\mathbf{k}$, respectively. Figure 7.1 depicts the geometry of the problem.

7.2 Synchronism in relative motion

Let $\mathbf{r} \in \mathbb{R}^3$ be the solution to the two-body problem

$$\frac{\mathrm{d}^2\mathbf{r}}{\mathrm{d}t^2} = -\frac{\mu}{r^3}\mathbf{r} + \mathbf{a}_p \tag{7.1}$$

in which $r = \|\mathbf{r}\|$ is the distance to the origin, μ is the gravitational parameter, and $\mathbf{a}_p$ is the perturbing acceleration. Introducing a characteristic length ℓ_c and a characteristic

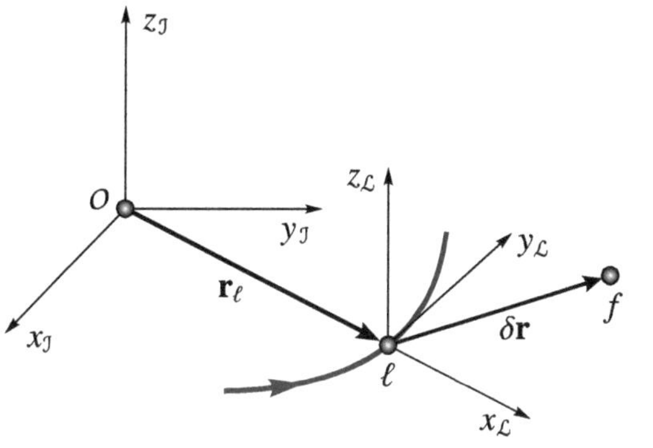

Fig. 7.1: Geometrical definition of the problem.

time $t_c = (\ell_c^3/\mu)^{1/2}$, variables are normalized according to

$$\tilde{\mathbf{r}} = \frac{\mathbf{r}}{\ell_c}, \qquad \tilde{\mathbf{v}} = \frac{\mathbf{v}}{\ell_c/t_c}, \qquad \tilde{t} = \frac{t}{t_c}, \qquad \tilde{\mathbf{a}}_p = \frac{\mathbf{a}_p}{\ell_c/t_c^2}$$

Here $\mathbf{v}$ denotes the velocity. Although ℓ_c can be chosen arbitrarily, in this chapter we shall make it equal to the semimajor axis of the reference orbit, $\ell_c \equiv a_\ell$, in order for the mean motion to simplify to unity, $\tilde{n}_\ell = 1$. To alleviate the notation all variables are assumed to be normalized, and the tilde $\tilde{\square}$ will be dropped from now on.

Equation (7.1) needs to be integrated from the initial conditions

$$\text{at} \quad t = t_0: \quad \mathbf{r}(t_0) = \mathbf{r}_0\,, \quad \mathbf{v}(t_0) = \mathbf{v}_0 \tag{7.2}$$

in order to obtain the state vector $\mathbf{x} = [\mathbf{r}^\top, \mathbf{v}^\top]^\top$, formed by the position and velocity vectors,

$$\mathbf{x} = \mathbf{f}(t, \boldsymbol{\text{œ}}_0) \tag{7.3}$$

in normalized variables. The vector $\boldsymbol{\text{œ}}_0$ denotes the initial values of the set of parameters used to describe the solution, i.e., the six integration constants. If the equations of motion are propagated in Cartesian coordinates $\boldsymbol{\text{œ}}_0$ reduces to the initial conditions in Eq. (7.2). If the orbit is modeled with a set of orbital elements, then $\boldsymbol{\text{œ}}_0$ are the osculating values of the elements at departure. A common choice is to take $\boldsymbol{\text{œ}} = [a, e, i, \omega, \Omega, \vartheta_0]^\top$, corresponding to the classical elements of the orbit. Alternative sets of elements have been discussed in Sect. 2.5. For Keplerian orbits ($\mathbf{a}_p = 0$), the elements $\boldsymbol{\text{œ}}$ are constant and they are solved from the initial conditions (7.2); for perturbed problems, however, they are functions of time.

In the following lines we will first derive the usual solution to linear relative motion, what we shall call the time-synchronous solution. Then, the asynchronous approach will be presented, and we will show how the asynchronous solution can be transformed into the time-synchronous one via a first-order correction of the time delay. Section 7.2.3 contains a simple example of application. Finally, Sect. 7.2.4 discusses how to take advantage of this construction to define a high-order correction that improves the linear solution easily. This last section includes a diagram sketching how the solution is built.

7.2.1 Time-synchronous approach

Let $\mathbf{x}_\ell$ and $\mathbf{x}_f$ be the inertial state vectors of the leader and follower spacecraft, respectively. Both states relate by means of

$$\mathbf{x}_f(t) = \mathbf{x}_\ell(t) + \delta\mathbf{x}(t) \tag{7.4}$$

where $\delta\mathbf{x}$ is the relative state vector. Similarly, if $\boldsymbol{\mathfrak{oe}}_\ell$ denotes the elements of the leader, then the elements of the follower read

$$\boldsymbol{\mathfrak{oe}}_f(t) = \boldsymbol{\mathfrak{oe}}_\ell(t) + \delta\boldsymbol{\mathfrak{oe}}(t) \tag{7.5}$$

and initially it is

$$\boldsymbol{\mathfrak{oe}}_{f,0} = \boldsymbol{\mathfrak{oe}}_{\ell,0} + \delta\boldsymbol{\mathfrak{oe}}_0$$

Vector $\delta\boldsymbol{\mathfrak{oe}}$ denotes the differential elements. In the absence of perturbations the elements will remain constant, to wit, $\delta\boldsymbol{\mathfrak{oe}}(t) \equiv \delta\boldsymbol{\mathfrak{oe}}_0$.

If the relative separation and velocity are small compared to the absolute states, the differential elements will be small compared to the absolute elements too. Under this assumption and considering the solution (7.3), the state vector of the follower can be obtained from the series expansion

$$\mathbf{x}_f = \mathbf{f}(t, \boldsymbol{\mathfrak{oe}}_{\ell,0} + \delta\boldsymbol{\mathfrak{oe}}_0) \approx \mathbf{f}(t, \boldsymbol{\mathfrak{oe}}_{\ell,0}) + \left.\frac{\partial \mathbf{f}_\ell}{\partial \boldsymbol{\mathfrak{oe}}_0}\right|_t \delta\boldsymbol{\mathfrak{oe}}_0 = \mathbf{x}_\ell + \left.\frac{\partial \mathbf{f}_\ell}{\partial \boldsymbol{\mathfrak{oe}}_0}\right|_t \delta\boldsymbol{\mathfrak{oe}}_0$$

Identifying this result with Eq. (7.4) proves that the variations on the leader position and velocity vectors yield the relative state vector, $\delta\mathbf{x}$:

$$\delta\mathbf{x} = \mathsf{J}_t\, \delta\boldsymbol{\mathfrak{oe}}_0\,, \qquad \text{where} \quad \mathsf{J}_t = \left.\frac{\partial \mathbf{f}_\ell}{\partial \boldsymbol{\mathfrak{oe}}_0}\right|_t \tag{7.6}$$

Here J_t represents the Jacobian matrix of the orbit of the leader spacecraft. The problem of relative motion reduces to computing this Jacobian matrix, defined by the partial derivatives of the leader state vector $\mathbf{x}_\ell = \mathbf{f}(t, \boldsymbol{\mathfrak{oe}}_{\ell,0})$ with respect to $\boldsymbol{\mathfrak{oe}}_0$.

Since the leader and follower states need to be computed at the same time, t is kept constant when deriving the Jacobian matrix, which justifies the subscript in J_t. This is the *time-synchronous* (or simply synchronous) Jacobian matrix. In this way, the relative orbit can be constructed geometrically by applying a set of small differences $\delta\boldsymbol{\mathfrak{oe}}_0$ to the reference orbit. When perturbations are present J_t is typically propagated numerically, following the usual procedure summarized in Sect. 6.1.4.

7.2.2 Asynchronous approach

The solution to linear relative motion presented in the previous section is based on the partial derivatives of the reference orbit, computed with constant time. Although this provides the physical solution to the problem, it is subject to the known Lyapunov instability of orbital motion. This section explores an alternative (asynchronous) approach, in which the partial derivatives are obtained keeping constant an independent variable different from time, often equivalent to an angle. Time is no longer constant, and the resulting time delay needs to be corrected in order to obtain the solution from the previous section exactly. As a result, thanks to the Poincaré stability of motion, all secular terms are confined to the time delay.

The problem can be formulated using any set of parameters $\mathbf{oe}$, and the physical time can be replaced by an alternative independent variable, ϕ, which behaves as the fictitious time. The solution can be expressed as

$$\mathbf{x} = \mathbf{g}(\phi, \mathbf{oe}_0) \tag{7.7}$$

$$t = T(\phi, \mathbf{oe}_0) \tag{7.8}$$

The physical time becomes a dependent variable defined by Eq. (7.8), which is equivalent to Kepler's equation in the corresponding variables. In any case, Eq. (7.8) is a bijective function in $\mathbb{R}$, whose inverse provides

$$\phi = \Phi(t, \mathbf{oe}_0)$$

The bijection defines the mapping $T\colon \phi \mapsto t$, and $\Phi\colon t \mapsto \phi$, which might involve an iterative process.

Now the inertial state vector of the follower at ϕ can be written

$$\mathbf{x}_f(\phi) = \mathbf{x}_\ell(\phi) + \delta\mathbf{x}_{\text{asyn}}(\phi)\,. \tag{7.9}$$

The relative state vector in the present asynchronous approach, $\delta\mathbf{x}_{\text{asyn}}$, is different from the one appearing in Eq. (7.4), $\delta\mathbf{x}$, as we will prove in what remains of this section.

Given Eq. (7.7), the state vector of the follower abides by

$$\begin{aligned}\mathbf{x}_f &= \mathbf{g}(\phi, \mathbf{oe}_{\ell,0} + \delta\mathbf{oe}_0) \approx \mathbf{g}(\phi, \mathbf{oe}_{\ell,0}) + \left.\frac{\partial \mathbf{g}_\ell}{\partial \mathbf{oe}_0}\right|_\phi \delta\mathbf{oe}_0 \\ &= \mathbf{x}_\ell + \left.\frac{\partial \mathbf{g}_\ell}{\partial \mathbf{oe}_0}\right|_\phi \delta\mathbf{oe}_0\end{aligned} \tag{7.10}$$

By virtue of Eq. (7.9) the variations on the leader position and velocity vectors result in the relative state vector, $\delta\mathbf{x}_{\text{asyn}}$, given by:

$$\delta\mathbf{x}_{\text{asyn}} = \mathsf{J}_\phi\, \delta\mathbf{oe}_0\,, \qquad \text{where} \quad \mathsf{J}_\phi = \left.\frac{\partial \mathbf{g}_\ell}{\partial \mathbf{oe}_0}\right|_\phi \tag{7.11}$$

Matrix J_ϕ is the Jacobian matrix of (7.7), particularized along the orbit of the leader. Defining the Jacobian matrix requires the partial derivatives of the leader state vector $\mathbf{x}_\ell = \mathbf{g}(\phi, \mathbf{oe}_{\ell,0})$ with respect to $\mathbf{oe}_0$. When perturbations are considered, the Jacobian is computed numerically; see Sect. 6.1.4 for details on the implementation. The states of the leader and follower are computed at the same value of independent variable. Thus, ϕ is kept constant when computing the partials leading to the Jacobian matrix. This provides the *ϕ-synchronous* (or asynchronous) Jacobian matrix, $\mathbf{J}_\phi$. Comparing Eqs. (7.6) and (7.11) clearly shows the difference between the synchronous, $\delta\mathbf{x}$, and asynchronous solutions, $\delta\mathbf{x}_{\text{asyn}}$, which comes from the different definitions of the Jacobian:

$$\delta\mathbf{x} = \mathbf{J}_t\,\delta\boldsymbol{\mathfrak{oe}}_0\,, \qquad \text{and} \qquad \delta\mathbf{x}_{\text{asyn}} = \mathbf{J}_\phi\,\delta\boldsymbol{\mathfrak{oe}}_0$$

In the asynchronous approach, the state vector of the leader $\mathbf{x}_\ell$ and the state vector of the follower $\mathbf{x}_f$ are defined at *different times*, t_ℓ and t_f, which come from Eq. (7.8):

$$t_f = T(\phi, \boldsymbol{\mathfrak{oe}}_{f,0}) \qquad \text{and} \qquad t_\ell = T(\phi, \boldsymbol{\mathfrak{oe}}_{\ell,0})$$

Following the same procedure that yielded Eq. (7.10), it is possible to relate these two different times thanks to

$$t_f = T(\phi, \boldsymbol{\mathfrak{oe}}_{\ell,0} + \delta\boldsymbol{\mathfrak{oe}}_0) \approx T(\phi, \boldsymbol{\mathfrak{oe}}_{\ell,0}) + \left.\frac{\partial T_\ell}{\partial \boldsymbol{\mathfrak{oe}}_0}\right|_\phi \cdot \delta\boldsymbol{\mathfrak{oe}}_0 = t_\ell + \left.\frac{\partial T_\ell}{\partial \boldsymbol{\mathfrak{oe}}_0}\right|_\phi \cdot \delta\boldsymbol{\mathfrak{oe}}_0$$

in which a time delay δt appears, defined as

$$\delta t = t_f - t_\ell = \left.\frac{\partial T_\ell}{\partial \boldsymbol{\mathfrak{oe}}_0}\right|_\phi \cdot \delta\boldsymbol{\mathfrak{oe}}_0 \tag{7.12}$$

This means that the relative state vector $\delta\mathbf{x}_{\text{asyn}}$ determines where the follower will be in a time δt in the future, or where it was $-\delta t$ ago. In general, the partials in Eq. (7.12) will be solved numerically, although in the Keplerian case they can be computed explicitly. It is worth noting that if $\delta\boldsymbol{\mathfrak{oe}}_0$ is small, the time delay will also be small. Figure 7.2 depicts the configuration of the solution. Three dots represent the position of the leader spacecraft at t_0, t_1, and t_2. Similar dots show the position of the follower spacecraft at those same times. The relative states between these pairs of points furnish the synchronous solution to the problem $\delta\mathbf{x}$ (in gray). The positions of the leader spacecraft at t_1 and t_2 correspond to the values ϕ_1 and ϕ_2. The stars mark the points in the follower orbit corresponding to these same values of ϕ_1 and ϕ_2. The resulting state vectors define the asynchronous solution $\delta\mathbf{x}_{\text{asyn}}$, and the time delay connects the synchronous and asynchronous solutions (points and stars, respectively).

Now that the asynchronous solution and the time delay are known, the question of how to recover the synchronous solution arises. That is, how can we refer the solution $\mathbf{x}_f(t_f)$ to the reference time of the leader spacecraft, $\mathbf{x}_f(t_\ell)$? The answer comes from the series expansion

$$\mathbf{x}_f(t_\ell) = \mathbf{x}_f(t_f - \delta t) = \mathbf{x}_{f,\text{asyn}} - \left.\frac{\partial \mathbf{f}_f}{\partial t}\right|_{\text{asyn}} \delta t + \mathcal{O}(\delta t^2) \tag{7.13}$$

The derivatives appearing in the expansion are

$$\begin{aligned} \left.\frac{\partial \mathbf{r}_f}{\partial t}\right|_{\text{asyn}} &= \mathbf{v}_\ell + \delta\mathbf{v}_{\text{asyn}} \\ \left.\frac{\partial \mathbf{v}_f}{\partial t}\right|_{\text{asyn}} &= -\left.\frac{\mathbf{r}_f}{r_f^3}\right|_{\text{asyn}} + \mathbf{a}_{p,f} = -\frac{\mathbf{r}_\ell}{r_\ell^3} + \mathbf{a}_{p,\ell} + \frac{1}{r_\ell^3}[\mathbf{I} - 3(\mathbf{i}\otimes\mathbf{i})]\delta\mathbf{r}_{\text{asyn}} + \delta\mathbf{a}_p \end{aligned} \tag{7.14}$$

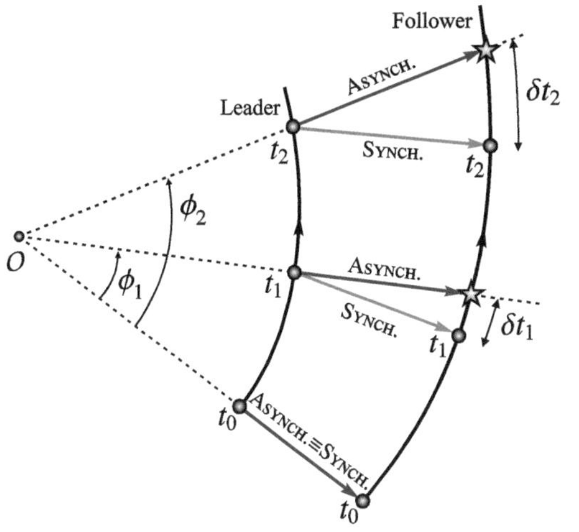

Fig. 7.2: Schematic representation of the synchronous and asynchronous solutions, together with the time delay.

The term between square brackets results in a three-by-three matrix, defined by the identity matrix $\mathbf{I}$ and the dyadic product $\otimes$ (p. 14). The unit vector $\mathbf{i}$, attached to the orbital frame, reads $\mathbf{i} = \mathbf{r}_\ell / r_\ell$.

The perturbations in the leader orbit are $\mathbf{a}_{p,\ell}$, the perturbations affecting the follower orbit will be written $\mathbf{a}_{p,f}$, and $\delta\mathbf{a}_p = \mathbf{a}_{p,f} - \mathbf{a}_{p,\ell}$ denotes the differential perturbations. When the separation is small, $\|\delta\mathbf{x}\|/\|\mathbf{x}_\ell\| \sim \varepsilon \ll 1$, the time delay and the differential perturbations will be small too, $\delta t/t_\ell \sim \|\delta\mathbf{a}_p\|/\|\mathbf{a}_{p,\ell}\| \sim \varepsilon \ll 1$. Subtracting the leader state vector $\mathbf{x}_\ell(t_\ell)$ from the expansion and retaining only first-order terms in ε results in

$$\delta\mathbf{x} = \delta\mathbf{x}_{\text{asyn}} - \frac{\partial \mathbf{f}_\ell}{\partial t}\,\delta t \tag{7.15}$$

Equation (7.15) can be separated to define the relative position and velocity vectors,

$$\delta\mathbf{r} = \delta\mathbf{r}_{\text{asyn}} - \mathbf{v}_\ell\,\delta t \tag{7.16}$$

$$\delta\mathbf{v} = \delta\mathbf{v}_{\text{asyn}} + \left(\frac{\mathbf{r}_\ell}{r_\ell^3} - \mathbf{a}_{p,\ell}\right)\delta t \tag{7.17}$$

The asynchronous solution can be transformed into the true, synchronous solution by applying an intuitive linear correction, which is equivalent to assuming locally rectilinear motion.

We will now derive the relation between the synchronous and the asynchronous Jacobian matrices, J_t and J_ϕ, respectively. Equations (7.3) and (7.7) are two alternative solutions to the same problem, meaning that the following relations hold:

$$\mathbf{f}(t, \boldsymbol{\text{œ}}_0) = \mathbf{g}(\Phi(t, \boldsymbol{\text{œ}}_0), \boldsymbol{\text{œ}}_0) \tag{7.18}$$

$$\mathbf{g}(\phi, \boldsymbol{\text{œ}}_0) = \mathbf{f}(T(\phi, \boldsymbol{\text{œ}}_0), \boldsymbol{\text{œ}}_0) \tag{7.19}$$

The partial derivatives of Eq. (7.18), particularized along the leader orbit, render

$$\left.\frac{\partial \mathbf{f}_\ell}{\partial \mathbf{œ}_0}\right|_t = \left.\frac{\partial \mathbf{g}_\ell}{\partial \mathbf{œ}_0}\right|_\phi + \left.\frac{\partial \mathbf{g}_\ell}{\partial \phi}\frac{\partial \Phi}{\partial \mathbf{œ}_0}\right|_t \implies \mathsf{J}_t = \mathsf{J}_\phi + \left.\frac{\partial \mathbf{g}_\ell}{\partial \phi}\frac{\partial \Phi}{\partial \mathbf{œ}_0}\right|_t \tag{7.20}$$

The first summand appearing in the definition of J_t, J_ϕ, is the asynchronous Jacobian matrix. The second summand is a correction that accounts for the variation in the independent variable ϕ due to the differences $\delta\mathbf{œ}_0$, keeping the time constant. It recovers the time-synchronism. The first term is usually straight-forward to compute. But the second term is cumbersome since Kepler's equation (or equivalent) needs to be differentiated. It is worth noting that this is the approach found in the literature (see for example Broucke, 2003), although no references to the synchronism of the solutions are made.

The partial derivatives of Eq. (7.19) furnish the relation

$$\left.\frac{\partial \mathbf{g}_\ell}{\partial \mathbf{œ}_0}\right|_\phi = \left.\frac{\partial \mathbf{f}_\ell}{\partial \mathbf{œ}_0}\right|_t + \left.\frac{\partial \mathbf{f}_\ell}{\partial t}\frac{\partial T}{\partial \mathbf{œ}_0}\right|_\phi \implies \mathsf{J}_\phi = \mathsf{J}_t + \left.\frac{\partial \mathbf{f}_\ell}{\partial t}\frac{\partial T}{\partial \mathbf{œ}_0}\right|_\phi \tag{7.21}$$

Subtracting Eqs. (7.20) and (7.21) yields the following identity relating the partials of t and ϕ:

$$\left.\frac{\partial \mathbf{g}_\ell}{\partial \phi}\frac{\partial \Phi}{\partial \mathbf{œ}_0}\right|_t = -\left.\frac{\partial \mathbf{f}_\ell}{\partial t}\frac{\partial T}{\partial \mathbf{œ}_0}\right|_\phi \tag{7.22}$$

Indeed, recovering Eq. (7.6) and making use of Eq. (7.20) and the identity (7.22) leads to

$$\begin{aligned}\delta\mathbf{x} = \mathsf{J}_t\,\delta\mathbf{œ}_0 &= \left(\mathsf{J}_\phi + \left.\frac{\partial \mathbf{g}_\ell}{\partial \phi}\frac{\partial \phi}{\partial \mathbf{œ}_0}\right|_t\right)\delta\mathbf{œ}_0 = \left(\mathsf{J}_\phi - \left.\frac{\partial \mathbf{f}_\ell}{\partial t}\frac{\partial T}{\partial \mathbf{œ}_0}\right|_t\right)\delta\mathbf{œ}_0 \\ &= \delta\mathbf{x}_{\text{asyn}} - \frac{\partial \mathbf{f}_\ell}{\partial t}\,\delta t\end{aligned}$$

which is none other than Eq. (7.15). This means that Eq. (7.15) is the exact solution to the linear equations of relative motion. Moreover, the synchronous Jacobian reads

$$\mathsf{J}_t = \mathsf{J}_\phi - \left.\frac{\partial \mathbf{f}_\ell}{\partial t}\frac{\partial T}{\partial \mathbf{œ}_0}\right|_\phi \tag{7.23}$$

and this form of the Jacobian is easier to compute than the one in Eq. (7.21).

The corrected form of the relative velocity referred to the rotating frame $\mathcal{L}$ is

$$\delta\dot{\mathbf{r}} = \delta\dot{\mathbf{r}}_{\text{asyn}} + \delta t\left[\left(\frac{1}{r_\ell} - v_\ell^2\right)\mathsf{I} + (\mathbf{v}_\ell \otimes \mathbf{v}_\ell)\right]\frac{\mathbf{r}_\ell}{r_\ell^2} - \mathbf{a}_{p,\ell}\,\delta t$$

The asynchronous solution is corrected with the instantaneous velocity and acceleration of the leader spacecraft, multiplied by the time delay.

7.2.3 A simple example

This example will demonstrate how to compute the asynchronous solution and the time delay, and how the first-order correction of the time delay transforms the asynchronous solution into the synchronous one. We will compute the radial separation between the leader and follower spacecraft, δr, due to a small difference in the semimajor axis, δa. The spacecraft share the same values for the rest of the orbital elements. No perturbations are considered in this example.

The radial distance is defined in time by the functional relation

$$r = f(t, a) \tag{7.24}$$

Following Eq. (7.6), the solution to the radial motion when only the semimajor axis changes is

$$\delta r = \mathrm{J}_t\, \delta a \equiv \left.\frac{\partial f}{\partial a}\right|_t \delta a \tag{7.25}$$

In practice, the radial motion is easily written in terms of the true anomaly ϑ,

$$r = g(\vartheta, a) \equiv \frac{a(1 - e^2)}{1 + e\cos\vartheta}$$

The true anomaly relates to physical time by means of the inverse Kepler equation

$$\vartheta = \Phi(t, a)$$

In order to differentiate Eq. (7.24) keeping time constant we apply the chain rule

$$\left.\frac{\partial f}{\partial a}\right|_t = \left.\frac{\partial g}{\partial a}\right|_\vartheta + \left.\frac{\partial g}{\partial \vartheta}\frac{\partial \Phi}{\partial a}\right|_t \tag{7.26}$$

as shown in Eq. (7.20). The first term reduces to:

$$\left.\frac{\partial g}{\partial a}\right|_\vartheta = \frac{r}{a}$$

and the first factor in the second term is

$$\frac{\partial g}{\partial \vartheta} = \frac{r^2 e \sin\vartheta}{a(1 - e^2)}$$

The last term requires the derivatives of Kepler's equation with constant time.

Differentiating Kepler's equation,

$$n(t - t_p) = E - e\sin E$$

(in which E is the eccentric anomaly and t_p is the time of periapsis passage) with respect to a yields

$$-\frac{3n}{2a}(t - t_p) = \left.\frac{\partial E}{\partial a}\right|_t (1 - e\cos E) \tag{7.27}$$

The eccentric anomaly relates to the true anomaly by means of:

$$\sin E = \frac{\sqrt{1-e^2}\sin\vartheta}{1+e\cos\vartheta}, \qquad \text{and} \qquad \cos E = \frac{e+\cos\vartheta}{1+e\cos\vartheta}$$

These expressions are differentiated to provide

$$\left.\frac{\partial E}{\partial a}\right|_t = \frac{\sqrt{1-e^2}}{1+e\cos\vartheta}\left.\frac{\partial \Phi}{\partial a}\right|_t$$

Introducing this result into Eq. (7.27) furnishes the relation

$$\left.\frac{\partial \Phi}{\partial a}\right|_t = -\frac{3n}{2a}\frac{(1+e\cos\vartheta)^2}{(1-e^2)^{3/2}}(t-t_p) = -\frac{3}{2}\frac{na\sqrt{1-e^2}}{r^2}(t-t_p)$$

which completes the second term from Eq. (7.26),

$$\frac{\partial g}{\partial \vartheta}\left.\frac{\partial \Phi}{\partial a}\right|_t = -\frac{3}{2}\frac{ne\sin\vartheta}{\sqrt{1-e^2}}(t-t_p) \tag{7.28}$$

Finally, Eq. (7.25) leads to the time-synchronous solution to the problem:

$$\delta r = \left[\frac{r}{a} - \frac{3}{2}\frac{ne\sin\vartheta}{\sqrt{1-e^2}}(t-t_p)\right]\delta a \tag{7.29}$$

This result can also be found in, for example, the work by Broucke (2003), Gim and Alfriend (2003), or Schaub (2004).

Now the same result will be derived applying the theory of asynchronous relative motion, according to Eq. (7.15):

$$\delta r = \delta r_{\text{asyn}} - \frac{\partial f}{\partial t}\delta t \tag{7.30}$$

in which

$$\delta r_{\text{asyn}} = \left.\frac{\partial g}{\partial a}\right|_\vartheta \delta a = \frac{r}{a}\delta a, \qquad \text{and} \qquad \frac{\partial f}{\partial t} \equiv \dot r = \frac{\mu}{h}e\sin\vartheta$$

are the asynchronous solution and the radial velocity, respectively. The time delay is solved from its definition, given in Eq. (7.12),

$$\delta t = \left.\frac{\partial T}{\partial a}\right|_\vartheta \delta a$$

The function $T(\vartheta, a)$ is Kepler's equation:

$$t = T(\vartheta, a) \equiv t_p + [E(\vartheta) - e\sin E(\vartheta)](a^3/\mu)^{1/2}$$

Differentiating with respect to a, keeping the anomaly ϑ constant, leads to

$$\left.\frac{\partial T}{\partial a}\right|_\vartheta = \frac{3}{2a}(t-t_p)$$

and the time delay takes the form

$$\delta t = \frac{3}{2a}(t - t_p)\,\delta a$$

Consequently, Eq. (7.30) becomes

$$\delta r = \frac{r}{a}\delta a - \frac{3}{2}\frac{ne\sin\vartheta}{\sqrt{1-e^2}}(t - t_p)\,\delta a = \left[\frac{r}{a} - \frac{3}{2}\frac{ne\sin\vartheta}{\sqrt{1-e^2}}(t - t_p)\right]\delta a$$

This is the same exact result provided by Eq. (7.29), although this result has been obtained from the first-order correction of the time delay applied to the asynchronous solution. Note also that

$$-\frac{\partial f}{\partial t}\frac{\partial T}{\partial a}\bigg|_{\vartheta} = -\frac{3}{2}\frac{ne\sin\vartheta}{\sqrt{1-e^2}}(t - t_p)$$

equals the expression in Eq. (7.28), which proves the identity (7.22).

7.2.4 Improving the accuracy with second-order corrections of the time delay

The time delay has been corrected thanks to the series expansion in Eq. (7.13). Retaining only first-order terms provides the exact solution to the linear equations of relative motion, Eq. (7.15). If higher order terms are considered, the accuracy of the correction will improve because nonlinearities are introduced in the solution. Extending Eq. (7.13) yields

$$\mathbf{x}_f(t_\ell) = \mathbf{x}_{f,\text{asyn}} - \frac{\partial \mathbf{f}_f}{\partial t}\bigg|_{\text{asyn}}\delta t + \frac{1}{2}\frac{\partial^2 \mathbf{f}_f}{\partial t^2}\bigg|_{\text{asyn}}\delta t^2 + O(\delta t^3) \tag{7.31}$$

Apart from the terms in Eq. (7.14), this expansion also involves

$$\frac{\partial^2 \mathbf{v}_f}{\partial t^2}\bigg|_{\text{asyn}} = -\frac{\mathbf{v}_\ell}{r_\ell^3} + 3\dot{r}_\ell\frac{\mathbf{r}_\ell}{r_\ell^4} + O(\varepsilon) \approx -\frac{1}{r_\ell^3}[\mathbf{I} - 3(\mathbf{i}\otimes\mathbf{i})]\mathbf{v}_\ell$$

The correction is an approximation of the solution, which tries to retain accuracy without complicating the computations excessively. A rigorous definition of the second-order correction of the time delay requires the explicit value of the differential perturbations $\delta\mathbf{a}_p$. Although it can be solved by evaluating the partial derivatives of the perturbing acceleration, this is not a usual feature of orbit propagators, and important modifications in the code would be required. For this reason and for the sake of simplicity, we neglect the contribution of the differential perturbations to the second-order correction of the time delay. It is worth emphasizing that we are only neglecting $\delta\mathbf{a}_p$ when computing this high-order correction, but not in the propagation of the linear solution. The partials leading to the linear solution and the time delay are integrated considering all the perturbations.

As a result, retaining terms up to second order in ε in Eq. (7.31), the relative position and velocity corrected to second order (denoted with a star $\square^\star$) read

$$\delta\mathbf{r}^\star = \delta\mathbf{r}_{\text{asyn}} - (\mathbf{v}_\ell + \delta\mathbf{v}_{\text{asyn}})\delta t - \left(\frac{\mathbf{r}_\ell}{r_\ell^3} - \mathbf{a}_{p,\ell}\right)\frac{\delta t^2}{2} \tag{7.32}$$

$$\delta\mathbf{v}^\star = \delta\mathbf{v}_{\text{asyn}} + \left(\frac{\mathbf{r}_\ell}{r_\ell^3} - \mathbf{a}_{p,\ell}\right)\delta t + \frac{\delta t}{2r_\ell^3}[\mathbf{I} - 3(\mathbf{i}\otimes\mathbf{i})](2\delta\mathbf{r}_{\text{asyn}} - \mathbf{v}_\ell\delta t) \tag{7.33}$$

This technique is a generic concept, not restricted to any specific formulation or perturbation model. In fact, combining Eqs. (7.16–7.17) and (7.32–7.33) shows how the linear solution ($\delta\mathbf{r}$ and $\delta\mathbf{v}$) can be improved by introducing nonlinear terms:

$$\delta\mathbf{r}^\star = \delta\mathbf{r} - \delta\mathbf{v}\,\delta t + \frac{\mathbf{r}_\ell}{2r_\ell^3}\delta t^2 \tag{7.34}$$

$$\delta\mathbf{v}^\star = \delta\mathbf{v} + \frac{\delta t}{2r_\ell^3}\,[\mathbf{I} - 3(\mathbf{i}\otimes\mathbf{i})]\,(2\delta\mathbf{r} + \mathbf{v}_\ell\delta t) \tag{7.35}$$

The only extra variable that we need, apart from the linear solution and the state of the leader, is the time delay. Recall that $\mathbf{i} = \mathbf{r}_\ell/r_\ell$. In a series of papers, Roa and Peláez

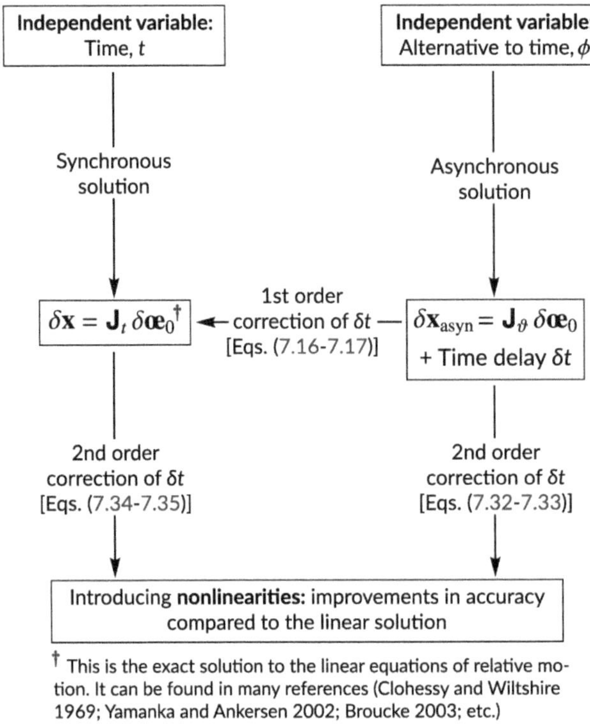

Fig. 7.3: Diagram showing the construction of the solution.

(2015b,f, 2017) and Roa et al. (2015a) arrived at different expressions for the time delay using regularized variables, which will be discussed in the next chapter. In the next section we will derive a simple explicit solution for the time delay using the set of equinoctial elements in the unperturbed case. When perturbations are present, the time delay needs to be propagated numerically. Figure 7.3 depicts the flowchart of the solution. The linear solution to relative motion (7.6) can be obtained with both the usual time-synchronous approach (in line with the references in the introduction) or with the asynchronous approach. In any case, from this solution and when the time delay is known, a corrected, more accurate solution can be built. See Sect. 7.5 for examples of the improvements in accuracy.

7.3 Generalizing the transformation

Consider that the linear, time-synchronous solution to the problem ($\delta\mathbf{r}$ and $\delta\mathbf{v}$) has been obtained using a certain formulation. The formulations discussed in the introduction are good examples. We will now see how the accuracy can be improved thanks to the second-order correction of the time delay. In what remains of the chapter the relative velocity is defined in the rotating reference $\mathcal{L}$, i.e., $\delta\mathbf{x} = [x, y, z, \dot{x}, \dot{y}, \dot{z}]^\top$. Recall that $\delta\mathbf{v} = \delta\dot{\mathbf{r}} + \boldsymbol{\omega}_{\mathcal{L}\mathcal{I}} \times \delta\mathbf{r}$, with $\delta\dot{\mathbf{r}} = \dot{x}\,\mathbf{i} + \dot{y}\,\mathbf{j} + \dot{z}\,\mathbf{k}$. Combining Eqs. (7.34–7.35) yields the corrected velocity in the rotating frame,

$$\delta\dot{\mathbf{r}}^\star = \delta\dot{\mathbf{r}} + \frac{\delta t}{r_\ell^4}\{[(r_\ell - h_\ell^2)\mathbf{I} + h_\ell^2(\mathbf{k}\otimes\mathbf{k})]\delta\mathbf{r} - r_\ell(\mathbf{i}\otimes\mathbf{i})(3\delta\mathbf{r} + \mathbf{v}_\ell\delta t) + r_\ell^2\mathbf{h}_\ell \times \delta\dot{\mathbf{r}}\}$$

As summarized in Fig. 7.3, the linear solution can be improved without the need for computing the asynchronous solution (only the time delay is required). Therefore, this section is devoted to deriving the generic definition of the time delay, so it can be applied to any existing solution directly. No perturbations are considered in this section ($\mathbf{a}_p = 0$, meaning that $\delta\boldsymbol{\alpha} \equiv \delta\boldsymbol{\alpha}_0$). The resulting expression is parameterized with the equinoctial orbital elements, in order to avoid singularities.

7.3.1 The time delay using equinoctial orbital elements

In order to derive a generic nonsingular expression for the time delay, the variational solution to relative motion is obtained using the equinoctial orbital elements. This set of elements, introduced by Brouwer and Clemence (1961), overcomes the singularities of the classical elements for equatorial and circular orbits. Gim and Alfriend (2005) obtained the time-synchronous solution to relative motion in curvilinear coordinates using these elements. The transformation proposed in this section is derived in the usual rectangular coordinates. All variables refer to the leader spacecraft.

Let $\mathbf{q} \in \mathbb{R}^7$ denote the equinoctial elements, ordered as $\mathbf{q} = [a, q_1, q_2, k_1, k_2, \lambda_0, t_p]^\top$, where:

$$q_1 = e\cos(\omega + \Omega)\,, \qquad k_1 = \tan\frac{i}{2}\cos\Omega$$

$$q_2 = e\sin(\omega + \Omega), \qquad k_2 = \tan\frac{i}{2}\sin\Omega$$

and t_p is the time of pericenter passage. The true anomaly ϑ is replaced by the true longitude

$$\lambda = \omega + \Omega + \vartheta$$

Introducing the auxiliary variables $\eta^2 = 1 - q_1^2 - q_2^2$ (equivalent to the angular momentum squared) and

$$s = 1 + q_1\cos\lambda + q_2\sin\lambda\,, \qquad u = q_1\sin\lambda - q_2\cos\lambda$$

the position and velocity vectors read:

$$\mathbf{r} = \frac{\eta^2}{s}\mathbf{i}\,, \qquad \text{and} \qquad \mathbf{v} = \frac{1}{\eta}(u\,\mathbf{i} + s\,\mathbf{j})$$

Recall that both the normalized semimajor axis and the mean motion are equal to one, so that $h \equiv \eta$. The equinoctial reference frame $\mathcal{Q}$ (Broucke and Cefola, 1972) is defined in the inertial reference $\mathcal{J}$ by means of: (i) a rotation about the z-axis of magnitude Ω, which defines the node, (ii) a rotation about the rotated x-axis of magnitude i, (iii) a rotation of magnitude $-\Omega$ about the rotated z-axis. The equinoctial frame is shown in Fig. 7.4.

The rotation $\mathcal{Q} \mapsto \mathcal{J}$ is represented by the matrix

$$\mathsf{P} = \frac{1}{\ell^2}\begin{bmatrix} 1 + k_1^2 - k_2^2, & +2k_1k_2, & +2k_2 \\ +2k_1k_2, & 1 - k_1^2 + k_2^2, & -2k_1 \\ -2k_2, & +2k_1, & 1 - k_1^2 - k_2^2 \end{bmatrix}$$

with $\ell^2 = 1 + k_1^2 + k_2^2$. It defines the transformation

$$\mathbf{r}_{\mathcal{J}} = \mathsf{P}\,\mathbf{r}_{\mathcal{Q}}$$

which maps the coordinates in the equinoctial frame, $\mathbf{r}_{\mathcal{Q}}$, to the coordinates in the inertial frame, $\mathbf{r}_{\mathcal{J}}$. The orbital frame is obtained through a rotation of magnitude λ about the $z_{\mathcal{Q}}$-axis.

The dimension of the relative state vector is extended to include the time delay as the seventh component, resulting in $\delta\mathbf{x}'_{\text{asyn}} = [x, y, z, \dot{x}, \dot{y}, \dot{z}, \delta t]^\top$. For computing the asynchronous solution it is more convenient to use the angular displacement in true longitude as the independent variable:

$$\gamma = \lambda - \lambda_0$$

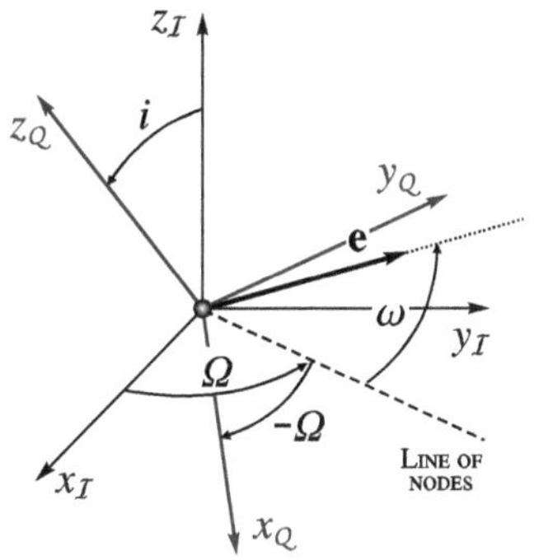

Fig. 7.4: Equinoctial frame.

so the initial condition λ_0 contained in $\mathbf{q}$ can be modified. According to the notation in Sect. 7.2, it is $\phi \equiv \gamma$. The asynchronous solution is

$$\delta\mathbf{x}'_{\text{asyn}}(\gamma) = \mathsf{J}_\gamma(\gamma)\,\delta\mathbf{q}$$

with $\delta\mathbf{œ} \equiv \delta\mathbf{q} = [\delta a, \delta q_1, \delta q_2, \delta k_1, \delta k_2, \delta\lambda_0, \delta t_p]^\top$. In this section $\delta\mathbf{q} \equiv \delta\mathbf{q}_0$. The nonzero elements of the Jacobian J_γ are

$$\begin{aligned}
\mathsf{J}_{\gamma,11} &= r & \mathsf{J}_{\gamma,16} &= \eta^2 u/s \\
\mathsf{J}_{\gamma,41} &= -u/(2\eta) & \mathsf{J}_{\gamma,46} &= (s-1)/\eta \\
\mathsf{J}_{\gamma,51} &= -3s/(2\eta) & \mathsf{J}_{\gamma,56} &= -u/\eta \\
\mathsf{J}_{\gamma,71} &= 3(t-t_p)/2 & \mathsf{J}_{\gamma,76} &= \eta^3/s^2 \\
\mathsf{J}_{\gamma,12} &= -(2q_1 + r\cos\lambda)/s & \mathsf{J}_{\gamma,13} &= -(2q_2 + r\sin\lambda)/s \\
\mathsf{J}_{\gamma,42} &= (uq_1 + \eta^2\sin\lambda)/\eta^3 & \mathsf{J}_{\gamma,43} &= (uq_2 - \eta^2\cos\lambda)/\eta^3 \\
\mathsf{J}_{\gamma,52} &= (3q_1 s + 2\eta^2\cos\lambda)/\eta^3 & \mathsf{J}_{\gamma,53} &= (3q_2 s + 2\eta^2\sin\lambda)/\eta^3 \\
\mathsf{J}_{\gamma,24} &= 2rk_2/\ell^2 & \mathsf{J}_{\gamma,25} &= -2rk_1/\ell^2 \\
\mathsf{J}_{\gamma,34} &= 2r\sin\lambda/\ell^2 & \mathsf{J}_{\gamma,35} &= -2r\cos\lambda/\ell^2 \\
\mathsf{J}_{\gamma,54} &= 2uk_2/(\eta\ell^2) & \mathsf{J}_{\gamma,55} &= -2uk_1/\ell^2 \\
\mathsf{J}_{\gamma,64} &= 2(q_1 + \cos\lambda)/(\eta\ell^2) & \mathsf{J}_{\gamma,65} &= 2(q_2 + \sin\lambda)/(\eta\ell^2) \\
\mathsf{J}_{\gamma,26} &= r & \mathsf{J}_{\gamma,77} &= 1 \\
\mathsf{J}_{\gamma,72} &= \frac{\eta}{s^2 e^2}[\eta^2 q_2 - q_1 u(s+1)] & \mathsf{J}_{\gamma,73} &= -\frac{\eta}{s^2 e^2}[\eta^2 q_1 + q_2 u(s+1)]
\end{aligned} \tag{7.36}$$

The normalized semimajor axis, $a = 1$, has been omitted for brevity. Its determinant is simply:

$$\det(\mathsf{J}_\gamma) = 2\eta^3/(s^2\ell^4)$$

so the matrix is regular even in the circular and equatorial cases. However, the time delay (given by the last row of the matrix, $\mathsf{J}_{\gamma,7i}$) becomes singular for $e \to 0$, due to the form of $\mathsf{J}_{\gamma,72}$ and $\mathsf{J}_{\gamma,73}$. The singularity can be overcome by replacing the second and third columns with a linear combination with the seventh column, which removes e^2 from the denominator. The only terms affected by this combination are those corresponding to the time delay. If col_i denotes the i-th column of matrix $\mathbf{J}_\gamma$, this matrix is transformed according to

$$\mathrm{col}_2 \mapsto \mathrm{col}_2 - \frac{\eta q_2}{e^2}\mathrm{col}_7\,, \qquad \text{and} \qquad \mathrm{col}_3 \mapsto \mathrm{col}_3 + \frac{\eta q_1}{e^2}\mathrm{col}_7$$

As a result, the terms $\mathrm{J}_{\gamma,72}$ and $\mathrm{J}_{\gamma,73}$ become regular,

$$\mathrm{J}_{\gamma,72} = -\frac{\eta}{s^2}[q_2 + (s+1)\sin\lambda], \quad \mathrm{J}_{\gamma,73} = +\frac{\eta}{s^2}[q_1 + (s+1)\cos\lambda] \tag{7.37}$$

The determinant of the matrix is not affected by the linear combination of the columns. The time delay is given explicitly by

$$\delta t = \mathbf{d} \cdot \delta\mathbf{q} \tag{7.38}$$

where vector $\mathbf{d}$ is the last row of matrix $\mathbf{J}_\gamma$. That is,

$$\delta t = \frac{3}{2}(t - t_p)\delta a - \frac{\eta}{s^2}(C_2\delta q_1 - C_1\delta q_2) + \frac{\eta^3}{s^2}\delta\lambda_0 + \delta t_p \tag{7.39}$$

with $C_1 = q_1 + (s+1)\cos\lambda$ and $C_2 = q_2 + (s+1)\sin\lambda$. The time delay introduces a secular term in the solution. This term appears due to a variation on the semimajor axis. In the case of co-orbital motion between leader and follower the semimajor axes coincide and $\delta a = 0$. The solution becomes periodic and the relative orbits are closed.

The linear transformation $\mathbf{J}_\gamma\colon \delta\mathbf{q} \mapsto \delta\mathbf{x}'_{\mathrm{asyn}}$ maps a set of differences on the equinoctial elements to the asynchronous relative state vector and the time delay. According to Fig. 7.3, the asynchronous solution can be corrected to define the synchronous one – Eqs. (7.16–7.17) – and then improved to account for nonlinearities – Eqs. (7.34–7.35). Equations (7.32) and (7.33) can be solved directly to go from the asynchronous solution to the improved one including nonlinearities.

No perturbations are considered in this section, so $\delta\mathbf{q} \equiv \delta\mathbf{q}_0$ is constant. This set of differences can be determined initially by means of the inverse transformation:

$$\delta\mathbf{q} = \mathbf{J}_\gamma^{-1}(0)\,\delta\mathbf{x}'_0$$

It is important to note that the initial value of the time delay is always zero, $\delta t_0 = 0$. The inverse matrix $\mathbf{J}_\gamma^{-1}$ can be found in Appendix D. From this expression, the state-transition matrix

$$\Xi(\gamma) = \mathbf{J}_\gamma(\gamma)\,\mathbf{J}_\gamma^{-1}(0) \tag{7.40}$$

can be constructed, providing the asynchronous relative state vector:

$$\delta\mathbf{x}'_{\mathrm{asyn}}(\gamma) = \Xi(\gamma)\,\delta\mathbf{x}'_0$$

The last element of this matrix product, which involves the state-transition matrix, defines the time delay in terms of the relative initial conditions, and $\delta\mathbf{x}'_0 = [x_0, y_0, z_0, \dot{x}_0, \dot{y}_0, \dot{z}_0, 0]^\top$. It reads:

$$\delta t = \delta t_x x_0 + \delta t_y y_0 + \delta t_{\dot{x}} \dot{x}_0 + \delta t_{\dot{y}} \dot{y}_0 \tag{7.41}$$

Note that the out-of-plane motion does not affect the time delay. The contribution of each term is solved from the matrix product in Eq. (7.40):

$$\begin{aligned}
\delta t_x &= \frac{3}{r_0^2}(s_0 + 1)(t - t_0) + \frac{u_0}{\eta} + \frac{s_0}{\eta s^2}[C_1(2C_{20} - s_0 \sin\lambda_0) \\
&\quad - C_2(2C_{10} - s_0 \cos\lambda_0)] \\
\delta t_y &= -\frac{3u_0}{r_0^2}(t - t_0) + \frac{1}{s^2\eta}[C_1(s_0^2 \cos\lambda_0 - C_{30}) \\
&\quad + C_2(s_0^2 \sin\lambda_0 + C_{40}) + s_0(\eta^2 - s^2)] \\
\delta t_{\dot{x}} &= \frac{3u_0}{\eta}(t - t_0) - \frac{r}{s}[(s + 1)\cos y + (s_0 - 1)] + 2r_0 \\
\delta t_{\dot{y}} &= \frac{3\eta}{r_0}(t - t_0) + \frac{r_0}{s^2}(C_1 C_{20} - C_2 C_{10})
\end{aligned}$$

with

$$C_1 = q_1 + (s + 1)\cos\lambda, \qquad C_2 = q_2 + (s + 1)\sin\lambda$$
$$C_3 = uq_2 + \cos\lambda + q_1, \qquad C_4 = uq_1 - \sin\lambda - q_2$$

These expressions provide the evolution of the time delay from a set of initial conditions. Once the time delay is obtained – evaluating Eq. (7.41) – it can be applied to the asynchronous solution to obtain the synchronous one. In addition, a second-order correction can be computed to improve the accuracy of the linear solution, according to Fig. 7.3.

7.3.2 Connection with the time-synchronous solution

The time-synchronous Jacobian matrix written in terms of the equinoctial elements, already provided by Gim and Alfriend (2003, 2005), can be recovered from matrix $\mathbf{J}_y$. First, let us call $\mathbf{d}$ the first six elements of the last row of $\mathbf{J}_y$. Second, we ignore the last row and last column of $\mathbf{J}_y$ so it reduces to a 6×6 matrix. Then, considering the identity in Eq. (7.23),

$$\mathbf{J}_t = \mathbf{J}_y - \frac{\partial \mathbf{f}_\ell}{\partial t} \otimes \mathbf{d} \tag{7.42}$$

(recall that vector $\mathbf{d}$ has already been introduced in Eq. (7.38) and here $\mathbf{a}_p = 0$) the partials $\partial \mathbf{f}/\partial t$ take the form

$$\frac{\partial \mathbf{f}_\ell}{\partial t} = \begin{bmatrix} \mathbf{v}_\ell \\ \left\{ \left(\frac{1}{r_\ell} - v_\ell^2 \right) \mathsf{I} + (\mathbf{v}_\ell \otimes \mathbf{v}_\ell) \right\} \frac{\mathbf{r}_\ell}{r_\ell^2} \end{bmatrix}$$

Evaluating Eq. (7.42) renders the synchronous Jacobian J_t, which is precisely the one derived by Gim and Alfriend (2003), although they interpreted the solution in curvilinear coordinates. The nonzero elements of the synchronous Jacobian matrix are:

$$\begin{aligned}
\mathsf{J}_{t,11} &= -3u(t-t_p)/(2\eta) + \eta^2/s & \mathsf{J}_{t,21} &= -3s(t-t_p)/(2\eta) \\
\mathsf{J}_{t,41} &= \frac{3s^2}{2\eta^4}(1-s)(t-t_p) - \frac{u}{2\eta} & \mathsf{J}_{t,51} &= \frac{3s^2u}{2\eta^4}(t-t_p) - \frac{3s}{2\eta} \\
\mathsf{J}_{t,12} &= -\cos\lambda & \mathsf{J}_{t,13} &= -\sin\lambda \\
\mathsf{J}_{t,22} &= [q_2 + (s+1)\sin\lambda]/s & \mathsf{J}_{t,23} &= -[q_1 + (s+1)\cos\lambda]/s \\
\mathsf{J}_{t,42} &= s^2\sin\lambda/\eta^3 & \mathsf{J}_{t,43} &= -s^2\cos\lambda/\eta^3 \\
\mathsf{J}_{t,52} &= [sq_1 + (\eta^2+s^2)\cos\lambda]/\eta^3 & \mathsf{J}_{t,53} &= [sq_2 + (\eta^2+s^2)\sin\lambda]/\eta^3 \\
\mathsf{J}_{t,24} &= 2\eta^2 k_2/(\ell^2 s) & \mathsf{J}_{t,25} &= -2\eta^2 k_1/(\ell^2 s) \\
\mathsf{J}_{t,34} &= 2\eta^2\sin\lambda/(\ell^2 s) & \mathsf{J}_{t,35} &= -2\eta^2\cos\lambda/(\ell^2 s) \\
\mathsf{J}_{t,54} &= 2uk_2/(\ell^2\eta) & \mathsf{J}_{t,55} &= -2uk_1/(\ell^2\eta) \\
\mathsf{J}_{t,64} &= 2(q_1+\cos\lambda)/(\ell^2\eta) & \mathsf{J}_{t,65} &= 2(q_2+\sin\lambda)/(\ell^2\eta) \\
\mathsf{J}_{t,26} &= 2\eta^2/s & \mathsf{J}_{t,56} &= 2u/\eta
\end{aligned} \tag{7.43}$$

The determinant

$$\det(\mathsf{J}_t) = 4\eta/\ell^4$$

confirms that there are no singularities in the formulation. The inverse matrix is given in Appendix D. An important difference between matrices J_y and J_t is that the only secular terms in matrix J_y appear in the time delay. However, in matrix J_t there are secular terms in the radial and along-track components of the solution. This is a consequence of the stability properties of orbital motion. The asynchronous solution is based on the geometry of the orbits and, consequently, exploits the Poincaré stability of motion. The effect of Lyapunov instability appears only when considering the time within each orbit, and affects only the time delay.

7.4 The circular case

The simplest case of relative motion is the motion about a circular reference orbit. In this case, the orbital radius of the leader spacecraft r_ℓ is constant and equal to the semimajor axis. The eccentricity is zero and the γ-synchronous Jacobian matrix is denoted by $\mathbf{J}_{\gamma,c}$. The state-transition matrix is obtained from the product $\Xi_c(\gamma) = \mathbf{J}_{\gamma,c}(\gamma)\,\mathbf{J}_{\gamma,c}^{-1}(0)$, resulting in:

$$\Xi_c(\gamma) = \begin{bmatrix} 4-3\cos\gamma & 0 & 0 & \sin\gamma & 2(1-\cos\gamma) & 0 & 0 \\ 0 & 1 & 0 & 0 & 0 & 0 & 0 \\ 0 & 0 & \cos\gamma & 0 & 0 & \sin\gamma & 0 \\ 3\sin\gamma & 0 & 0 & \cos\gamma & 2\sin\gamma & 0 & 0 \\ 6(\cos\gamma-1) & 0 & 0 & -2\sin\gamma & 4\cos\gamma-3 & 0 & 0 \\ 0 & 0 & -\sin\gamma & 0 & 0 & \cos\gamma & 0 \\ 6(\gamma-\sin\gamma) & 0 & 0 & 2(1-\cos\gamma) & 3\gamma-4\sin\gamma & 0 & 1 \end{bmatrix}$$

The angular displacement γ relates to the normalized time t through the linear equation:

$$\gamma = t - t_0$$

because the normalized mean motion equals unity.

From the state-transition matrix it is possible to derive in closed form the asynchronous relative state vector, namely $\delta\mathbf{x}_{\text{asyn}}(\gamma) = \Xi_c(\gamma)\,\delta\mathbf{x}_0$:

$$x_{\text{asyn}}(\gamma) = 2(2x_0 + \dot{y}_0) - (3x_0 + 2\dot{y}_0)\cos\gamma + \dot{x}_0\sin\gamma \tag{7.44}$$

$$y_{\text{asyn}}(\gamma) = y_0 \tag{7.45}$$

$$z_{\text{asyn}}(\gamma) = z_0\cos\gamma + \dot{z}_0\sin\gamma \tag{7.46}$$

$$\dot{x}_{\text{asyn}}(\gamma) = \dot{x}_0\cos\gamma + (3x_0 + 2\dot{y}_0)\sin\gamma \tag{7.47}$$

$$\dot{y}_{\text{asyn}}(\gamma) = -3(2x_0 + \dot{y}_0) + 2(3x_0 + 2\dot{y}_0)\cos\gamma - 2\dot{x}_0\sin\gamma \tag{7.48}$$

$$\dot{z}_{\text{asyn}}(\gamma) = \dot{z}_0\cos\gamma - z_0\sin\gamma \tag{7.49}$$

This system of equations is the asynchronous equivalent to the CW solution. The time delay becomes:

$$\delta t = 2\dot{x}_0 + 3(2x_0 + \dot{y}_0)\gamma - 2\dot{x}_0\cos\gamma - 2(3x_0 + 2\dot{y}_0)\sin\gamma$$

The secular term vanishes when $\dot{y}_0 = -2x_0$ (recall that variables are nondimensional). This is in fact the condition for co-orbital motion from the CW solution (Alfriend et al., 2009, p. 86).

Following the diagram in Fig. 7.3, a first-order correction of the time delay will transform Eqs. (7.44–7.49) into the true CW solution. In the circular case, the linear corrections to the position and velocity, given in Eqs. (7.16–7.17), reduce to

$$\delta\mathbf{r} = \delta\mathbf{r}_{\text{asyn}} - \delta t\,\mathbf{j}\,, \quad \text{and} \quad \delta\dot{\mathbf{r}} = \delta\dot{\mathbf{r}}_{\text{asyn}} + \delta t\,\mathbf{i} - \mathbf{k}\times(-\delta t\,\mathbf{j}) = \delta\dot{\mathbf{r}}_{\text{asyn}}$$

The term $\mathbf{k} \times (-\delta t\,\mathbf{j})$ is the correction to the inertia terms due to defining the velocity in the rotating frame. This factor cancels the first term in the correction of the time delay. Hence, the asynchronous velocity coincides with the synchronous velocity. The only component of the relative position vector that is affected by the correction is the along-track component. The equations corrected up to first order become:

$$
\begin{aligned}
x(\gamma) &= 2(2x_0 + \dot{y}_0) - (3x_0 + 2\dot{y}_0)\cos\gamma + \dot{x}_0 \sin\gamma \\
y(\gamma) &= (y_0 - 2\dot{x}_0) - 3(2x_0 + \dot{y}_0)\gamma + 2\dot{x}_0 \cos\gamma + 2(3x_0 + 2\dot{y}_0)\sin\gamma \\
z(\gamma) &= z_0 \cos\gamma + \dot{z}_0 \sin\gamma
\end{aligned}
$$

$$
\begin{aligned}
\dot{x}(\gamma) &= \dot{x}_0 \cos\gamma + (3x_0 + 2\dot{y}_0)\sin\gamma \\
\dot{y}(\gamma) &= -3(2x_0 + \dot{y}_0) + 2(3x_0 + 2\dot{y}_0)\cos\gamma - 2\dot{x}_0 \sin\gamma \\
\dot{z}(\gamma) &= \dot{z}_0 \cos\gamma - z_0 \sin\gamma
\end{aligned}
$$

These equations are, indeed, the solution to relative motion about a circular reference orbit already provided by Clohessy and Wiltshire (1960).

7.4.1 Second-order correction

The CW solution can be improved using the second-order correction of the time delay introduced in Sect. 7.2.4. According to Eqs. (7.34–7.35), the corrected state vector becomes:

$$
\begin{aligned}
\delta\mathbf{r}^\star &= \delta\mathbf{r} - \frac{\delta t}{2}\left(2\delta\dot{\mathbf{r}}_{\text{asyn}} + 2\mathbf{k} \times \delta\mathbf{r}_{\text{asyn}} - \delta t\,\mathbf{i}\right) \\
\delta\dot{\mathbf{r}}^\star &= \delta\dot{\mathbf{r}} + \delta t\left\{\left[(\mathbf{k}\otimes\mathbf{k}) - 3(\mathbf{i}\otimes\mathbf{i})\right]\delta\mathbf{r}_{\text{asyn}} + \mathbf{k} \times \delta\dot{\mathbf{r}}_{\text{asyn}}\right\}
\end{aligned}
$$

Expanding these expressions yields the explicit form of the improved solution. The in-plane relative motion in normalized variables is given by:

$$x^\star(\gamma) = 2A_1 + \frac{1}{2}A_4A_5 + (\dot{x}_0 - A_4A_2)\sin\gamma - (A_2 + A_4\dot{x}_0)\cos\gamma \tag{7.50}$$

$$\begin{aligned} y^\star(\gamma) = {} & A_8A_5 + y_0 + (\dot{x}_0A_5 - 2A_2A_8)\sin\gamma + A_6^- \sin 2\gamma + 2\dot{x}_0A_2\cos 2\gamma \\ & - (2\dot{x}_0A_8 + A_2A_5)\cos\gamma \end{aligned} \tag{7.51}$$

$$\begin{aligned} \dot{x}^\star(\gamma) = {} & -3A_1A_5 + (A_2A_7 - \dot{x}_0A_5)\sin\gamma + (\dot{x}_0A_7 + A_2A_5)\cos\gamma - A_6^- \sin 2\gamma \\ & - 2\dot{x}_0A_2\cos 2\gamma \end{aligned} \tag{7.52}$$

$$\begin{aligned} \dot{y}^\star(\gamma) = {} & -A_6^+ - 3A_1 + (A_2A_5 - 2\dot{x}_0)\sin\gamma + (2A_2 + \dot{x}_0A_5)\cos\gamma - 2\dot{x}_0A_2\sin 2\gamma \\ & + A_6^- \cos 2\gamma \end{aligned} \tag{7.53}$$

which has been written in terms of the following coefficients:

$$
\begin{aligned}
A_1 &= (2x_0 + \dot{y}_0)\,, & A_2 &= (3x_0 + 2\dot{y}_0)\,, & A_3 &= (y_0 - \dot{x}_0) \\
A_4 &= (2A_3 - 3A_1\gamma)\,, & A_5 &= (2\dot{x}_0 + 3A_1\gamma)\,, & A_6^\pm &= \left(A_2^2 \pm \dot{x}_0^2\right) \\
A_7 &= 1 + 6A_1\,, & A_8 &= A_1 - 1
\end{aligned}
$$

The +/− sign is chosen according to the sign that appears in Eqs. (7.50–7.53).

The out-of-plane motion is governed by the equations:

$$z^{\star}(\gamma) = -B_4^- + B_1 \sin\gamma - B_2 \cos\gamma + B_3^- \sin 2\gamma + B_4^+ \cos 2\gamma \tag{7.54}$$

$$\dot{z}^{\star}(\gamma) = -B_3^+ + B_2 \sin\gamma + B_1 \cos\gamma - B_4^+ \sin 2\gamma + B_3^- \cos 2\gamma \tag{7.55}$$

where

$$B_1 = (z_0 A_5 + \dot{z}_0), \qquad B_2 = (\dot{z}_0 A_5 - z_0)$$

$$B_3^{\pm} = (\dot{z}_0 A_2 \pm z_0 \dot{x}_0), \quad B_4^{\pm} = (z_0 A_2 \pm \dot{x}_0 \dot{z}_0)$$

The time delay is the only source of secular terms. The condition for closing the relative orbits when using the corrected solution is still the same condition that applies to the original CW solution, $\dot{y}_0 = -2x_0$. For $A_1 = 0$ the corrected solution is periodic, just like the CW solution. The out-of-plane motion is now coupled with the in-plane motion.

The expressions provided in this section introduce nonlinear terms to the CW solution in a simple way, which does not involve the second derivatives of the relative state vector. Nevertheless, it should be noted that Eqs. (7.50–7.55) are neither a solution to the second or to the third-order equations of circular relative motion. The third-order equations are (see, for example, Richardson and Mitchell, 2003):

$$\ddot{x} - 3x - 2\dot{y} + \frac{3}{2}(2x^2 - y^2 - z^2) - 2x(2x^2 - 3y^2 - 3z^2) = 0$$

$$\ddot{y} + 2\dot{x} - 3xy + \frac{3y}{2}(4x^2 - y^2 - z^2) = 0$$

$$\ddot{z} + z - 3xz + \frac{3z}{2}(4x^2 - y^2 - z^2) = 0$$

Figure 7.5 compares the accuracy of the improved CW solution – Eqs. (7.50–7.55) – with the numerical integration of the first-, second-, and third-order equations of circular relative motion. Note that the solution to the first-order equations is simply the CW solution. The error is measured as the norm of the difference with respect to the integration of the exact nonlinear equations of relative motion. It is observed that the solutions to the second and third-order equations are more accurate at first, but they diverge from the exact solution after several revolutions. The theory of asynchronous relative motion will eventually provide a more accurate description of the dynamics after sufficient revolutions. No divergence in the corrected solution has been observed. The initial conditions are small to ensure that the linear approach is valid.

7.5 Numerical evaluation

This section evaluates the performance of the new theory of asynchronous relative motion with a series of examples. First, we consider a circular and an elliptic Keplerian orbit. Next, we proceed to the perturbed case and propagate the relative dynamics of spacecraft orbiting the Earth and Jupiter.

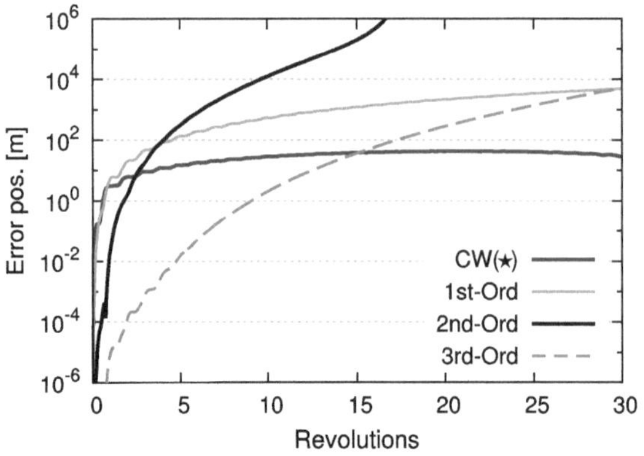

(a) a = 6900 km, $\|\delta \mathbf{r}_0\|$ = 137.5 m, $\|\delta \dot{\mathbf{r}}_0\|$ = 1.1 m/s

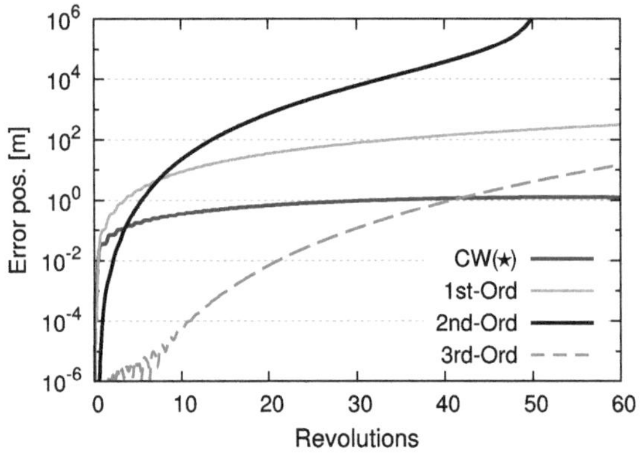

(b) a = 8500 km, $\|\delta \mathbf{r}_0\|$ = 65.6 m, $\|\delta \dot{\mathbf{r}}_0\|$ = 0.1 m/s

Fig. 7.5: Error comparison between the improved CW solution ("CW(⋆)"), and the exact solution to the first- ("1st-Ord"), the second- ("2nd-Ord"), and the third-order ("3rd-Ord") equations of circular relative motion.

The error in determining the relative position at time j, ϵ_j, is measured as:

$$\epsilon_j = \|\delta \mathbf{r}_j - \delta \mathbf{r}_{\text{ref}}\|$$

where $\delta \mathbf{r}_{\text{ref}}$ is the reference solution. To compute this solution the orbits of the leader and the follower are solved separately, then projected in the rotating Euler–Hill frame and subtracted. The same definition of error applies to the relative velocity, $\delta \dot{\mathbf{r}}$.

7.5.1 Keplerian motion

The reference orbits for the two examples of Keplerian orbits are defined in Table 7.1. They are propagated for 15 revolutions. The first example refers to a circular reference orbit, whereas the second corresponds to elliptic relative motion. The initial relative state vector is defined in the rotating reference.

Tab. 7.1: Definition of the reference orbits and relative initial conditions in the $\mathcal{L}$ frame.

Case	a [km]	e	i [deg]	ω [deg]	Ω [deg]	ϑ_0 [deg]
1	8500	0.0	0	10	20	0
2	30000	0.6	70	20	60	50
	x [m]	y [m]	z [m]	$\dot{x}$ [m/s]	$\dot{y}$ [m/s]	$\dot{z}$ [m/s]
1	−30	−30	50	0.0	0.1	0.0
2	−60	−100	1000	0.2	0.2	−1.0

The linear solution to relative motion is computed using the Clohessy–Wiltshire solution in the circular case, the Yamanaka–Ankersen state-transition matrix in the elliptic case, and the variational solution using equinoctial elements (Sect. 7.3.2) is applied to both cases. All these approaches yield the same exact results. Then, the accuracy of these linear solutions will be improved using the second-order correction of the time delay described in Sect. 7.2.4. Finally, the solution in curvilinear coordinates provided by Gim and Alfriend (2005) is also included in the comparison.

Figure 7.6 presents the evolution of the propagation errors both in position and velocity. Figure 7.6a corresponds to the circular case. The nonlinear correction improves the accuracy of the relative position by more than one order of magnitude after 15 revolutions, and of the relative velocity by almost three orders of magnitude. It is interesting to note that the curvilinear approach exhibits similar advantages when propagating the relative position, but when dealing with the velocity it is clearly outperformed by the new theory. In the elliptic case (Fig. 7.6b) the use of curvilinear coordinates does not improve the accuracy of the linear solution noticeably, whereas applying the new nonlinear corrections reduces the error by over one order of magnitude both in position and velocity. The exact values of the final errors can be compared in Table 7.2. The numbers in the table correspond to the errors divided by the magnitude of the relative separation and relative velocity, computed with the exact solution to the problem. In this way, it is easy to evaluate the impact of the error in each case and, more importantly, the significance of the improvements in accuracy achieved with the nonlinear correction of the solutions. All linear approaches yield the same result.

7.5.2 Perturbed motion

Having discussed the accuracy of the method in the Keplerian case, we now consider the effect of generic perturbations on the relative dynamics. The synchronous and the asynchronous Jacobian matrices are propagated numerically together with the time delay: the synchronous solution is computed by direct propagation of the variational equations of orbital motion in Cartesian coordinates; next, integrating the variational equations attached to the equinoctial elements, using $\gamma = \lambda - \lambda_0$ as the independent

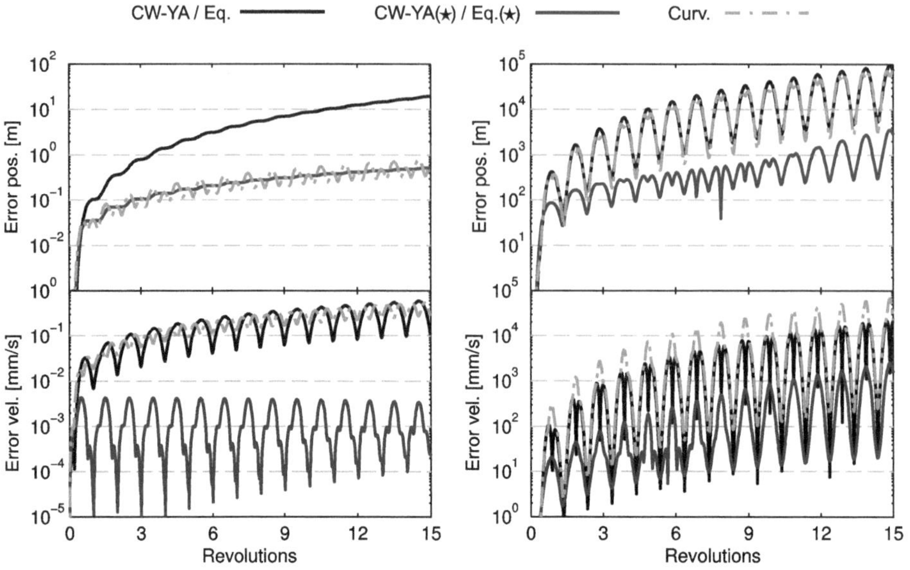

(a) Case 1 (circular) (b) Case 2 (elliptic)

Fig. 7.6: Error in the relative position and velocity using the linear solutions ("CW-YA/Eq."), the improved version of these formulations ("CW-YA(⋆)/Eq.(⋆)"), and the solution using curvilinear coordinates ("Curv").

Tab. 7.2: Final error in position and velocity relative to the magnitude of the relative separation and velocity between the spacecraft. Cases 1 and 2. The star ⋆ denotes the improved solutions.

	Case 1		**Case 2**	
	$\epsilon_f/\|\boldsymbol{\delta}\mathbf{r}_f\|$	$\dot{\epsilon}_f/\|\boldsymbol{\delta}\dot{\mathbf{r}}_f\|$	$\epsilon_f/\|\boldsymbol{\delta}\mathbf{r}_f\|$	$\dot{\epsilon}_f/\|\boldsymbol{\delta}\dot{\mathbf{r}}_f\|$
CW	1.07×10^{-3}	1.04×10^{-3}	–	–
CW (⋆)	2.85×10^{-5}	2.28×10^{-6}	–	–
YA	1.07×10^{-3}	1.04×10^{-3}	4.12×10^{-2}	6.80×10^{-2}
YA (⋆)	2.85×10^{-5}	2.28×10^{-6}	1.63×10^{-3}	5.02×10^{-3}
Eq.	1.07×10^{-3}	1.04×10^{-3}	4.12×10^{-2}	6.80×10^{-2}
Eq. (⋆)	2.85×10^{-5}	2.28×10^{-6}	1.63×10^{-3}	5.02×10^{-3}
Curv.	3.08×10^{-5}	2.85×10^{-3}	3.96×10^{-2}	7.22×10^{-2}

variable, renders the asynchronous solution and the time delay (Chap. 6 has details on the implementation). In these examples, we will focus on the difference between the purely linear solution and the one improved with the second-order correction of the time delay. All computations are carried out with PERFORM.

The first example (Case 3) corresponds to the motion of a spacecraft around the Earth integrated for 12 h (roughly seven revolutions). The model includes luni-solar perturbations, a nonuniform Earth gravity field, atmospheric drag, and solar radiation pressure. The gravitational attraction from the Sun and the Moon are computed using the DE431 ephemeris solution. The Earth gravity field is reduced to a 4×4 grid using the GGM03S model. For such a short propagation the nutation and precession of the Earth is ignored and its orientation is computed with the simplified model from Archinal et al. (2011). The leader and follower spacecraft are assigned an area-to-mass ratio of $1/500\,\mathrm{m}^2/\mathrm{kg}$, a drag coefficient $c_D = 2.0$, and the atmospheric density is modeled with the MSISE90 standard. The reflectivity coefficient for the two spacecraft is the same and equal to $c_R = 1.2$. The values of the osculating elements at the initial epoch (MJD 55198 ET) can be found in Table 7.3.

Tab. 7.3: Definition of the reference orbits and relative initial conditions in the $\mathcal{L}$ frame. The reference orbits are defined in the ICRF/J2000 system, using the Earth mean equator and equinox as the reference plane.

Case	a [km]	e	i [deg]	ω [deg]	Ω [deg]	ϑ_0 [deg]
3	7400	0.100	30.00	90.00	0.00	0.00
4	673,048	0.222	25.55	114.53	−0.03	0.60
	x [m]	y [m]	z [m]	$\dot{x}$ [m/s]	$\dot{y}$ [m/s]	$\dot{z}$ [m/s]
3	0	0	0	1	1	1
4	500,000	−10,000	0	0	10	0

Figure 7.7a displays the error in the propagation of the relative position and velocity in Case 3. The nonlinear correction reduces the error-growth rate and, as a result, the error in the final position is reduced from 1561 m down to 181 m. The oscillations in the error profile come from the orbital frequency, as the magnitude of the relative perturbations is small compared to the differential gravity. The error in velocity is also reduced by one order of magnitude, although the improvements are smaller (in relative terms) than in the relative position. The reason is that for the second-order correction in velocity we neglected the contribution of the differential perturbations.

The second example (Case 4) corresponds to a formation around Jupiter propagated for 13.45 days (approximately four revolutions). More specifically, the leader

Tab. 7.4: Final error in position and velocity relative to the magnitude of the relative separation and velocity between the spacecraft. Cases 3 and 4.

	Case 3		Case 4	
	$\epsilon_f/\|\delta\mathbf{r}_f\|$	$\dot{\epsilon}_f/\|\delta\dot{\mathbf{r}}_f\|$	$\epsilon_f/\|\delta\mathbf{r}_f\|$	$\dot{\epsilon}_f/\|\delta\dot{\mathbf{r}}_f\|$
Linear	1.02×10^{-2}	2.17×10^{-2}	1.33×10^{-1}	2.46×10^{-1}
Corr.	1.17×10^{-3}	2.97×10^{-3}	2.52×10^{-2}	7.53×10^{-2}

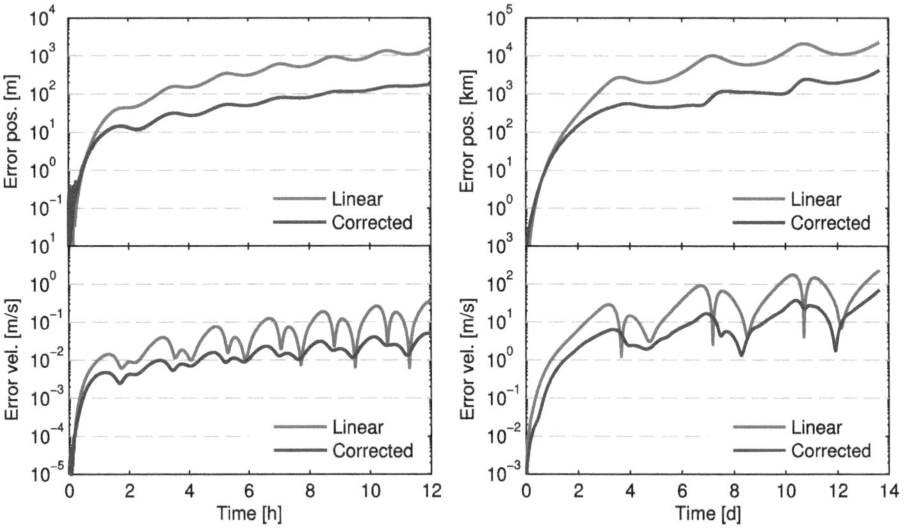

(a) Case 3 (geocentric orbit)

(b) Case 4 (Jupiter-Europa system)

Fig. 7.7: Propagation error for Cases 3 and 4.

spacecraft describes a distant retrograde orbit around Europa. The initial conditions for this orbit were first computed in the circular restricted three-body problem, and they were then refined using a Levenberg–Marquardt algorithm in order to minimize the separation in phase space between the initial crossing of the Poincaré section and the crossing after four revolutions. The initial values (MJD 55255 ET) of the osculating elements with respect to Jupiter are presented in Table 7.3. The model accounts for the gravitational perturbations from the Sun, Io, Europa, Ganymede, and Callisto, using their true ephemeris. The accuracy of the propagation is shown in Fig. 7.7b. The improvements in accuracy introduced by the nonlinear correction are of about one order of magnitude after four revolutions. In this case the relative initial conditions are larger than the ones considered in Case 3, as the relative separation is close to 0.1% of the radius of the reference orbit. This explains the large errors in position observed in this case. Like in Case 3, the improvements in accuracy when propagating the relative velocities are more subtle, because the differential perturbations were omitted in the correction. Table 7.4 presents the exact value of the final propagation errors, divided by the magnitude of the relative separation and velocity at that time. The error in position for Case 3 with the linear solution is about 10% of the relative separation, and the improved solution reduces the error to 1.2%. Similarly, in Case 4 the error in position is reduced from 13% down to 2.5%.

7.6 Conclusions

The theory of asynchronous relative motion has two main applications. First, it provides a systematic technique for propagating the partial derivatives of any formulation using independent variables different from time. Second, it is easy to introduce nonlinear terms in the linear solution to improve its accuracy. As a result, the error in the propagation of the relative orbits between spacecraft is reduced significantly.

The method presented in this chapter is not restricted to any particular formulation or perturbation model. The proposed method can be applied eventually to any of the existing solutions to the linear equations of relative motion, and to the motion under any perturbation. For example, the corrected form of the Clohessy–Wiltshire solution has been provided explicitly. The corrected solution reduces the propagation error by several orders of magnitude. Such improvements may have a positive impact on the design of GNC and filtering algorithms. The fact that one only needs to propagate the reference trajectory makes the method appealing for modeling the dynamics of swarms including hundreds or thousands of spacecraft.

Regularization methods introduce time transformations motivated by the structural stability of orbital motion, trying to escape from the known Lyapunov instability. Similarly, the theory of asynchronous relative motion shows how the secular terms can be confined to the definition of a time delay, without affecting the rest of the variables in the so-called asynchronous solution. The time delay plays a key role in the variational solution and, because of its physical meaning, the mechanism for introducing nonlinearities is simple and does not require the computation of the second partial derivatives of the state vector, nor deriving multiple scale expansions. The improved equations of motion are not the exact solution to the second-order equations of motion. The solution may be, in fact, more accurate than the second and third-order solutions after sufficient revolutions, because it is not affected by the divergence of these equations.

This approach has implications beyond spacecraft relative motion. On the one hand, it shows how to propagate the state-transition matrix (which maps the initial value of the separation in position and velocity up to a given time) when using formulations different from the Cartesian coordinates and, more importantly, when the physical time is no longer the independent variable. This result is useful when the variational equations of motion need to be integrated numerically. Thanks to this theory it is now possible to extend the use of regularized formulations, which behave very well for numerical integration, to the problem of propagating the partial derivatives. On the other hand, it presents a generic method for increasing the accuracy of the propagation of the variational equations through simple corrections applied a posteriori.

8 Universal and regular solutions to relative motion

By definition, the relative motion between two particles is the difference between their respective absolute motions. When their relative separation is small this difference is linearized and the relative state vector is computed directly without the need for solving the dynamics of the two particles independently. The outcome of the linearization process depends on the parameterization of the dynamics, i.e., on the variables chosen to write the analytic solution to the problem.

Cartesian coordinates are the most intuitive variables. This form of the equations of relative motion dates back to Laplace (1799, book II, chap. II, §14). But the problem can be formulated using any set of variables or elements. Finding the most adequate representation of the equations of motion for a given application may provide a deep insight into the dynamics of the problem, simplify its analysis, and even allow one to obtain solutions that would be intractable otherwise. In Chap. 7 we have already presented a more detailed survey of existing solutions.

Carter (1990) found the connection between the equations in Lawden's primer vector theory and the linearization by De Vries. He recovered the integral (Lawden, 1963, pag. 85, eq. 5.51)

$$I(\vartheta) = \int_{\vartheta_0}^{\vartheta} \frac{\mathrm{d}\chi}{(1 + e\cos\chi)^2 \sin^2\chi} \tag{8.1}$$

and succeeded in solving it in closed-form by replacing the true anomaly ϑ by the eccentric anomaly. Yamanaka and Ankersen (2002) advanced on Carter's work and arrived at a simplified state-transition matrix for solving the elliptic rendezvous problem. The resulting state-transition matrix is explicit both in time and true anomaly. Because of how compact the solution is, and the fact that it is valid for moderate eccentricities, this method has been applied in many practical scenarios. Alfriend et al. (2009, chap. 5) presented a detailed overview of a number of state-transition matrices that can be found in the literature, and Casotto (2014) referred to an interesting classification of them.

The formulations discussed in the previous lines and in Chap. 7 become singular as the eccentricity goes to one. Not only are they not valid for describing relative motion about parabolic or hyperbolic orbits, but the performance may also be affected when the elliptic orbit is quasiparabolic. Little attention has been paid to this particular issue. Carter (1990) introduced the eccentric anomaly for solving Lawden's integral in the elliptic case, meaning that the result will only be valid for elliptic orbits. He advanced on this result and was able to find the solution to Eq. (8.1) for the case of open orbits, although $I(\vartheta)$ takes different values depending on the eccentricity of the orbit.

There is a renewed interest in relative motion about hyperbolic orbits. Missions designed to fly by a certain asteroid, comet, or planet and to deploy a landing probe may fall into this category. The concept of the Aldrin cycler (Byrnes et al., 1993) is a

https://doi.org/10.1515/9783110559125-008

good example. Landau and Longuski (2007) proposed a solution to hyperbolic rendezvous based on impulsive maneuvers and geometrical constructions. Carter (1996) analyzed optimal impulsive strategies to rendezvous with highly eccentric orbits.

An important part of the theory of regularization deals precisely with finding suitable parameterizations of orbital motion. On the basis of this theory, we aim for a fully *regular* and *universal* solution to the relative dynamics. First, regular means that there are no singularities; in this scenario the typical singularity corresponds to $e \to 1$, and not so much to $r \to 0$. Second, a solution is said to be universal if there is a unique form of the equations that is valid for any type of reference orbit (circular, elliptic, parabolic, and hyperbolic). Section 2.4.3 already presented some universal solutions to Kepler's problem using regularization. Everhart and Pitkin (1983) explained in detail how introducing the Stumpff functions leads to universal solutions to the two-body problem, no matter the eccentricity of the reference orbit. Danby (1987) discussed exhaustively the role of the Stumpff functions in solving Kepler's equation. Folta and Quinn (1998) investigated the use of universal variables in the problem of relative motion.

In the present chapter we will solve the problem of linear Keplerian relative motion using the Dromo formulation, the Kustaanheimo–Stiefel (KS) transformation, and Sperling–Burdet (SB) regularization. The goal is to provide formulations that are completely free of singularities, and valid for any type of reference orbit. The KS and SB regularizations use the same time transformation, and the solution to relative motion admits a compact tensorial representation. In order to simplify the derivations, the solution is built using the theory of asynchronous relative motion (see Chap. 7 and Roa et al., 2015a; Roa and Peláez, 2014, 2015b, 2017). Thanks to this theory, the linear solution can also be corrected to account for nonlinear terms. This correction depends strongly on the form of the time delay, and consequently on the definition of the time transformation. Thus, it is interesting to compare the performance of the Dromo-based solution (based on $\mathrm{d}t/\mathrm{d}s = r^2/h$) with the performance of the KS and SB solutions (with $\mathrm{d}t/\mathrm{d}s = r$).

Section 8.1 solves the problem of relative motion using the Dromo elements. Next, Sects. 8.2 and 8.3 present the solution using the SB and KS variables. The orbits are assumed to be Keplerian. Practical comments about the fictitious time can be found in Sect. 8.4. Numerical examples showing the accuracy of the methods and the impact of the nonlinear correction appear in Sect. 8.5. An additional result obtained by applying the theory of asynchronous relative motion is a generic method for propagating numerically the variational equations of formulations using an independent variable different from time. This method is presented in Sect. 8.6, and admits any kind of orbital perturbations.

8.1 Relative motion in Dromo variables

Let $\mathbf{q} = [\zeta_1, \zeta_2, \zeta_3, \eta_1, \eta_2, \eta_3, \eta_4, \sigma_0]^\top$ denote the set of Dromo elements. The orbit of the follower spacecraft can be constructed in terms of a set of differences on the elements defining the leader orbit, $\mathbf{q}_f = \mathbf{q}_\ell + \delta\mathbf{q}$. Assuming that the difference between the elements is small (or equivalently that the relative separation is small), the first-order asynchronous solution is given by the linear equation

$$\delta\mathbf{x}_{\text{asyn}} = \mathbf{D}_{\text{asyn}}(\sigma)\,\delta\mathbf{q} \tag{8.2}$$

Here $\delta\mathbf{x} = [\delta x, \delta y, \delta z, \delta v_x, \delta v_y, \delta v_z]^\top$ is the relative state vector from the perspective of the inertial frame. In this chapter no perturbations are considered, meaning that vector $\delta\mathbf{q}$ is constant. The velocity in the rotating Euler–Hill frame $\mathcal{L}$, $\delta\dot{\mathbf{r}}$, follows from

$$\delta\dot{\mathbf{r}} = \delta\mathbf{v} - \boldsymbol{\omega}_{\mathcal{L}\mathcal{I}} \times \delta\mathbf{r}$$

The matrix $\mathbf{D}_{\text{asyn}}(\sigma)$ contains the partial derivatives of the state vector with respect to Dromo variables, and reduces to (Sect. 4.4):

$$\begin{bmatrix}
-\dfrac{\cos\sigma}{\zeta_3^2 s^2} & -\dfrac{\sin\sigma}{\zeta_3^2 s^2} & -\dfrac{2}{\zeta_3^3 s} & 0 & 0 & 0 & 0 & \dfrac{u}{\zeta_3^2 s^2} \\
0 & 0 & 0 & +\dfrac{2\eta_2}{\zeta_3^2 s} & -\dfrac{2\eta_1}{\zeta_3^2 s} & +\dfrac{2\eta_4}{\zeta_3^2 s} & -\dfrac{2\eta_3}{\zeta_3^2 s} & \dfrac{1}{\zeta_3^2 s} \\
0 & 0 & 0 & +\dfrac{2N_{34}^+}{\zeta_3^2 s} & -\dfrac{2N_{43}^-}{\zeta_3^2 s} & -\dfrac{2N_{12}^+}{\zeta_3^2 s} & +\dfrac{2N_{21}^-}{\zeta_3^2 s} & 0 \\
\zeta_3\sin\sigma & -\zeta_3\cos\sigma & u & -2\zeta_3 s\eta_2 & +2\zeta_3 s\eta_1 & -2\zeta_3 s\eta_4 & +2\zeta_3 s\eta_3 & -\zeta_3 \\
\zeta_3\cos\sigma & +\zeta_3\sin\sigma & s & +2\zeta_3 u\eta_2 & -2\zeta_3 u\eta_1 & +2\zeta_3 u\eta_4 & -2\zeta_3 u\eta_3 & 0 \\
0 & 0 & 0 & 2\zeta_3 M_{34}^+ & -2\zeta_3 M_{43}^- & -2\zeta_3 M_{12}^+ & 2\zeta_3 M_{21}^- & 0
\end{bmatrix}$$

We have introduced the auxiliary terms $M_{ij}^{\pm} = uN_{ij}^{\pm} \pm sN_{ji}^{\mp}$ and $N_{ij}^{\pm} = \eta_i\cos\sigma \pm \eta_j\sin\sigma$ to simplify the notation. Recall that $s = 1 + \zeta_1\cos\sigma + \zeta_2\sin\sigma$ and $u = \zeta_1\sin\sigma - \zeta_2\cos\sigma$.

The elements $(\eta_1, \eta_2, \eta_3, \eta_4)$ are the components of a unit quaternion, $|\mathfrak{n}| = 1$. Thus, the set of differences $\delta\eta_i$ are not independent as they must satisfy the constraint (Sect. 4.5)

$$\eta_1\,\delta\eta_1 + \eta_2\,\delta\eta_2 + \eta_3\,\delta\eta_3 + \eta_4\,\delta\eta_4 = 0 \tag{8.3}$$

This means that one of the four components $\delta\eta_i$ cannot be decided arbitrarily. For example, choosing η_4 as the dependent variable $\delta\eta_4$ takes the form

$$\delta\eta_4 = -\frac{1}{\eta_4}(\eta_1\,\delta\eta_1 + \eta_2\,\delta\eta_2 + \eta_3\,\delta\eta_3)$$

If $\eta_4 = 0$, then another component should be chosen as the dependent variable.

The physical time is obtained by integrating the Sundman transformation in Eq. (4.9),

$$t = T(\sigma; \mathbf{q}) \equiv t_0 + \int_{\sigma_0}^{\sigma} \frac{1}{\zeta_3^3 (1 + \zeta_1 \cos\chi + \zeta_2 \sin\chi)^2} \,\mathrm{d}\chi$$

Differentiating this equation with respect to each Dromo element while keeping σ constant yields the time delay

$$\delta t = I_1\,\delta\zeta_1 + I_2\,\delta\zeta_2 + I_3\,\delta\zeta_3 + I_4\,\delta\sigma_0$$

defined in terms of

$$I_1 = \left.\frac{\partial T}{\partial \zeta_1}\right|_{\sigma} = \int_{\sigma_0}^{\sigma} \frac{2\cos\chi}{\zeta_3^3 (1 + \zeta_1 \cos\chi + \zeta_2 \sin\chi)^3} \,\mathrm{d}\chi$$

$$I_2 = \left.\frac{\partial T}{\partial \zeta_2}\right|_{\sigma} = \int_{\sigma_0}^{\sigma} \frac{2\sin\chi}{\zeta_3^3 (1 + \zeta_1 \cos\chi + \zeta_2 \sin\chi)^3} \,\mathrm{d}\chi$$

$$I_3 = \left.\frac{\partial T}{\partial \zeta_3}\right|_{\sigma} = \int_{\sigma_0}^{\sigma} \frac{3}{\zeta_3^4 (1 + \zeta_1 \cos\chi + \zeta_2 \sin\chi)^2} \,\mathrm{d}\chi$$

$$I_4 = \left.\frac{\partial T}{\partial \sigma_0}\right|_{\sigma} = \frac{s_0^2 - s^2}{\zeta_3^3 s_0^2 s^2}$$

In the Keplerian case the eccentricity vector is constant so $\zeta_1 \equiv e$, $\zeta_2 = 0$, and $\delta\zeta_2 = 0$. Therefore

$$\delta t = I_1\,\delta\zeta_1 + I_3\,\delta\zeta_3 + I_4\,\delta\sigma_0 \tag{8.4}$$

Matrix **D** is valid for any type of orbit and it is regular, except for the case when the angular momentum vanishes or there is a collision. A dedicated analysis of the singularities in Dromo can be found in Sect. 4.6. Although these limitations exist, in practice the formation flight scenarios are far from the singularities. The formulation is universal in form although the explicit solutions to the integrals I_i are different depending on the type of orbit. It is worth noting that a similar difficulty was encountered by Carter (1990) when solving Lawden's integral. In fact, in the elliptic and hyperbolic cases it is

$$I_1 = -\frac{3\,\zeta_1 (t - t_0)}{1 - \zeta_1^2} + \frac{1}{\zeta_3^3 \left(1 - \zeta_1^2\right)} \left(\frac{1 + s}{s^2} \sin\sigma - \frac{1 + s_0}{s_0^2} \sin\sigma_0 \right)$$

$$I_3 = \frac{3}{\zeta_3} (t - t_0)$$

whereas in the parabolic case the integrals reduce to

$$I_1 = \frac{1}{10\zeta_3^3}\left[5\left(\tan\frac{\sigma}{2} - \tan\frac{\sigma_0}{2}\right) - \tan^5\frac{\sigma}{2} + \tan^5\frac{\sigma_0}{2}\right]$$

$$I_3 = \frac{1}{2\zeta_3^4}\left[3\left(\tan\frac{\sigma}{2} - \tan\frac{\sigma_0}{2}\right) + \tan^3\frac{\sigma}{2} - \tan^3\frac{\sigma_0}{2}\right]$$

under the assumption $\zeta_2 = 0$.

The transformation from the relative Dromo elements to the relative state vector needs to be inverted at departure in order to compute $\delta\mathbf{q}$ from the initial conditions $\delta\mathbf{x}_0$, defining the linear map

$$\delta\mathbf{q} = \mathbf{Q}(\sigma_0)\,\delta\mathbf{x}_0$$

Combining matrix $\mathbf{Q}$ and matrix $\mathbf{D}_{\text{asyn}}$ provides the asynchronous state-transition matrix

$$\delta\mathbf{x}(\sigma) = \Xi_{\text{asyn}}(\sigma, \sigma_0)\,\delta\mathbf{x}_0$$

with $\Xi_{\text{asyn}}(\sigma, \sigma_0) = \mathbf{D}_{\text{asyn}}(\sigma)\mathbf{Q}(\sigma_0)$. Details on the inverse transformation can be found in Appendix D. The orbit is defined in Dromo variables in terms of the eccentricity vector. When the orbit is circular the eccentricity vector vanishes and the Dromo differential elements $\delta\mathbf{q}$ cannot be initialized properly. This singularity in the transformation translates into matrix $\mathbf{Q}$ becoming singular for circular reference orbits.

Summary

When using the Dromo formulation the solution is obtained according to the following steps:

1. Use matrix $\mathbf{Q}$ (defined in Appendix D) to transform the initial relative state vector to the relative Dromo elements, $\delta\mathbf{q} = \mathbf{Q}\,\delta\mathbf{x}_0$. The velocity needs to be referred to the inertial frame, $\delta\mathbf{v}_0 = \delta\dot{\mathbf{r}}_0 + \boldsymbol{\omega}_{\mathcal{L}\mathcal{I},0} \times \delta\mathbf{r}_0$, with $\boldsymbol{\omega}_{\mathcal{L}\mathcal{I},0} = \mathbf{h}/r_0^2$.
2. Propagate the state vector using Eq. (8.2) and matrix $\mathbf{D}_{\text{asyn}}$.
3. Solve the time delay from Eq. (8.4).
4. Recover the synchronism through the first-order correction defined in Eqs. (7.32–7.33).
5. (Optional) If more accuracy is needed, the linear solution can be refined by introducing nonlinear terms according to Eqs. (7.34–7.35). They give the improved solution $\delta\mathbf{r}^\star$ and $\delta\mathbf{v}^\star$.

8.2 Relative motion in Sperling–Burdet variables

The explicit solution to Kepler's problem via the SB regularization is given in Eq. (2.19). Assuming $\mu = 1$, introducing the auxiliary vector

$$\mathbf{d} = -(\omega^2 \mathbf{r}_0 + \mathbf{e}) \tag{8.5}$$

and considering the modified argument $z = \omega^2 s^2$, the dynamics of the leader spacecraft are governed by the equations

$$\mathbf{r}_\ell(s) = \mathbf{r}_{\ell,0} + s\, r_{\ell,0}\mathfrak{C}_1(z)\mathbf{v}_{\ell,0} + s^2\mathfrak{C}_2(z)\,\mathbf{d}_\ell \tag{8.6}$$

$$\mathbf{r}'_\ell(s) = r_{\ell,0}\mathfrak{C}_0(z)\mathbf{v}_{\ell,0} + s\mathfrak{C}_1(z)\,\mathbf{d}_\ell \tag{8.7}$$

Recall that $\mathbf{v}_\ell = \mathbf{r}'_\ell / r_\ell$. They are defined in terms of the initial conditions $\mathbf{r}_{\ell,0}$ and $\mathbf{v}_{\ell,0}$. The radial distance evolves like

$$r(s) = r_0\mathfrak{C}_0(z) + r'_0 s\,\mathfrak{C}_1(z) + s^2\mathfrak{C}_2(z) \tag{8.8}$$

Similarly, Eq. (2.20) defines the universal form of Kepler's equation:

$$t = T(t; \mathbf{r}, \mathbf{r}') \equiv t_0 + s r_0 \mathfrak{C}_1 + s^2 r'_0 \mathfrak{C}_2 + s^3 \mathfrak{C}_3 \tag{8.9}$$

The solution to the problem of relative motion is obtained by computing how vectors $\mathbf{r}_\ell$ and $\mathbf{v}_\ell$ change given a set of differences in the initial conditions $\delta\mathbf{r}_0$ and $\delta\mathbf{v}_0$. The relative velocity will be solved from the fictitious velocity $\mathbf{r}'_\ell$. In what remains of the chapter all variables are referred to the leader spacecraft; to alleviate the notation the subscript ℓ will be omitted. The asynchronous solution reads

$$\delta\mathbf{x}_{\text{asyn}} = \Xi_{\text{asyn}}(s)\,\delta\mathbf{x}_0$$

Here $\Xi_{\text{asyn}}(s)$ denotes the asynchronous state-transition matrix. It can be written in blocks as:

$$\Xi_{\text{asyn}}(s) = \begin{bmatrix} \nabla_{\mathbf{r}_0}\big|_s \mathbf{r}, & \nabla_{\mathbf{v}_0}\big|_s \mathbf{r} \\ \nabla_{\mathbf{r}_0}\big|_s \mathbf{v}, & \nabla_{\mathbf{v}_0}\big|_s \mathbf{v} \end{bmatrix} \tag{8.10}$$

The blocks correspond to the partial derivatives of the position and velocity vectors with respect to the initial conditions, keeping constant the fictitious time. Each block is computed as the gradient of the corresponding vector field. The resulting rank-two tensors are:

$$\nabla_{\mathbf{r}_0}\big|_s \mathbf{r} = \mathsf{I} + s r_0 (\mathbf{v}_0 \otimes \nabla_{\mathbf{r}_0}\big|_s \mathfrak{C}_1) + s^2\big[(\mathbf{d} \otimes \nabla_{\mathbf{r}_0}\big|_s \mathfrak{C}_2) + \mathfrak{C}_2 \nabla_{\mathbf{r}_0}\big|_s \mathbf{d}\big] \tag{8.11}$$

$$\nabla_{\mathbf{v}_0}\big|_s \mathbf{r} = s r_0 \big[\mathfrak{C}_1 \mathsf{I} + (\mathbf{v}_0 \otimes \nabla_{\mathbf{v}_0}\big|_s \mathfrak{C}_1)\big] + s^2\big[(\mathbf{d} \otimes \nabla_{\mathbf{v}_0}\big|_s \mathfrak{C}_2) + \mathfrak{C}_2 \nabla_{\mathbf{v}_0}\big|_s \mathbf{d}\big] \tag{8.12}$$

$$\nabla_{\mathbf{r}_0}\big|_s \mathbf{v} = \frac{1}{r^2}\big[r \nabla_{\mathbf{r}_0}\big|_s \mathbf{r}' - (\mathbf{r}' \otimes \nabla_{\mathbf{r}_0}\big|_s r)\big] \tag{8.13}$$

$$\nabla_{\mathbf{v}_0}\big|_s \mathbf{v} = \frac{1}{r^2}\big[r \nabla_{\mathbf{v}_0}\big|_s \mathbf{r}' - (\mathbf{r}' \otimes \nabla_{\mathbf{v}_0}\big|_s r)\big] \tag{8.14}$$

The gradients of the Stumpff functions are obtained attending to the derivation rules (C.3–C.4) established in Appendix C:[1]

$$\nabla_{\mathbf{r}_0}\big|_s \mathfrak{C}_k = (\partial_\omega \mathfrak{C}_k)\,\nabla_{\mathbf{r}_0}\big|_s \omega = \frac{s^2}{r_0^3}\mathfrak{C}^*_{k+2}\mathbf{r}_0$$

$$\nabla_{\mathbf{v}_0}\big|_s \mathfrak{C}_k = (\partial_\omega \mathfrak{C}_k)\,\nabla_{\mathbf{v}_0}\big|_s \omega = s^2\mathfrak{C}^*_{k+2}\mathbf{v}_0$$

Equations (8.11–8.12) then become:

$$\nabla_{\mathbf{r}_0}\big|_s \mathbf{r} = \mathbf{I} + \frac{s^3}{r_0^2}\mathfrak{C}^*_3(\mathbf{v}_0 \otimes \mathbf{r}_0) + \frac{s^4}{r_0^3}\mathfrak{C}^*_4(\mathbf{d} \otimes \mathbf{r}_0) + s^2\mathfrak{C}_2\nabla_{\mathbf{r}_0}\big|_s \mathbf{d} \tag{8.15}$$

$$\nabla_{\mathbf{v}_0}\big|_s \mathbf{r} = s r_0 \mathfrak{C}_1 \mathbf{I} + s^3 r_0 \mathfrak{C}^*_3(\mathbf{v}_0 \otimes \mathbf{v}_0) + s^4\mathfrak{C}^*_4(\mathbf{d} \otimes \mathbf{v}_0) + s^2\mathfrak{C}_2\nabla_{\mathbf{v}_0}\big|_s \mathbf{d} \tag{8.16}$$

The gradients of vector **d** are required. Differentiating Eq. (8.5) renders

$$\nabla_{\mathbf{r}_0}\big|_s \mathbf{d} = -\omega^2\mathbf{I} + \frac{2}{r_0^3}(\mathbf{r}_0 \otimes \mathbf{r}_0) - \nabla_{\mathbf{r}_0}\big|_s \mathbf{e}$$

$$\nabla_{\mathbf{v}_0}\big|_s \mathbf{d} = 2(\mathbf{r}_0 \otimes \mathbf{v}_0) - \nabla_{\mathbf{v}_0}\big|_s \mathbf{e}$$

The gradients of the eccentricity vector are obtained directly from its definition:

$$\nabla_{\mathbf{r}_0}\big|_s \mathbf{e} = \left(v_0^2 - \frac{1}{r_0}\right)\mathbf{I} + \frac{1}{r_0^3}(\mathbf{r}_0 \otimes \mathbf{r}_0) - \mathbf{v}_0 \otimes \mathbf{v}_0$$

$$\nabla_{\mathbf{v}_0}\big|_s \mathbf{e} = 2(\mathbf{r}_0 \otimes \mathbf{v}_0) - \mathbf{v}_0 \otimes \mathbf{r}_0 - r_0'\mathbf{I}$$

Recall that $r_0' = (\mathbf{r}_0 \cdot \mathbf{v}_0)$. At this point the solution is completely referred to the initial conditions $\mathbf{r}_0$ and $\mathbf{v}_0$. The gradients of vector **d** reduce to:

$$\nabla_{\mathbf{r}_0}\big|_s \mathbf{d} = -\frac{1}{r_0^3}[r_0^2\mathbf{I} - (\mathbf{r}_0 \otimes \mathbf{r}_0)] + \mathbf{v}_0 \otimes \mathbf{v}_0\,, \quad \nabla_{\mathbf{v}_0}\big|_s \mathbf{d} = r_0'\mathbf{I} + \mathbf{v}_0 \otimes \mathbf{r}_0$$

Considering the dyadic products

$$\mathbf{d} \otimes \mathbf{r}_0 = \left(r_0'\mathbf{v}_0 - \frac{\mathbf{r}_0}{r_0}\right) \otimes \mathbf{r}_0 \quad \text{and} \quad \mathbf{d} \otimes \mathbf{v}_0 = \left(r_0'\mathbf{v}_0 - \frac{\mathbf{r}_0}{r_0}\right) \otimes \mathbf{v}_0$$

1 We abbreviate the notation of partial derivatives according to

$$\frac{\partial}{\partial y} \equiv \partial_y$$

in order to simplify the notation.

Equations (8.15–8.16) transform into:

$$\nabla_{\mathbf{r}_0}\big|_s \mathbf{r} = \mathbf{I} + \frac{s}{r_0^4}\left[r_0^2(r_0\mathfrak{C}_1 + s^2\mathfrak{C}_3^*)\mathbf{v}_0 + r_0 s\mathfrak{C}_2\mathbf{r}_0 + s^3\mathfrak{C}_4^*(r_0 r_0'\mathbf{v}_0 - \mathbf{r}_0)\right] \otimes \mathbf{r}_0$$
$$- \mathfrak{C}_2\frac{s^2}{r_0}[\mathbf{I} - r_0(\mathbf{v}_0 \otimes \mathbf{v}_0)]$$
$$\nabla_{\mathbf{v}_0}\big|_s \mathbf{r} = s r_0\mathfrak{C}_1\mathbf{I} + s^2\mathfrak{C}_2\left[r_0'\mathbf{I} + (\mathbf{v}_0 \otimes \mathbf{r}_0)\right] - \frac{s^4}{r_0}\mathfrak{C}_4^*(\mathbf{r}_0 - r_0 r_0'\mathbf{v}_0) \otimes \mathbf{v}_0$$
$$+ s^3 r_0\mathfrak{C}_3^*(\mathbf{v}_0 \otimes \mathbf{v}_0)$$

The first two blocks of the state-transition matrix $\mathbf{M}_{\text{asyn}}(s)$ are given by the initial conditions and the fictitious time.

The gradients of the velocity vector require the gradients of both the fictitious velocity vector $\mathbf{r}'$ and the radial distance r. From the solution to the radial distance $r(s)$ – Eq. (8.8) –, it follows that:

$$\nabla_{\mathbf{r}_0}\big|_s r = s\mathfrak{C}_1\mathbf{v}_0 + \frac{1}{r_0^3}(r_0^2\mathfrak{C}_0 + s^2 r_0\mathfrak{C}_1 + s^3 r_0'\mathfrak{C}_3^* + s^4\mathfrak{C}_4^*)\mathbf{r}_0$$
$$\nabla_{\mathbf{v}_0}\big|_s r = s\mathfrak{C}_1\mathbf{r}_0 + s^2(r_0\mathfrak{C}_1 + r_0' s\mathfrak{C}_3^* + s^2\mathfrak{C}_4^*)\mathbf{v}_0$$

These equations are already referred to $\mathbf{r}_0$ and $\mathbf{v}_0$ and involve no additional terms. Computing the partial derivatives of Eq. (8.7) renders

$$\nabla_{\mathbf{r}_0}\big|_s \mathbf{r}' = \frac{1}{r_0^4}\left[s r_0\mathfrak{C}_1\mathbf{r}_0 + r_0^2(r_0\mathfrak{C}_0 + s^2\mathfrak{C}_1)\mathbf{v}_0 - s^3\mathfrak{C}_3^*(\mathbf{r}_0 - r_0 r_0'\mathbf{v}_0)\right] \otimes \mathbf{r}_0$$
$$+ \frac{s}{r_0}\mathfrak{C}_1\left[r_0(\mathbf{v}_0 \otimes \mathbf{v}_0) - \mathbf{I}\right]$$
$$\nabla_{\mathbf{v}_0}\big|_s \mathbf{r}' = r_0\mathfrak{C}_0\mathbf{I} + s\mathfrak{C}_1\left[r_0'\mathbf{I} + \mathbf{v}_0 \otimes (\mathbf{r}_0 + r_0 s\mathbf{v}_0)\right] - \frac{s^3}{r_0}\mathfrak{C}_3^*(\mathbf{r}_0 - r_0 r_0'\mathbf{v}_0) \otimes \mathbf{v}_0$$

These results complete the gradients of the velocity. The four blocks in matrix $\mathbf{M}_{\text{asyn}}$, given in Eqs. (8.11–8.14), take the form:

$$\nabla_{\mathbf{r}_0}\big|_s \mathbf{r} = \mathbf{I} + \frac{s}{r_0^4}\left[r_0(r_0 a_{13}^+ + s^3 r_0'\mathfrak{C}_4^*)\mathbf{v}_0 + s a_{24}^-\mathbf{r}_0\right] \otimes \mathbf{r}_0$$
$$- \mathfrak{C}_2\frac{s^2}{r_0}[\mathbf{I} - r_0(\mathbf{v}_0 \otimes \mathbf{v}_0)] \tag{8.17}$$

$$\nabla_{\mathbf{v}_0}\big|_s \mathbf{r} = s b_{12}^+\mathbf{I} + s^2\mathfrak{C}_2(\mathbf{v}_0 \otimes \mathbf{r}_0) - \frac{s^4}{r_0}\mathfrak{C}_4^*(\mathbf{r}_0 - r_0 r_0'\mathbf{v}_0) \otimes \mathbf{v}_0$$
$$+ s^3 r_0\mathfrak{C}_3^*(\mathbf{v}_0 \otimes \mathbf{v}_0) \tag{8.18}$$

$$
\begin{aligned}
r^2\nabla_{\mathbf{r}_0}\big|_s\mathbf{v} = & -\frac{sr}{r_0}\mathfrak{C}_1\mathbf{I} + \frac{s^2}{r_0}(\mathfrak{C}_1^2\mathbf{r}_0 + r_0 s\mathfrak{C}_1\mathfrak{C}_2\mathbf{v}_0)\otimes\mathbf{v}_0 \\
& + \frac{s}{r_0^4}\{[r_0 b_{01}^+\mathfrak{C}_1 + s^2 a_{14}^+\mathfrak{C}_1 + a_{02}^+ a_{13}^-]\mathbf{r}_0 \\
& + r_0 s[r_0^2\mathfrak{C}_0\mathfrak{C}_2 + s^2(b_{13}^+\mathfrak{C}_2 - b_{01}^+\mathfrak{C}_4^*)]\mathbf{v}_0\}\otimes\mathbf{r}_0
\end{aligned}
\tag{8.19}
$$

$$
\begin{aligned}
r^2\nabla_{\mathbf{v}_0}\big|_s\mathbf{v} = & \, r b_{01}^+\mathbf{I} + \frac{s^2}{r_0}\left(\mathfrak{C}_1^2\mathbf{r}_0 + r_0 s\mathfrak{C}_1\mathfrak{C}_2\mathbf{v}_0\right)\otimes\mathbf{r}_0 + \frac{s^3}{r_0}[(a_{14}^+\mathfrak{C}_1 - a_{02}^+\mathfrak{C}_3^*)\mathbf{r}_0 \\
& + r_0 s(b_{13}^+\mathfrak{C}_2 - b_{01}^+\mathfrak{C}_4^*)\mathbf{v}_0]\otimes\mathbf{v}_0
\end{aligned}
\tag{8.20}
$$

having introduced the auxiliary terms:

$$
a_{ij}^{\pm} = r_0\mathfrak{C}_i \pm s^2\mathfrak{C}_j^* \quad \text{and} \quad b_{ij}^{\pm} = r_0\mathfrak{C}_i \pm s r_0'\mathfrak{C}_j^*
$$

Recall that r, r', $\mathbf{r}$, and $\mathbf{v}$ refer to the leader spacecraft.

The physical time is given explicitly by Eq. (8.9), the generalized form of Kepler's equation. The gradients of the time with respect to the initial state vector are obtained by deriving this equation, and result in

$$
\begin{aligned}
\nabla_{\mathbf{r}_0}\big|_s T &= \frac{s}{r_0^3}(r_0 a_{13}^+ + s^3 r_0'\mathfrak{C}_4^* + s^4\mathfrak{C}_5^*)\mathbf{r}_0 + s^2\mathfrak{C}_2\mathbf{v}_0 \\
\nabla_{\mathbf{v}_0}\big|_s T &= s^2\mathfrak{C}_2\mathbf{r}_0 + s^3(r_0\mathfrak{C}_3^* + s\mathfrak{C}_4^* r_0' + s^2\mathfrak{C}_5^*)\mathbf{v}_0
\end{aligned}
$$

The time delay δt reduces to

$$
\delta t = (\nabla_{\mathbf{r}_0}\big|_s T\cdot\delta\mathbf{r}_0) + (\nabla_{\mathbf{v}_0}\big|_s T\cdot\delta\mathbf{v}_0)
\tag{8.21}
$$

and it is completely determined by the equations given above. Having derived the asynchronous state-transition matrix and the time delay, the solution is complete.

Summary

The method presented in this section is summarized as follows:

1. Given the initial relative state vector $\delta\mathbf{r}_0$ and $\delta\mathbf{v}_0$, compute the gradients of the state vector, defined in Eqs. (8.17–8.20). The velocity needs to be referred to the inertial frame, $\delta\mathbf{v}_0 = \delta\dot{\mathbf{r}}_0 + \boldsymbol{\omega}_{\mathcal{LI},0}\times\delta\mathbf{r}_0$, with $\boldsymbol{\omega}_{\mathcal{LI},0} = \mathbf{h}/r_0^2$.
2. Build the asynchronous state-transition matrix $\Xi_{\text{asyn}}(s)$ defined in Eq. (8.10) and find the asynchronous relative state vector, $\delta\mathbf{x}_{\text{asyn}} = \Xi_{\text{asyn}}(s)\delta\mathbf{x}_0$.
3. Solve the time delay from Eq. (8.21).
4. Recover the synchronism through the first-order correction defined in Eqs. (7.32–7.33).
5. (Optional) If more accuracy is needed, the linear solution can be refined by introducing nonlinear terms according to Eqs. (7.34–7.35). They give the improved solution $\delta\mathbf{r}^\star$ and $\delta\mathbf{v}^\star$.

8.3 Relative motion in Kustaanheimo–Stiefel variables

In KS language the position and velocity vectors are written in terms of the coordinates $\mathbf{u}$ and $\mathbf{u}'$ in $\mathbb{R}^4$. Combining $[\mathbf{u}^\top, \mathbf{u}'^\top]$ into vector $\mathbf{y}$, the relative state vector $\delta\mathbf{x}$ can be defined in terms of $\delta\mathbf{y}$ thanks to

$$\delta\mathbf{x}_{\mathrm{asyn}}(s) = \mathbf{M}_{\mathrm{asyn}}(s)\,\delta\mathbf{y}(s) \tag{8.22}$$

The relative state vector in KS space $\delta\mathbf{y}$ evolves in fictitious time, meaning that

$$\delta\mathbf{y}(s) = \mathbf{N}_{\mathrm{asyn}}(s)\,\delta\mathbf{y}_0 \tag{8.23}$$

Finally, vector $\delta\mathbf{y}_0$ relates to the initial conditions by means of

$$\delta\mathbf{y}_0 = \mathbf{T}(\mathbf{x}_0)\,\delta\mathbf{x}_0$$

Matrix $\mathbf{T}(\mathbf{x}_0)$ is solved from the linear form of the Hopf fibration in Appendix D. It connects the relative state vector in Cartesian space with the relative state vector in KS space.

Differentiating Eq. (3.6), $\mathbf{r} = \mathbf{L}(\mathbf{u})\,\mathbf{u}$, provides the relative position vector:[2]

$$\delta\mathbf{r} = \delta\mathbf{L}(\mathbf{u})\,\mathbf{u} + \mathbf{L}(\mathbf{u})\,\delta\mathbf{u} = \mathbf{L}(\delta\mathbf{u})\,\mathbf{u} + \mathbf{L}(\mathbf{u})\,\delta\mathbf{u} = 2\mathbf{L}(\mathbf{u})\,\delta\mathbf{u} \tag{8.24}$$

The fact that the fourth component of $\delta\mathbf{r}$ is zero proves that the bilinear relation $\ell(\mathbf{u}, \delta\mathbf{u}) = 0$ holds, which motivates the last simplification. Equation (8.24) defines only the relative position vector. The relative velocity results in

$$\delta\mathbf{v} = \frac{2}{r}\left[\mathbf{L}(\mathbf{u}')\,\delta\mathbf{u} + \mathbf{L}(\mathbf{u})\,\delta\mathbf{u}'\right] - \frac{4}{r^2}(\mathbf{u}\cdot\delta\mathbf{u})\,\mathbf{L}(\mathbf{u})\,\mathbf{u}' \tag{8.25}$$

Equations (8.24) and (8.25) can be written in matrix form in order to define $\mathbf{M}_{\mathrm{asyn}}$,

$$\mathbf{M}_{\mathrm{asyn}}(s) = \begin{bmatrix} 2\mathbf{L}(\mathbf{u}), & \mathbf{0}_4 \\ \dfrac{2}{r}\mathbf{L}(\mathbf{u}'), & \dfrac{2}{r^2}\left\{r\mathbf{L}(\mathbf{u}) - 2[\{\mathbf{L}(\mathbf{u})\,\mathbf{u}'\}\otimes\mathbf{u}]\right\} \end{bmatrix} \tag{8.26}$$

which has already been introduced in Eq. (8.22). Matrix $\mathbf{0}_4$ is the zero-matrix of dimension four.

Next, we focus on matrix $\mathbf{N}_{\mathrm{asyn}}(s)$ – Eq. (8.23) –. In the Keplerian case Eq. (3.31) can be solved explicitly to provide

$$\mathbf{u}(s) = \mathbf{u}_0\,\mathfrak{D}_0(z) + \quad s\,\mathbf{u}_0'\mathfrak{D}_1(z) \tag{8.27}$$

$$\mathbf{u}'(s) = \mathbf{u}_0'\,\mathfrak{D}_0(z) - \psi^2 s\,\mathbf{u}_0\mathfrak{D}_1(z) \tag{8.28}$$

2 We have relaxed the notation by denoting $\mathbf{r}$ and $\mathbf{v}$ the extensions to $\mathbb{R}^4$ of the position and velocity vectors, $[x, y, z, 0]^\top$ and $[v_x, v_y, v_z, 0]^\top$, respectively.

with $\psi^2 = -\mathcal{E}/2$ and having introduced the modified Stumpff functions $\mathfrak{D}_k(z) = \mathfrak{C}_k(z/4) \equiv \mathfrak{D}_k$. This is the solution to Kepler's problem in KS variables. The $\mathfrak{C}_k$ and $\mathfrak{D}_k$ functions relate through the half-angle formulas, given in Appendix C. Computing the partial derivatives of these equations yields

$$\mathbf{N}_{\text{asyn}}(s) = \begin{bmatrix} \nabla_{\mathbf{u}_0}\big|_s \mathbf{u}, & \nabla_{\mathbf{u}'_0}\big|_s \mathbf{u} \\ \nabla_{\mathbf{u}_0}\big|_s \mathbf{u}', & \nabla_{\mathbf{u}'_0}\big|_s \mathbf{u}' \end{bmatrix} \tag{8.29}$$

Kriz (1978) also relied on the partial derivatives of $\mathbf{u}$ and $\mathbf{u}'$ with respect to $\mathbf{u}_0$ and $\mathbf{u}'_0$ when solving the perturbed two-point boundary value problem using the KS transformation. In the following lines a new derivation is presented in tensorial form. See Shefer (2007) and the references therein for an overview of some works on this topic that appeared in the Russian literature.

Applying the gradients $\nabla_{\mathbf{u}_0}, \nabla_{\mathbf{u}'_0} : \mathbb{R}^4 \to \mathbb{R}^4 \times \mathbb{R}^4$ to Eqs. (8.27–8.28) yields:

$$\begin{aligned}
\nabla_{\mathbf{u}_0}\big|_s \mathbf{u} &= \mathfrak{D}_0\, \mathbf{I}_4 + \mathbf{u}_0 \otimes \nabla_{\mathbf{u}_0}\big|_s \mathfrak{D}_0 + s\left(\mathbf{u}'_0 \otimes \nabla_{\mathbf{u}_0}\big|_s \mathfrak{D}_1\right) \\
\nabla_{\mathbf{u}'_0}\big|_s \mathbf{u} &= s\mathfrak{D}_1\, \mathbf{I}_4 + \mathbf{u}_0 \otimes \nabla_{\mathbf{u}'_0}\big|_s \mathfrak{D}_0 + s\left(\mathbf{u}'_0 \otimes \nabla_{\mathbf{u}'_0}\big|_s \mathfrak{D}_1\right) \\
\nabla_{\mathbf{u}_0}\big|_s \mathbf{u}' &= -\psi^2 s\mathfrak{D}_1\, \mathbf{I}_4 + \mathbf{u}'_0 \otimes \nabla_{\mathbf{u}_0}\big|_s \mathfrak{D}_0 - \psi^2 s\left(\mathbf{u}_0 \otimes \nabla_{\mathbf{u}_0}\big|_s \mathfrak{D}_1\right) \\
\nabla_{\mathbf{u}'_0}\big|_s \mathbf{u}' &= \mathfrak{D}_0\, \mathbf{I}_4 + \mathbf{u}'_0 \otimes \nabla_{\mathbf{u}'_0}\big|_s \mathfrak{D}_0 - \psi^2 s\left(\mathbf{u}_0 \otimes \nabla_{\mathbf{u}'_0}\big|_s \mathfrak{D}_1\right)
\end{aligned}$$

From the properties of the Stumpff functions it follows that:

$$\begin{aligned}
\nabla_{\mathbf{u}_0}\big|_s \mathfrak{D}_0 &= \partial_\psi\, \mathfrak{D}_0\, \nabla_{\mathbf{u}_0}\big|_s \psi = s^2 \frac{\psi}{r_0}\, \mathfrak{D}_1\, \mathbf{u}_0\,, \\
\nabla_{\mathbf{u}_0}\big|_s \mathfrak{D}_1 &= s^3 \frac{\psi}{r_0}\, \mathfrak{D}_3^*\, \mathbf{u}_0 \\
\nabla_{\mathbf{u}'_0}\big|_s \mathfrak{D}_0 &= \partial_\psi \mathfrak{D}_0\, \nabla_{\mathbf{u}'_0}\big|_s \psi = \frac{s^2}{r_0}\, \mathfrak{D}_1\, \mathbf{u}'_0 \\
\nabla_{\mathbf{u}'_0}\big|_s \mathfrak{D}_1 &= \frac{s^3}{r_0}\, \mathfrak{D}_3^*\, \mathbf{u}'_0
\end{aligned}$$

The gradients of vectors $\mathbf{u}$ and $\mathbf{u}'$ then become

$$\nabla_{\mathbf{u}_0}\big|_s \mathbf{u} = \mathfrak{D}_0\, \mathbf{I}_4 + \psi^2 \frac{s^2}{r_0}(\mathfrak{D}_1 \mathbf{u}_0 + s\mathfrak{D}_3^* \mathbf{u}'_0) \otimes \mathbf{u}_0 \tag{8.30}$$

$$\nabla_{\mathbf{u}'_0}\big|_s \mathbf{u} = s\mathfrak{D}_1\, \mathbf{I}_4 + \frac{s^2}{r_0}(\mathfrak{D}_1 \mathbf{u}_0 + s\mathfrak{D}_3^* \mathbf{u}'_0) \otimes \mathbf{u}'_0 \tag{8.31}$$

$$\nabla_{\mathbf{u}_0}\big|_s \mathbf{u}' = +\psi^2 \frac{s}{r_0}[(2\mathfrak{D}_1 - \psi^2 s^2 \mathfrak{D}_3^*)\mathbf{u}_0 + s\mathfrak{D}_1 \mathbf{u}'_0] \otimes \mathbf{u}_0 - \psi^2 s\mathfrak{D}_1\, \mathbf{I}_4 \tag{8.32}$$

$$\nabla_{\mathbf{u}'_0}\big|_s \mathbf{u}' = \frac{s}{r_0}[(2\mathfrak{D}_1 - \psi^2 s^2 \mathfrak{D}_3^*)\mathbf{u}_0 + s\mathfrak{D}_1 \mathbf{u}'_0] \otimes \mathbf{u}'_0 + \mathfrak{D}_0\, \mathbf{I}_4 \tag{8.33}$$

These expressions complete matrix $\mathbf{N}_{\text{asyn}}$ according to its definition given in Eq. (8.29).

The universal Kepler equation is written in terms of vectors $\mathbf{u}_0$ and $\mathbf{u}_0'$ as

$$t = T\left(s; \mathbf{u}_0, \mathbf{u}_0'\right) \equiv \frac{s}{2} r_0 (1 + \mathfrak{C}_1) + 2s^2 \left(\mathbf{u}_0 \cdot \mathbf{u}_0'\right) \mathfrak{C}_2 + 2s^3 \left(\mathbf{u}_0' \cdot \mathbf{u}_0'\right) \mathfrak{C}_3$$

Note that it involves the Stumpff functions $\mathfrak{C}_k$ and not $\mathfrak{D}_k$. The time delay is determined through

$$\delta t = \nabla_{\mathbf{u}_0}\big|_s T \cdot \delta \mathbf{u}_0 + \nabla_{\mathbf{u}_0'}\big|_s T \cdot \delta \mathbf{u}_0' \tag{8.34}$$

where the gradients of time are

$$\nabla_{\mathbf{u}_0}\big|_s T = \frac{8}{r_0} s^3 \psi^2 [s(\mathbf{u}_0 \cdot \mathbf{u}_0')\, \mathfrak{C}_4^* + s^2 (\mathbf{u}_0' \cdot \mathbf{u}_0') \mathfrak{C}_5^* + \frac{r_0}{4}\, \mathfrak{C}_3^*] \mathbf{u}_0$$
$$+ s(1 + \mathfrak{C}_1)\mathbf{u}_0 + 2s^2 \mathfrak{C}_2\, \mathbf{u}_0'$$

$$\nabla_{\mathbf{u}_0'}\big|_s T = \frac{2s^2}{r_0} \{4s^2 [(\mathbf{u}_0' \cdot \mathbf{u}_0') s \mathfrak{C}_5^* + (\mathbf{u}_0 \cdot \mathbf{u}_0') \mathfrak{C}_4^*] \mathbf{u}_0' + r_0 [\mathbf{u}_0' (\mathfrak{C}_2 + \mathfrak{C}_3^*) s + \mathbf{u}_0 \mathfrak{C}_2]\}$$

Equation (8.34) is now defined in terms of the initial state vector in KS variables and the solution is complete.

8.3.1 Summary

The following steps summarize the algorithm for computing the relative state vector from the KS transformation.

1. Transform the initial separation $\delta \mathbf{x}_0$ to the differences $\delta \mathbf{y}_0$ using matrix $\mathsf{T}(\mathbf{x}_0)$ (Appendix D). The velocity needs to be referred to the inertial frame, $\delta \mathbf{v}_0 = \delta \dot{\mathbf{r}}_0 + \boldsymbol{\omega}_{\mathcal{L}\mathcal{I},0} \times \delta \mathbf{r}_0$, with $\boldsymbol{\omega}_{\mathcal{L}\mathcal{I},0} = \mathbf{h}/r_0^2$.
2. Use Eqs. (8.30–8.33) to compute the matrix N_{asyn}.
3. Get matrix $\mathsf{M}_{\text{asyn}}(s)$ from Eq. (8.26) (all variables are referred to the leader spacecraft).
4. The asynchronous solution to the problem is:
$$\delta \mathbf{x}_{\text{asyn}} = \mathsf{M}_{\text{asyn}}(s) \mathsf{N}_{\text{asyn}}(s) \mathsf{T}(\mathbf{x}_0)\, \delta \mathbf{x}_0$$
5. Compute the time delay δt from Eq. (8.34).
6. Recover the synchronism through the first-order correction defined in Eqs. (7.32–7.33).
7. (Optional) If more accuracy is needed, the linear solution can be refined by introducing nonlinear terms according to Eqs. (7.34–7.35). They give the improved solution $\delta \mathbf{r}^\star$ and $\delta \mathbf{v}^\star$.

8.4 On the fictitious time

In practice, the solution needs to be propagated across a certain time interval, $t \in [t_0, t_f]$. The fictitious time is initially zero, spanning across $s \in [0, s_f]$. The final value of the fictitious time, s_f, is solved from the universal Kepler equation:

$$t_f - t_0 = s_f r_0 \mathfrak{C}_1(z_f) + s_f^2 r_0' \mathfrak{C}_2(z_f) + s_f^3 \mathfrak{C}_3(z_f) \tag{8.35}$$

Battin (1999, §4.5) wrote this equation in terms of the universal functions $\mathfrak{U}_k(z)$, which relate to the Stumpff functions by means of $\mathfrak{U}_k(z) = s^k \mathfrak{C}_k(z)$. The universal functions are also referred to as generalized conic functions (Everhart and Pitkin, 1983). Special attention has been paid to the numerical resolution of this form of Kepler's equation (Burkardt and Danby, 1983; Danby, 1987).

Battin (1999, p. 179) credited Charles M. Newman from MIT with deriving an explicit expression for the final value of s in terms of

$$s_f = \omega^2(t_f - t_0) + r_f' - r_0', \qquad \text{with } r' = (\mathbf{r} \cdot \mathbf{v})$$

This expression is useful when the relative state vector is to be computed from the position and velocity of the leader spacecraft. Given the initial and final times, the final value of the fictitious time is solved from Eq. (8.35), which is not affected by the eccentricity of the orbit.

The fictitious time can be related to the true anomaly. However, since how the true anomaly relates to the time depends on the type of orbit, the transformation from the true anomaly to the fictitious time will also be eccentricity-dependent. But this does not affect the universality of the solution. From the Sundman transformation it follows that

$$\frac{\mathrm{d}s}{\mathrm{d}\vartheta} = \frac{\mathrm{d}s}{\mathrm{d}t}\frac{\mathrm{d}t}{\mathrm{d}\vartheta} = \frac{r}{h}$$

In the limit case $e \to 1$, this differential equation leads to

$$s_f = h\left(\tan\frac{\vartheta}{2} - \tan\frac{\vartheta_0}{2}\right)$$

Simple expressions can be obtained for the elliptic and hyperbolic cases:

$$e < 1: \qquad s_f = \sqrt{a}(E - E_0)$$
$$e > 1: \qquad s_f = \sqrt{|a|}(H - H_0)$$

in which E and H are the eccentric and hyperbolic anomaly, respectively.

8.5 Numerical examples

This section is devoted to testing the performance of the second-order correction of the time delay for the four different types of reference orbits: circular, elliptic, parabolic, and hyperbolic. The accuracy of the proposed formulations is analyzed by comparing them with the exact solution to the problem. The error at each step is measured as:

$$\varepsilon = ||\delta\mathbf{r} - \delta\mathbf{r}_{\text{ref}}||, \qquad \dot{\varepsilon} = ||\delta\dot{\mathbf{r}} - \delta\dot{\mathbf{r}}_{\text{ref}}||$$

where $\delta\mathbf{r}_{\mathrm{ref}}$ and $\delta\dot{\mathbf{r}}_{\mathrm{ref}}$ are the exact relative position and velocity vectors. The exact solution is constructed by solving the nonlinear two-body problem for the leader and the follower, and then subtracting the absolute state vectors. Vector $\delta\dot{\mathbf{r}} = [\delta\dot{x}, \delta\dot{y}, \delta\dot{z}]^{\top}$ denotes the relative velocity from the perspective of the Euler–Hill reference frame that rotates with the leader spacecraft.

Table 8.1 defines four test cases, giving both the reference orbit and the initial relative conditions. In the circular and elliptic cases the solution is propagated for 15 complete revolutions. The Clohessy–Wiltshire (CW) and Yamanaka–Ankersen (YA) solutions will be included as references. In the parabolic and hyperbolic cases the propagation spans from ϑ_0 to $-\vartheta_0$. Examples of equatorial-retrograde and polar orbits are selected to show that the KS and SB formulations are not affected by typical singularities such as $i = 0$ or $e = 0$, and that the Dromo-based solution is valid for any noncircular orbit.

Figure 8.1a displays the relative orbit and the error in position and velocity for Case 1, an equatorial, retrograde, circular orbit. The solution is computed only with the KS and SB formulations, because the Dromo-based solution is singular in this case. The linear solution ("Linear") obtained with the KS and SB methods coincides exactly with the CW and YA results, because they are all the exact solution to the same problem (linearized relative motion). The improved solution ("KS-SB (⋆)") adequately captures the nonlinear behavior of the dynamics and improves the accuracy of the propagation by one order of magnitude both in position and velocity. This improved formulation introduces nonlinear terms using the time delay. Consequently, since the KS and SB schemes use the same time transformation, they yield the same exact accuracy.

The results for Case 2 are presented in Figure 8.1b. This case is an example of a polar, highly eccentric reference orbit. It is propagated with the YA solution, and the Dromo, SB, and KS formulations. The linear solution (with the first-order correction) coincides in practice for the four methods. However, the second-order correction de-

Tab. 8.1: Definition of the reference orbits and relative initial conditions in the $\mathcal{L}$ frame.

Case	a [km]	e	i [deg]	ω [deg]	Ω [deg]	ϑ_0 [deg]
1	7500	0.0	180	50	0	−45
2	9000	0.7	90	100	50	10
3	∞	1.0	20	250	10	−120
4	−20000	1.4	20	100	60	−130
	δx [m]	δy [m]	δz [m]	$\delta\dot{x}$ [m/s]	$\delta\dot{y}$ [m/s]	$\delta\dot{z}$ [m/s]
1	−150	50	−200	−0.6	−0.1	−0.1
2	0	−50	150	0.1	0.1	0.5
3	−170	160	−20	−1.7	−1.0	0.5
4	−60	50	100	−0.2	0.2	−0.5

Note: the reference orbit for Case 3 is defined by its angular momentum, which is $h = 60{,}000\ \mathrm{km^2/s}$

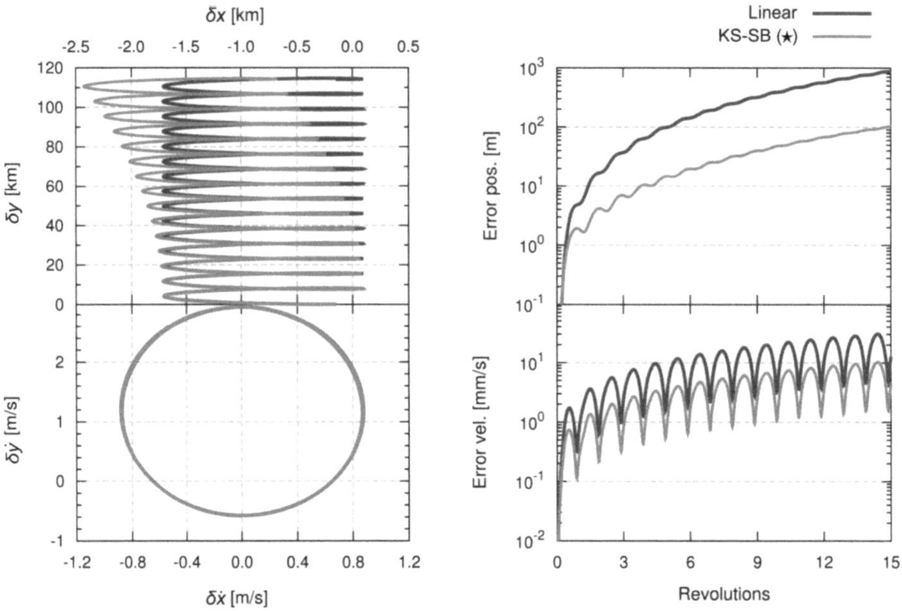

(a) Case 1: circular reference orbit

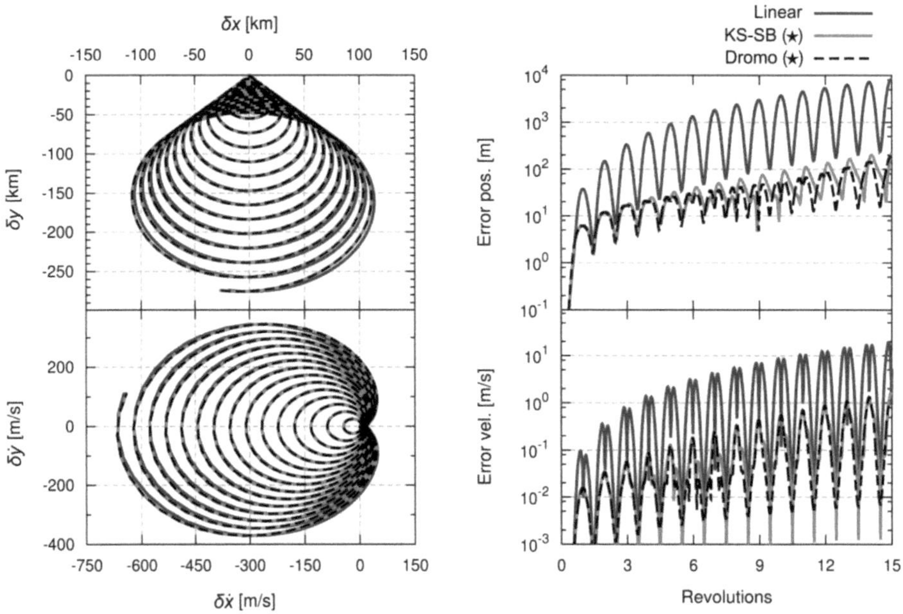

(b) Case 2: elliptic reference orbit

Fig. 8.1: Relative orbit and propagation error for the circular and elliptic cases.

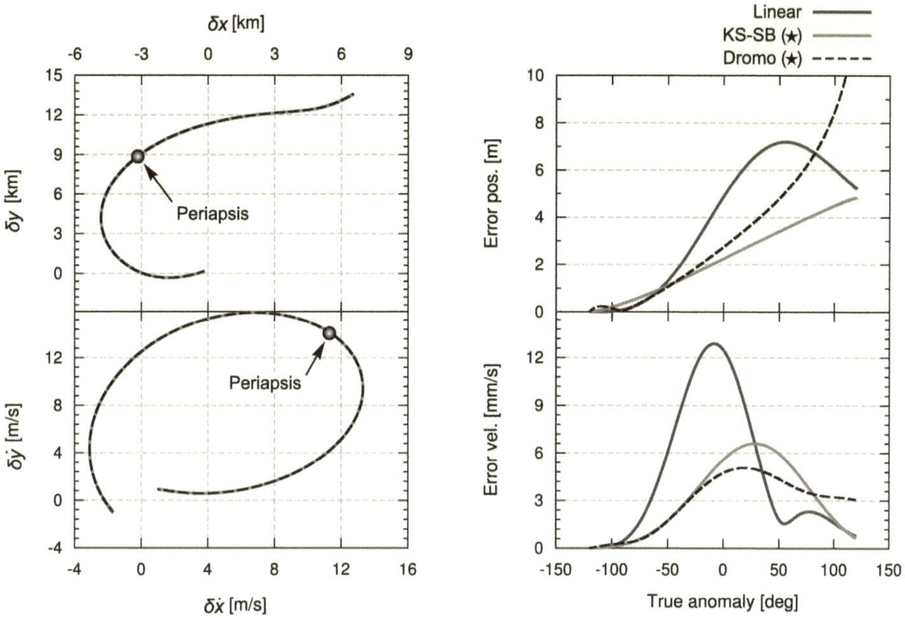

(a) Case 3: parabolic reference orbit

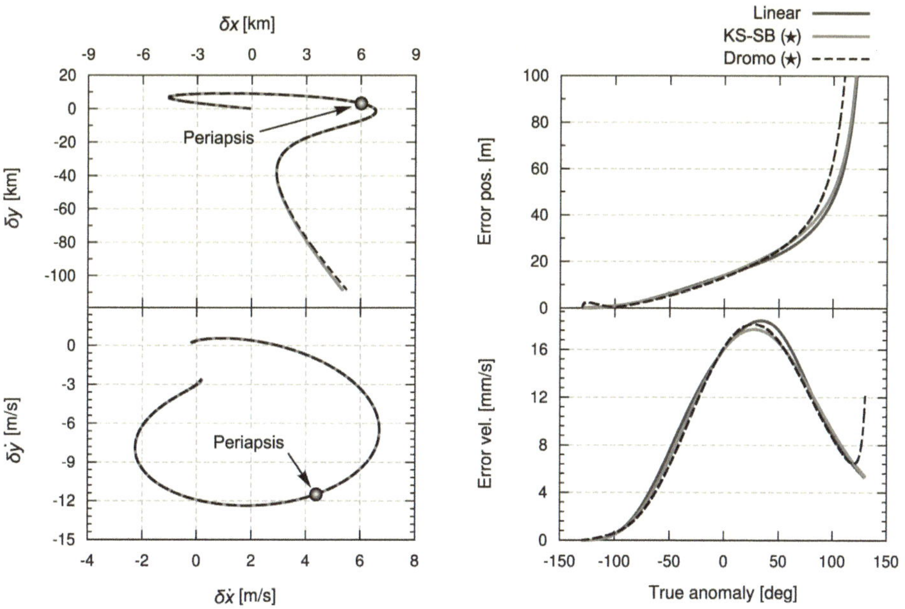

(b) Case 4: hyperbolic reference orbit

Fig. 8.2: Relative orbit and propagation error for the parabolic and hyperbolic cases.

pends on the time transformation and the results are slightly different. The Dromo formulation relies on a second-order Sundman transformation, whereas the time transformation in the SB and KS schemes is of first order. The discretization of the orbit is different: the former is equivalent to the true anomaly, whereas the latter corresponds to the eccentric anomaly. In this particular case, the solution corrected to second order in Dromo variables ("Dromo (⋆)") is slightly more accurate than the equivalent solution using the KS or the SB ("KS-SB (⋆)"), which coincide exactly. These differences will grow with the eccentricity of the reference orbit. The second-order correction of the time delay improves the accuracy of the propagation by almost two orders of magnitude in position and one order of magnitude in velocity after 15 revolutions.

Figures 8.2a and 8.2b correspond to parabolic and hyperbolic reference orbits, respectively. The three formulations presented in this chapter can be applied directly to these type of orbits, although the Dromo-based formulation requires changing the form of the integrals I_i (p. 175). The results in both cases are qualitatively similar. It is observed that the error in velocity is maximum around perigee no matter the formulation. This is caused by the strong divergence of the dynamics around periapsis. The nonlinear correction partially mitigates this phenomenon in the parabolic case, although no clear improvements are observed in the hyperbolic case. In the parabolic case, the second-order time transformation used in the Dromo formulation exhibits

Tab. 8.2: Absolute and relative errors in position and velocity for the proposed formulations (final value).

	Case 1		Case 2		Case 3		Case 4	
Units	ε_f [m]	$\dot{\varepsilon}_f$ [mm/s]	ε_f [m]	$\dot{\varepsilon}_f$ [mm/s]	ε_f [m]	$\dot{\varepsilon}_f$ [mm/s]	ε_f [m]	$\dot{\varepsilon}_f$ [mm/s]
CW	875.46	12.68	–	–	–	–	–	–
YA	875.46	12.68	8224.9	5309.0	–	–	–	–
SB	875.46	12.68	8224.9	5309.0	5.22	0.07	231.6	5.28
SB(⋆)	103.78	4.29	221.7	1475.5	4.84	0.07	249.0	5.33
KS	875.46	12.68	8224.9	5309.0	5.22	0.07	231.6	5.28
KS(⋆)	103.78	4.29	221.7	1475.5	4.84	0.07	249.0	5.33
Dromo	–	–	8224.9	5309.0	5.22	0.07	231.6	5.28
Dromo(⋆)	–	–	188.9	1542.4	12.35	3.05	238.5	19.73
	Relative error (Error/Final value)							
CW	7.68E−3	2.03E−2	–	–	–	–	–	–
YA	7.68E−3	2.03E−2	2.99E−2	8.14E−3	–	–	–	–
SB	7.68E−3	2.03E−2	2.99E−2	8.14E−3	3.43E−4	4.70E−4	9.93E−4	8.89E−4
SB(⋆)	9.11E−4	6.90E−3	8.05E−4	2.26E−3	3.18E−4	4.01E−4	1.07E−3	8.97E−4
KS	7.68E−3	2.03E−2	2.99E−2	8.14E−3	3.43E−4	4.70E−4	9.93E−4	8.89E−4
KS(⋆)	9.11E−4	6.90E−3	8.05E−4	2.26E−3	3.18E−4	4.01E−4	1.07E−3	8.97E−4
Dromo	–	–	2.99E−2	8.14E−3	3.43E−4	4.70E−4	9.93E−4	8.89E−4
Dromo(⋆)	–	–	6.86E−4	2.36E−3	8.11E−4	1.83E−3	6.17E−3	2.05E−3

small advantages in the propagation of the velocity, whereas the propagation of the position is less accurate. The Dromo-based solution seems to be less accurate in the case of open orbits. The loss of accuracy is attributed to the issues related to parabolic and hyperbolic orbits reported in Sect. 4.6.

Table 8.2 summarizes the previous discussions on the accuracy of the methods. The final errors in position and velocity are presented, showing both their absolute magnitude and the value relative to the final relative separation and velocity. In this way, the real impact of the error on the solution can be quantified, and the significance of the error reductions when introducing nonlinearities is easier to appreciate. The improved solutions are denoted with a star ($\star$). This table clearly shows that the linear solutions coincide identically, as they are the exact solution to the same system of equations. The nonlinear correction, conversely, depends on the time transformation and therefore the accuracy of the KS and SB methods coincides (both use $\mathrm{d}t/\mathrm{d}s = r$), but it is different from that of Dromo (which uses $\mathrm{d}t/\mathrm{d}s = r^2/h$).

8.6 Generic propagation of the variational equations

This section summarizes the procedure for propagating the state-transition matrix using an arbitrary formulation, subject to any source of perturbations $\mathbf{a}_p$. The special set of variables $\boldsymbol{\mathfrak{oe}}$ relates to the state vector $\mathbf{x}$ by means of an invertible transformation $\mathbf{q}\colon \mathbf{x} \mapsto \boldsymbol{\mathfrak{oe}}$:

$$\boldsymbol{\mathfrak{oe}}(\phi) = \mathbf{q}(t(\phi); \mathbf{x}) \,, \qquad \mathbf{x}(t) = \mathbf{q}^{-1}(\phi(t); \boldsymbol{\mathfrak{oe}})$$

The linear form of these transformations is given in terms of the matrices $\mathbf{Q}$ and $\mathbf{Q}^{\dagger}$:

$$\delta\boldsymbol{\mathfrak{oe}}(\phi) = \mathbf{Q}(t(\phi); \mathbf{x})\, \delta\mathbf{x} \,, \qquad \delta\mathbf{x}(t) = \mathbf{Q}^{\dagger}(\phi(t); \boldsymbol{\mathfrak{oe}})\, \delta\boldsymbol{\mathfrak{oe}}$$

Here $\mathbf{Q}^{\dagger}$ is the pseudo-inverse of $\mathbf{Q}$, because this is not necessarily a square matrix. The physical time and the independent variable ϕ relate by means of

$$\phi = \Phi(t; \mathbf{x}) \,, \qquad t = T(\phi; \boldsymbol{\mathfrak{oe}})$$

with $T \equiv \Phi^{-1}$. Under this notation the generalized Sundman transformation reads

$$\frac{\partial t}{\partial \phi} = T'(\phi; \boldsymbol{\mathfrak{oe}}) \tag{8.36}$$

The evolution of the set $\boldsymbol{\mathfrak{oe}}(\phi)$ obeys the differential equation

$$\frac{\partial \boldsymbol{\mathfrak{oe}}}{\partial \phi} = \mathbf{g}(\phi; \boldsymbol{\mathfrak{oe}}, \mathbf{a}_p) \tag{8.37}$$

This equation is propagated together with Eq. (8.36), which provides the physical time, and from the initial conditions

$$\phi = 0\colon \qquad \boldsymbol{\mathfrak{oe}}(0) = \boldsymbol{\mathfrak{oe}}_0 \,, \quad t(0) = t_0$$

The solution to this initial value problem determines the evolution of the set $\mathbf{œ}$, namely

$$\mathbf{œ}(\phi) = \mathbf{G}(\phi; \mathbf{œ}_0)\,, \qquad \text{with} \quad \mathbf{G}(\phi; \mathbf{œ}_0) = \int_{\phi_0}^{\phi} \mathbf{g}(\phi; \mathbf{œ}_0)\, d\phi$$

Following Eq. (7.15), a small separation $\delta\mathbf{œ}_0$ can be propagated in time according to

$$\delta\mathbf{œ}(t) = \delta\mathbf{œ}_{\text{asyn}}(\phi(t)) - \frac{\partial \mathbf{œ}}{\partial t}\, \delta t$$

This renders the synchronous variation on the set $\mathbf{œ}$ at time t. Using matrix notation it is

$$\delta\mathbf{œ}(t) = \mathbf{M}(t, t_0)\, \delta\mathbf{œ}_0 \tag{8.38}$$

in which matrix $\mathbf{M}(t, t_0)$ can be referred to the asynchronous transition matrix

$$\mathbf{M}_{\text{asyn}}(\phi, \phi_0) = \nabla_{\mathbf{œ}_0}\big|_{\phi}\mathbf{G}$$

by means of

$$\mathbf{M}(t, t_0) = \mathbf{M}_{\text{asyn}}(\phi(t), \phi_0) - \frac{1}{T'}[\mathbf{g}(\phi; \mathbf{œ}, \mathbf{a}_p) \otimes \nabla_{\mathbf{œ}_0} T]$$

Considering the transformation between the set $\mathbf{œ}$ and the state vector $\mathbf{x}$ the previous expressions can be extended to define the state-transition matrix $\Xi(t, t_0)$. This matrix can be used to propagate the relative state vector

$$\delta\mathbf{x}(t) = \Xi(t, t_0)\, \delta\mathbf{x}_0$$

Indeed, from Eq. (8.38) it follows that

$$\delta\mathbf{x}(t) = \mathbf{Q}^{\dagger}(\phi(t), \mathbf{œ})\, \delta\mathbf{œ}(t) = \left[\mathbf{Q}^{\dagger}(\phi(t), \mathbf{œ})\, \mathbf{M}(t, t_0)\, \mathbf{Q}(t_0, \mathbf{x}_0)\right] \delta\mathbf{x}_0$$

This identity proves that the state-transition matrix is built from

$$\Xi(t, t_0) = \mathbf{Q}^{\dagger}\left\{\mathbf{M}_{\text{asyn}}(\phi(t), \phi_0) - \frac{1}{T'}[\mathbf{g}(\phi; \mathbf{œ}, \mathbf{a}_p) \otimes \nabla_{\mathbf{œ}_0} T]\right\} \mathbf{Q}_0 \tag{8.39}$$

Matrices $\mathbf{Q}^{\dagger}$ and $\mathbf{Q}$ contain the partial derivatives of the state vector with respect to $\mathbf{œ}$, and vice-versa. These two matrices can be computed analytically. The term T' is the right-hand side of the Sundman transformation (8.36), and $\mathbf{g}(\phi; \mathbf{œ}, \mathbf{a}_p)$ is the right-hand side of the evolution equations (8.37). If t is considered part of the set $\mathbf{œ}$, as is the usual practice, then $\nabla_{\mathbf{œ}_0} T$ is simply the row of $\mathbf{M}_{\text{asyn}}$ corresponding to the physical time. Similarly, T' will be one of the components of $\mathbf{g}(\phi; \mathbf{œ}, \mathbf{a}_p)$. Matrix $\mathbf{M}_{\text{asyn}}$ is the solution to the initial value problem

$$\left\{\begin{aligned} &\frac{\partial \Xi_{\text{asyn}}}{\partial \phi} = \nabla_{\mathbf{œ}}\mathbf{g}(\phi; \mathbf{œ}, \mathbf{a}_p)\, \Xi_{\text{asyn}} \\ &\Xi_{\text{asyn}}(0) = \mathbf{I} \end{aligned}\right. \tag{8.40}$$

Here $\nabla_{\mathbf{oe}}\mathbf{g}(\phi;\mathbf{oe},\mathbf{a}_p)$ is the Jacobian matrix of the function (8.37). The initial value problem (8.40) is integrated together with Eqs. (8.36) and (8.37). Thus, at every integration step the solutions $\mathbf{oe}(\phi)$, $\Xi_{\text{asyn}}(\phi,\phi_0)$, and $t(\phi)$ are available. In addition, the right-hand sides $\mathbf{g}(\phi;\mathbf{oe},\mathbf{a}_p)$ and T' are also known. Since $\nabla_{\mathbf{oe}_0}T$ is a row of $\Xi_{\text{asyn}}(\phi,\phi_0)$, all the terms in Eq. (8.39) are known and this equation furnishes the state-transition matrix. As shown in Sect. 6.1.4, PERFORM can be used for propagating the variational equations following this procedure.

When propagating the variational equations of a formulation different from Cowell's method, two factors should be considered. First, the initialization of the independent variable. Second, the correction of the time delay when using a time element.

8.6.1 Initializing the independent variable

The initial value of the new independent variable is given in a specific way, and it may be $\phi_0 \neq 0$. For example, if the independent variable is the true anomaly its initial value is solved from the projections of the position vector onto the perifocal frame.

When the initial conditions change, as is the case when the variational formulation is integrated, then the initial value of the independent variable will change too. The gradient of the set $\mathbf{oe}$ does not account for the change in ϕ_0, meaning that in the current setup this information will be lost and the differentiation will fail. To solve this issue, the formulation must be adapted so that the independent variable is always zero at departure. This is achieved by introducing a modified independent variable, $\phi^* = \phi - \phi_0$. There are two systematic ways of modifying the formulation accordingly (Roa and Peláez, 2015a):

1. To attach ϕ_0 to the vector of elements or coordinates $\mathbf{oe}$. It remains constant throughout the integration process, and the original value of the independent variable is recovered simply by means of $\phi = \phi^* + \phi_0$. This approach requires little modification of the equations of motion, but increases the size of the system.
2. To reformulate the problem in terms of the modified independent variable ϕ^*. This approach preserves the dimension of the system, at the cost of having to derive the modified equations of motion.

Examples of formulations requiring this correction are Dromo (Chap. 4), the Minkowskian formulation for hyperbolic orbits (Chap. 5), the Stiefel–Scheifele method (Sect. 3.6), the equinoctial elements with the longitude as the independent variable (Sect. 7.3), etc.

8.6.2 The time element

Time elements separate the physical time t in a term that depends on the perturbations (the time element per se, t_{te}) and a term not affected by perturbations, t_{np},

$$t(\phi;\mathbf{œ}) = t_{\text{te}}(\phi;\mathbf{œ}) + t_{\text{np}}(\phi;\mathbf{œ}) \tag{8.41}$$

The time element vanishes if there are no perturbations. The derivative of the time element scales with the perturbation, and yields a smoother evolution in weakly perturbed problems.

The presence of a time element complicates slightly the definition of the time delay: the gradient $\nabla_{\mathbf{œ}_0} t$ is no longer given explicitly by the state-transition matrix Ξ_{asyn}, since this matrix propagates the time element and not the physical time. In fact, taking the gradient $\nabla_{\mathbf{œ}_0}$ in Eq. (8.41) provides

$$\nabla_{\mathbf{œ}_0} t = \nabla_{\mathbf{œ}_0} t_{\text{te}}(\phi;\mathbf{œ}) + \nabla_{\mathbf{œ}_0} t_{\text{np}}(\phi;\mathbf{œ})$$

where $\nabla_{\mathbf{œ}_0} t_{\text{te}}(\phi;\mathbf{œ})$ is the corresponding row of the matrix Ξ_{asyn}. The second term is required for adjusting the physical time from the variation on the time element. For convenience, the following sections present the explicit form of the partial derivatives of the formulations discussed in this book.

8.6.2.1 Time-element in Minkowskian variables

The time element improves the numerical performance of the method defined in Chap. 5, for which it is $\mathbf{œ} = [t, \lambda_1, \lambda_2, \lambda_3, \chi_1, \chi_2, \chi_3, \chi_4]^\top$. The physical time decomposes in

$$t = t_{\text{te}} + \lambda_3^{3/2}\left[\left(\lambda_1^2 - \lambda_2^2\right)^{1/2}\sinh H - H\right] - \lambda_{30}^{3/2}(\lambda_{10}\sinh H_0 - H_0)$$

The time delay is obtained from the gradient $\nabla_{\mathbf{œ}_0} t$. Considering the function

$$d(u,x) = \frac{\partial t_{\text{te}}}{\partial x} + \lambda_3^{3/2}\left(\sinh u\,\frac{\partial \lambda_1}{\partial \xi_0} + \cosh u\,\frac{\partial \lambda_2}{\partial x}\right) + \frac{3}{2}\sqrt{\lambda_3}(\hat{r} + 1 - u)\frac{\partial \lambda_3}{\partial x}$$

the partial derivatives forming the gradient take the form:

$$\begin{aligned}
\frac{\partial t}{\partial \lambda_{10}} &= d(u,\lambda_{10}) - \lambda_{30}^{3/2}\sinh H_0\\
\frac{\partial t}{\partial \lambda_{30}} &= d(u,\lambda_{10}) - \frac{3}{2}\sqrt{\lambda_{30}}\,(\hat{r}_0 + 1 - H_0)\\
\frac{\partial t}{\partial H_0} &= d(u,H_0) - \frac{3}{2}\sqrt{\lambda_{30}}\,(\lambda_{10}\cosh H_0 - 1)\\
\frac{\partial t}{\partial \xi_{i,0}} &= d(u,\xi_{i,0}), \qquad \boldsymbol{\xi}_0 = [t_{\text{te}0}, \lambda_{20}, \chi_{10}, \chi_{20}, \chi_{30}, \chi_{40}]^\top
\end{aligned}$$

The partial derivatives with respect to the initial values of the elements are given by the integration procedure described in the previous lines.

8.6.2.2 Time-element in KS variables

In KS variables it is $\boldsymbol{\alpha}\mathbf{e}^\top = [t, \mathbf{u}^\top, \mathbf{u}'^\top]$. The time delay is given by

$$\delta t = \nabla_{\mathbf{u}_0} t \cdot \delta\mathbf{u}_0 + \nabla_{\mathbf{u}_0'} t \cdot \delta\mathbf{u}_0'$$

The derivatives of time t with respect to the initial conditions $\mathbf{u}_0$ and $\mathbf{u}_0'$ read

$$\begin{aligned}
\nabla_{\mathbf{u}_0} t &= \nabla_{\mathbf{u}_0} t_{\mathrm{te}} - \frac{1}{2r\alpha}\left(r\mathbf{u}' + 2r'\mathbf{u}\right) \circ \nabla_{\mathbf{u}_0}\mathbf{u} - \frac{1}{2r\alpha^2}(2r'\mathbf{u}' + r\alpha\mathbf{u}) \circ \nabla_{\mathbf{u}_0}\mathbf{u}' \\
\nabla_{\mathbf{u}_0'} t &= \nabla_{\mathbf{u}_0'} t_{\mathrm{te}} - \frac{1}{2r\alpha}\left(r\mathbf{u}' + 2r'\mathbf{u}\right) \circ \nabla_{\mathbf{u}_0'}\mathbf{u} - \frac{1}{2r\alpha^2}(2r'\mathbf{u}' + r\alpha\mathbf{u}) \circ \nabla_{\mathbf{u}_0'}\mathbf{u}'
\end{aligned}$$

Typically, the energy α is integrated together with the time and the coordinates $\mathbf{u}$ and $\mathbf{u}'$. In such a case the gradients $\nabla_{\mathbf{u}_0}\alpha$ and $\nabla_{\mathbf{u}_0'}\alpha$ are given by the corresponding row of the state-transition matrix. The previous equations then reduce to

$$\begin{aligned}
\nabla_{\mathbf{u}_0} t &= \nabla_{\mathbf{u}_0} t_{\mathrm{te}} - \frac{1}{2\alpha}(\mathbf{u} \circ \nabla_{\mathbf{u}_0}\mathbf{u}' + \mathbf{u}' \circ \nabla_{\mathbf{u}_0}\mathbf{u}) + \frac{(\mathbf{u}\cdot\mathbf{u}')}{2\alpha^2}\nabla_{\mathbf{u}_0}\alpha \\
\nabla_{\mathbf{u}_0'} t &= \nabla_{\mathbf{u}_0'} t_{\mathrm{te}} - \frac{1}{2\alpha}(\mathbf{u} \circ \nabla_{\mathbf{u}_0'}\mathbf{u}' + \mathbf{u}' \circ \nabla_{\mathbf{u}_0'}\mathbf{u}) + \frac{(\mathbf{u}\cdot\mathbf{u}')}{2\alpha^2}\nabla_{\mathbf{u}_0'}\alpha
\end{aligned}$$

Finally, the last term required for solving the t-synchronous state-transition matrix is t':

$$t' = t_{\mathrm{te}}' - \frac{1}{2r\alpha}\left(r\mathbf{u}' + 2r'\mathbf{u}\right)\cdot\mathbf{u}' - \frac{1}{2r\alpha^2}(2r'\mathbf{u}' + r\alpha\mathbf{u})\cdot\mathbf{u}''$$

8.6.2.3 Time-element in Sperling–Burdet variables

Both the Sperling–Burdet regularization and the stabilized Cowell method (Sect. 2.3) introduce a fictitious time $\phi \equiv s$ by means of the Sundman transformation $\mathrm{d}t = r\,\mathrm{d}s$. For the former method with a time element it is $\boldsymbol{\alpha}\mathbf{e} = [t_{\mathrm{te}}, \mathbf{r}^\top, \mathbf{r}'^\top, r, r', \alpha, \boldsymbol{\mu}\mathbf{e}^\top]^\top$, and for the latter $\boldsymbol{\alpha}\mathbf{e} = [t_{\mathrm{te}}, \mathbf{r}^\top, \mathbf{r}'^\top, \alpha]^\top$. Considering the auxiliary terms

$$r' = \frac{\mathbf{r}\cdot\mathbf{r}'}{r}, \qquad \alpha = \frac{1}{r}\left(2 - \frac{||\mathbf{r}'||^2}{r}\right)$$

the gradient $\nabla_{\boldsymbol{\alpha}\mathbf{e}_0} t$ consists of

$$\begin{aligned}
\nabla_{\mathbf{r}_0} t &= \nabla_{\mathbf{r}_0} t_{\mathrm{te}} + \frac{1}{r^4\alpha^2}\left[(r'||\mathbf{r}'||^2\mathbf{r} - r^3\alpha\,\mathbf{r}') \circ \nabla_{\mathbf{r}_0}\mathbf{r} - r^2(2r'\mathbf{r}' + \alpha r\,\mathbf{r}) \circ \nabla_{\mathbf{r}_0}\mathbf{r}'\right] \\
\nabla_{\mathbf{r}_0'} t &= \nabla_{\mathbf{r}_0'} t_{\mathrm{te}} + \frac{1}{r^4\alpha^2}\left[(r'||\mathbf{r}'||^2\mathbf{r} - r^3\alpha\,\mathbf{r}') \circ \nabla_{\mathbf{r}_0'}\mathbf{r} - r^2(2r'\mathbf{r}' + \alpha r\,\mathbf{r}) \circ \nabla_{\mathbf{r}_0'}\mathbf{r}'\right]
\end{aligned}$$

and the derivative with respect to fictitious time reads

$$t' = t_{\mathrm{te}}' + \frac{1}{r^4\alpha^2}\left[(r'||\mathbf{r}'||^2\mathbf{r} - r^3\alpha\,\mathbf{r}')\cdot\mathbf{r}' - r^2(2r'\mathbf{r}' + \alpha r\,\mathbf{r})\cdot\mathbf{r}''\right]$$

8.7 Conclusions

The theory of asynchronous relative motion allows one to solve the relative dynamics using regularized formulations of orbital motion. The theory provides a systematic technique that can be applied eventually to any formulation relying on an independent variable different from time. When the motion is formulated using fully regular schemes, like the Sperling–Burdet or the Kustaanheimo–Stiefel methods, the result is a universal formulation that is valid for any kind of reference orbit: circular, elliptic, parabolic, and hyperbolic. The use of Dromo elements also yields a solution valid for any type of reference orbit, although there is a singularity arising from the indeterminacy of the eccentricity vector for circular orbits.

The nonlinear correction of the time delay can be applied easily and the accuracy of the linear solution is increased significantly. For moderately to highly elliptical orbits the discretization of the reference orbit in terms of the true anomaly seems more effective than the discretization in terms of the eccentric anomaly. This conclusion translates into the Dromo-based solution being more accurate than the KS and SB schemes in the elliptic case. However, following the discussion in Chap. 2, the full regularization of the problem will only be possible if the Sundman transformation is of order $n < 3/2$.

Another outcome of the theory is a general procedure for propagating the variational equations of any formulation that uses an independent variable different from time. The typical numerical procedure defines the transition matrix at every integration step. But the corresponding partial derivatives are computed with constant fictitious time (or a given angle), rather than with constant time. This introduces an intrinsic time delay in the solution. It can be corrected easily using the theory of asynchronous relative motion.

9 Generalized logarithmic spirals: A new analytic solution with continuous thrust

Element-based formulations have proven efficient for propagating weakly perturbed orbits (Sect. 2.5). The right-hand side of the equations of motion scales with the magnitude of the perturbation, and evolves in a similar time scale. A problem of fundamental importance in astrodynamics that involves small perturbations is the use of low-thrust propulsion systems. The low acceleration from the engines results in trajectories that spiral away or toward the central body. When the spacecraft spirals around the primary for hundreds or thousands of revolutions, the slow time scale of the thrust acceleration couples with different frequencies coming from, for example, the high-order gravity harmonics of the central body. This phenomenon slows down the integration significantly due to an inefficient step-size control. It might become an issue when many propagations are required, as is the case for trajectory optimization. Finding a suitable set of elements for integrating such trajectories may improve the design process.

The evolution of the elements is typically formulated applying the variation of parameters technique to an analytic solution. The key to deriving an efficient representation is finding an analytic solution to a problem that is sufficiently close to the perturbed one. This is why special perturbation methods are constructed by perturbing Keplerian orbits. However, Keplerian orbits might not be a good base solution to model low-thrust orbits. To improve the dynamical model and the meaning of the orbital elements, a new, more representative analytic solution is required. Finding an analytic solution that at least approximates low-thrust trajectories will be the first step in constructing a dedicated formulation. In the field of low-thrust preliminary trajectory design, many authors have devoted themselves to finding special trajectories that can be defined in closed form. The most common approach is the so-called *shape-based* method: instead of integrating the equations of motion given a thrust profile, the shape of the trajectory is assumed a priori and then the thrust required to follow such a curve is obtained. This chapter will focus on presenting the family of *generalized logarithmic spirals*, an analytic solution that represents the orbital motion of a particle under a continuous acceleration (Roa and Peláez, 2015c; Roa et al., 2016a). Interestingly, the system admits two integrals of motion equivalent to the conservation of the energy and angular momentum. A control parameter will be introduced in Sect. 10.2 to improve the flexibility of the solution, and a collection of useful formulas can be found in Appendix F.

The shape-based approach was conceived to provide suitable initial guesses for more sophisticated algorithms for trajectory optimization. Bacon (1959) and Tsu (1959) explored the equations of motion for the case of logarithmic spirals, the latter considering the use of solar sails. However, the flight-path angle is constant along a logarithmic spiral so they turn out not to be practical for trajectory design. Petropoulos and

https://doi.org/10.1515/9783110559125-009

Longuski (2004) proposed a more sophisticated curve, the exponential sinusoid,

$$r(\theta) = k_0 \exp[k_1 \sin(k_2\theta + \phi)]$$

which describes the trajectory by means of four constants (k_0, k_1, k_2, ϕ) and resulted in a celebrated mission design technique. They derived analytic expressions for the required tangential acceleration and for the angular rate. Practical applications of this solution to optimization problems have also been discussed by McConaghy et al. (2003), Vasile et al. (2006), and Schütze et al. (2009). Izzo (2006) made a significant contribution by formulating and solving Lambert's problem with exponential sinusoids, and presented a systematic treatment of the boundary conditions. Wall and Conway (2009) suggested generalized forms of the exponential sinusoid together with inverse polynomial functions. Alternative families of solutions have been discussed by Petropoulos and Sims (2002). Other solutions include that of Pinkham (1962), who published a particular representation of the logarithmic spiral and improved its versatility by introducing a scaling factor of the semilatus-rectum, whereas Lawden (1963) established the connection between spirals and optimality conditions. Perkins (1959) obtained interesting analytic results when studying the motion of a spacecraft leaving a circular orbit under a constant thrust acceleration. We refer to the work by McInnes (2004, chap. 5) for a detailed review of solutions arising from the design of solar sails. The design method presented by De Pascale and Vasile (2006), which was later improved by Novak and Vasile (2011), relies on the use of pseudoequinoctial elements for shaping the trajectory, relaxing the Lagrange constraint, and considering a nonzero gauge. A dedicated analysis of the optimality of the method can be found in Vasile et al. (2007). Taheri and Abdelkhalik (2012, 2015a,b) followed a different approach by modeling the trajectory using Fourier series. The method proves flexible for different preliminary design scenarios. Gondelach and Noomen (2015) published an ingenious technique based on working in the hodograph plane and modeling the trajectory using the velocities.

A different problem is that of solving the equations of motion given the thrust profile. Analytic solutions are only available for very specific cases, but due to their theoretical and practical importance it is worth exploring the conditions that yield closed-form solutions. Possibly the first comprehensive approach to the dynamics of a spacecraft under a constant radial and circumferential acceleration is due to Tsien (1953). He solved explicitly the radial case and proposed an approximate solution for the circumferential part. The explicit solution to the constant radial thrust problem is found in terms of elliptic integrals (Battin, 1999, pp. 408–418), and alternative solutions involving the Weierstrass elliptic functions (Izzo and Biscani, 2015), approximate methods (Quarta and Mengali, 2012), and asymptotic expansions (Gonzalo and Bombardelli, 2014) have been proposed. Urrutxua et al. (2015b) solved the Tsien problem in closed form relying on the Dromo formulation (Chap. 4).

Benney (1958) analyzed for the first time the escape trajectories of a spacecraft subject to a constant tangential thrust and provided approximate solutions for the motion, while Lawden (1955) had already found the optimal direction of the thrust to minimize propellant consumption. Boltz (1992) advanced on these pioneering works by assuming that the ratio between the thrust and the local gravity is constant, instead of assuming purely constant thrust. He derived approximate solutions for both the high- and low-thrust cases. Exact solutions to the constant tangential thrust problem have eluded researchers but explicit solutions to certain variables can be found. For instance, the expressions defining the escape conditions or the amplitude of the bounded motion have been provided by different authors (Prussing and Coverstone-Carroll, 1998; Mengali and Quarta, 2009). Bombardelli et al. (2011) derived an alternative asymptotic solution using regular sets of elements. A more didactic approach to perturbation methods applied to the motion of a continuously accelerated spacecraft is presented by Kevorkian and Cole (1981), who took the atmospheric drag as the perturbing acceleration. In a series of works, Zuiani et al. (2012) and Zuiani and Vasile (2015) exploited the advantages of approximate analytical solutions based on equinoctial elements. Colombo et al. (2009) adopted a thrust profile that decreases with the square of the radial distance and arrived at a semianalytical solution. They applied the method to the design of asteroid deflection missions.

The existence of integrals of motion enriches the dynamics of any system. The analysis of the constant radial thrust problem benefits from the fact that the force is conservative. The perturbing potential determines the total energy, which is conserved. The angular momentum is also conserved. Prussing and Coverstone-Carroll (1998) and Akella and Broucke (2002) approached the problem from the energy perspective and discussed the corresponding integrals of motion. General considerations on the integrability of the system can be found in the work by San-Juan et al. (2012). For the tangential thrust problem Battin (1999, p. 418) arrived at an equation with separate variables that is integrated to define a first integral of the motion.

This chapter is organized as follows. In Sect. 9.1 we recover the tangential thrust profile that Bacon (1959) related to the logarithmic spiral and formulate the dynamics. The first reason for recovering the logarithmic spiral is how simple the thrust profile is. It decreases with the square of the radial distance, which seems convenient for solar sails or solar electric propulsion. The second reason is the number of mathematical properties of the logarithmic spirals.[1] The equations of motion are then solved with no prior assumptions about the shape of the trajectory. Two first integrals appear naturally. If the equations of motion under such acceleration are solved rigorously one finds that the logarithmic spiral is not the only solution, but a particular case of an entire family of curves that we call generalized logarithmic spirals, $\mathcal{S}$. The integrals

1 This curve captivated Jakob Bernoulli in the 17th century, so much so that he wanted a logarithmic spiral to be engraved on his tombstone as his coat of arms. His epitaph reads *"Eadem mutata resurgo"*, which translates to "although changed, I arise the same" and refers to the selfsimilarity properties of this curve. More details on its construction can be found in Lockwood (1967). Anecdotally, the craftsman made a mistake and carved an Archimedean spiral instead.

of motion derived in this section yield a natural classification of the trajectories in elliptic, parabolic, and hyperbolic spirals. Sections 9.2, 9.3, and 9.4 present the fully analytic solution for each type of spiral. Closed-form expressions for the trajectory, velocity, and time of flight are provided. They are summarized in Sect. 9.5. Section 9.6 analyzes the evolution of the orbital elements. Properties related to the osculating orbits are highlighted. The departure conditions for spirals emanating from a Keplerian orbit are analyzed in Sect. 9.7. Practical comments can be found in Sect. 9.8, together with a numerical example comparing the new spirals with a solar sail trajectory. The continuity of the solution is proved in Sect. 9.9.

9.1 The equations of motion. First integrals

Let $\mathbf{r} \in \mathbb{R}^3$ denote the position vector of a particle. The dynamics under the action of a central gravitational acceleration and a perturbation $\mathbf{a}_p$ abide by

$$\frac{\mathrm{d}^2\mathbf{r}}{\mathrm{d}t^2} + \frac{\mu}{r^3}\mathbf{r} = \mathbf{a}_p\,, \qquad \text{with} \quad r = \|\mathbf{r}\|$$

Shape-based approaches first assume that the trajectory of the particle can be described by a certain curve, and then the required acceleration to generate such a trajectory is obtained. In particular, Bacon (1959) proved that the thrust that renders a logarithmic spiral can be represented by the perturbing acceleration

$$\mathbf{a}_p = \frac{\mu}{2r^2}\cos\psi\,\mathbf{t} \tag{9.1}$$

where ψ is the flight direction angle (complementary to the flight-path angle) and $\mathbf{t}$ denotes the vector tangent to the trajectory. Bacon considered ψ constant, but we shall make no assumptions and let ψ change. In what follows the problem is normalized so that $\mu = 1$. By definition it is

$$\mathbf{t} = \frac{\mathbf{v}}{v}\,, \qquad \text{and} \quad (\mathbf{t}\cdot\mathbf{r}) = r\cos\psi$$

Here $\mathbf{v} \in \mathbb{R}^3$ is the velocity vector and $v = \|\mathbf{v}\|$. If the perturbing term is confined to the orbital plane the problem is planar and can be described by the polar coordinates (r, θ). Figure 9.1 depicts the geometry of the problem.

The orbital frame $\mathcal{L} = \{\mathbf{i}, \mathbf{j}, \mathbf{k}\}$ is given the usual definition

$$\mathbf{i} = \frac{\mathbf{r}}{r}\,, \qquad \mathbf{k} = \frac{\mathbf{h}}{h}\,, \qquad \mathbf{j} = \mathbf{k}\times\mathbf{i}$$

The angular momentum vector is $\mathbf{h} = \mathbf{r}\times\mathbf{v}$. The velocity of the particle projected in $\mathcal{L}$ reads

$$\mathbf{v} = \dot{r}\,\mathbf{i} + r\dot{\theta}\,\mathbf{j}$$

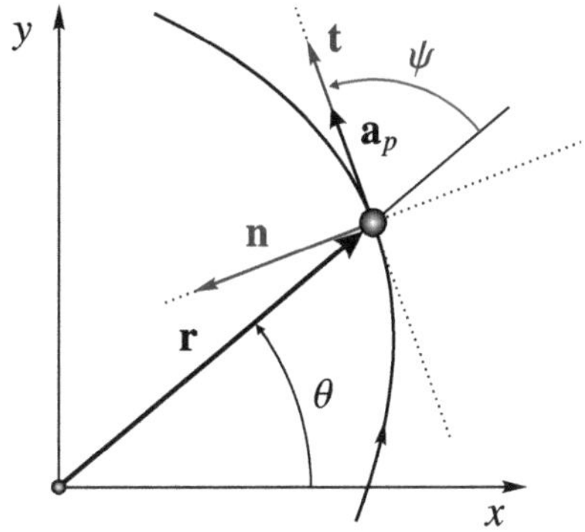

Fig. 9.1: Geometry of the problem. The velocity vector **v** follows the direction of **t**.

where $\dot{\square}$ denotes derivatives with respect to time. The problem can be formulated using intrinsic coordinates in terms of the tangent and normal vectors,

$$\mathbf{t} = \frac{1}{v}(\dot{r}\,\mathbf{i} + r\dot{\theta}\,\mathbf{j}), \quad \text{and} \quad \mathbf{n} = \frac{1}{v}(\dot{r}\,\mathbf{j} - r\dot{\theta}\,\mathbf{i}) \tag{9.2}$$

which define the intrinsic frame $\mathcal{T} = \{\mathbf{t}, \mathbf{n}, \mathbf{b}\}$ with $\mathbf{b} \equiv \mathbf{k}$. The tangent vector can be alternatively defined in terms of the flight direction angle ψ:

$$\mathbf{t} = \cos\psi\,\mathbf{i} + \sin\psi\,\mathbf{j}$$

Recall that ψ is not necessarily constant. Comparing this expression to Eq. (9.2a) it follows that

$$\cos\psi = \frac{\dot{r}}{v}, \quad \sin\psi = \frac{r\dot{\theta}}{v}, \quad \tan\psi = r\frac{\mathrm{d}\theta}{\mathrm{d}r} \tag{9.3}$$

The specific forces acting on the particle are the gravitational attraction of the central body, $\mathbf{a}_g$, and the perturbing acceleration defined in Eq. (9.1), $\mathbf{a}_p$. These accelerations are written in $\mathcal{T}$ as

$$\mathbf{a}_g = -\frac{1}{r^2}(\cos\psi\,\mathbf{t} - \sin\psi\,\mathbf{n})$$
$$\mathbf{a}_p = \frac{\cos\psi}{2r^2}\,\mathbf{t}$$

The intrinsic equations of motion turn out to be

$$\frac{\mathrm{d}v}{\mathrm{d}t} = -\frac{\cos\psi}{2r^2} \tag{9.4}$$

$$v^2 \kappa = +\frac{\sin\psi}{r^2} \tag{9.5}$$

where κ is the curvature of the trajectory, namely

$$\kappa = \frac{\mathrm{d}}{\mathrm{d}s}(\psi + \theta) \tag{9.6}$$

Here s denotes an arclength. It relates to the velocity by means of

$$v = \frac{\mathrm{d}s}{\mathrm{d}t} \tag{9.7}$$

Combining this identity with Eqs. (9.4–9.5) and the geometrical relations in Eq. (9.3) yields the final form of the equations of motion:

$$\frac{\mathrm{d}v}{\mathrm{d}t} = -\frac{\cos\psi}{2r^2} \tag{9.8}$$

$$v\frac{\mathrm{d}}{\mathrm{d}t}(\psi+\theta) = +\frac{\sin\psi}{r^2} \tag{9.9}$$

$$\cos\psi = \frac{\dot{r}}{v} \tag{9.10}$$

$$\sin\psi = \frac{r\dot{\theta}}{v} \tag{9.11}$$

and they must be integrated from the initial conditions

$$t = 0: \qquad v = v_0\,, \quad \psi = \psi_0\,, \quad r = r_0\,, \quad \theta = \theta_0$$

An additional equation relates the polar angle and the radius,

$$\tan\psi = r\,\frac{\mathrm{d}\theta}{\mathrm{d}r} \tag{9.12}$$

If the inertial reference frame $\mathcal{I} = \{\mathbf{i}_\mathcal{I}, \mathbf{j}_\mathcal{I}, \mathbf{k}_\mathcal{I}\}$ is defined so that the angular momentum is $\mathbf{h} = h\,\mathbf{k}_\mathcal{I}$, then $\dot{\theta} > 0$ and from Eq. (9.11) it follows that $\sin\psi > 0 \Rightarrow \psi \in (0, \pi)$. This assumption forces the angular velocity to be $\dot{\theta} > 0$. The retrograde case could be derived in an analogous way, but it is obviated for the sake of clarity. The limits $\psi = 0$ and $\psi = \pi$ are also omitted since they yield rectilinear orbits crossing the origin.

9.1.1 The equation of the energy

Let $\mathcal{E}_k$ denote the specific Keplerian energy of the system, defined by the vis-viva equation

$$\mathcal{E}_k = \frac{v^2}{2} - \frac{1}{r}$$

Its time evolution is given by the power performed by the perturbing acceleration,

$$\frac{\mathrm{d}\mathcal{E}_k}{\mathrm{d}\,t} = \mathbf{a}_p \cdot \mathbf{v}$$

From Eq. (9.1) it follows that

$$\frac{\mathrm{d}\mathcal{E}_k}{\mathrm{d}\,t} = \frac{\cos\psi}{2r^2}(\mathbf{t}\cdot\mathbf{v}) = \frac{v\cos\psi}{2r^2}$$

The geometrical relation between the radial velocity and $\cos\psi$ – Eq. (9.10) – yields an equation with separate variables,

$$\frac{\mathrm{d}\mathcal{E}_k}{\mathrm{d}t} = \frac{1}{2r^2}\frac{\mathrm{d}r}{\mathrm{d}t} \Longrightarrow \frac{\mathrm{d}\mathcal{E}_k}{\mathrm{d}r} = \frac{1}{2r^2}$$

which is integrated to provide a first integral of motion:

$$\mathcal{E}_k = -\frac{1}{2r} + \frac{K_1}{2} \Longrightarrow v^2 - \frac{1}{r} = K_1 \tag{9.13}$$

Here K_1 is a constant of integration, determined by

$$K_1 = v_0^2 - \frac{1}{r_0} \tag{9.14}$$

in terms of the initial values of the radius and velocity, r_0 and v_0. It can take positive or negative values. The integral of motion (9.13) is the generalized integral of the energy.

For the velocity to be real it must be

$$K_1 \geq -\frac{1}{r} \Longrightarrow 1 + K_1 r \geq 0 \tag{9.15}$$

which sets a lower bound on the values that K_1 can take.

Combining the integral of motion (9.13) with the equation of the Keplerian energy furnishes a local property that holds for any value of r:

$$2v^2 - v_k^2 = 2(K_1 - \mathcal{E}_k)$$

This expression relates the velocity at r of the spiral trajectory with the velocity of the Keplerian orbit with energy $\mathcal{E}_k$.

9.1.2 The equation of the angular momentum

In this section we prove the existence of an additional first integral related to the angular momentum. Dividing Eq. (9.8) by Eq. (9.9) and introducing Eq. (9.12) it transforms into an equation with separate variables,

$$\frac{\mathrm{d}v}{v} = -\frac{\cot\psi}{2}(\mathrm{d}\psi + \mathrm{d}\theta) = -\frac{\mathrm{d}r}{2r} - \frac{1}{2}\cot\psi\,\mathrm{d}\psi$$

which is integrated easily,

$$2\ln v = -\ln r - \ln(\sin\psi) + \ln K_2 \Longrightarrow v^2 r\sin\psi = K_2 \tag{9.16}$$

considering that $\sin\psi > 0$. This result defines an integral of motion written in terms of the constant K_2, which is determined from the initial values r_0, v_0, ψ_0:

$$K_2 = v_0^2 r_0 \sin\psi_0 = v_0 r_0^2 \dot{\theta}_0 \tag{9.17}$$

The integral of motion (9.16) is the generalization of the angular momentum equation. It relates to the angular momentum of the Keplerian orbit at r, h_k, by means of

$$\frac{v_k \sin \psi_k}{v^2 \sin \psi} = \frac{h_k}{K_2}$$

If the orbit is prograde, $(\mathbf{h} \cdot \mathbf{k}_{\mathcal{I}}) > 0$, and nondegenerate, $h \neq 0$, then $K_2 > 0$. The condition for the orbit to be prograde can be satisfied by an adequate choice of the inertial reference so that $\dot{\theta} > 0$. The limit $K_2 \to 0$ yields degenerate rectilinear orbits for which $\theta = \theta_0$ and $\psi = 0$ or $\psi = \pi$.

The form of Eq. (9.17) yields an interesting property of the generalized spirals: since $\sin \psi_0$ is symmetric with respect to $\psi_0 = \pi/2$, two different trajectories that emanate from the same radius and with the same value of K_1 and K_2 can be found. The first trajectory will depart in the lowering regime ($\psi_0^+ = \pi/2 + \delta$) and the second in the raising regime ($\psi_0^- = \pi/2 - \delta$), while sharing the same value of K_2. These two trajectories relate to the two different *regimes* that characterize the generalized spirals considered in this chapter. The radius of the orbit will increase/decrease depending on the sign of $\dot{r}$. Since $\psi \in (0, \pi)$, Eq. (9.10) shows that there are two possibilities:

Raising regime: corresponding to $\cos \psi > 0$ ($\dot{r} > 0$).
Lowering regime: corresponding to $\cos \psi < 0$ ($\dot{r} < 0$).

In the raising regime the perturbing acceleration $\mathbf{a}_p$ moves the particle away from the attracting body (the origin). In the lowering regime the opposite happens. These two regimes are separated by the classical circular Keplerian orbit. In this analysis, circular Keplerian orbits appear when $\psi = \pi/2$ and the perturbing acceleration $\mathbf{a}_p$ vanishes. When ψ crosses $\pi/2$ the direction of the thrust vector changes although it remains tangential to the velocity. Hence, it is important to understand the evolution of the flight direction angle.

9.1.3 The flight direction angle ψ

The value of $\sin \psi$ can be solved from Eq. (9.16) and the equation of the energy:

$$\sin \psi = \frac{K_2}{r v^2} = \frac{K_2}{1 + K_1 r} \tag{9.18}$$

This relation is the solution to the differential equation

$$\frac{\mathrm{d}\psi}{\mathrm{d}r} = -\frac{K_1}{r v^2} \tan \psi \qquad \text{or} \qquad \frac{\mathrm{d}\psi}{\mathrm{d}t} = -\frac{K_1}{r v} \sin \psi \tag{9.19}$$

that governs the evolution of the flight direction angle ψ. Note that for $K_1 = 0$ the angle ψ will be constant. The trajectory in this case reduces to a logarithmic spiral. Logarithmic spirals are not of much interest in mission design for this reason. Nonzero values of the constant K_1 yield an entire family of *generalized spirals*. If $K_1 < 0$ the flight direction angle will always grow in time, and for $K_1 > 0$ it will decrease. Note that, since $0 < \sin\psi \le 1$, it must be $K_2 \le |1 + K_1 r| = 1 + K_1 r$. The last simplification comes from Eq. (9.15), because $K_1 \ge -1/r$.

The term $\cos\psi$ can be defined from Eq. (9.18), resulting in

$$\cos\psi = \pm\sqrt{1 - \frac{K_2^2}{(1+K_1 r)^2}} = \pm\frac{\sqrt{(1+K_1 r)^2 - K_2^2}}{1+K_1 r} \tag{9.20}$$

The condition from Eq. (9.15) is required for the last simplification. The +/– sign of the cosine relates to the raising/lowering regime of the trajectory. In what remains of the chapter the first sign will always correspond to the raising regime, and the second to the lowering regime. Considering Eqs. (9.18) and (9.20), $\tan\psi$ can be written as

$$\tan\psi = \pm\frac{K_2}{\sqrt{(1+K_1 r)^2 - K_2^2}} \tag{9.21}$$

9.1.4 The fundamental theorem of curves

Let $\kappa(s)$ and $\tau(s)$ be two singled-valued functions of the arclength. The fundamental theorem of curves states that any curve with torsion $\tau(s)$ and nonzero curvature $\kappa(s)$ is uniquely determined except for a Euclidean transformation (rotation, translation, and/or reflection). The arclength, curvature, and torsion of the curve are invariant under such actions.

Equation (9.7) defines the velocity along the trajectory as the time derivative of the arclength. Combining this result with Eq. (9.10), it follows that

$$\mathrm{d}s = v\,\mathrm{d}t = \sec\psi\,\mathrm{d}r \tag{9.22}$$

The arclength of the spiral connecting r_1 and r_2 is obtained from the integral of the previous expression, taking Eq. (9.20) into account. It yields

$$\begin{aligned} s &= \frac{1}{|K_1|}\left|\sqrt{(1+K_1 r_2)^2 - K_2^2} - \sqrt{(1+K_1 r_1)^2 - K_2^2}\right| \\ &= \frac{K_2}{|K_1|}(|\cot\psi_2| - |\cot\psi_1|) \end{aligned}$$

If there are transitions between the raising and lowering regimes the arclength has to be computed in separate arcs. Once the trajectory is solved for each type of spiral the arclength will be given as a function of the polar angle, $s = s(\theta)$.

The curvature $\kappa(s)$ has already been introduced by Eq. (9.5), and reduces to

$$\kappa = \frac{\sin\psi}{(rv)^2} = \frac{K_2}{r(1+K_1 r)^2}$$

The curvature depends only on the radial distance and the constants K_1 and K_2 by means of Eqs. (9.13) and (9.18). For $K_2 = 0$, the curvature vanishes and the trajectory becomes rectilinear. This case will not be considered in this chapter.

An important result in differential geometry is that the torsion of any planar curve is zero. Hence, two spirals with the same values of K_1 and K_2 have the same curvature and torsion. By the fundamental theorem of curves the resulting curves are equal except perhaps for a Euclidean transformation. This result suggests the concept of the *generating spiral*:

Definition 1. (Generating spiral Γ) The generalized logarithmic spiral defined by K_1 and K_2 that spans from $\theta \to -\infty$ to $\theta \to +\infty$ with an arbitrary orientation is called the generating spiral, and it is written Γ-spiral. It contains all the possible solutions to Eqs. (9.8–9.11) given K_1 and K_2, except perhaps for a rotation or reflection depending on the initial conditions.

Two spirals that originate from the same Γ-spiral have the same K_1 and K_2. If they are initially in the same regime but defined by different initial positions (r_0, θ_0), then they relate through a single rotation. On the contrary, if the initial positions are the same but the departure regime changes, then the trajectories relate by a reflection $\mathscr{R}$. This last reflecting transformation is particularly interesting because of its inherent symmetry:

Definition 2. ($\mathscr{C}$-symmetry) Two curves are said to be $\mathscr{C}$-symmetric if they originate from the same Γ-spiral (they share the same values of K_1 and K_2) and relate by a reflection about a polar axis, $\mathscr{R}: \mathcal{S} \mapsto \mathcal{S}^\dagger$, where $\mathcal{S}^\dagger$ is the $\mathscr{C}$-symmetric of $\mathcal{S}$. The spirals are reflected about the axis defined by θ_0. One spiral is initially in the raising regime, whereas the other is in the lowering regime.

The reflection $\mathscr{R}$ simply transforms a spiral which is initially in the raising regime into another one initially in the lowering regime (or vice-versa), while keeping constant K_1, K_2 and (r_0, θ_0). The reflection $\mathscr{R}$ transforms the trajectory but, since the orbit is assumed to be prograde, the initial velocity vectors are not reflected about θ_0. They are reflected about the perpendicular to the initial radius vector. It is natural to consider the following concept:

Definition 3. ($\mathscr{R}$-invariance) A spiral $\mathcal{S}$ is $\mathscr{R}$-invariant if it coincides with the $\mathscr{C}$-symmetric spiral, i.e., $\mathcal{S} \equiv \mathcal{S}^\dagger$.

The concept of $\mathscr{C}$-symmetry involves two spirals that originate from the same Γ-spiral. It should not be confused with an important intrinsic property of a single spiral, which is:

Definition 4. ($\mathscr{T}$-symmetry) A spiral $r = r(\theta)$ is said to be $\mathscr{T}$-symmetric if $r(\theta_m + \delta) = r(\theta_m - \delta)$, with $\delta \geq 0$. The polar angle θ_m is the axis of $\mathscr{T}$-symmetry.

The concept of $\mathscr{T}$-symmetry yields two important properties of the transformation $\mathscr{R}$. First, if a spiral is $\mathscr{T}$-symmetric then the reflection $\mathscr{R}$ can be represented by a rotation. Second, if a spiral is $\mathscr{R}$-invariant then the spiral is $\mathscr{T}$-symmetric. The opposite is not true, in general. This statement follows from the fact that the definition of a reflection

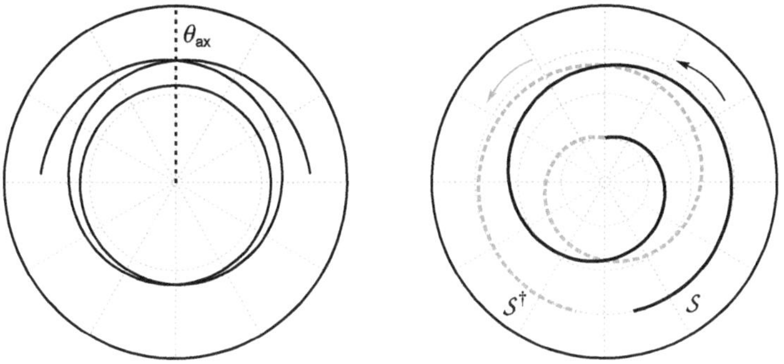

(a) A $\mathscr{T}$-symmetric spiral **(b)** A pair of $\mathscr{C}$-symmetric spirals

Fig. 9.2: Graphical representation of the $\mathscr{T}$- and $\mathscr{C}$-symmetries.

with respect to a polar axis θ_{ax} is $\mathscr{R}$: $r(\theta_{ax} + \delta) \mapsto r(\theta_{ax} - \delta)$. If the Γ-spiral is $\mathscr{R}$-invariant then $r(\theta_{ax} + \delta) = r(\theta_{ax} - \delta)$, which is exactly the definition of $\mathscr{T}$-symmetry for $\theta_m \equiv \theta_{ax}$. Conversely, a $\mathscr{T}$-symmetric spiral is not $\mathscr{R}$-invariant unless the reflection is applied about the axis of symmetry, $\theta_{ax} \equiv \theta_m$. Figure 9.2 helps in visualizing the $\mathscr{T}$-symmetry of a spiral and a pair of $\mathscr{C}$-symmetric spirals.

9.1.5 Families of solutions

Like $\mathcal{E}_k$ in the Keplerian case, different values of the constant of the generalized energy K_1 yield different types of solutions. The resulting families are now introduced briefly, and will be analyzed in detail in Secs. 9.2, 9.3, and 9.4, respectively:

Elliptic spirals: ($K_1 < 0$) There is a physical limit to the radius, r_ℓ, set by the condition $v^2 \geq 0$ in Eq. (9.14):

$$r_\ell = -\frac{1}{K_1}, \qquad \text{with} \quad K_1 < 0 \tag{9.23}$$

Elliptic spirals are dynamically bounded and symmetric. In Section 9.2 an additional, more restrictive limit to the radius is found. Since ψ grows continuously in time – see Eq. (9.19b) – the sign of $\dot{r}$ changes when crossing the value $\psi = \pi/2$ and, in such a case, the trajectory evolves from the raising to the lowering regime.

Parabolic spirals: ($K_1 = 0$) In this case it is $v^2 = 1/r$ meaning that the velocity at every point of the spiral matches the local circular velocity. A particle in the raising regime will reach infinity with $v \to 0$. The angle ψ remains constant and the trajectory reduces to a logarithmic spiral.

Hyperbolic spirals: ($K_1 > 0$) A particle in the raising regime will spiral away and reach infinity with nonzero finite velocity, $v \to v_\infty \equiv \sqrt{K_1}$ as $r \to \infty$. The constant

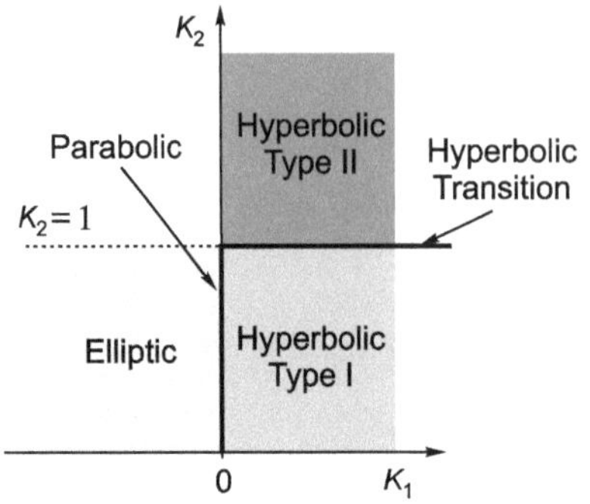

Fig. 9.3: Types of spirals in the parametric space (K_1, K_2).

K_1 equals the characteristic energy C_3. There are two subfamilies of hyperbolic spirals:
- Spirals of *Type I*, defined by $K_2 < 1$. They only have one asymptote.
- Spirals of *Type II*, defined by $K_2 > 1$. They have two symmetric asymptotes.

Both subfamilies are separated by an asymptotic case, $K_2 = 1$, as will be proved later on.

The discussed families of solutions can be distinguished in Fig. 9.3. The figure depicts the regions corresponding to each type of spiral on the (K_1, K_2) plane. The spirals are elliptic for $K_1 < 0$, parabolic for $K_1 = 0$, and hyperbolic for $K_1 > 0$. It has already been stated that two subfamilies of hyperbolic spirals exist: Type I ($K_2 < 1$) and Type II ($K_2 > 1$). The transition corresponds to the limit case $K_2 = 1$. Elliptic spirals exist only for $K_2 \in (0, 1)$.

Figure 9.4 depicts the three different families of generalized spirals, with the elliptic and hyperbolic spirals bifurcating from the parabolic (logarithmic) one. Although all trajectories are in the raising regime ($\dot{r} > 0$), the angle ψ only grows for the case of elliptic spirals. For parabolic (logarithmic) spirals it remains constant, and decreases in the hyperbolic case.

In the following sections, Eqs. (9.8–9.11) are solved in closed form to determine the trajectory and time of flight. The solution takes different forms depending on the type of spiral. The continuity of the solution is proved in Sect. 9.9.

It is worth mentioning that the solutions can be unified using the Weierstrass elliptic functions. In this formalism there is no need to distinguish the different types of orbits. The most general form of the referred universal solution will be derived in Sect. 12.8, after the properties of each family have been characterized.

9.1.6 Form of the solution

The problem has been reduced to the planar case. Thus, there are four constants of motion defining the shape of the trajectory, and relating the position in the orbit with

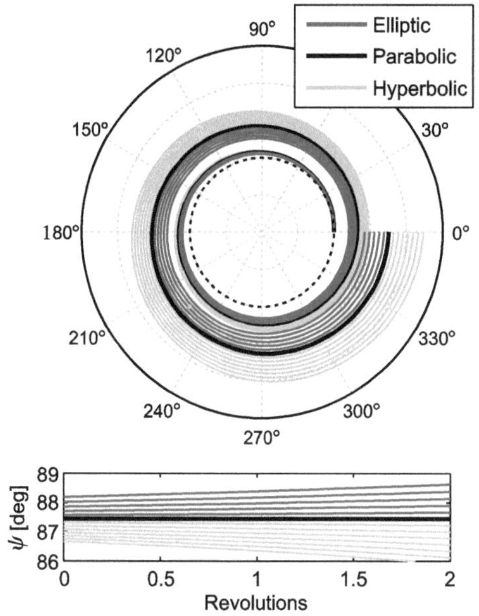

Fig. 9.4: Families of generalized spirals: elliptic ($K_1 < 0$), parabolic ($K_1 = 0$), and hyperbolic ($K_1 > 0$).

the time. In order to integrate the equations of motion (9.8–9.11) we shall first obtain the relation between the polar angle and the radial distance, $\theta(r)$. Inverting this equation yields the trajectory, $r(\theta)$. Finally, the time evolution needs to be accounted for. Typically one seeks a relation $\theta(t)$ that yields $r(\theta(t))$, usually obtained as the inverse of $t(\theta)$. In some cases, like Keplerian orbits, the equation $t(\theta)$ cannot be inverted explicitly and $\theta(t)$ needs to be solved numerically. This is none other than Kepler's equation. In the present problem, we adopt a slightly different approach by writing $t(r)$. The explicit expressions for $t(r)$ derived in the following sections are the spiral form of Kepler's equation, keeping in mind that $t(r) \equiv t(r(\theta))$. Like Kepler's equation, the equation for the time of flight cannot be inverted analytically.

9.2 Elliptic spirals ($K_1 < 0$)

The family of elliptic spirals is defined by $K_1 < 0$, meaning that the velocity is bounded by the circular velocity

$$v^2 < \frac{1}{r}$$

As $v^2 \to 1/r$ or $K_1 \to 0$ elliptic spirals converge to parabolic (logarithmic) spirals.

Since $\sin\psi, r, v > 0$, and $K_1 < 0$, Eq. (9.19b) shows that the time derivative of ψ is always positive. That is, the angle ψ always grows in an elliptic spiral. For these kinds of orbits the constant K_2 always belongs to the open interval $K_2 \in (0, 1)$. Indeed, from Eq. (9.18)

$$K_2 = \sin\psi_0(1 + K_1 r_0) < \sin\psi_0 \le 1 \implies K_2 \in (0, 1)$$

If the orbit is initially in the lowering regime, ($\psi_0 > \pi/2$ and $\cos\psi_0 < 0$), the orbit will reach the origin $r = 0$, which is a singular point in this formulation. Therefore, the final fate for an orbit that is initially in the lowering regime is $r = 0$ and $\psi = \pi - \arcsin(K_2)$.

If the orbit is initially in the raising regime ($\psi < \pi/2$ and $\cos\psi > 0$) it will eventually reach $\psi = \pi/2$, so $\cos\psi$ will become negative, changing the sign of the radial velocity. The radius reaches a maximum when $\psi = \pi/2$. By Eq. (9.18) the apoapsis radius reads

$$r_{\max} = \frac{1 - K_2}{-K_1} \tag{9.24}$$

Once the maximum radius is reached, $\cos\psi$ changes its sign and the spiral enters the lowering regime, $\dot{r} < 0$. Thus, elliptic spirals transition from the raising regime to the lowering regime naturally. The opposite is not true, i.e., the trajectory will never change from the lowering to the raising regime. The limit $r_{\max}$ will always be smaller than $r_\ell = -1/K_1$.

Elliptic spirals cannot escape to infinity. Once they transition to the lowering regime ($\cos\psi < 0$), they will remain in that regime and fall toward the origin. The spiral can only be in the raising regime ($\cos\psi > 0$) if it is initially in the raising regime, i.e., $\cos\psi_0 > 0$. If initially it is $r_0 = r_{\max}$, then the spiral immediately transitions to the lowering regime and falls toward the origin.

When $r = r_{\max}$ the velocity vector is normal to the radius vector ($\psi = \pi/2$) and from Eq. (9.13) the velocity will be

$$v_m = \sqrt{K_1 + \frac{1}{r_{\max}}} = \sqrt{\frac{-K_1 K_2}{1 - K_2}} = \sqrt{\frac{K_2}{r_{\max}}} \tag{9.25}$$

The velocity in an elliptic spiral is minimum at apoapsis ($\psi = \pi/2$ and $r = r_{\max}$), i.e., $\min(v) = v_m$.

9.2.1 The trajectory

In what remains of the section the equations of motion are solved for the elliptic case. When substituting Eq. (9.21) in Eq. (9.12) it yields

$$\mathrm{d}\theta = \pm \frac{K_2\,\mathrm{d}r}{r\sqrt{(1 + K_1 r)^2 - K_2^2}} \tag{9.26}$$

Recall that the first/second sign corresponds to the raising/lowering regime. This expression can be integrated to provide the evolution of the polar angle,

$$\begin{aligned}\theta - \theta_m &= \mp \frac{K_2}{\ell}\left|\operatorname{arccosh}\left[\frac{r_{\max}}{r}\left(1 + \frac{1}{K_2}\right) - \frac{1}{K_2}\right]\right| \\ &= \mp \frac{K_2}{\ell}\left|\operatorname{arccosh}\left[-\frac{1}{K_2}\left(1 + \frac{\ell^2}{K_1 r}\right)\right]\right|\end{aligned} \tag{9.27}$$

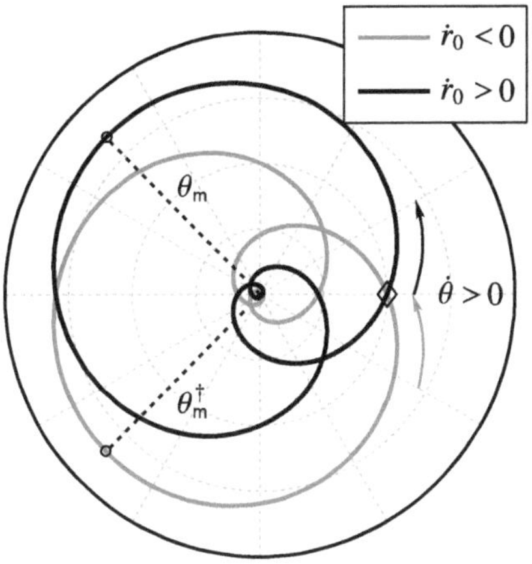

Fig. 9.5: A pair of $\mathscr{C}$-symmetric elliptic spirals departing from (◊), propagated forward and backward.

with $\ell = (1 - K_2^2)^{1/2}$. Here θ_m defines the orientation of the apoapsis, $\theta_m = \theta(r_{\max})$, which is equivalent to the apse-line. When implementing this method the value of θ_m is solved initially from Eq. (9.27) particularized at (r_0, θ_0). Depending on the initial regime of the spiral a pair of $\mathscr{C}$-symmetric elliptic spirals appear, defined by

$$\theta_m = \theta_0 + \frac{K_2}{\ell}\left|\operatorname{arccosh}\left[-\frac{1}{K_2}\left(1 + \frac{\ell^2}{K_1 r_0}\right)\right]\right| \tag{9.28}$$

$$\theta_m^\dagger = \theta_0 - \frac{K_2}{\ell}\left|\operatorname{arccosh}\left[-\frac{1}{K_2}\left(1 + \frac{\ell^2}{K_1 r_0}\right)\right]\right| \tag{9.29}$$

Equation (9.28) applies to spirals initially in the raising regime, and Eq. (9.29) corresponds to spirals initially in the lowering regime. The pair of resulting spirals are $\mathscr{C}$-symmetric and relate simply by a reflection about θ_0.

A pair of $\mathscr{C}$-symmetric spirals generated by the same Γ-spiral is depicted in Fig. 9.5. One curve is initially in the lowering regime, whereas the other is initially in the raising regime. The angular momentum is positive, so $\dot{\theta} > 0$. Being $\theta_0 = 0$, $\mathscr{R}$ defines a reflection with respect to the horizontal axis. The direction of the initial velocity vector determines which trajectory is the one followed by the particle.

The equation for the trajectory is obtained upon inversion of Eq. (9.27) and yields:

$$\frac{r(\theta)}{r_{\max}} = \frac{1 + K_2}{1 + K_2 \cosh\beta} \tag{9.30}$$

where β denotes the spiral anomaly:

$$\beta = \frac{\ell}{K_2}(\theta - \theta_m)\,, \qquad \beta^\dagger = \frac{\ell}{K_2}(\theta - \theta_m^\dagger)$$

The first argument, β, corresponds to a spiral initially in the raising regime. The second, $\beta^\dagger$, defines the trajectory of a spiral initially in the lowering regime.

Elliptic spirals are $\mathscr{T}$-symmetric with respect to the axis defined by θ_m, i.e., $r(\theta_m + \delta) = r(\theta_m - \delta)$ with $\delta \geq 0$. This property follows from the equation of the trajectory,

Eq. (9.30). Thus, the reflection $\mathscr{R}$ can alternatively be represented by a rotation[2] of magnitude ϑ, with

$$\vartheta = \frac{2K_2}{\ell}\left|\operatorname{arccosh}\left[-\frac{1}{K_2}\left(1+\frac{\ell^2}{K_1 r_0}\right)\right]\right|$$

An elliptic spiral in the raising regime will always intersect itself at least once when propagated for sufficient revolutions and with adequate initial conditions. Intersections occur on the axis of symmetry, $\theta = \theta_m$.

From Eq. (9.30) and the symmetry properties of the trajectory, it can be verified that $\lim_{\theta\to+\infty} r(\theta) = \lim_{\theta\to-\infty} r(\theta) = 0$, which means that all elliptic spirals fall toward the origin for sufficiently large times, and also when propagated backward. The Γ-spirals depart from the origin, reach $r_{\max}$, and fall to the origin again.

The arclength is measured from θ_m so it takes positive values for $\theta > \theta_m$ and negative values for $\theta < \theta_m$. It can be solved in terms of the polar angle by introducing Eq. (9.30) into Eq. (9.22), and then integrating from θ_m to θ thanks to Eq. (9.10). It results in

$$s(\theta) = \frac{\ell K_2 \sinh\beta}{(-K_1)(1+K_2\cosh\beta)}$$

For spirals initially in the raising regime, $\dot{r}_0 > 0$, $s(\theta)$ relates to the polar angle through the spiral anomaly β. Conversely, if $\dot{r}_0 < 0$ then β is replaced by $\beta^\dagger$. The total arclength of the Γ-spiral, s_Γ, reduces to

$$s_\Gamma = \frac{2\ell}{(-K_1)}$$

It only depends on the constants K_1 and K_2, equivalent to the energy and angular momentum. Given $K_1 < 0$ and K_2 the corresponding Γ-spiral contains all the possible elliptic spirals defined by K_1 and K_2. In addition, s_Γ is the maximum arclength that the particle can travel along the spiral.

9.2.2 The time of flight

In order to solve completely the equations of motion – Eqs. (9.8–9.11) – a relation between the position in the orbit and the time is required. The radial velocity, defined in

2 The trajectory being $\mathscr{T}$-symmetric is a necessary condition for $\mathscr{R}$-invariance. An elliptic spiral will be $\mathscr{R}$-invariant if both the axes of $\mathscr{T}$-symmetry and $\mathscr{C}$-symmetry coincide. This is equivalent to $\theta_0 = \theta_m + n\pi$, with $n = 0, 1, 2, \ldots$. From Eq. (9.30) it follows

$$\mathscr{R}\text{-invariance} \iff \frac{r_{0,\text{inv}}}{r_{\max}} = \frac{1+K_2}{1+K_2\cosh\left(\frac{n\pi\ell}{K_2}\right)}$$

If initially $r_0 \equiv r_{0,\text{inv}}$ then the axis of symmetry θ_m coincides with the initial direction given by θ_0. The axis of $\mathscr{T}$-symmetry is $\mathscr{R}$-invariant.

Eq. (9.10), can be inverted to provide

$$\frac{\mathrm{d}t}{\mathrm{d}r} = \frac{\sec\psi}{v} = \pm\sqrt{\frac{r(1+K_1 r)}{(1+K_1 r)^2 - K_2^2}} \tag{9.31}$$

Integrating this expression from $r_{\min}$ to r yields the time of flight as a function of the radial distance and referred to the time of apoapsis passage, t_m:

$$t(r) - t_m = \pm\left(\frac{rv}{K_1}\sqrt{\frac{1-\sin\psi}{1+\sin\psi}} + \frac{K_2 \Delta\,\mathrm{E} - k'^2 \Delta\Pi}{v_m K_1 k}\right) \tag{9.32}$$

Here $\Delta\,\mathrm{E}$ and $\Delta\Pi$ are the difference between the incomplete and the complete elliptic integrals of the second and third kinds,

$$\Delta\,\mathrm{E} = \mathrm{E}(\phi, k) - \mathrm{E}(k)\,, \qquad \Delta\Pi = \Pi(n;\phi,k) - \Pi(n;k)$$

Their argument, modulus, and parameter are

$$\sin\phi = \frac{v_m}{v}\sqrt{\frac{2}{1+\sin\psi}}\,, \qquad k = \sqrt{\frac{1-K_2}{2}}\,, \qquad n = \frac{K_2 - 1}{2K_2}$$

and k' denotes the complementary modulus, which relates to the modulus k by means of $k^2 + k'^2 = 1$ (for a discussion about the notation and required properties see Appendix E). This form of the equation for the time of flight resembles Kepler's equation, provided that t_m can be seen as the spiral equivalent to the time of periapsis passage. Its value is solved initially from Eq. (9.32) particularized at $t(r_0) = t_0$,

$$t_m = t_0 \mp \left(\frac{r_0 v_0}{K_1}\sqrt{\frac{1-\sin\psi_0}{1+\sin\psi_0}} + \frac{K_2 \Delta\,\mathrm{E}_0 - k'^2 \Delta\Pi_0}{v_m K_1 k}\right) \tag{9.33}$$

The sign is chosen according to the initial regime of the spiral.

The time of flight to describe the complete generating spiral, t_Γ, is

$$t_\Gamma = \frac{2\left[K_2\,\mathrm{E}(k) - k'^2\,\Pi(n;k)\right]}{v_m K_1 k}$$

This expression depends only on K_1 and K_2. It is the maximum time that any elliptic spiral generated by Γ will require to connect two points.

9.3 Parabolic spirals ($K_1 = 0$)

Parabolic spirals are equivalent to the well-known logarithmic spirals. The velocity coincides with the local circular velocity,

$$v = \frac{1}{\sqrt{r}}$$

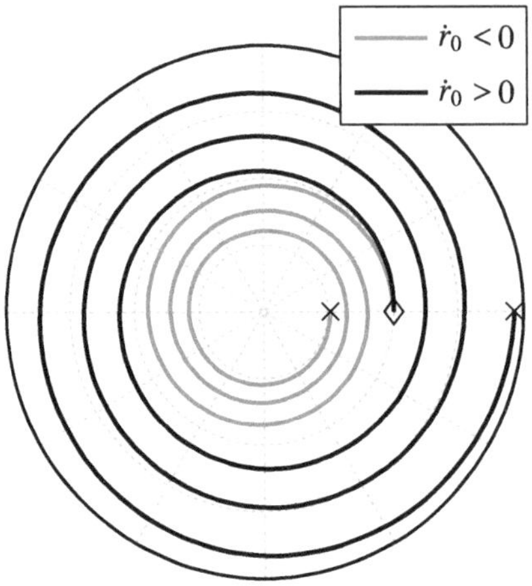

Fig. 9.6: Pair of $\mathscr{C}$-symmetric parabolic spirals, with $\psi = 88°$ and $\psi^\dagger = 92°$, respectively.

and shows that parabolic (logarithmic) spirals reach infinity with zero velocity, $\lim_{r\to\infty} v = 0$, just like parabolic orbits in the Keplerian case. The evolution of the angle ψ is governed by Eq. (9.19). For $K_1 = 0$ it vanishes and proves that the angle ψ is constant along a parabolic spiral.

The definition of K_2 provided in Eq. (9.16) becomes

$$\sin\psi = K_2\,, \quad \text{and} \quad \cos\psi = \pm\sqrt{1 - K_2^2} = \pm\ell$$

With ψ constant, the regime is defined initially by the value of ψ_0. In the limit case $\psi = \pi/2$ ($K_2 = 1$) parabolic spirals degenerate into circular Keplerian orbits. The thrust vanishes and, since ψ is constant, it will remain zero. The constant K_2 is restricted to $K_2 \in (0, 1]$. Figure 9.6 shows a pair of $\mathscr{C}$-symmetric parabolic spirals. The corresponding flight direction angles make $\sin\psi = \sin\psi^\dagger$ and therefore the $\mathscr{C}$-symmetry.

9.3.1 The trajectory

This section proves that the solution to the system of equations (9.8–9.11) when $K_1 = 0$ is indeed a logarithmic spiral. Equation (9.12) is integrated to provide

$$\theta - \theta_0 = \pm\frac{K_2}{\ell}\ln(r/r_0) \tag{9.34}$$

The fact that $\lim_{r\to\infty}\theta(r) = \infty$ shows that the particle follows a spiral branch when reaching infinity.

Equation (9.34) is inverted to define the trajectory of the particle,

$$r(\theta) = r_0\,e^{(\theta-\theta_0)\cot\psi} \tag{9.35}$$

which is none other than the equation of a logarithmic spiral.

If measured from the initial position θ_0, the arclength[3] takes the form

$$s(\theta) = \frac{r_0}{\cos\psi}\left[e^{(\theta-\theta_0)\cot\psi} - 1\right]$$

This definition yields positive values of the arclength for $\theta > \theta_0$, and negative values if propagated backward ($\theta < \theta_0$).

9.3.2 The time of flight

The time of flight is solved from the inverse of the radial velocity and results in:

$$t - t_0 = \pm\frac{2}{3\ell}\left(r^{3/2} - r_0^{3/2}\right) \tag{9.36}$$

It takes an infinite time to reach $r \to \infty$, as shown by the limit $\lim_{r\to\infty} t(r) = \infty$. On the contrary, parabolic spirals in the lowering regime reach the attractive center in finite time:

$$\lim_{r\to 0} t(r) = \frac{2r_0^{3/2}}{3\ell}$$

Of course for $\psi = \pi/2$ ($\ell = 0$) the spiral never reaches the origin, since the spiral degenerates into a circular Keplerian orbit.

9.4 Hyperbolic spirals ($K_1 > 0$)

Hyperbolic spirals are generalized logarithmic spirals with positive constant of the generalized energy. When in the raising regime spirals of this family will reach infinity with a finite, nonzero velocity v_∞. The hyperbolic excess velocity can be solved from Eq. (9.14):

$$v_\infty = \lim_{r\to\infty}\sqrt{K_1 + \frac{1}{r}} = \sqrt{K_1} \implies K_1 \equiv C_3 \tag{9.37}$$

This result is analogous to the Keplerian case, where the hyperbolic excess velocity is defined by the characteristic energy, $C_3 = v_\infty^2$. The time derivative of ψ, given by Eq. (9.19b), proves that the angle ψ always decreases along hyperbolic spirals.

While parabolic spirals reach infinity along a spiral branch, hyperbolic spirals reach infinity along an asymptotic branch. In the limit case $r \to \infty$ the polar angle converges to a finite value θ_{as}. Since ψ decreases in time the position and velocity vectors will become parallel at infinity. The impact parameter of the resulting asymptotes,

3 The original expression for the arclength of the logarithmic spiral is attributed to Evangelista Torricelli back in the 17th century. He derived the first rectification of a transcendental curve and was able to integrate the length of the spiral by evaluating the sum of small values of $d\theta$.

c, can be solved from the equation of the angular momentum, Eq. (9.16),

$$h = v_\infty c \implies c = \frac{K_2}{K_1} \tag{9.38}$$

which defines the minimum distance from the asymptote to the center of attraction.

Equation (9.18) is valid for all types of spirals. It relates the constants K_1 and K_2 with the flight direction angle ψ, and of course $\sin\psi \leq 1$. For $K_2 < 1$ this condition is satisfied naturally, provided that $K_1 r > 0$. However, in the case $K_2 > 1$ this condition holds only if

$$r \geq r_{\min} = \frac{K_2 - 1}{K_1}$$

A minimum radius exists for hyperbolic spirals with $K_2 > 1$, so they will never reach the center of attraction. This radius is referred to as the periapsis of the spiral. The expression for $r_{\min}$ is equal to that for $r_{\max}$ (for elliptic spirals), the only difference being the values that K_1 and K_2 can take.

There are two different types of hyperbolic spirals: those with $K_2 < 1$ (Type I) and those with $K_2 > 1$ (Type II). Figure 9.7 shows the evolution of the radius and the angle ψ for the two types of hyperbolic spirals. Type II spirals in the lowering regime reach $r_{\min}$, then transition to the raising regime and escape. Type I spirals in the lowering regime simply converge to the origin, and escape to infinity if in the raising regime. Also, $\cos\psi$ can change its sign only for Type II spirals, which means that the trajectory transitions from the lowering to the raising regime. The two types of hyperbolic spirals are separated by the asymptotic limit $K_2 \to 1$.

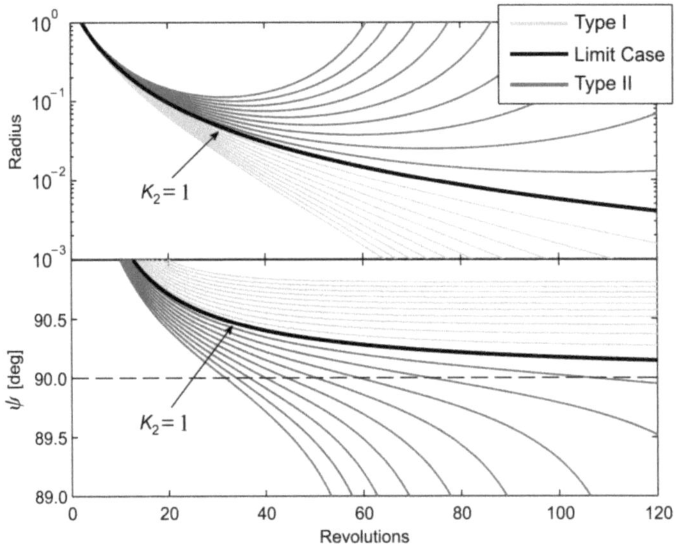

Fig. 9.7: Evolution of the radius and angle ψ along hyperbolic spirals of Types I and II initially in the lowering regime.

This figure also unveils a fundamental difference between hyperbolic spirals of Types I and II: a hyperbolic spiral of Type I only has one asymptote, whereas spirals of Type II have two asymptotes.

9.4.1 Type I hyperbolic spirals

Spirals of this type are defined by $(K_1, K_2) \in (0, \infty) \times (0, 1)$. There are no limitations to the values the radius can take, since the condition $\sin\psi \le 1$ holds naturally. If the trajectory is initially in the lowering regime ($\cos\psi_0 < 0$) then the particle falls toward the origin. The dynamics of the particle when reaching the origin can be understood by taking the limit $r \to 0$ in Eq. (9.18),

$$\lim_{r\to 0} \psi(r) = \pi - \arcsin K_2 \tag{9.39}$$

which only depends on the angular momentum. Since $K_2 < 1$ it is $\arcsin K_2 < \pi/2$, meaning that ψ will never cross $\psi = \pi/2$ as shown in Fig. 9.7. That is, the trajectory remains in the lowering regime. If initially it is $\cos\psi_0 > 0$ then the spiral will be in the raising regime forever. The asymptotic escape is defined by $\lim_{r\to\infty}\psi(r) = 0$ so the position and velocity vectors become parallel. In this case $\psi_0 < \pi/2$ and, since the flight direction angle always decreases, it is not possible for the trajectory to transition to the lowering regime.

A particle in the lowering regime will always fall to the origin, whereas particles in the raising regime will always escape to infinity. Natural transitions between regimes are not possible since $\cos\psi$ can never change its sign.

9.4.1.1 The trajectory

When assuming $K_2 < 1$, the integration of Eq. (9.26) yields

$$\theta - \theta_0 = \pm\frac{K_2}{\ell} \ln\left[\frac{r\sin\psi}{r_0\sin\psi_0}\left(\frac{1 - K_2\sin\psi_0 + \ell\,|\cos\psi_0|}{1 - K_2\sin\psi + \ell\,|\cos\psi|}\right)\right] \tag{9.40}$$

There are two possible $\mathscr{C}$-symmetric solutions, given by the choice of the sign. Having obtained the solution for the polar angle, taking the limit $r \to \infty$ in Eq. (9.40) provides the direction to the asymptote, $\theta_{as} = \lim_{r\to\infty}\theta(r)$:

$$\theta_{as} = \theta_0 + \frac{K_2}{\ell} \ln\left[\frac{K_2(\ell\,|\cos\psi_0| + 1 - K_2\sin\psi_0)}{(K_2 - \sin\psi_0)(1 + \ell)}\right] \tag{9.41}$$

This equation is valid for spirals initially in the raising regime. The asymptote of the $\mathscr{C}$-symmetric spiral, initially in the lowering regime, follows a different direction given by

$$\theta_{as}^{\dagger} = \theta_0 - \frac{K_2}{\ell} \ln\left[\frac{K_2(\ell\,|\cos\psi_0| + 1 - K_2\sin\psi_0)}{(K_2 - \sin\psi_0)(1 + \ell)}\right] \tag{9.42}$$

Once K_1 and K_2 are fixed, two different possible trajectories are obtained: one departs in the raising regime, and the other in the lowering regime. They are $\mathscr{C}$-symmetric and relate by a reflection $\mathscr{R}$ about θ_0. Type I hyperbolic spirals can never be $\mathscr{R}$-invariant, since they are not $\mathscr{T}$-symmetric.

The equation for the trajectory is obtained upon inversion of Eq. (9.40) and introducing the direction of the asymptote. After some simplifications it takes the form

$$r(\theta) = \frac{\ell^2/K_1}{2\sinh\frac{\beta}{2}\left(\sinh\frac{\beta}{2} + \ell\cosh\frac{\beta}{2}\right)} \tag{9.43}$$

with

$$\beta = \frac{\ell}{K_2}(\theta_{as} - \theta)\,, \qquad \beta^\dagger = -\frac{\ell}{K_2}(\theta^\dagger_{as} - \theta)$$

The argument β applies to spirals initially in the raising regime, and $\beta^\dagger$ defines the $\mathscr{C}$-symmetric spiral initially in the lowering regime. The direction of the initial velocity vector determines which one to choose.

Figure 9.8 shows two $\mathscr{C}$-symmetric hyperbolic spirals of Type I. The corresponding asymptotes are denoted by θ_{as} and $\theta^\dagger_{as}$. They intersect on the axis of $\mathscr{C}$-symmetry, at a distance d from the origin, which follows from the impact parameter defined in Eq. (9.38):

$$d = \frac{K_2}{K_1\,|\sin(\theta_{as} - \theta_0)|} = \frac{K_2}{K_1\,|\sin(\theta^\dagger_{as} - \theta_0)|}$$

These types of spirals have only one asymptote, so the Γ-spiral connects the origin with infinity.

The arclength measured from the direction of the asymptote θ_{as} is obtained by integrating Eq. (9.22) and results in

$$s(\theta) = \frac{\left[(1 - 2\ell^2)\cosh\beta - \ell\sinh\beta - 1\right]\operatorname{csch}^2\frac{\beta}{2}}{4K_1(1 + \ell\coth\frac{\beta}{2})}$$

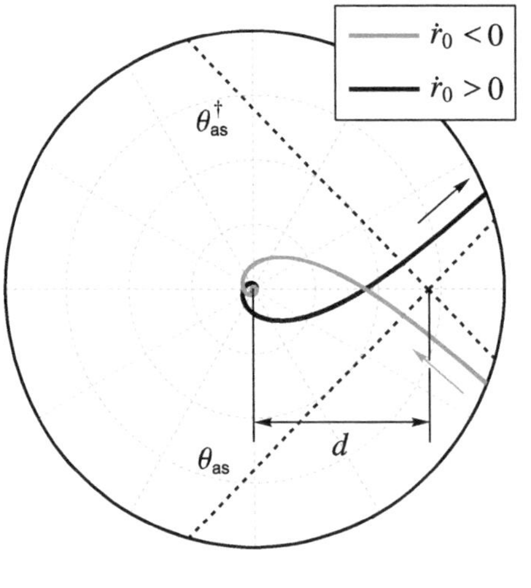

Fig. 9.8: Two $\mathscr{C}$-symmetric hyperbolic spirals of Type I.

This expression yields positive values of the arclength. If the spiral is in the lowering regime, then the argument β is replaced by $\beta^\dagger$ and the arclength takes negative values.

9.4.1.2 The time of flight

Integration of Eq. (9.31) for $K_2 < 1$ requires the use of the incomplete elliptic integrals of the second, $\mathrm{E} = \mathrm{E}(\phi, k)$, and third kinds, $\Pi = \Pi(n; \phi, k)$:

$$t(r) = K_4 \pm \left(\frac{rv}{K_1} \sqrt{\frac{1 + \sin\psi}{1 - \sin\psi}} - \frac{K_2\,\mathrm{E} + k'^2 \Pi}{K_1 \sqrt{K_1 K_2 / 2}} \right) \tag{9.44}$$

with

$$\sin\phi = \sqrt{\frac{2rK_1 \sin\psi}{(1 + K_2)(1 - \sin\psi)}}, \qquad k = \sqrt{\frac{1 + K_2}{2}}, \qquad n = \frac{1 + K_2}{2K_2}$$

The complementary modulus is $k'^2 = (1 - K_2)/2$. The constant of integration K_4 is obtained from the initial conditions:

$$K_4 = t_0 \mp \left(\frac{r_0 v_0}{K_1} \sqrt{\frac{1 + \sin\psi_0}{1 - \sin\psi_0}} - \frac{K_2\,\mathrm{E}_0 + k'^2 \Pi_0}{K_1 \sqrt{K_1 K_2 / 2}} \right) \tag{9.45}$$

The time to reach the origin derives from the limit $\lim_{r\to 0} t(r) = K_4$. It is finite and depends on the initial conditions by means of Eq. (9.45).

9.4.2 Type II hyperbolic spirals

A particle moving along a Type II hyperbolic spiral in the lowering regime ($\cos\psi < 0$) will reach a limit radius $r_{\min}$ with $\psi = \pi/2$ in finite time. Then, it enters the raising regime and escapes to infinity. It can never reach the origin. Once the particle is in the raising regime ($\cos\psi > 0$) it will stay in that regime forever and escape to infinity. This behavior follows from the fact that ψ decreases in time. It can only be in the lowering regime ($\cos\psi < 0$) if it is initially in the lowering regime, i.e., $\cos\psi_0 < 0$. If initially it is $r_0 = r_{\min}$ then the spiral immediately transitions to the raising regime and escapes to infinity.

The velocity of the particle when reaching the periapsis is obtained from Eq. (9.14), and reads

$$v_m = \sqrt{\frac{K_1 K_2}{K_2 - 1}} = \sqrt{\frac{K_2}{r_{\min}}} \tag{9.46}$$

This expression coincides with the one obtained for the elliptic case in Eq. (9.25). The velocity is maximum at periapsis.

9.4.2.1 The trajectory

The evolution of the polar angle for Type II hyperbolic spirals comes from solving Eq. (9.26) assuming that $K_2 > 1$:

$$\theta - \theta_m = \pm\frac{K_2}{\ell}\left\{\frac{\pi}{2} + \arctan\left[\frac{1 + K_1 r - K_2^2}{\ell\sqrt{(1 + K_1 r)^2 - K_2^2}}\right]\right\} \tag{9.47}$$

and with $\ell = \left(K_2^2 - 1\right)^{1/2}$. The value of θ_m is solved initially from

$$\theta_m = \theta_0 - \frac{K_2}{\ell}\left\{\frac{\pi}{2} + \arctan\left[\frac{1 + K_1 r_0 - K_2^2}{\ell\sqrt{(1 + K_1 r_0)^2 - K_2^2}}\right]\right\} \tag{9.48}$$

$$\theta_m^{\dagger} = \theta_0 + \frac{K_2}{\ell}\left\{\frac{\pi}{2} + \arctan\left[\frac{1 + K_1 r_0 - K_2^2}{\ell\sqrt{(1 + K_1 r_0)^2 - K_2^2}}\right]\right\} \tag{9.49}$$

Like in the case of elliptic spirals these expressions define two different trajectories, depending on the initial regime. Equation (9.48) is valid when $\dot{r}_0 > 0$, and it is replaced by Eq. (9.49) if $\dot{r}_0 < 0$. The adequate choice is given by the direction of the initial velocity vector.

From Eq. (9.47) it is possible to solve for the radial distance,

$$\frac{r(\theta)}{r_{\min}} = \frac{1 + K_2}{1 + K_2\cos\beta} \tag{9.50}$$

where the spiral anomaly reads

$$\beta = \frac{\ell}{K_2}(\theta - \theta_m)\,, \qquad \beta^{\dagger} = \frac{\ell}{K_2}(\theta - \theta_m^{\dagger})$$

When the spiral is initially in the raising regime the spiral anomaly is β, and if initially in the lowering regime it is replaced by $\beta^{\dagger}$. The solution takes the same values for $\theta > \theta_m$ and for $\theta < \theta_m$, meaning that hyperbolic spirals of Type II are $\mathscr{T}$-symmetric with respect to the axis defined by θ_m, i.e., $r(\theta_m + \delta) = r(\theta_m - \delta)$ with $\delta \geq 0$.

The reflection $\mathscr{R}$ connecting the $\mathscr{C}$-symmetric spirals $\mathcal{S}$ and $\mathcal{S}^{\dagger}$ can alternatively be represented by a rotation[4] of magnitude ϑ, with

$$\vartheta = \frac{2K_2}{\ell}\left\{\frac{\pi}{2} + \arctan\left[\frac{1 + K_1 r_0 - K_2^2}{\ell\sqrt{(1 + K_1 r_0)^2 - K_2^2}}\right]\right\}$$

---

4 Since these types of spirals are $\mathscr{T}$-symmetric the spirals can be $\mathscr{R}$-invariant. The condition for $\mathscr{R}$-invariance reduces to $\theta_0 = \theta_m + n\pi$, or

$$\mathscr{R}\text{-invariance} \iff \frac{r_{0,\mathrm{inv}}}{r_{\min}} = \frac{1 + K_2}{1 + K_2\cos\left(\frac{n\pi\ell}{K_2}\right)}$$

If $r_0 = r_{0,\mathrm{inv}}$ then the axis of $\mathscr{T}$-symmetry coincides with the axis of $\mathscr{C}$-symmetry.

Type II hyperbolic spirals initially in the lowering regime will always intersect themselves at least once when propagated for sufficient revolutions and with adequate initial conditions. Intersections occur on the axis of symmetry, $\theta = \theta_m$. If initially in the lowering regime, the spiral will reach $r_{\min}$ and then enter the raising regime, following a $\mathscr{T}$-symmetric path.

How the spiral reaches infinity is studied from the limit $r \to \infty$. The existence of two asymptotes can be proved from Eq. (9.50): the denominator cancels for

$$\beta = \pm \arccos\left(-\frac{1}{K_2}\right) \implies \theta_{\text{as}}^{\pm} = \theta_m \pm \frac{K_2}{\ell} \arccos\left(-\frac{1}{K_2}\right)$$

This equation defines the orientation of the two asymptotes, that are symmetric with respect to θ_m. The $\mathscr{C}$-symmetric spiral is defined by $\theta_m^{\dagger}$, given in Eq. (9.49). Its two asymptotes correspond to

$$\theta_{\text{as}}^{\pm\,\dagger} = \theta_m^{\dagger} \pm \frac{K_2}{\ell} \arccos\left(-\frac{1}{K_2}\right)$$

The difference between the asymptotes $\theta_{\text{as}}^{\pm}$ and $\theta_{\text{as}}^{\pm\,\dagger}$ comes from the fact that $\theta_m \neq \theta_m^{\dagger}$. The geometry of the spirals and the asymptotes can be analyzed in Fig. 9.9. Each spiral is $\mathscr{T}$-symmetric with respect to its corresponding θ_m. In addition, the spirals are $\mathscr{C}$-symmetric with respect to the initial polar angle, $\theta_0 = 0$.

Like in the elliptic case, the arclength is measured from θ_m, so that $s(\theta) > 0$ for $\theta > \theta_m$ and $s(\theta) < 0$ when $\theta < \theta_m$. Considering Eq. (9.50) the arclength takes the form

$$s(\theta) = \frac{\ell K_2 \sin\beta}{K_1(1 + K_2 \cos\beta)}$$

Depending on the initial regime of the spiral the argument β might be replaced by $\beta^{\dagger}$. The arclength of the Γ-spiral is infinite, as deduced from the limit $\lim_{\theta\to\pm\infty} s_m = \pm\infty$.

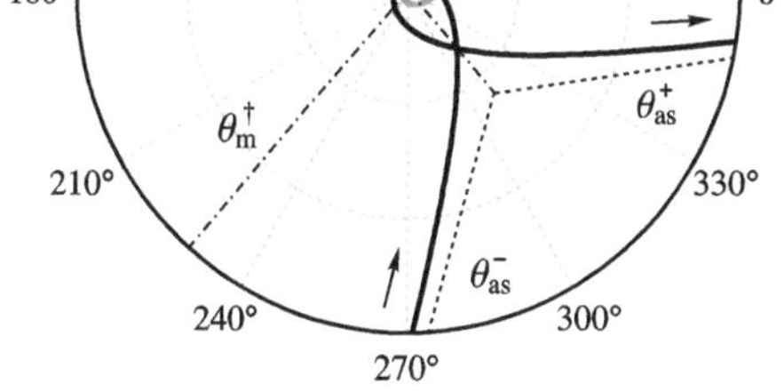

Fig. 9.9: Pair of $\mathscr{C}$-symmetric hyperbolic spirals of Type II, with their corresponding asymptotes and axes of $\mathscr{T}$-symmetry.

9.4.2.2 The time of flight

Integrating Eq. (9.31) and introducing the time of periapsis passage t_m (following the same procedure applied to the elliptic spirals) defines the time of flight. It reads

$$t(r) - t_m = \pm \frac{1}{K_1^{3/2}} \left\{ \sqrt{\frac{z\left(z^2 - K_2^2\right)}{(z-1)}} - \operatorname{arcsinh} \sqrt{\frac{(z-1)(z-K_2)}{z + K_2(2z-1)}} + \frac{\sqrt{2}\,[k'^2(\Pi + K_2\,\mathrm{F}) - K_2\,\mathrm{E}]}{k\sqrt{K_2}} \right\} \tag{9.51}$$

with $z = 1 + K_1 r$. In this case the arguments of the elliptic integrals are

$$\sin\phi = \sqrt{\frac{n\,(z-K_2)}{k^2(z-1)}}, \qquad k = \sqrt{\frac{2}{1+K_2}}, \qquad n = \frac{1}{K_2}$$

The time of flight to $r_{\min}$ is solved from Eq. (9.51) particularized at $r = r_0$:

$$t_m = t_0 \mp \frac{1}{K_1^{3/2}} \left\{ \sqrt{\frac{z_0\left(z_0^2 - K_2^2\right)}{(z_0-1)}} - \operatorname{arcsinh} \sqrt{\frac{(z_0-1)(z_0-K_2)}{z_0 + K_2(2z_0-1)}} + \frac{\sqrt{2}\,[k'^2(\Pi_0 + K_2\,\mathrm{F}_0) - K_2\,\mathrm{E}_0]}{k\sqrt{K_2}} \right\} \tag{9.52}$$

and it is defined in terms of the constants K_1 and K_2.

9.4.3 Transition between Type I and Type II spirals

The limit $K_2 \to 1$ defines the transition regime between Type I and Type II hyperbolic spirals. The continuity of the transition is discussed in Sect. 9.9. As shown in the previous sections, the flight direction angle for Type I spirals in the lowering regime reaches a limit value given in Eq. (9.39). For $K_2 = 1$ it is $\lim_{r\to 0} \psi(r) = \pi/2$. This result, together with the fact that $\lim_{K_2\to 1} r_{\min} = 0$ shows that the limit case $K_2 = 1$ can be understood as a Type II hyperbolic spiral with a periapsis radius equal to zero. Finally, from Eq. (9.47) it is $\lim_{K_2\to 1} \theta_m = \pm\infty$. The spiral can never reach the axis of symmetry and transitions from the lowering regime to the raising regime are not possible. The spiral approaches the origin along a spiral branch. These spirals are not $\mathscr{T}$-symmetric.

9.4.3.1 The trajectory

Imposing $K_2 = 1$ in Eq. (9.26) and integrating the result yields

$$\theta = K_3 \mp \sqrt{1 + \frac{2}{K_1 r}}, \quad \text{with} \quad K_3 = \theta_0 \pm \sqrt{1 + \frac{2}{K_1 r_0}} \tag{9.53}$$

The inverse of this equation defines the trajectory for $K_2 = 1$,

$$r(\theta) = \frac{2}{K_1\left[(\theta - K_3)^2 - 1\right]} \tag{9.54}$$

The definition of K_3 in Eq. (9.53b) contains the information on the initial regime of the spiral.

The asymptote in this case is given by the simple relation:

$$\theta_{as} = \lim_{r\to\infty} \theta(r) = K_3 - 1 = \theta_0 - \left(1 - \sqrt{1 + \frac{2}{K_1 r_0}}\right) \tag{9.55}$$

if the spiral is initially in the raising regime. In the opposite case it is

$$\theta_{as}^{\dagger} = K_3^{\dagger} + 1 = \theta_0 + \left(1 - \sqrt{1 + \frac{2}{K_1 r_0}}\right) \tag{9.56}$$

These results provide a more compact description of the trajectory

$$r(\theta) = \frac{2/K_1}{\beta(\beta + 2)} \qquad \text{or} \qquad r^{\dagger}(\theta) = \frac{2/K_1}{\beta^{\dagger}(\beta^{\dagger} - 2)} \tag{9.57}$$

with $\beta = (\theta_{as} - \theta)$ and $\beta^{\dagger} = (\theta_{as}^{\dagger} - \theta)$. The first equation defines the trajectory of a spiral initially in the raising regime, and the second corresponds to a spiral initially in the lowering regime. The asymptote is defined by Eqs. (9.55) and (9.56), depending on the regime of the spiral.

Transitions from the lowering to the raising regimes are not possible, so the spirals of this type have one asymptote. Figure 9.10 shows an example of two $\mathscr{C}$-symmetric spirals with $K_2 = 1$.

When in the raising regime, the length of the curve measured from the direction of the asymptote, θ_{as} takes the form

$$s(\theta) = \frac{2(\beta + 1)}{K_1\beta(\beta + 2)}, \qquad \beta = (\theta_{as} - \theta)$$

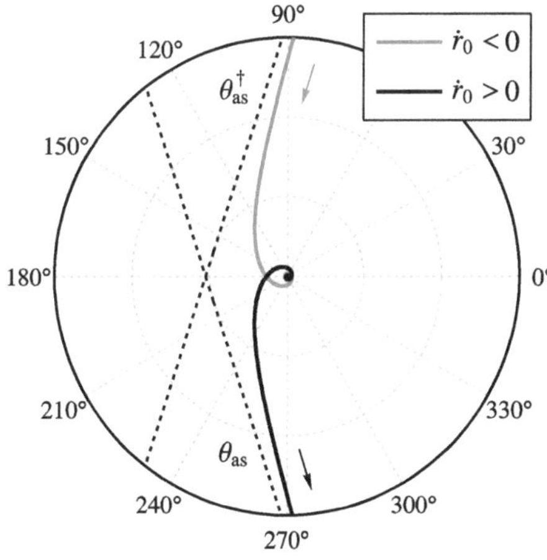

Fig. 9.10: Pair of $\mathscr{C}$-symmetric hyperbolic spirals in the limit case $K_2 = 1$.

whereas for a spiral in the lowering regime $s^\dagger(\theta)$ is obtained by replacing β with $\beta^\dagger = (\theta_{as}^\dagger - \theta)$ in the previous expression. The sign criterion has been chosen so that $s(\theta) > 0$ and $s^\dagger(\theta) < 0$. Because of the $\mathscr{C}$-symmetry of the spirals, when $|\theta - \theta_{as}| = |\theta - \theta_{as}^\dagger|$ then it is $s(\theta) = -s^\dagger(\theta)$.

9.4.3.2 The time of flight

For $K_2 = 1$ the integration of Eq. (9.31) reduces to

$$t - t_0 = \pm\frac{1}{2K_1^{3/2}}\left\{2(\Xi - \Xi_0) - \ln\left[\frac{1 + 2(z + \Xi)}{1 + 2(z_0 + \Xi_0)}\right]\right\} \tag{9.58}$$

where $z = 1 + K_1 r$ and $\Xi = \sqrt{z(1 + z)}$. The time for reaching the origin is solved from the limit

$$\lim_{r\to 0} t(r) = \frac{1}{2K_1^{3/2}}\left\{2(\Xi_0 - \sqrt{2}) + \ln\left[\frac{3 + 2\sqrt{2}}{1 + 2(z_0 + \Xi_0)}\right]\right\}$$

Although the number of revolutions is infinite, it is compensated by a decreasing period that yields a finite time to reach the origin.

9.5 Summary

Table 9.1 summarizes the different types of solutions that can be found depending on the values of the parameters K_1 and K_2. The equations defining the trajectory and the time of flight are referenced in this table for convenience.

Taking the limit $K_1 \to 0$ in the equations of the trajectory for elliptic and hyperbolic spirals yields the equation of the trajectory for parabolic (logarithmic) spirals. Similarly, the limit $K_2 \to 1$ for hyperbolic spirals of Types I and II is continuous and the transition between both types is given by the solution for $K_2 = 1$.

9.6 Osculating elements

The orbital elements defining the osculating orbit can be related to the constants of motion K_1 and K_2. The eccentricity of the osculating orbit is obtained from the defini-

Tab. 9.1: Equations describing the different families of generalized logarithmic spirals.

Type of spiral	K_1	K_2	Trajectory	TOF
Elliptic	< 0	$\in (0, 1)$	Eq. (9.30)	Eq. (9.32)
Parabolic	$= 0$	$\in (0, 1]$	Eq. (9.35)	Eq. (9.36)
Hyperb. Type I	> 0	$\in (0, 1)$	Eq. (9.43)	Eq. (9.44)
Hyperb. Transition	> 0	$= 1$	Eq. (9.57)	Eq. (9.58)
Hyperb. Type II	> 0	> 1	Eq. (9.50)	Eq. (9.51)

tion,

$$\mathbf{e} = \mathbf{v} \times \mathbf{h} - \frac{\mathbf{r}}{r} = (K_2 \sin\psi - 1)\,\mathbf{i} - K_2 \cos\psi\,\mathbf{j} \tag{9.59}$$

which provides

$$e(r) = \sqrt{1 - K_2^2 \left(\frac{1 - K_1 r}{1 + K_1 r}\right)} \tag{9.60}$$

It has been proven that elliptic spirals ($K_1 < 0$) are bounded by a maximum radius, $r_{\max}$, defined in Eq. (9.24). Similarly, for hyperbolic spirals of Type II ($K_1 > 0$ and $K_2 > 1$) a minimum radius $r_{\min}$ exists. The eccentricity of the osculating orbit at those points is

$$e_m = 1 - K_2\,, \qquad \text{for } K_1 < 0 \tag{9.61}$$

$$e_m = K_2 - 1\,, \qquad \text{for } K_1 > 0, K_2 > 1 \tag{9.62}$$

These expressions show that for $K_2 \to 1$ the final orbit is quasicircular, and that the eccentricity at $r = r_{\max}$ and $r = r_{\min}$ only depends on the angular momentum K_2. The osculating orbit will never be perfectly circular except for the case of degenerate parabolic spirals. In fact such orbits turn out to be circular Keplerian orbits. All elliptic spirals, together with parabolic and Type I hyperbolic spirals in the lowering regime fall toward the origin after sufficient time. The osculating eccentricity becomes $\lim_{r\to 0} e(r) = \left(1 - K_2^2\right)^{1/2}$.

The eccentricity of the osculating orbit can be rewritten in terms of the flight direction angle:

$$e = \sqrt{K_2^2 + 1 - 2K_2 \sin\psi} \tag{9.63}$$

It only depends on the constant K_2 and the angle ψ. This expression, together with Eqs. (9.61–9.62), shows that the eccentricity of the osculating orbit reaches a minimum value at $r = r_{\min}$ or $r = r_{\max}$.

The eccentricity always decreases along an elliptic spiral in the raising regime. It grows in the lowering regime. Conversely, the eccentricity grows for all hyperbolic spirals in the raising regime, and decreases in the lowering regime. The eccentricity of the osculating orbit at infinity is defined by

$$e_\infty = \lim_{r\to\infty} e(r) = \sqrt{1 + K_2^2} \tag{9.64}$$

so the osculating orbit is always hyperbolic. The eccentricity of the osculating orbit of a parabolic (logarithmic) spiral is constant and equals $e = (1 - K_2^2)^{1/2}$. Equation (9.62) relates the eccentricity at $r = r_{\min}$ with the value of K_2. A simple analysis of this expression shows that hyperbolic spirals with $K_2 = 2$ yield parabolic osculating orbits at $r = r_{\min}$. If $K_2 > 2$, then the osculating orbit is hyperbolic for every r.

The angular momentum reduces to

$$\mathbf{h} = \mathbf{r} \times \mathbf{v} = rv \sin\psi\,\mathbf{k} = \frac{K_2}{v}\,\mathbf{k} = K_2 \sqrt{\frac{r}{1 + K_1 r}}\,\mathbf{k} \tag{9.65}$$

The semimajor axis can be solved from Eqs. (9.60) and (9.65):

$$a(r) = \frac{h^2}{1 - e^2} = \frac{r}{1 - K_1 r} \tag{9.66}$$

The semimajor axis grows in the raising regime and decreases in the lowering regime, no matter the type of spiral. It only depends on the energy and the radial distance. This property leads to a relation between the osculating semimajor axes at two different points of the spiral, r_1 and r_2:

$$\frac{1}{r_2} - \frac{1}{r_1} = \frac{1}{a_2} - \frac{1}{a_1} \tag{9.67}$$

If two isoenergetic spirals intersect, the corresponding osculating orbits have the same semimajor axis at the intersection point. The semimajor axis of the osculating orbit at $r = r_{\max}$ and $r = r_{\min}$ is given by a single expression:

$$a_m = \frac{K_2 - 1}{K_1(2 - K_2)}$$

It defines the maximum value the semimajor axis can reach. Conversely, for spirals reaching infinity ($K_1 \geq 0$) it is

$$a_\infty = \lim_{r \to \infty} a(r) = -\frac{1}{K_1} \tag{9.68}$$

and this result confirms a beautiful connection with Keplerian orbits: the osculating orbit at $r \to \infty$ is parabolic if the spiral is parabolic, and hyperbolic if the spiral is hyperbolic.

Note that the denominator in Eq. (9.66) vanishes for $r = 1/K_1$ and only when $K_1 > 0$. Equation (9.60) shows that this singularity corresponds to the point where the osculating orbit becomes parabolic, $e = 1$. That is, when the particle escapes from the gravitational field of the central body,

$$r^* = 1/K_1 \implies e(r^*) = e^* = 1$$

In this case it is $v^* = (2K_1)^{1/2}$ and $\psi^* = \arcsin(K_2/2)$. It is interesting to note that the flight direction angle at escape, ψ^*, only depends on the generalized angular momentum K_2. In the limit case $K_2 = 1$, which separates Type I and Type II hyperbolic spirals, it is $\psi^* = \pi/6$. Hyperbolic spirals of Type I ($K_2 < 1$) are defined by $\psi^* < \pi/6$, whereas those of Type II ($K_2 > 1$) satisfy $\psi^* > \pi/6$.

Combining the limits from Eqs. (9.64) and (9.68) the expression for the impact parameter c, already given by Eq. (9.38), follows

$$c = |a_\infty| \sqrt{e_\infty^2 - 1} = \frac{K_2}{K_1}$$

The argument of periapsis, ω, can be solved from the projection of the eccentricity vector in the inertial frame $\mathfrak{I}$. From Eq. (9.59) it is

$$\omega = \arctan\left[\frac{\sin\theta + K_2\cos(\theta + \psi)}{\cos\theta - K_2\sin(\theta + \psi)}\right] \tag{9.69}$$

Equation (9.69) determines the evolution of the apse line. The argument of periapsis ω always increases, no matter the type of spiral. The apse line rotates counterclockwise in the inertial reference.

9.7 In-orbit departure point

The constant K_1 is defined initially by r_0 and v_0, and K_2 also requires ψ_0. Although an initial $\Delta\mathbf{v}$ can be applied to provide the adequate energy and angular momentum, it is interesting to study the spirals that emanate naturally from a Keplerian orbit with no additional maneuvers. That is, the thrust given in Eq. (9.1) initially has the direction of the velocity in the departure orbit. The eccentricity of this orbit is e_0, a_0 is the semimajor axis, and h_0 is the angular momentum.

The velocity is expressed in terms of the orbital elements and the true anomaly, ϑ, as

$$v = \frac{1}{h_0}\sqrt{1 + e_0^2 + 2e_0\cos\vartheta}$$

and the angle ψ is obtained from the expressions:

$$\sin\psi = \frac{1 + e_0\cos\vartheta}{\sqrt{1 + 2e_0\cos\vartheta + e_0^2}}, \qquad \cos\psi = \frac{e_0\sin\vartheta}{\sqrt{1 + 2e_0\cos\vartheta + e_0^2}}$$

Since the eccentricity of the departure orbit is fixed, the angle ψ only depends on the departure point, ϑ_0. From the definitions of K_1 and K_2, given in Eqs. (9.14) and (9.16), it follows that

$$K_1 = \frac{e_0}{h_0^2}(e_0 + \cos\vartheta) \quad \text{and} \quad K_2 = \sqrt{1 + 2e_0\cos\vartheta + e_0^2} \tag{9.70}$$

The constants K_1 and K_2 define the type of spiral. The previous equations corroborate that all generalized spirals that emanate from circular orbits are degenerate parabolic spirals ($K_1 = 0$ and $K_2 = 1$). Equation (9.70) proves that, once the initial orbit is given by a_0 and e_0, K_1 and K_2 only depend on the departure point $\vartheta(t_0) = \vartheta_0$. In addition, if the departure orbit is elliptic the constant K_2 is constrained to $K_2 \in (0, 2)$, as deduced from Eq. (9.70b). If $K_2 = 2$ the departure orbit is parabolic, and for $K_2 > 2$ it is hyperbolic.

It is clear from Eq. (9.70) that the eccentricity of the departure orbit limits the values that K_1 and K_2 can take. The thrust profile defined in Eq. (9.1) may not be adequate for departing from a circular orbit. If it is not possible to meet the design requirements with that thrust, then an initial maneuver is required. A solution to this problem without the need for impulsive maneuvers can be found in Sect. 10.2, and it is later applied in Chap. 11 for the design of orbit transfers (Roa and Peláez, 2016a). Given the design values of K_1 and K_2, together with the energy of the departure orbit (a_0), Eq. (9.70) can be inverted to define the required eccentricity and departure point:

$$e_0 = \sqrt{\frac{A_0 - K_2^2}{A_0}} \quad \text{and} \quad \cos\vartheta_0 = \frac{K_2^2(1 + K_1 a_0) - A_0}{\sqrt{A_0\left(A_0 - K_2^2\right)}}$$

with $A_0 = 1 + 2K_1 a_0$.

Let $\varphi_1 = \arccos(-e_0)$ and $\varphi_2 = 2\pi - \varphi_1$, with $e_0 < 1$. The angles $\vartheta = \varphi_1$ and $\vartheta = \varphi_2$ correspond to the semiminor axis of the ellipse, since $r(\varphi_1) = r(\varphi_2) = a_0$. Parabolic (logarithmic) spirals in the raising regime originate from $\vartheta_0 = \varphi_1$, and those in the lowering regime depart from $\vartheta_0 = \varphi_2$. Elliptic spirals in the raising regime correspond to $\vartheta_0 \in (\varphi_1, \pi)$, and for the lowering regime it is $\vartheta_0 \in (\pi, \varphi_2)$. If the particle departs from apoapsis ($\vartheta_0 = \pi$) then $r_0 = r_{max}$ and it immediately enters the lowering regime. The spirals departing from $\vartheta_0 \in (0, \varphi_1)$ are hyperbolic spirals in the raising regime, whereas $\vartheta_0 \in (\varphi_2, 2\pi)$ corresponds to hyperbolic spirals in the lowering regime. For hyperbolic spirals departing from $\vartheta_0 = 0$ it is $r_0 = r_{min}$, so the spiral immediately enters the raising regime.

Figure 9.11 depicts the different regions found in an elliptic Keplerian orbit. If the thrust starts between periapsis and apoapsis then the spiral is in the raising regime. Between apoapsis and periapsis the resulting spirals are in the lowering regime.

The maximum radius that an elliptic spiral can reach, r_{max}, relates to the departure point by means of

$$r_{max} = \frac{h_0^2(e_0 + 2\cos\vartheta_0)}{(e_0 + \cos\vartheta_0)\left(\sqrt{1 + 2e_0\cos\vartheta_0 + e_0^2} + 1\right)} \tag{9.71}$$

with $\vartheta_0 \in (\varphi_1, \pi)$. In order to analyze how much the radius of the orbit can be raised from a certain departure orbit the limit of Eq. (9.71) shall be considered. In particular,

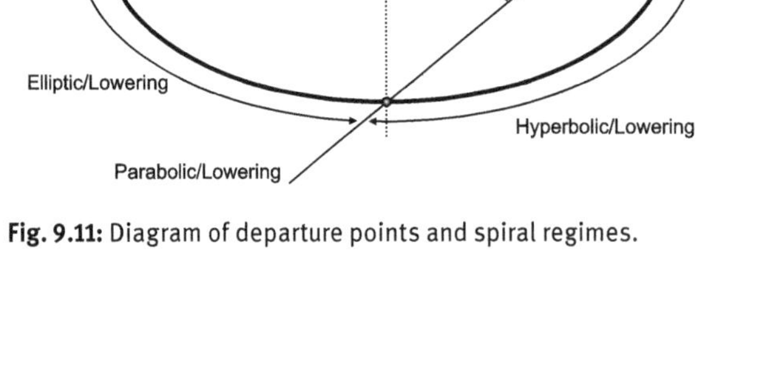

Fig. 9.11: Diagram of departure points and spiral regimes.

for an elliptic spiral in the raising regime the final radius maximizes in the limit case

$$r_{max}\big|_{max} = \lim_{\vartheta_0 \to \varphi_1^-} r_{max}(\vartheta_0) = +\infty$$

Obviating practical limitations such as maximum admissible thrust, there is no theoretical limit to the maximum radius an elliptic spiral can reach. There is, however, a minimum value for the maximum reachable radius:

$$r_{max}\big|_{min} = \lim_{\vartheta_0 \to \pi} r_{max}(\vartheta_0) = a_0(1 + e_0) \tag{9.72}$$

For a transfer to apoapsis, the apoapsis of the target orbit must be higher than that of the departure orbit.

For the hyperbolic case the type of spiral (I or II) comes from the value of K_2. From Eq. (9.70b) it follows that $K_2 = 1$ for $\vartheta_0 = \varphi_3$ or $\vartheta_0 = \varphi_4$, with $\varphi_3 = \arccos(-e_0/2)$ and $\varphi_4 = 2\pi - \varphi_3$. If $\vartheta_0 \in (\varphi_3, \varphi_1) \cup (\varphi_2, \varphi_4)$ then $K_2 < 1$, and for $K_2 > 1$ it is $\vartheta_0 \in [0, \varphi_3) \cup (\varphi_4, 2\pi]$. Figure 9.12 depicts the discussed intervals.

The limit radius for a hyperbolic spiral with $K_2 > 1$ is defined by an expression equivalent to that for the elliptic case – Eq. (9.71) –:

$$r_{min} = \frac{h_0^2(e_0 + 2\cos\vartheta_0)}{(e_0 + \cos\vartheta_0)\left(\sqrt{1 + 2e_0\cos\vartheta_0 + e_0^2} + 1\right)} \tag{9.73}$$

with $\vartheta_0 \in (\varphi_4, 2\pi)$. The minimum radius r_m is reduced for $K_1 \to 0$ and $K_2 \to 1$. But the previous analysis, as summarized in Fig. 9.12, shows that the interval where $K_2 > 1$ is limited by φ_4. The limit $K_1 \to 1$ ($\vartheta_0 \to \varphi_2$) cannot be reached with $K_2 > 1$. Hence, the minimum possible value of r_m to be reached by the hyperbolic spiral is nil. In analogy with the elliptic case there is no theoretical limit to how much the orbit can be lowered by a hyperbolic spiral. But there exists an upper limit,

$$r_{min}\big|_{max} = \lim_{\vartheta_0 \to 0} r_{min}(\vartheta_0) = a_0(1 - e_0) \tag{9.74}$$

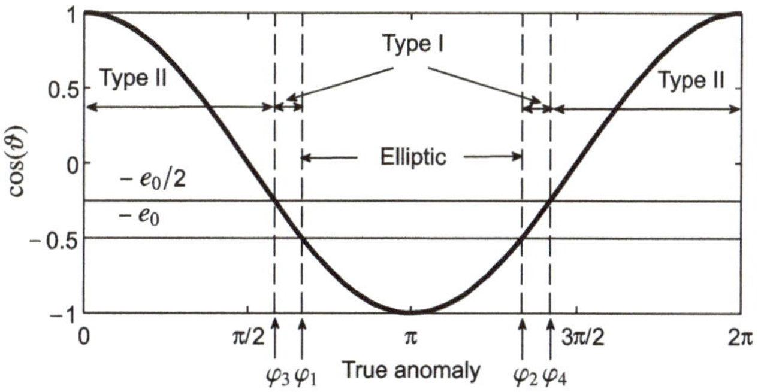

Fig. 9.12: Hyperbolic spirals emanating from an elliptic orbit ($e_0 = 0.5$).

for the limit radius. That is, for transfers to periapsis the periapsis of the target orbit must be lower than that of the departure orbit.

The upper limit for r_{max} and the lower limit for r_{min} are a direct consequence of the continuity of the solution. The limit $\vartheta_0 \rightarrow \varphi_1^-$ shows that the maximum radius grows unboundedly, and for $\vartheta_0 = \varphi_1$ the spiral is parabolic. Parabolic spirals are in fact elliptic spirals with infinite r_{max}. Conversely, the limit $\vartheta_0 \rightarrow \varphi_4^-$ in the case of Type II hyperbolic spirals leads to $r_{min} = 0$, which defines the limit case $K_2 = 1$ for $\vartheta_0 = \varphi_4^-$.

9.8 Practical considerations

The fact that the acceleration in Eq. (9.1) decreases with $1/r^2$ suggests that it might be similar to that originated by a solar sail. Figure 9.13 shows an example of a spiral transfer between the Earth and Mars, assuming circular orbits with radii 1 and 1.524 au, respectively, and neglecting the phasing of the planets. The acceleration from the solar sail reads

$$\mathbf{a}_p = \beta \frac{\mu}{r^2} \cos^2 \alpha \, \mathbf{n}_s$$

where β denotes the lightness number, $\mathbf{n}_s$ is the unit vector normal to the sail, and the pitch angle α is the angle between $\mathbf{n}_s$ and $\mathbf{r}$. Using a solar sail with constant pitch angle $\alpha = 76.95^\circ$ and lightness number $\beta = 0.35$ the spacecraft reaches the orbit of Mars along a logarithmic spiral with $\psi = 88^\circ$ after 980.43 days (see McInnes, 2004, pp. 129–136 for details). The same exact trajectory can be obtained with a parabolic generalized logarithmic spiral, and the total time of flight is 978.75 days. The acceleration profiles are not the same meaning that the velocities will be different too, causing the differ-

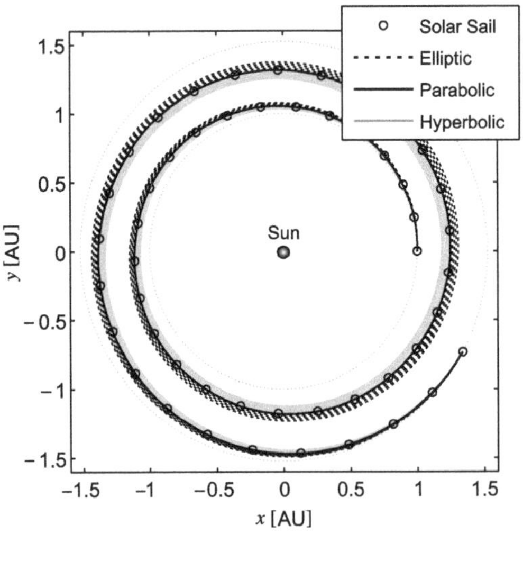

Fig. 9.13: Spiral transfers from the Earth to Mars.

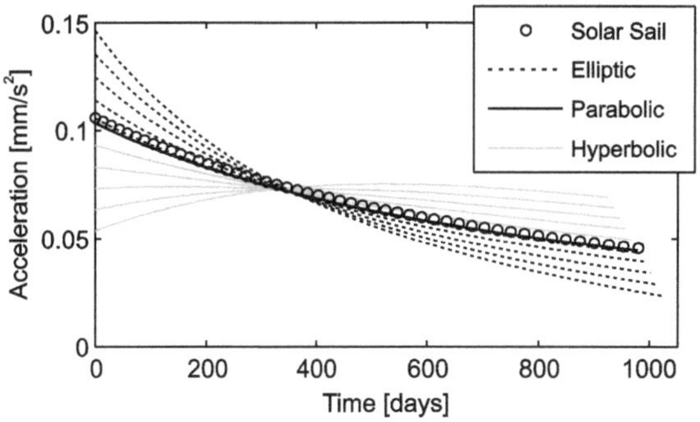

Fig. 9.14: Acceleration profile along the spiral transfers.

ence observed in the times of flight. The figure also depicts a number of elliptic and hyperbolic spirals connecting the same departure and arrival points in different times.

The acceleration profiles are plotted in Fig. 9.14. First, the figure shows that the magnitude of the acceleration exerted by the sail coincides almost exactly with the acceleration required by a parabolic spiral. Changing the value of K_1 the solutions to the two-point boundary-value problem are elliptic and hyperbolic spirals. The acceleration profile separates from that of the solar sail as K_1 increases in magnitude. Elliptic spirals require a high acceleration at first but it decreases rapidly. On the contrary, the acceleration along hyperbolic spirals grows at first and then decreases after reaching a maximum. The times of flight vary accordingly. Hyperbolic spirals are faster than the parabolic one for a given transfer geometry, whereas elliptic spirals are slower. In practice, small values of K_1 will lead to long transfer times with small accelerations, whereas high-energy transfers yield short times and high acceleration levels. Parabolic spirals are very limited due to having a constant flight direction angle. Thus, the mission analyst seeking low-energy spirals should focus on the elliptic and hyperbolic cases with K_1 small. Chapter 11 presents a design strategy using generalized logarithmic spirals, satisfying constraints not only on the position and velocity vectors, but also on the time of flight. The technique is based on matching the values of the constants of motion, just like the patched conics method in the Keplerian case.

9.9 Continuity of the solution

Different expressions for the trajectory have been provided depending on the type of spiral. The solution is continuous, as proved in this section. Proofs are developed for the raising regime to simplify the formulation. The definition of the family of generalized spirals will be continuous as long as $\lim_{K_1 \to 0^-} r_{\mathrm{ell}} = \lim_{K_1 \to 0^+} r_{\mathrm{hyp}} = r_{\mathrm{par}}$, where r_{ell}, r_{par}, and r_{hyp} denote the equations for the trajectory in the elliptic, parabolic, and hyperbolic (Type I) cases. They correspond to Eqs. (9.30), (9.35), and (9.43), respectively.

In addition, the continuity of the solution for the hyperbolic case is determined by the condition $\lim_{K_2\to 1^-} r_{\mathrm{I}} = \lim_{K_2\to 1^+} r_{\mathrm{II}} = r_{\lim}$. Here r_{I} and r_{II} denote the trajectory for Type I and Type II hyperbolic spirals, corresponding to Eqs. (9.43) and (9.50). The trajectory in the asymptotic limit $K_2 \to 1$, $r_{\lim}$, is defined in Eq. (9.57).

9.9.1 Elliptic to parabolic

The transition from elliptic to parabolic spirals is defined by the limit $K_1 \to 0^-$ in Eq. (9.30). Considering the argument

$$\beta = \frac{\ell}{K_2}(\theta - K_3) + \ln(-K_1K_2) = \alpha + \ln(-K_1K_2)$$

the limit $K_1 \to 0^-$ shows that

$$\lim_{K_1\to 0^-} \frac{2\ell^2}{(-K_1)(2 - K_1K_2^2\,\mathrm{e}^{\alpha} - \mathrm{e}^{-\alpha}/K_1)} = 2\ell^2\,\mathrm{e}^{\alpha} \tag{9.75}$$

The argument $\alpha = \ell(\theta - K_3)/K_2$ depends on the value of K_3, which is given by the definition of θ_m. The limit $K_1 \to 0^-$ yields

$$\alpha = \frac{\ell}{K_2}(\theta - \theta_0) - \ln\left(\frac{2\ell^2}{r_0}\right)$$

With this result Eq. (9.75) readily becomes the equation for the trajectory of parabolic (logarithmic) spirals, $r = r_0\,\mathrm{e}^{\alpha'}$ with $\alpha' = \ell(\theta - \theta_0)/K_2$.

9.9.2 Hyperbolic to parabolic

Hyperbolic spirals of Type I define the transition from hyperbolic to parabolic spirals, so the constant K_2 is compatible. Taking the limit $K_1 \to 0^+$ in Eq. (9.43) yields $\lim_{K_1\to 0^+} r(\theta) = 2\ell^2\,\mathrm{e}^{\alpha}$, with $\alpha = \ell(\theta - K_3)/K_2$. In the limit case the constant K_3 reduces to

$$\lim_{K_1\to 0^+} K_3 = \theta_0 + \frac{K_2}{\ell}\ln\left(\frac{2\ell^2}{r_0}\right)$$

and, as proved for the elliptic case, the equation for the trajectory becomes $\lim_{K_1\to 0^+} r(\theta) = r_0\,\mathrm{e}^{\alpha'}$.

9.9.3 Asymptotic limit from Type I hyperbolic spirals

Let $\delta > 0$ be a parameter such that $K_2 = 1 - \delta$. The limit $K_2 \to 1^-$ is equivalent to $\delta \to 0^+$. It can be obtained from the series expansion of the numerator and denominator in Eq. (9.43) and retaining only first-order terms, i.e.,

$$\lim_{\delta\to 0^+} r(\theta) = \frac{-4\delta/K_1 + O(\delta^{3/2})}{-2\{[\theta-\theta_0-\sqrt{1+2/(K_1r_0)}]^2-1\}\delta + O(\delta^{3/2})}$$
$$= \frac{2}{K_1[(\theta-C)^2-1]}$$

where $C = \theta_0 + \sqrt{1+2/(K_1r_0)}$ is equivalent to K_3 in the limit case $K_2 \to 1^-$, as defined in Eq. (9.53b). The result is indeed the equation of the trajectory for $K_2 = 1$ given in Eq. (9.54).

9.9.4 Asymptotic limit from Type II hyperbolic spirals

Consider the expression defining the axis of symmetry, θ_m, given in Eq. (9.48). The trajectory – Eq. (9.50) – can be rewritten as

$$r = \frac{\ell^2}{K_1}\left(\frac{1+K_2\sin\beta'}{K_2^2\cos^2\beta'-\ell^2}\right), \qquad \beta' = \frac{\ell}{K_2}(\theta-K_3)$$

where K_3 is defined initially by the initial conditions. Introducing $\delta > 0$ so that $K_2 = 1+\delta$, the limit $K_2 \to 1^+$ transforms into $\delta \to 0^+$. The numerator and denominator in Eq. (9.50) are expanded to provide

$$\lim_{\delta\to 0^+} r(\theta) = \frac{-4\delta + O(\delta^{3/2})}{-2K_1\{[\theta-\theta_0-\sqrt{1+2/(K_1r_0)}]^2-1\}\delta + O(\delta^{3/2})}$$
$$= \frac{2}{K_1[(\theta-C)^2-1]}$$

Again, $C \equiv K_3$ as defined in Eq. (9.53b).

9.10 Conclusions

While seeking new sets of elements attached to a certain dynamical model, novel analytic solutions might be obtained. The elements are the integration constants upon which the solution is built. When dealing with continuous-thrust engines, shape-based methods become useful for defining closed-form solutions.

A priori solutions for shape-based approaches might only represent a particular case of all the possible trajectories rendered by a given thrust profile. A rigorous approach to the case of logarithmic spirals shows that entire families of solutions were missing, and that the system admits two integrals of motion. One of the first integrals is a generalization of the equation of the energy and allows an energy-based analysis, analogous to the Keplerian case. Three different families of solutions are found: elliptic, parabolic, and hyperbolic. Parabolic spirals turn out to be pure logarithmic spirals. Transitions between the families of spirals have been proven to be continuous.

The existence of upper/lower limit radii yields bounded and symmetric trajectories. The second integral of motion relates to the angular momentum.

The definitions here provided are purely dynamical. The parameters defining the spirals are constants of motion with a clear physical interpretation, and not shape parameters introduced artificially. This is particularly useful for parametric analyses. Although mathematically interesting, the logarithmic spiral has received little attention for practical applications in the past due to its constant flight direction angle. If one thinks about its Keplerian equivalent, the parabola, this is also the least practical conic section from the mission design perspective. Just like Keplerian ellipses and hyperbolas are of much more interest, the new elliptic and hyperbolic spirals prove more flexible than the parabolic spiral, and are still fully integrable.

10 Lambert's problem with generalized logarithmic spirals

Lambert's problem is solved for the case of a spacecraft accelerated by a continuous thrust. The solution is based on the family of generalized logarithmic spirals, which provides a fully analytic description of the dynamics including the time of flight and involves two conservation laws (Chap. 9). Before solving the boundary-value problem an extended version of the generalized logarithmic spirals is presented, including a control parameter. The structure of the solution yields a collection of properties that are closely related to those of the Keplerian case. A minimum-energy spiral is found, with pairs of conjugate spirals bifurcating from it. Thanks to the integral of motion related to the energy the solutions are classified into elliptic, parabolic, and hyperbolic. The maximum acceleration reached along the transfer can be solved in closed form. The problem of designing a low-thrust trajectory between two bodies reduces to solving two equations with two unknowns. Double-time opportunity transfers appear naturally thanks to the symmetry properties of the generalized logarithmic spirals. Comparing the Keplerian and spiral pork-chop plots in an Earth-Mars example shows that the spiral solution might increase the mass fraction delivered to the final orbit, thanks to reducing the magnitude of the impulsive maneuvers at departure and arrival.

10.1 Introduction to Lambert's problem

Originally formulated by Leonhard Euler and Johann H. Lambert in the 18th century, the problem of finding the orbit that connects two position vectors in a certain time has occupied many authors throughout the years. Because of Lambert's pioneering contributions the problem is typically referred to as Lambert's problem. Even Gauss, the Prince of Mathematicians, said that this particular problem is "to be considered among the most important in the theory of the motions of the heavenly bodies". He published the first formal solution to the problem from the mathematical point of view in his treaty on the motion of celestial bodies in 1809 (Gauss, 1809, §3). He devoted the entire third section of his book to the problem of relating several places in orbit. Gauss rewrote Kepler's equation by means of Cotes' formulas, and established the connection with previous work from Euler. He then furnished a number of transformations to be solved by series expansions represented by continuous fractions (see Gauss, 1809, pp. 112–117). It is interesting to note that Bate et al. (1971) referred to Lambert's problem as Gauss' problem instead, as a tribute to his contributions to orbit determination.

Battin was particularly captivated by this classical problem and is responsible for a large collection of properties of the solutions and reviews of existing methods (see Battin, 1999, §6 and §7). He greatly refined and tuned Gauss' method for numerical

https://doi.org/10.1515/9783110559125-010

implementation. In fact, his formulation of Gauss' method (Battin, 1977) was implemented in the onboard guidance computer of the NASA space shuttle. On the basis of his previous discovery of the invariance of the mean or normal point (Battin et al., 1978), Battin and Vaughan (1984) derived an algorithm that successfully removed the singularity in Gauss' method when the transfer angle equals π. A simplified version of Battin and Vaughan's method has been proposed by Avanzini (2008). He recovered an interesting property of the eccentricity vector of the families of solutions: its component along the chord remains constant (Battin, 1999, p. 256). This property motivated the author to parameterize the problem in terms of the component of the eccentricity vector normal to the chord, which yields a monotonic evolution of the time. This algorithm competes with the original method in accuracy and computational cost, but it is simpler to implement.

Finding a universal description of the problem occupied Lancaster and Blanchard (1968), Battin (1977), and Bate et al. (1971). The method from Lancaster and Blanchard (1968), originally presented in a short note (Lancaster et al., 1966), was further improved by Gooding (1990), who focused on the computational aspects. He relied on Halley's cubic iteration process (Gander, 1985) to solve for the universal variable and showed that typically only three iterations are required for convergence. In the work of Peterson et al. (1991), the performance of different algorithms is assessed. Sarnecki (1988) investigated the along-track component of the motion and found a physical interpretation for the universal variable introduced by Lancaster and Blanchard (1968), making use of the components of the velocity. Later improvements in Lancaster–Blanchard's method were proposed by Izzo (2014). He introduced a new Lambert-invariant variable to parameterize the problem. His main contribution is that the curve of the time of flight exhibits two asymptotes, and it can be approximately inverted to provide adequate initial guesses for the iterative procedure. The overall performance of the method matches Gooding's, although the algorithm is simpler. Similarly, Arora and Russell (2013) have recently recovered Bate's algorithm in order to improve the computational efficiency by introducing a new iteration parameter, which relates to the cosine of the difference in eccentric anomaly. The formulation has been extended to account for the partial derivatives (Arora et al., 2015). Starting from Bate's method Luo et al. (2011) tackled the quasi-Lambert problem, where the constraint on the transfer angle is replaced by a constraint on the departure flight-direction angle. The pseudostate method was also exploited by Senent (2010) for finding abort return trajectories for manned missions to the Moon. No numerical optimizers are required and flyby return trajectories are also found by the algorithm. Sun (1981) followed a different approach by finding a universal differential equation to describe the problem. He made extensive use of hypergeometric functions and analyzes carefully the differences with respect to Battin's method.

The rise of interplanetary exploration has brought renewed interest in Lambert's problem for mission design applications. For example, it is the cornerstone of the automated method for gravity-assist trajectory design proposed by Longuski and Williams

(1991). The transfer legs form a patched-conic solution and the algorithm admits constraints in both the number of admissible revolutions and the time of flight. Sims et al. (1997) proposed techniques for maximizing the energy gains from gravity-assist maneuvers. Vasile and De Pascale (2006) posed the problem of designing interplanetary trajectories with multiple gravity-assists as a global optimization problem. Izzo et al. (2007) followed a heuristic approach and published a search space pruning method for finding optimal solutions.

Low thrust is a powerful technique for the design of interplanetary trajectories due to the high specific impulse of the propulsive systems. For this reason, Izzo (2006) explored the potential of exponential sinusoids for solving the accelerated Lambert problem. A different solution is that of Avanzini et al. (2015), who wrote the problem in equinoctial elements and derived a series solution. The present chapter formulates the spiral Lambert problem using the generalized logarithmic spirals presented in Chap. 9. The structure of the solution is surprisingly similar to that of the Keplerian case. The approach followed in this chapter is motivated by Battin's ideas (1999, p. 237):

> "[I] have been fascinated by this subject for many years and have collected (almost as a hobby as others would collect stamps) a number of delightful and often useful properties of the two-body, two-point, boundary-value problem."

Special attention is paid to formally stating the properties of the solution. The families of solutions are parameterized using the constant appearing in the generalized equation of the energy. A minimum-energy transfer is found with pairs of conjugate spirals emanating from it. Alternative discussions can be found in Roa et al. (2016b) and Roa and Peláez (2016b).

This chapter is organized as follows. Section 10.3 solves the two-point boundary-value problem with a continuous acceleration and free transfer time. The minimum-energy spiral, the families of solutions, and their properties are discussed. Section 10.4 introduces the constraint on the time of flight to tackle Lambert's problem. The design of repetitive transfers is addressed in Sect. 10.5. Finally, Sect. 10.6 analyzes the practical advantages of using the spiral solution in place of the Keplerian one when designing transfers to Mars. Additional properties and dynamical constraints are analyzed in Sects. 10.7 and 10.8.

10.2 Controlled generalized logarithmic spirals

It is worth recalling that generalized logarithmic spirals are the solution to the dynamics under the acceleration in Eq. (9.1):

$$\mathbf{a}_p = \frac{\mu}{2r^2} \cos\psi\, \mathbf{t} \tag{10.1}$$

Mission analysts will notice the limitation of this thrust profile: there are no control parameters. Because there are no degrees of freedom the spirals are simply the analytic solution to an initial value problem. In addition, if the departure orbit is circular the thrust will always be zero unless an impulsive maneuver is considered.

In order to overcome this limitation we shall consider the more general perturbation

$$\mathbf{a}_p = \frac{\mu}{r^2}[\xi \cos\psi\,\mathbf{t} + (1-2\xi)\sin\psi\,\mathbf{n}] \tag{10.2}$$

involving both tangential and normal components of the thrust vector. Here ξ is a constant that behaves as a control parameter. The normalized gravitational parameter $\mu = 1$ will be omitted for brevity. Note that Eq. (10.2) reduces to Eq. (10.1) when $\xi = 1/2$ (Roa and Peláez, 2016a). The angle between the radial direction and the thrust vector will be denoted φ. Figure 10.1 sketches the geometry of the problem.

The choice of the thrust in Eq. (10.2) is not arbitrary. It is the natural generalization of the acceleration (10.1), because both problems share the same exact properties.

Combining the thrust profile in Eq. (10.2) with the gravitational acceleration

$$\mathbf{a}_g = -\frac{1}{r^2}(\cos\psi\,\mathbf{t} - \sin\psi\,\mathbf{n})$$

yields the equations of motion projected in the intrinsic frame:

$$\frac{\mathrm{d}v}{\mathrm{d}t} = \frac{\xi-1}{r^2}\cos\psi \tag{10.3}$$

$$v\frac{\mathrm{d}}{\mathrm{d}t}(\psi+\theta) = \frac{2(1-\xi)}{r^2}\sin\psi \tag{10.4}$$

$$\frac{\mathrm{d}r}{\mathrm{d}t} = v\cos\psi \tag{10.5}$$

$$\frac{\mathrm{d}\theta}{\mathrm{d}t} = \frac{v}{r}\sin\psi \tag{10.6}$$

In the limit case $\xi \to 1$ the thrust compensates the gravitational attraction of the central body and the velocity remains constant. We consider this a natural limit to the control parameter, so $\xi < 1$. The control parameter ξ can be bounded from limitations on the ratio between the thrust acceleration and the local gravity.

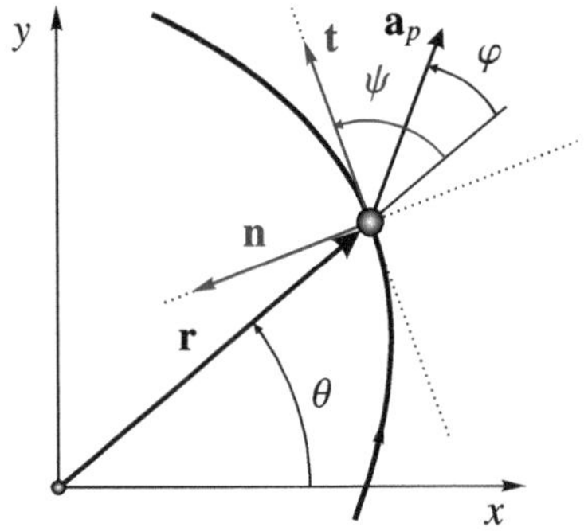

Fig. 10.1: Geometry of the problem.

The problem can be transformed by means of a linear application $\mathscr{S}$ of the form

$$\mathscr{S}: (t, v, r, \theta, \psi) \to (\tau, \tilde{v}, \tilde{r}, \theta, \psi)$$

involving the constant parameters α, β, and δ:

$$\tau = \frac{t}{\beta}, \qquad \tilde{r} = \frac{r}{\alpha}, \qquad \tilde{v} = \frac{v}{\delta}$$

The parameter α is simply a dilation of the solution, so it can be set to unity without loss of generality. In order for the parameters to be physically compatible it must be

$$\beta = \sqrt{\frac{\alpha^3}{2(1-\xi)}} \qquad \text{and} \qquad \delta = \frac{\alpha}{\beta} = \sqrt{\frac{2(1-\xi)}{\alpha}}$$

Under the action of $\mathscr{S}$ Eqs. (10.3–10.6) transform into

$$\begin{aligned}
\frac{d\tilde{v}}{d\tau} &= -\frac{\cos\psi}{2\tilde{r}^2} \\
\tilde{v}\frac{d}{d\tau}(\psi + \theta) &= +\frac{\sin\psi}{\tilde{r}^2} \\
\frac{d\tilde{r}}{d\tau} &= \tilde{v}\cos\psi \\
\frac{d\theta}{d\tau} &= \frac{\tilde{v}}{\tilde{r}}\sin\psi
\end{aligned}$$

These equations are none other than the equations of motion resulting from the acceleration in Eq. (10.1), Eqs. (9.8–9.11). That is, $\mathscr{S}$ is a *similarity* transformation that transforms the problem perturbed by the thrust acceleration in Eq. (10.2) to the simplified problem governed by Eq. (10.1). The solution to the more general problem can be easily obtained by mapping the inverse transformation $\mathscr{S}^{-1}$ over the simplified solution $\tilde{r}(\theta)$ and $\tau(\tilde{r})$, which corresponds to that in Chap. 9. There is no need to re-derive the solution including the control parameter ξ. The time of flight, trajectory, and velocity of the general problem is obtained from the simplified one thanks to

$$t = \beta\,\tau, \qquad r = \alpha\,\tilde{r}, \qquad v = \delta\,\tilde{v}$$

Because of the importance and reach of such an intriguing transformation the entire Chap. 12 is devoted to analyzing this transformation, and new families of spirals will be obtained. For practical reasons in this chapter we only recover the basic properties required for designing orbit transfers. The complete reformulation of the solution with the control parameter can be found in Appendix F.

The constants of motion K_1 and K_2 relate to the transformed values, κ_1 and κ_2, by means of

$$K_1 = \delta^2\kappa_1 = v^2 - \frac{2(1-\xi)}{r} \tag{10.7}$$

and also

$$K_2 = \alpha\delta^2\kappa_2 = rv^2 \sin\psi \tag{10.8}$$

The fact that the integrals of motion (10.8) and (9.16) have the same form justifies the choice of the acceleration (10.2).

The flight-direction angle is defined in terms of

$$\sin\psi = \frac{K_2}{2(1-\xi) + K_1 r} \tag{10.9}$$

It is important to note that the constant κ_2 is not affected by the value of the control parameter ξ. The apoapsis radius for elliptic spirals is

$$r_{\max} = \frac{2(1-\xi) - K_2}{(-K_1)}$$

and the periapsis of a hyperbolic spiral of Type II reads

$$r_{\min} = \frac{K_2 - 2(1-\xi)}{K_1}$$

Hyperbolic spirals of Type I correspond to $K_2 < 2(1-\xi)$, equivalent to $\kappa_2 < 1$, and spirals of Type II are those with $K_2 > 2(1-\xi)$ or $\kappa_2 > 1$.

The tangent and normal vectors, $\mathbf{t}$ and $\mathbf{n}$, can be referred to the orbital frame $\mathcal{L} = \{\mathbf{i}, \mathbf{j}, \mathbf{k}\}$,

$$\mathbf{i} = \frac{\mathbf{r}}{r}, \qquad \mathbf{k} = \frac{\mathbf{h}}{h}, \qquad \mathbf{j} = \mathbf{k} \times \mathbf{i}$$

by means of

$$\mathbf{t} = \cos\psi\,\mathbf{i} + \sin\psi\,\mathbf{j} \qquad \text{and} \qquad \mathbf{n} = -\sin\psi\,\mathbf{i} + \cos\psi\,\mathbf{j}$$

The thrust acceleration from Eq. (10.2) becomes

$$\mathbf{a}_p = \frac{1}{r^2}\{[\xi\cos^2\psi - (1-2\xi)\sin^2\psi]\,\mathbf{i} + (1-\xi)\sin\psi\cos\psi\,\mathbf{j}\}$$

The angle between the thrust direction and the radial direction, φ, is defined by

$$\tan\varphi = \frac{(1-\xi)\sin\psi\cos\psi}{\xi - (1-\xi)\sin^2\psi} = \frac{(1-\xi)\sin 2\psi}{3\xi - 1 + (1-\xi)\cos 2\psi} \tag{10.10}$$

For $\xi = 1/2$ it follows $\varphi = \psi$, which is the case of purely tangential thrust discussed in Chap. 9. For $\xi = 1/3$ this equation reduces to $\varphi = 2\psi$. That is, the velocity lies on the bisection between $\mathbf{r}$ and $\mathbf{a}_p$.

As discussed above, the original family of generalized logarithmic spirals ($\xi = 1/2$) requires an impulsive Δv for departing from circular orbits. This comes from the combination of two factors: First, the velocity on the circular orbit is $v = r^{-1/2}$, which makes $K_1 = 0$, and the trajectory is a pure logarithmic spiral. In this case the flight-direction angle ψ is constant. Second, on a circular orbit it is always $\psi = \pi/2$ so the thrust vanishes. Since ψ remains constant along the logarithmic spiral the thrust acceleration will always be zero and the particle describes a Keplerian circular orbit. In the general case $\xi \neq 1/2$, the presence of a normal acceleration overcomes this limitation, and impulsive maneuvers can be avoided. When applying the similarity transformation $\mathscr{S}^{-1}$ it is

$$v(r) = \delta\tilde{v} = \sqrt{\frac{2(1-\xi)}{r}}$$

meaning that it is possible to describe the same trajectory with different velocities. When $\xi \neq 1/2$ logarithmic spirals are suitable for departing from circular orbits.

10.3 The two-point boundary-value problem

Consider the problem of finding a generalized logarithmic spiral $\mathcal{S}$ that connects a departure point $\mathcal{P}_1$, defined by (r_1, θ_1), with a final point $\mathcal{P}_2$, defined by (r_2, θ_2), with $\theta_2 > \theta_1$. The position vectors for $\mathcal{P}_1$ and $\mathcal{P}_2$ are $\mathbf{r}_1$ and $\mathbf{r}_2$, respectively. In order to account for the number of revolutions n, θ_2 is decomposed in $\theta_2 = \tilde{\theta}_2 + 2n\pi$, with $n = 0, 1, 2\ldots$ and $\tilde{\theta}_2 \in [0, 2\pi)$. Changing the number of revolutions is equivalent to modifying the geometry of the boundary-value problem (BVP). No constraints are imposed on the time of flight in this section. In order for the trajectory to depart from $\mathcal{P}_1$ and arrive at $\mathcal{P}_2$ the following equation must be satisfied:

$$r_2 = r(\theta_2; \xi, K_1, K_2, r_1, \theta_1) \tag{10.11}$$

The function $r(\theta; \xi, K_1, K_2, r_1, \theta_1)$ is the equation of the trajectory and depends on the type of spiral. The polar angle θ is the independent variable, the constants K_1 and K_2 define the shape of the spiral, (r_1, θ_1) are the initial conditions at $\mathcal{P}_1$, and ξ is the control parameter.

In the Keplerian case the minimum-energy ellipse plays a key role in the configuration of the transfers (Battin, 1999, pp. 240–241). The Keplerian energy (or equivalently the semimajor axis of the orbit) yields an intuitive parameterization of the solutions. An analogous parameterization can be introduced for the spiral case in terms of the constant K_1. With this technique the solutions are easily classified into elliptic, parabolic, or hyperbolic spiral transfers.

Given the boundary conditions (r_1, θ_1) and (r_2, θ_2), the control parameter ξ, and a certain value of K_1 the problem reduces to solving for K_2 in Eq. (10.11). Note that once the value of K_1 is fixed, the departure and terminal velocities are known:

$$v_1 = \sqrt{K_1 + \frac{2(1-\xi)}{r_1}} \quad \text{and} \quad v_2 = \sqrt{K_1 + \frac{2(1-\xi)}{r_2}}$$

If the initial and final radii are the same, $r_1 = r_2$, then the departure and terminal velocities in a generalized spiral trajectory are also equal, $v_1 = v_2$.

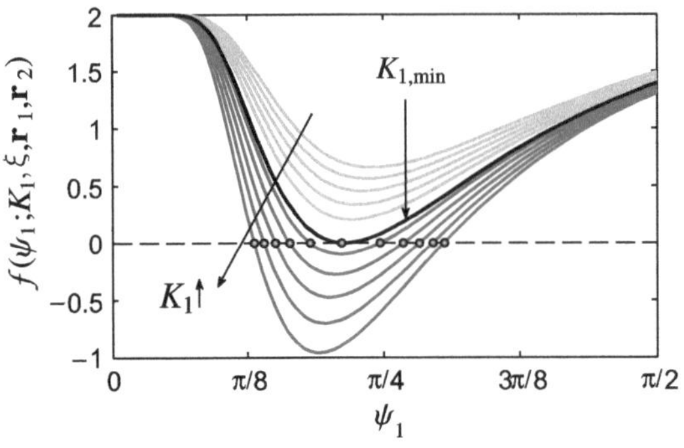

Fig. 10.2: Zeros of the function $f(\psi_1)$ for increasing values of $K_1 < 0$. In this example $r_2/r_1 = 2$, $\theta_2 - \theta_1 = 2\pi/3$, and $\xi = 1/2$.

Since K_1, ξ and r_1 are fixed Eq. (10.9) shows that solving for K_2 is equivalent to solving for the initial flight-direction angle, ψ_1. In fact, because of the $\mathscr{C}$-symmetry of the spirals it is more convenient to solve for the departure angle ψ_1 in order not to lose any solution, and then compute the value of K_2 from Eq. (10.9).

Elliptic spirals are those with the minimum value of the constant K_1. In this case the boundary-value problem reduces to finding the zeros of the function

$$f(\psi_1; \xi, K_1, \mathbf{r}_1, \mathbf{r}_2) \equiv \frac{r_2}{r_{\max}} - \frac{2(1-\xi) + K_2}{2(1-\xi) + K_2 \cosh \beta_2} = 0 \tag{10.12}$$

where β_2 is the spiral anomaly at $\mathcal{P}_2$ and $r_{\max}$ is the apoapsis of the spiral, defined in Eq. (F.10).

Figure 10.2 shows the effect of the constant K_1 on the zeros of the function $f(\psi_1)$ for an example transfer geometry. This figure proves that a minimum value of K_1 exists: for $K_1 > K_{1,\min}$ there are two solutions, that collapse to one double solution when $K_1 = K_{1,\min}$. If $K_1 < K_{1,\min}$ the function has no roots: the BVP has no solution.

10.3.1 The minimum-energy spiral

Let $\mathbb{S}$ denote the set of all generalized logarithmic spirals $\mathcal{S}$ that are solutions to the BVP. The elliptic spiral for which $K_1 = K_{1,\min}$ is referred to as the *minimum-energy spiral*, $\mathcal{S}_m \in \mathbb{S}$. The constant of the energy relates to the specific Keplerian energy $\mathcal{E}_k$ by means of

$$K_1 = 2\left(\mathcal{E}_k + \frac{\xi}{r}\right) \tag{10.13}$$

Consider two spirals $\mathcal{S}_i, \mathcal{S}_j \in \mathbb{S}$, with $i \neq j$. Thanks to Eq. (10.13) it is possible to compute the value of K_1 for both spirals from the values of the energy of the osculating orbits at r_1, $\mathcal{E}_{k,i}$ and $\mathcal{E}_{k,j}$. If $\mathcal{S}_i$ is the minimum-energy spiral ($i \equiv m$), then the osculating Keplerian energy at r_1 is also minimum. The osculating orbits at $\mathcal{P}_1$ and $\mathcal{P}_2$ for the minimum-energy spiral are those with the minimum Keplerian energy and semimajor axis. When the thrust vanishes the osculating orbits at $\mathcal{P}_1$ and $\mathcal{P}_2$ coincide, and correspond to the transfer orbit. If they are the minimum-energy orbits they will reduce to the minimum-energy Keplerian transfer. Thus, the minimum-energy spiral can be seen as the generalization of the minimum-energy ellipse.

The value of $K_{1,\text{min}}$ can be obtained through the following numerical procedure. The function $f(\psi_1)$ defined in Eq. (10.12) reaches a minimum at ψ_1^*, so that $f'(\psi_1^*; \xi, K_1, \mathbf{r}_1, \mathbf{r}_2) = 0$. The prime $\square'$ denotes the derivative with respect to ψ_1. The minimum value of K_1 for which a solution to the boundary-value problem exists corresponds to the value that yields only one root of the function $f(\psi_1)$. That is, when the minimum of $f(\psi_1)$ is exactly zero. Given the control parameter and the boundary conditions $\mathbf{r}_1$ and $\mathbf{r}_2$ the minimum-energy transfer, defined by $K_{1,\text{min}}$ and ψ_1^*, is the spiral that satisfies

$$f\,(\psi_1^*; \xi, K_{1,\text{min}}, \mathbf{r}_1, \mathbf{r}_2) = 0 \tag{10.14}$$

$$f'(\psi_1^*; \xi, K_{1,\text{min}}, \mathbf{r}_1, \mathbf{r}_2) = 0 \tag{10.15}$$

The first condition forces the minimum to be zero, so there is only one root, and the second equation determines where the minimum is located. Note that $K_{1,\text{min}}$ is defined in terms of $\theta_2 = \tilde{\theta}_2 + 2n\pi$ and takes different values depending on the number of revolutions n, even if $\tilde{\theta}_2$ remains the same.

Figure 10.3 shows how the value of $K_{1,\text{min}}$ depends on the geometry of the transfer. As the transfer angle increases the value of $K_{1,\text{min}}$ decreases in magnitude, meaning that increasing the number of revolutions reduces the energy of the transfer. The evolution of the values of $K_{1,\text{min}}$ proves that $\lim_{\theta_2\to\infty} K_{1,\text{min}} = 0^-$; transfers with negative K_1 always exist, and the magnitude of the constant K_2 can be arbitrarily small. This yields an important result, which is that the BVP defined by $\mathbf{r}_1$ and $\mathbf{r}_2$ will always admit three types of solutions: elliptic, parabolic, and hyperbolic spiral transfers.

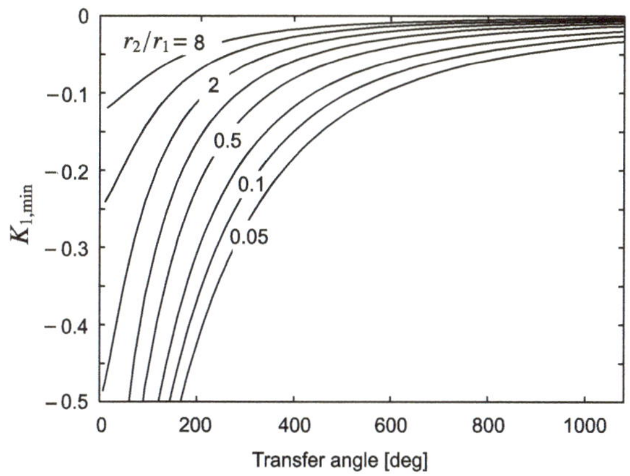

Fig. 10.3: Minimum K_1 for different transfer geometries. In this figure each curve corresponds to a different value of r_2/r_1.

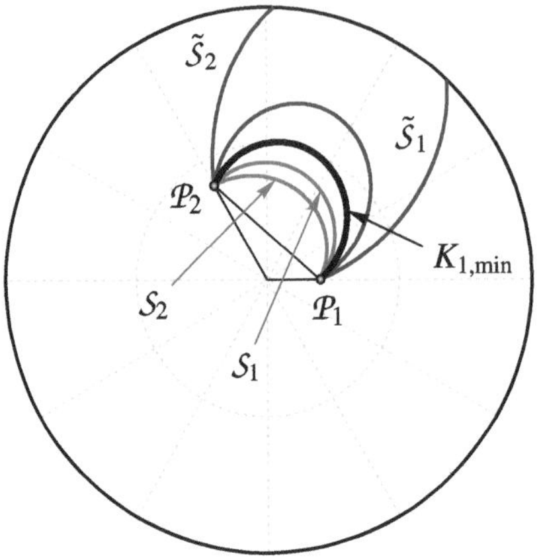

Fig. 10.4: Examples of pairs of conjugate spiral transfers and the minimum-energy spiral.

10.3.2 Conjugate spirals

There are two solutions to Eq. (10.12) for each value of the constant of the energy and control parameter. This behavior is intimately related to the conjugate Keplerian orbits that appear in the unperturbed form of Lambert's problem (Battin, 1999, p. 244):

Definition 5. (Conjugate spirals) Two generalized logarithmic spirals $\mathcal{S}, \tilde{\mathcal{S}} \in \mathbb{S}$ are said to be conjugate if they share the same value of the constant of the energy K_1 and control parameter ξ. The conjugate of $\mathcal{S}$ is denoted $\tilde{\mathcal{S}}$.

Figure 10.4 shows the spiral transfer corresponding to $K_{1,\min}$ and two pairs of conjugate spirals, $\mathcal{S}_i$ and $\tilde{\mathcal{S}}_i$. A spiral and its conjugate are always separated by the minimum-energy spiral. As K_1 becomes more negative both trajectories come closer, and they collapse into one single trajectory, $\mathcal{S}_m$, for $K_1 = K_{1,\min}$. In this example both the spiral $\mathcal{S}_i$ and its conjugate $\tilde{\mathcal{S}}_i$ reach $\mathcal{P}_2$ in the lowering regime, after having crossed the maximum radius $r_{\max}$.

Given a pair of conjugate solutions to the BVP, one will be faster than the minimum-energy transfer, whereas its conjugate will be slower. Conjugate spirals can be characterized by a collection of properties connecting the values that certain variables take at departure and arrival, summarized in Sect. 10.7. For example, the conservation of Eq. (10.7) proves that the departure and arrival velocities are the same for two conjugate spirals, i.e., $v_1 = \tilde{v}_1$ and $v_2 = \tilde{v}_2$.

10.3.3 Families of solutions

Parameterizing the solutions of the BVP in terms of the constant K_1 yields a natural classification of the orbits: starting from $K_{1,\min} < 0$, the spirals are elliptic in the interval $K_1 \in [K_{1,\min}, 0)$; solutions with $K_1 = 0$ are parabolic spirals, and for $K_1 > 0$ the spirals are hyperbolic.

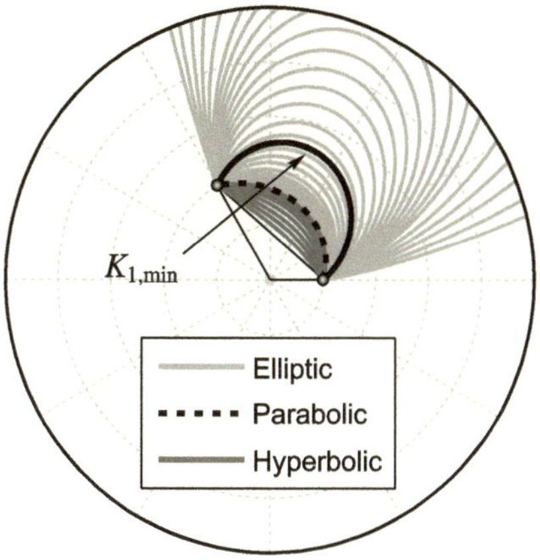

Fig. 10.5: Families of solutions for $r_2/r_1 = 2$, $\theta_2 - \theta_1 = 2\pi/3$, and fixed ξ.

Figure 10.5 shows an illustrative example of the families of solutions. First, note that the minimum-energy spiral separates the fast and slow transfers along elliptic spirals. Reducing the value of K_1 makes the pairs of conjugate spirals converge to the minimum-energy spiral. Conversely, increasing K_1 reduces the size of the fast transfers and increases the size of the slow ones. The limits are set by the corresponding parabolic spirals ($K_1 = 0$). The solution to the transfer along a parabolic (logarithmic) spiral can be given in closed form. The constraint in Eq. (10.11) combined with Eq. (F.19) provides

$$r_2 = r_1\, e^{(\theta_2-\theta_1)\cot\psi_1} \implies \psi_1 = \arctan\left[\frac{\theta_2-\theta_1}{\ln(r_2/r_1)}\right] \tag{10.16}$$

and defines the departure flight-direction angle. If the spiral is in the lowering regime, $r_2 < r_1$, the solution to Eq. (10.16) is negative and then the departure angle takes the value $\pi + \psi_1$. Equivalently, the value of K_2 results in

$$K_2 = \frac{b}{\sqrt{1+b^2}} \qquad \text{with} \qquad b = \frac{\theta_2-\theta_1}{\ln(r_2/r_1)} \tag{10.17}$$

This value of K_2 is unique no matter the regime of the spiral, thanks to the $\mathscr{C}$-symmetry of the trajectory. Note that Eq. (10.16) has only one solution: the second conjugate solution is not real, since it connects both points through infinity. That is equivalent to an elliptic spiral with $r_{\max} \to \infty$. If K_1 increases it becomes positive and yields hyperbolic spirals. Pairs of conjugate solutions also exist in this case, one of them being a fictitious solution through infinity. The fast transfers are below the parabolic transfer and connect the two points directly. In the limit $K_1 \to \infty$ the hyperbolic spiral degenerates into a rectilinear orbit along the chord connecting $\mathbf{r}_1$ and $\mathbf{r}_2$. The conjugate solutions connect the two points by reaching infinity, so the solution does not exist in practice. In the multirevolution case the transfer for $K_1 \to \infty$ decomposes into two rectilinear segments connecting $\mathcal{P}_1$ with the origin and the origin with $\mathcal{P}_2$.

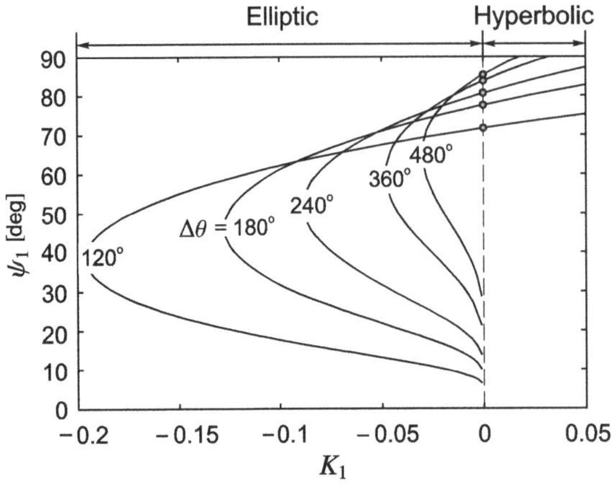

Fig. 10.6: Departure flight-direction angle as a function of the constant of the energy K_1 for fixed values of ξ. Different transfer geometries are considered, keeping the radii $r_2/r_1 = 2$ constant.

The existence of conjugate spirals and the discussed behavior can be understood from Fig. 10.6. There are two elliptic spiral transfers with the same value of K_1. For $K_1 = K_{1,\min}$ the pair of conjugate spirals converge to the minimum-energy spiral, and for $K_1 \to 0^-$ the apoapsis of the slow transfer becomes infinite. The branch of fast transfers ($\psi_1 > \psi_{1,m}$) exists for both the parabolic and hyperbolic cases, but the slow transfers ($\psi_1 < \psi_{1,m}$) exist only for the elliptic case. The branch of conjugate slow hyperbolic solutions corresponds to fictitious transfers through infinity, just like Keplerian hyperbolas. Despite the fact that the minimum-energy spiral $\mathcal{S}_m$ always departs with $\psi_1 \in (0, \pi/2)$, this constraint does not apply to generic fast transfers: arbitrary solutions with $\psi_1 \geq \pi/2$ may be found.

The trajectory of a particle following a hyperbolic spiral ($K_1 > 0$) takes different forms depending on the value of K_2, the transition being $K_2 = 2(1 - \xi)$. It is possible to find a limit value of K_1, $K_{1,\text{tr}}$, such that if $K_1 < K_{1,\text{tr}}$ then $\mathcal{S} \in \mathbb{S}$ is hyperbolic of Type I, and of Type II for $K_1 > K_{1,\text{tr}}$. It reads

$$K_{1,\text{tr}} = \frac{4(1-\xi)[2\sqrt{(r_2-r_1)^2 + r_1 r_2 \Delta\theta^2} - (r_2+r_1)\Delta\theta]}{r_1 r_2 \Delta\theta(4-\Delta\theta^2)} \tag{10.18}$$

with $\Delta\theta = \theta_2 - \theta_1$. When $K_{1,\text{tr}} \to 0$ all hyperbolic spirals in $\mathbb{S}$ are of Type II. It is interesting to note that for transfers where $r_2 = r_1$ it is $K_{1,\text{tr}} = 0$, which proves that for $r_1 = r_2$ all hyperbolic spirals $\mathcal{S} \in \mathbb{S}$ are of Type II.

For $r_1 = r_2$ the direction of the axis of $\mathcal{T}$-symmetry θ_m reduces to $\theta_m = (\theta_2 + \theta_1)/2$. In his book, Battin (1999, pp. 250–256) discusses a number of properties involving the bisection of the transfer angle. In the spiral case we found that for transfers with $r_1 = r_2$ all spiral solutions are $\mathcal{T}$-symmetric and the axis of $\mathcal{T}$-symmetry coincides with the bisection of the transfer angle.

The dynamics of the transfer depend on whether the spiral transitions between regimes or not. For a formal treatment of the solutions we introduce the following definition:

Definition 6. (Direct and indirect transfers) A spiral transfer $\mathcal{S} \in \mathbb{S}$ is *direct* if the evolution of the radius r from r_1 to r_2 is monotonic. Conversely, if r reaches a minimum or maximum value during the transfer, then the solution is an *indirect* transfer. Indirect transfers can only be elliptic or hyperbolic spirals of Type II, and the axis of $\mathscr{T}$-symmetry lies between θ_1 and θ_2, $\theta_m \in [\theta_1, \theta_2]$.

To determine whether a solution $\mathcal{S} \in \mathbb{S}$ is direct or indirect one should solve for θ_m in Eq.(F.24) and check if it lies between θ_1 and θ_2. Similarly, these relations are useful when looking explicitly for direct or indirect transfers. All spiral transfers connecting $r_1 = r_2$ are indirect. The case $K_1 = 0$ yields a degenerate indirect transfer, corresponding to a circular Keplerian orbit.

10.3.4 The thrust acceleration

The thrust acceleration profile is defined in Eq. (10.2), and depends on the radial distance and the flight-direction angle. Figure 10.7 depicts the evolution of the acceleration due to the tangential thrust ($\xi = 1/2$) for different types of spirals. Initially, the acceleration is maximum for slow elliptic transfers. It decreases rapidly and vanishes for $r = r_{\max}$. When the spiral transitions to the lowering regime the acceleration changes its sign. The figure shows how the acceleration profile resembles the typical thrust-coast-thrust sequence. The minimum-energy transfer separates the slow and the fast elliptic spiral transfers. The acceleration along hyperbolic spirals of Type II is zero for $r = r_{\min}$, and also $a_p \to 0$ as $r \to \infty$. That is, the magnitude of the thrust accelera-

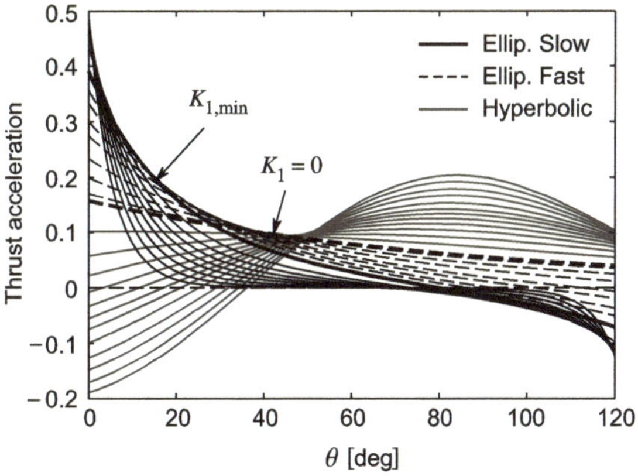

Fig. 10.7: Dimensionless thrust acceleration along different types of spiral transfers with $\xi = 1/2$.

tion decreases along the asymptotes. The thrust along parabolic or Type I hyperbolic spirals never vanishes, because there are no transitions between regimes.

In practice, knowing a priori the maximum value of the thrust greatly helps in the design process, since solutions that require a propulsive acceleration over the admissible maximum can be easily discarded. The magnitude of the acceleration due to the thrust $\mathbf{a}_p$ takes the form

$$a_p(r) = \frac{1}{r^2}\sqrt{\xi^2\cos^2\psi + (1-2\xi)^2\sin^2\psi} = \frac{\sqrt{\xi^2[2(1-\xi)+K_1 r]^2 + K_2^2(1-4\xi+3\xi^2)}}{r^2[2(1-\xi)+K_1 r]}$$

The thrust depends on K_1, K_2, the radial distance, and the control parameter ξ. Once the transfer spiral is selected the problem of finding the maximum value of $a_p(r)$ reduces to finding the radius r^* that maximizes $a_p(r)$, so that

$$a_{p,\max} = \frac{\sqrt{\xi^2[2(1-\xi)+K_1 r^*]^2 + K_2^2(1-4\xi+3\xi^2)}}{{r^*}^2[2(1-\xi)+K_1 r^*]} \tag{10.19}$$

The maximum acceleration exerted by the thrust $\mathbf{a}_p$ between $\mathcal{P}_1$ to $\mathcal{P}_2$ occurs at $r^* = \min(r_1, r_2)$, except for the case of hyperbolic spirals of Type II, where r^* is given by:

$$\xi \in [\xi^-, \xi^+]: \quad \begin{cases} \min(r_1, r_2), & r_Q \le \min(r_1, r_2),\ \mathrm{D} \\ \max(r_1, r_2), & r_Q \ge \max(r_1, r_2),\ \mathrm{D} \\ r_Q, & \text{rest} \end{cases} \tag{10.20}$$

$$\xi \notin [\xi^-, \xi^+]: \quad \begin{cases} r_{\min}, & \mathrm{I} \\ \min(r_1, r_2), & \text{rest} \end{cases} \tag{10.21}$$

Here D and I denote direct and indirect transfers, respectively. The value of r_Q is:

$$r_Q = \frac{6K_1K_2(1-\xi)}{R^{1/3}}\left[(1-3\xi)K_2\cos\left(\frac{\vartheta+2\pi}{3}\right) - \xi\sqrt{2(1-\xi)(3\xi-1)}\right]$$

where

$$R = 54K_1^6K_2^3\xi^3\sqrt{2(3\xi-1)^3(1-\xi)^3}$$

$$\vartheta = \arctan\left[\frac{\sqrt{2\left[K_2^2(3\xi-1) - 2\xi^2(1-\xi)\right]}}{2\xi\sqrt{1-\xi}}\right]$$

The limits ξ^+ and ξ^- are the solutions to the equation

$$\frac{|1-2\xi|}{r_{\min}^2} = \frac{\sqrt{\xi^2[2(1-\xi)+K_1 r_Q]^2 + K_2^2(1-4\xi+3\xi^2)}}{{r_Q}^2[2(1-\xi)+K_1 r_Q]}$$

and determine the values of ξ that make the acceleration at the periapsis of the spiral match the acceleration at r_Q. The constraint on K_2 becomes:

$$K_2^2 \leq \frac{a_{p,\max}^2 r^{*4}[2(1-\xi)+K_1 r^*]^2 - \xi^2[2(1-\xi)+K_1 r^*]^2}{1-4\xi+3\xi^2}$$

This expression leads to a criterion for bounding the search in the space of solutions.

10.3.5 The Δv due to the thrust

The total Δv imparted to the particle by the continuous thrust is defined by the integral

$$\Delta v = \int_{t_1}^{t_2} ||\mathbf{a}_p||\,dt$$

which, in general, needs to be evaluated numerically. However, for the special case $\xi = 1/2$ (purely tangential thrust) this integral can be solved in closed form, admitting two different solutions depending on whether the spiral transfer is direct or indirect. For the direct case and $\xi = 1/2$ it is

$$\Delta v = \left| \sqrt{K_1 + \frac{2(1-\xi)}{r_1}} - \sqrt{K_1 + \frac{2(1-\xi)}{r_2}} \right| = |v_1 - v_2| \tag{10.22}$$

The minimum-energy spiral involves the maximum change between the departure and arrival velocities. Thus, if the minimum-energy spiral defines a direct transfer, then the resulting Δv is the maximum among all the possible direct transfers with $\xi = 1/2$. If the spiral transfer is indirect, then it reaches a maximum or minimum radius r_m during the transfer, and the value of Δv is:

$$\Delta v = |v_1 - v_m| + |v_2 - v_m| = |v_1 + v_2 - 2v_m| \tag{10.23}$$

10.3.6 The locus of velocities

Figure 10.8 depicts the locus of velocity vectors projected on skewed axes (Battin, 1999, p. 244). The departure velocity vector $\mathbf{v}_1$ decomposes in

$$\mathbf{v}_1 = v_\rho\, \mathbf{u}_1 + v_c\, \mathbf{u}_c$$

where $\mathbf{u}_1 = \mathbf{r}_1/r_1$ and $\mathbf{u}_c$ is the unit vector along the chord connecting $\mathcal{P}_1$ and $\mathcal{P}_2$. The figure shows that the locus of minimum departure velocities, $v_m = ||\mathbf{v}_m||$, is tangent to the locus of solutions at the point where $K_1 = K_{1,\min}$: the departure velocity for $\mathcal{S}_m$ is the minimum among all possible transfers. Increasing the departure velocity yields

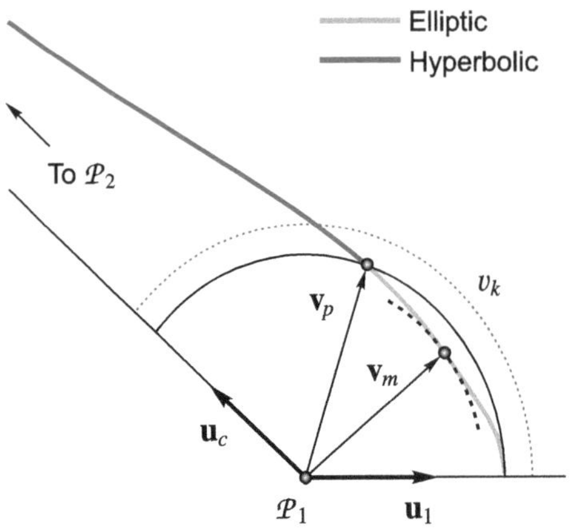

Fig. 10.8: Locus of velocity vectors projected on skewed axes for $\xi = 1/2$.

two intersection points with the locus of solutions. When the velocity becomes equal to $[2(1-\xi)/r]^{1/2}$ for $r = r_1$ the solutions are parabolic spirals, with

$$\mathbf{v}_p = \sqrt{\frac{2(1-\xi)}{r_1}}\,(\cos\psi_1\,\mathbf{u}_1 + \sin\psi_1\,\mathbf{u}_2)$$

Here, vector $\mathbf{u}_2$ is defined by the inplane perpendicular to $\mathbf{u}_1$, in counter-clockwise direction, and ψ_1 is given in Eq. (10.16). The velocity v_k corresponds to the magnitude of the departure velocity for the minimum-energy ellipse in the Keplerian case. It is observed that a spiral transfer allows one to reduce the minimum velocity required for the transfer, $v_m < v_k$. The Δv required for leaving the departure orbit will be smaller than the one in the ballistic case. This is an important improvement with respect to the logarithmic spirals (Petropoulos et al. (1999) and McInnes (2004) showed that logarithmic spirals typically require higher v_∞ at departure than the ballistic case). Increasing the magnitude of the departure velocity yields hyperbolic spiral transfers. This branch of solutions converges asymptotically to the chord connecting $\mathcal{P}_1$ and $\mathcal{P}_2$.

10.4 Fixing the time of flight

Generalized logarithmic spirals admit closed-form solutions for the time of flight. They depend on the type of spiral. This section analyzes the spiral transfers from $\mathbf{r}_1$ to $\mathbf{r}_2$ given a constraint on the time of flight.

The simplest solution is the transfer along a parabolic spiral ($K_1 = 0$), which corresponds to a logarithmic spiral. Combining Eqs. (F.20) and (10.16) the time of flight reduces to

$$t_2 = \frac{2\sqrt{2(1-\xi)\left[\ln^2(r_2/r_1) + (\theta_2-\theta_1)^2\right]}}{3\ln(r_2/r_1)}\left(r_2^{3/2} - r_1^{3/2}\right) \tag{10.24}$$

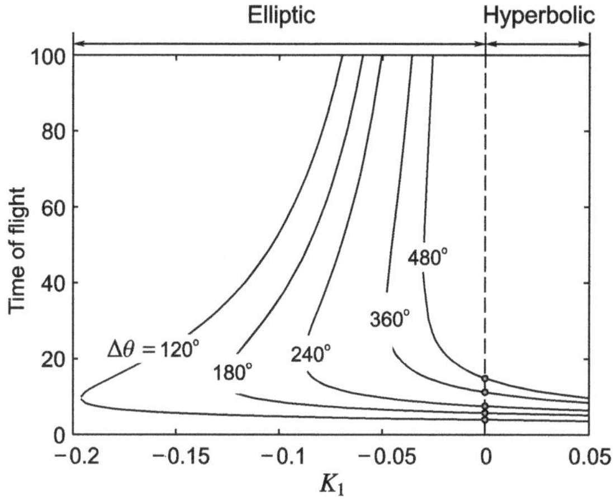

Fig. 10.9: Dimensionless time of flight parameterized in terms of the constant of the energy, K_1.

This expression depends only on the boundary conditions. In the limit case $r_2 = r_1$ and $\xi = 1/2$ this expression provides

$$\lim_{r_2 \to r_1} t_2 = \lim_{r_2 \to r_1} \frac{2}{3}(\theta_2 - \theta_1)\left[\frac{r_2^{3/2} - r_1^{3/2}}{\ln(r_2/r_1)}\right] = (\theta_2 - \theta_1)\, r_1^{3/2}$$

which is in fact the normalized time of flight corresponding to a circular Keplerian orbit of radius r_1.

Figure 10.9 shows the influence of the constant of the energy in the time of flight from $\mathcal{P}_1$ to $\mathcal{P}_2$. As $K_1 \to 0^-$ the branch of fast solutions converges to the fast parabolic transfer. It defines the transition from elliptic to hyperbolic spirals. Note that $K_1 = 0$ behaves as a vertical asymptote: the parabolic solution along the slow branch requires an infinite time to reach $\mathcal{P}_2$. This is due to the fact that the slow parabolic transfer connects both points through infinity, as discussed in regard to Fig. 10.5. Increasing K_1 from that point and along the slow branch yields the fictitious set of hyperbolic solutions that connect $\mathcal{P}_1$ and $\mathcal{P}_2$ through infinity. Given the time of flight, increasing the number of revolutions of the spiral transfer increases the value of K_1 of the solution. Consequently, increasing the number of revolutions of direct transfers always reduces the total Δv due to the thrust.

The parameterization used to construct Fig. 10.9 shows that the time of flight does not have any minima. Given the time of flight, the geometry of the transfer, and the control parameter ξ there is only one value of K_1 that satisfies the constraints. This statement holds for a given number of revolutions (defined as part of the geometry of the transfer) and for prograde orbits. There is an equivalent retrograde spiral transfer, $\psi_1 \in (\pi, 2\pi)$, with a different time of flight, unless the transfer is symmetric with respect to θ_1. Similarly, from Fig. 10.9 it follows that different solutions to Lambert's problem can be obtained by changing the number of revolutions, i.e., the transfer angle.

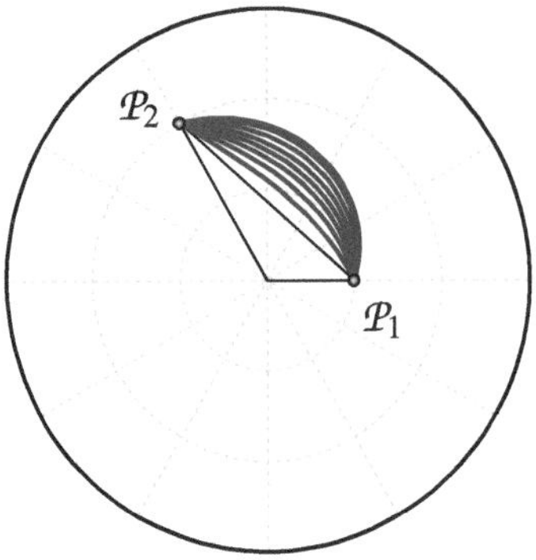

Fig. 10.10: Family of solutions to the spiral Lambert problem parameterized in terms of the control parameter ξ.

The constraint on the time of flight leaves one degree of freedom associated to the control parameter ξ, apart from the number of revolutions and the selection of prograde/retrograde motion. The spiral Lambert problem translates into solving for K_1 and ψ_1 in

$$r_2 = r(\theta_2; \xi, K_1, \psi_1, r_1, \theta_1) \tag{10.25}$$

$$t_2 = t(r(\theta_2); \xi, K_1, \psi_1, r_1, \theta_1) \tag{10.26}$$

Recall that the transfer angle θ_2 accounts for the number of revolutions n, so that $\theta_2 = \tilde{\theta}_2 + 2n\pi$ with $\theta_2 \in [0, 2\pi)$. The retrograde solutions can be found by solving a complementary problem where θ_2 is replaced by $\theta_{2,\text{ret}}$, defined as $\theta_{2,\text{ret}} = 2(n+1)\pi - \tilde{\theta}_2$. The solution to Eqs. (10.25) and (10.26) obtained for the complementary problem, $\psi_{1,\text{ret}}$, is modified so that it yields a retrograde orbit, i.e., $\psi_1 = 2\pi - \psi_{1,\text{ret}}$. Figure 10.10 depicts a family of solutions to the spiral Lambert problem obtained by changing the value of the control parameter ξ. The time of flight is the same for all the spirals in the figure.

An initial guess (K_1, ψ_1) is required in order to initialize the iterative procedure. Thus, given the geometry of the transfer, the time of flight, and the control parameter, the first step is to compute $K_{1,\text{min}}$ and ψ^* from Eqs. (10.14–10.15). For the parabolic case ψ_1 has been solved in closed form in Eq. (10.16). The corresponding times of flight follow naturally and it is now possible to determine the type of solution and provide a consistent initial guess. The next step is to solve for K_1 and ψ_1 in Eqs. (10.25–10.26). The method of generating an initial guess can be obviated if a better estimate is available.

10.5 Repetitive transfers

Assume that an elliptic spiral trajectory intersects the axis of symmetry in the raising regime. Because of the $\mathscr{T}$-symmetry of the trajectory it will always intersect the axis again at that exact point in the lowering regime after sufficient time. Hence, if the axis of symmetry is aligned with $\mathbf{r}_2$ and the spiral connects $\mathcal{P}_1$ and $\mathcal{P}_2$ in m revolutions, then it will always pass through $\mathcal{P}_2$ again after n revolutions. We refer to these kinds of transfers as m:n repetitive transfers. The departure flight-direction angle can be selected so that the integers m and n take prescribed values.

When defining these kinds of transfers, K_1 is fixed and cannot be chosen arbitrarily: it depends on the geometry of the transfer. The simplest way to formulate the problem is to solve for K_1 and ψ_1 in the system:

$$r_2 = r(\tilde{\theta}_2 + 2m\pi; \xi, K_1, \psi_1, r_1, \theta_1)$$
$$r_2 = r(\tilde{\theta}_2 + 2n\pi; \xi, K_1, \psi_1, r_1, \theta_1)$$

The solution depends on the configuration of the m:n sequence. Solutions only exist for elliptic spirals and hyperbolic spirals of Type II.

Figure 10.11 depicts two examples of different repetitive transfers. The solutions are indirect, by definition. If $m + n$ is even then the maximum is oriented in the direction of $\tilde{\theta}_2$, whereas if $m + n$ is odd then the maximum corresponds to $\tilde{\theta}_2 + \pi$. The magnitude of the maximum radius depends on $n - m$ and on the value of m, which indicates how soon the point $\mathcal{P}_2$ is reached for the first time.

During the transfer and along the spiral the values of $\{K_1, K_2, \xi\}$ are constant. This means that the only difference between crossings is the regime of the spiral, and the velocities after m and n revolutions (respectively v_2 and v_2') are the same, $v_2 = v_2'$. Similarly, the conservation of K_2 shows that the pair of solutions are $\mathscr{C}$-symmetric. It then follows that $\psi_2 + \psi_2' = \pi$. Equation (10.29) proves that the thrust acceleration at $\mathcal{P}_2$ only depends on $\{K_1, K_2\}$ and r_2. This means that the magnitude of the thrust accelerations at $\mathcal{P}_2$ for consecutive passes is the same, but one favors the velocity whereas the other opposes it.

The search for repetitive transfers is an additional technique for controlling the time of flight. Although the configuration of the transfer is given by discrete values

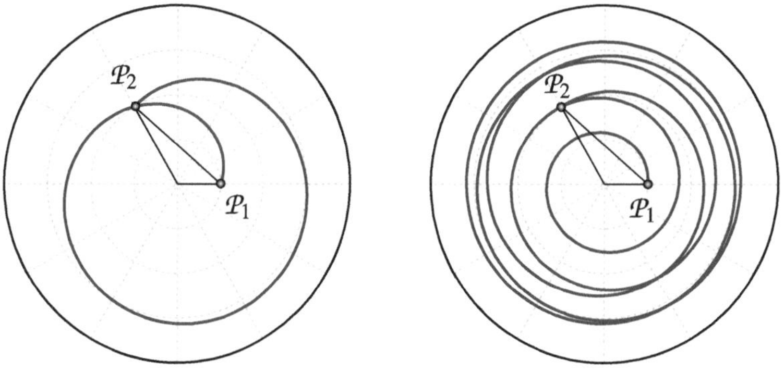

(a) $m = 0, n = 1$

(b) $m = 1, n = 5$

Fig. 10.11: Examples of $m:n$ repetitive transfers for $r_2/r_1 = 2$ and $\tilde{\theta}_2 - \theta_1 = 2\pi/3$.

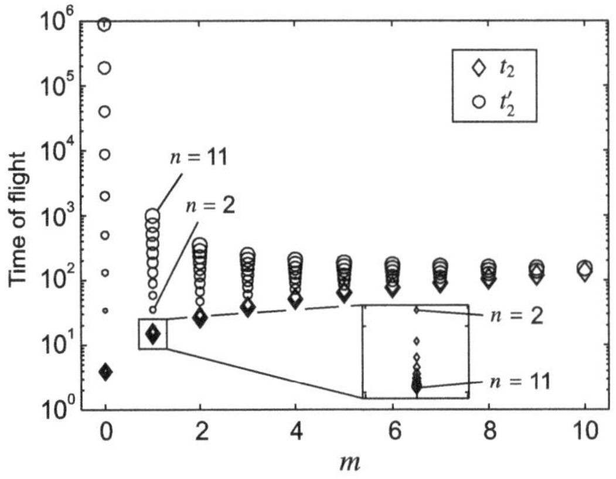

Fig. 10.12: Dimensionless time of flight for different $m{:}n$ repetitive configurations. The size of the markers scales with the values of the index $n > m$. Diamonds represent the time to the first pass, and circles correspond to the second pass.

of m and n, it might be useful for different mission scenarios. Figure 10.12 shows the time of flight for the two repetitive passes, t_2 and t_2', for different $m{:}n$ sequences. It is interesting to note that increasing the number of revolutions between the two passes reduces t_2, while the time of flight t_2' increases. For constant n increasing m yields slower transfers. Similarly, for constant m larger values of n increase the time of flight. When $m \to \infty$ the difference in the time of flight for the two passes decreases, and the time of flight approaches a limit value. The existence of a finite limit relates to the fact that increasing the number of revolutions makes $K_{1,\mathrm{min}}$ smaller in magnitude until $\mathcal{S}_m$ becomes parabolic in the theoretical limit (Fig. 10.3).

10.6 Evaluating the performance

The simplest approach to the design of an orbit transfer between two bodies is to connect them using a Keplerian arc. This means solving the ballistic Lambert problem given the state of the first body at departure, and the state of the second body at arrival. Two impulsive maneuvers are required (one at departure and one at arrival) in order to meet the conditions on the velocity. In this chapter, we explored the potential of replacing the Keplerian arc with a continuous-thrust arc. This allows us to reduce the magnitude of the impulsive maneuvers, which leads to more efficient transfers (Table 1.2 shows a comparison of specific impulses). In this section, we will evaluate the performance of the new design strategy by comparing it to the purely ballistic approach, using the 2016–2018 Mars launch campaign as an example. They are compared in terms of the required propellant mass fraction, the characteristic departure energy C_3, and the arrival v_∞. The ephemeris of the Earth and Mars are retrieved from the DE430 ephemeris, using SPICE.

The overall performance of the ballistic and accelerated transfers can be compared in Figs. 10.13a and 10.13d. The pork-chop plot in Fig. 10.13a shows the two launch windows corresponding to the 2016 and 2018 period. The figures show the propellant

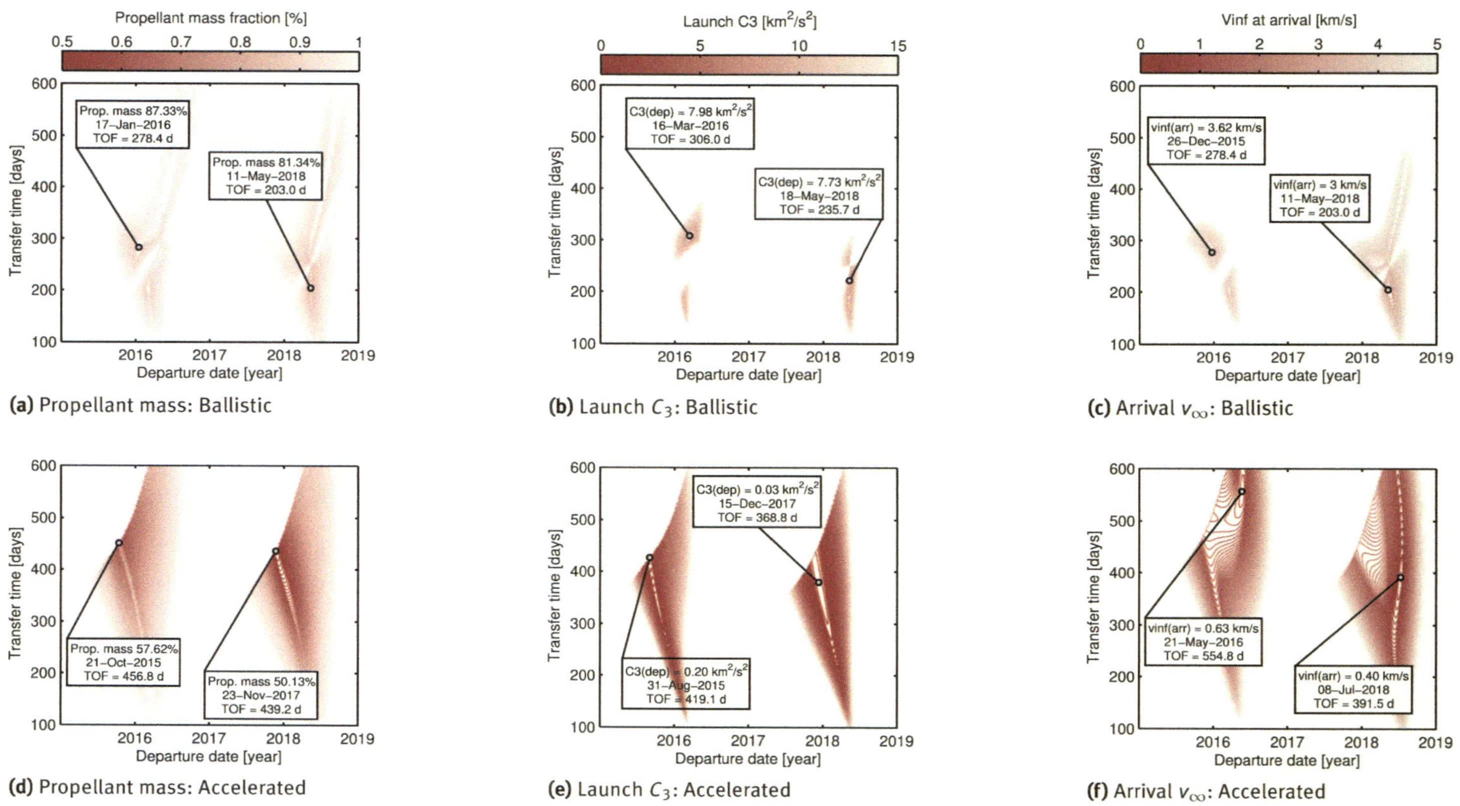

(a) Propellant mass: Ballistic

(b) Launch C_3: Ballistic

(c) Arrival v_∞: Ballistic

(d) Propellant mass: Accelerated

(e) Launch C_3: Accelerated

(f) Arrival v_∞: Accelerated

Fig. 10.13: Comparison of the ballistic and accelerated pork-chop plots for the Earth to Mars transfers.

mass fraction required for the transfer. This budget includes the impulsive maneuvers required at launch and to rendezvous with Mars. In the accelerated case, the propellant expenditures associated to the continuous-thrust phase are also accounted for. The specific impulses for the impulsive and continuous-thrust maneuvers are I_{sp} = 350 s and I_{sp} = 3500 s, respectively. The control parameter is fixed to $\xi = 1/2$ (purely tangential thrust). Equations (10.25–10.26) are solved with a Levenberg–Marquardt algorithm. The spiral transfers are restricted to the zero-revolutions case ($\theta_2 - \theta_1 < 2\pi$). With the ballistic solution more than 80% of the total mass budget will be occupied by propellant. The spiral transfer is able to reduce this figure to about 50%, which brings a significant increment in the payload mass delivered to the final orbit. Although the optimal spiral transfers require twice the time of flight compared to the Keplerian ones, the launch windows are broader and, for the same time of flight, the spiral solution still outperforms the ballistic approach.

Figures 10.13b and 10.13e depict the departure C_3 required for the transfer. The spiral solution extends the launch opportunities again, with transfers with C_3 as low as 0.03 km^2/s^2. This is a significant reduction compared to the 7.73 km^2/s^2 expected for May, 2018.

Finally, Figs. 10.13c and 10.13f focus on the arrival v_∞ when reaching Mars. The main advantage of the accelerated approach is its flexibility, as it extends the launch opportunities. In addition, the optimum solutions reduce the v_∞ at Mars by about one order of magnitude.

10.7 Additional properties

In this section, additional properties of the spiral two-point boundary-value problem are presented. They have been grouped in this section to simplify the reading of this chapter.

Property 1. *(Change in the semimajor axis) Given ξ, the change in the inverse of the semimajor axis between two points of the spiral depends on r_1 and r_2 alone, and reduces to:*

$$\frac{1}{a_2} - \frac{1}{a_1} = 2\xi\left(\frac{1}{r_2} - \frac{1}{r_1}\right)$$

Property 2. *(Minimum velocity) The departure and terminal velocities, v_1 and v_2, are minimum on the minimum-energy spiral.*

Property 3. *(Change in the velocity and semimajor axis) The change in the velocity from $\mathcal{P}_1$ to $\mathcal{P}_2$, $\Delta v = |v_2 - v_1|$, is maximum along the minimum-energy spiral. Similarly, the change in the semimajor axis, $|\Delta a| = |a_2 - a_1|$, is minimum along the minimum-energy spiral.*

In order to establish the properties of conjugate spirals it is important to note that:

Lemma 1. If a certain variable y depends only on K_1 and r, $y = y(r; \xi, K_1)$, then $y_1 = \tilde{y}_1$ and $y_2 = \tilde{y}_2$ for all pairs of conjugate spirals. Here $y_1 = y(r_1)$ and $y_2 = y(r_2)$.

This Lemma is a powerful contrivance for finding variables that take the same values on a spiral and its conjugate, and provides

Property 4. *(Departure and terminal velocities) Two conjugate generalized logarithmic spirals $\mathcal{S}$ and $\tilde{\mathcal{S}}$ have the same departure and terminal velocity, i.e., $v_1 = \tilde{v}_1$ and $v_2 = \tilde{v}_2$.*

Property 5. *(Departure and terminal semimajor axes) Two conjugate spirals $\mathcal{S}$ and $\tilde{\mathcal{S}}$ have the same departure and terminal osculating semimajor axes, i.e., $a_1 = \tilde{a}_1$ and $a_2 = \tilde{a}_2$.*

Property 6. *(Angular velocities) The ratio between the initial and terminal angular velocities is the same for two conjugate spirals, $\dot{\theta}_1/\dot{\theta}_2 = \dot{\tilde{\theta}}_1/\dot{\tilde{\theta}}_2$.*

Property 7. *(Flight-direction angle) The ratio between the initial and terminal values of $\sin\psi$ is the same for two conjugate spirals, $\sin\psi_1/\sin\psi_2 = \sin\tilde{\psi}_1/\sin\tilde{\psi}_2$.*

Property 8. *(Circumferential velocities) The ratio between the initial and terminal circumferential velocities $v_\theta = v\sin\psi$ is the same for two conjugate spirals, $v_{\theta 1}/v_{\theta 2} = \tilde{v}_{\theta 1}/\tilde{v}_{\theta 2}$.*

Property 9. *(Angular momentum) The ratio between the initial and terminal angular momenta (or equivalently the orbital parameter, p) is the same for two conjugate spirals, $h_1/h_2 = \tilde{h}_1/\tilde{h}_2$ and $p_1/p_2 = \tilde{p}_1/\tilde{p}_2$.*

Property 10. *(Eccentricity) The initial and terminal values of the eccentricity of two conjugate spirals satisfy the relation $(1 - e_1^2)/(1 - e_2^2) = (1 - \tilde{e}_1^2)/(1 - \tilde{e}_2^2)$.*

10.8 Additional dynamical constraints

Apart from the boundary conditions (r_1, θ_1), (r_2, θ_2) and fixing the time of flight, additional limitations might be imposed on the solutions. They typically originate from technical limitations or mission requirements. The goal of this section is to provide the conditions that $\{K_1, K_2\}$ must satisfy in order for the transfer to meet certain requirements. They are useful for rapidly rejecting solutions that violate the constraints, without having to integrate the trajectory. The conditions are similar to those related to the maximum thrust acceleration, presented in Sect. 10.3.

10.8.1 Arrival conditions

Once the transfer is determined by means of $\{K_1, K_2, \xi\}$ the arrival conditions can be solved explicitly. The velocity is given by

$$v_2 = \sqrt{K_1 + \frac{2(1-\xi)}{r_2}} \tag{10.27}$$

Property 4 showed that the arrival velocity for a pair of conjugate spirals is the same. The flight-direction angle at r_2 is solved from the definition of K_2:

$$\psi_2 = \arcsin\left[\frac{K_2}{2(1-\xi) + K_1 r_2}\right] \tag{10.28}$$

The two solutions to this equation are $\mathscr{C}$-symmetric. When the spiral reaches $\mathcal{P}_2$ in the raising regime it is $\psi_2 \in [0, \pi/2]$. This corresponds to direct transfers where $r_2 > r_1$ or indirect hyperbolic transfers of Type II. Conversely, if the spiral is in the lowering regime at $\mathcal{P}_2$ it is $\psi_2 \in [\pi/2, \pi]$. This occurs for direct transfers where $r_2 < r_1$ and for indirect elliptic transfers.

The magnitude of the thrust acceleration is

$$a_{p,2} = \pm\frac{\sqrt{\xi^2[2(1-\xi) + K_1 r_2]^2 + K_2^2(1 - 4\xi + 3\xi^2)}}{r_2^2[2(1-\xi) + K_1 r_2]} \tag{10.29}$$

The +/– is selected depending on whether the spiral is in the raising/lowering regime at $\mathcal{P}_2$.

10.8.2 Radius

Consider that the trajectory is bounded by a minimum and a maximum radius, so that

$$r_{\min} \le r(\theta) \le r_{\max}$$

where of course $[r_{\min}, r_{\max}] \supset [r_1, r_2]$. For direct transfers the previous consideration makes the condition on the radii hold naturally. Only for the case of indirect transfers can $r(t)$ take values outside the interval $[r_1, r_2]$, when $r_{\max} > \max(r_1, r_2)$ or $r_{\min} < \min(r_1, r_2)$.

Indirect transfers can only be elliptic spirals or hyperbolic spirals of Type II. A parameterization in terms of the constant K_1 yields a straightforward classification of the solutions. For the case of elliptic spirals the condition $r(\theta) \ge r_{\min}$ is satisfied naturally, whereas $r(\theta) \le r_{\max}$ requires

$$r_m \le r_{\max} \implies K_2 \ge 2(1-\xi) + K_1 r_{\max}$$

The solution $\mathcal{S} \in \mathbb{S}$ will be a Type II hyperbolic spiral as long as $K_1 > K_{1,\text{tr}}$, where the limit value $K_{1,\text{tr}}$ is given in Eq. (10.18). It then follows that

$$r_m \ge r_{\min} \implies K_2 \ge 2(1-\xi) + K_1 r_{\min}$$

10.8.3 Eccentricity

Assume that the transfer orbit must satisfy a constraint of the form $e(r) \in [e_{\text{min}}, e_{\text{max}}]$. The eccentricity of the osculating orbit is defined explicitly in Eq. (F.26). For elliptic spirals it always decreases in the raising regime. It grows in the lowering regime. The eccentricity at $r = r_{\text{max}}$ is minimum, and denoted $e(r_{\text{max}}) = e_m$. The eccentricity of the osculating orbit of a parabolic (logarithmic) spiral is constant. The eccentricity grows for all hyperbolic spirals in the raising regime. It decreases in the lowering regime. The eccentricity at $r = r_{\text{min}}$ is maximum, and denoted $e(r_{\text{min}}) = e_m$.

The minimum and maximum eccentricity for indirect transfers is given by the same expression, $e_m = |1 - K_2|$, which depends only on the constant of the angular momentum K_2. For the case of indirect elliptic transfers the spiral is initially in the raising regime, it reaches the maximum radius and then transitions to the lowering regime. In this case the minimum eccentricity occurs at $r = r_{\text{max}}$, and corresponds to $e_m = 1 - K_2$. A constraint on the minimum eccentricity translates into $K_2 \le 1 - e_{\text{min}}$. The maximum eccentricity occurs at r_1 or r_2, whichever is smaller. An upper limit on the osculating eccentricity translates into

$$K_2^2 \ge (1 - e_{\text{max}}^2) Q(r^*) \qquad \text{with} \quad r^* = \min(r_1, r_2)$$

and

$$Q(r) = \frac{2(1-\xi) + K_1 r}{2\xi - K_1 r}$$

Note that in the case $r_1 = r_2$ it is indifferent which value is chosen. Transfers to the same radius are always indirect. In addition, the axis of symmetry coincides with the bisection of the transfer vectors due to the $\mathscr{T}$-symmetry of the spiral.

For direct transfers there are no transitions between raising/lowering regimes. If $r_2 < r_1$ the spiral is in the lowering regime, and if $r_2 > r_1$ the spiral is in the raising regime. Note that the transfer $r_2 = r_1$ is not possible along direct elliptic spirals. To satisfy the constraints on the eccentricity it suffices that K_2 verifies

$$r_2 > r_1: \quad \left(1 - e_{\text{max}}^2\right) Q(r_1) \le K_2^2 \le \left(1 - e_{\text{min}}^2\right) Q(r_2) \tag{10.30}$$

$$r_2 < r_1: \quad \left(1 - e_{\text{max}}^2\right) Q(r_2) \le K_2^2 \le \left(1 - e_{\text{min}}^2\right) Q(r_1) \tag{10.31}$$

For the case of parabolic spirals ($K_1 = 0$) the eccentricity remains constant, meaning that the constraints on the values of the eccentricity reduce to $(1 - e_{\text{max}}^2)^{1/2} \le K_2 \le (1 - e_{\text{min}}^2)^{1/2}$. Recall that the required value of K_2 for transfers along parabolic spirals admits a closed-form solution, given in Eq. (10.17). The constraint on the eccentricity can be referred directly to the geometry of the problem

$$\frac{1 - e_{\text{max}}^2}{e_{\text{max}}^2} \le \left[\frac{\theta_2 - \theta_1}{\ln(r_2/r_1)}\right]^2 \le \frac{1 - e_{\text{min}}^2}{e_{\text{min}}^2}$$

The maximum and minimum eccentricity of the osculating orbit along Type I hyperbolic spirals occurs at r_1 and r_2: it is minimum at the smallest radius, and maximum at the largest. Note that solutions of this type are always direct. The constraint on K_2 is obtained by switching r_1 and r_2 in Eqs. (10.30–10.31).

The constraints on Type II hyperbolic spirals require an analysis of the orientation of the axis of $\mathcal{T}$-symmetry. For the case of indirect transfers the maximum eccentricity corresponds to $r = r_{\min}$, and reads $e_m = K_2 - 1$. The constraint on the maximum eccentricity reduces to $K_2 \leq 1 + e_{\max}$. The eccentricity will always be above the lowest bound as long as

$$K_2^2 \leq \left(1 - e_{\min}^2\right) Q(r^*)$$

For the case of direct transfers the conditions on K_2 are the same as for the case of hyperbolic spirals of Type I.

10.8.4 Semimajor axis

Consider that the semimajor axis of the osculating orbit is constrained to $a \in [a_{\min}, a_{\max}]$. It grows in the raising regime and decreases in the lowering regime, no matter the type of spiral. The semimajor axis will reach its minimum value when the radius is minimum, and its maximum value when the radius is maximum.

The semimajor axis only depends on the energy of the spiral, K_1, and the control parameter, ξ. Provided that the families of solutions to the transfer have been parameterized in terms of the value of K_1 the adequate selection of K_1 ensures that the constraints are satisfied:

$$\frac{2\xi}{r_{\min}} - \frac{1}{a_{\min}} \leq K_1 \leq \frac{2\xi}{r_{\max}} - \frac{1}{a_{\max}}$$

Here $r_{\min}$ and $r_{\max}$ denote the minimum and maximum radii that the spiral reaches. For indirect elliptic transfers the maximum radius corresponds to r_m, and for indirect hyperbolic spirals of Type II the minimum radius is r_m.

10.8.5 Periapsis and apoapsis radii

Consider the constraints on the radius at periapsis and apoapsis, $r_p \in [r_{p,\min}, r_{p,\max}]$ and $r_a \in [r_{a,\min}, r_{a,\max}]$. They are defined explicitly by combining Eqs. (F.26) and (F.28):

$$r_a = \frac{r}{2\xi - K_1 r}\left[1 + \sqrt{1 - \frac{K_2^2}{Q(r)}}\right]$$

$$r_p = \frac{r}{2\xi - K_1 r}\left[1 - \sqrt{1 - \frac{K_2^2}{Q(r)}}\right]$$

They grow in the raising regime, and diminish in the lowering regime. Therefore, for the case of direct transfers the constraints on the apoapsis radius reduce to

$$K_2^2 \le Q(r^*)\left\{1-\left[r_{a,\min}\left(\frac{2\xi}{r^*}-K_1\right)-1\right]^2\right\}, \qquad r^* = \min(r_1, r_2)$$

$$K_2^2 \ge Q(r^*)\left\{1-\left[r_{a,\max}\left(\frac{2\xi}{r^*}-K_1\right)-1\right]^2\right\}, \qquad r^* = \max(r_1, r_2)$$

The constraint on the periapsis radius takes the same form, simply inverting the inequality signs, and changing r_a to r_p.

Differentiating the definition of r_p and r_a with respect to r shows that both the apoapsis and periapsis radii have an extreme point which corresponds to $r_{\max}$ if $K_1 < 0$, and $r_{\min}$ if $K_1 > 0$ and $K_2 > 2(1-\xi)$. For the case of elliptic spirals the osculating orbit at $r = r_{\max}$ is defined by means of

$$r_p = \frac{K_2[2(1-\xi)-K_2]}{(-K_1)(2-K_2)} \qquad \text{and} \qquad r_a = \frac{2(1-\xi)-K_2}{(-K_1)}$$

These are the maximum values that r_p and r_a can take. They are reached only for the case of indirect transfers. In such a case the constraints translate into

$$K_2^2 - K_2[2(1-\xi) - K_1 r_{p,\max}] \ge 2K_1 r_{p,\max} \tag{10.32}$$

$$K_2 \ge 2(1-\xi) + K_1 r_{a,\max} \tag{10.33}$$

The conditions related to $r_{p,\min}$ and $r_{a,\min}$ are equal to those for direct transfers. Similarly, for indirect hyperbolic transfers of Type II the minimum values of r_p and r_a are the opposite of those for the elliptic case: the constraints are obtained by replacing the minimum/maximum values with the maximum/minimum ones in Eqs. (10.32–10.33), and inverting the inequality signs. The conditions for $r_{p,\max}$ and $r_{a,\max}$ are those for the direct transfers.

10.9 Conclusions

The connections with the Keplerian case found in the definition of the family of generalized logarithmic spirals simplify the study of Lambert's problem under a continuous acceleration. The existence of two integrals of motion provides a clear structure for the solution and yields dynamical properties of theoretical and practical interest. Classical geometrical and dynamical properties of the Keplerian Lambert problem have an equivalent expression in spiral form. There is a minimum-energy spiral transfer, with pairs of conjugate spirals bifurcating from it. Conjugate spirals are endowed with a collection of properties similar to those of conjugate Keplerian orbits.

All the numerical computations reduce to solving a system of two equations with two unknowns. Once the solution to the system is known, no other iterative processes are required, provided that all the relevant variables have been defined analytically. The resolution is simplified thanks to having located special cases for which the transfer can be solved analytically (parabolic spirals) or there is a specific procedure to compute them (minimum-energy spiral). When dealing with an arbitrary transfer knowing a priori two points on the curve for the time of flight helps to determine the range where the solution is to be found.

The generalized logarithmic spirals might improve the efficiency of the transfers computed with the Keplerian Lambert problem. Depending on the mission requirements the launch windows can be stretched, or the mass delivered to the final orbit might increase because the magnitude of the initial and final impulsive maneuvers can be reduced. The time of flight will typically increase, although special cases where the total Δv is reduced while maintaining the time of flight can be found.

11 Low-thrust trajectory design with controlled generalized logarithmic spirals

Chapter 9 introduced the family of generalized logarithmic spirals, a new analytic solution with continuous thrust. One of the possible practical applications of this kind of solution is the preliminary design of low-thrust trajectories. The pioneering work by Petropoulos (2001) and Petropoulos and Longuski (2004) showed the full potential of shape-based methods for exploring large spaces of solutions in mission design. They exploited their own analytic solution with continuous thrust in order to derive low-thrust gravity-assist mission design tools.

In this chapter, we shall explore the potential of generalized logarithmic spirals for low-thrust mission design. Thanks to having introduced a control parameter in Sect. 10.2, there is now a degree of freedom that can be adjusted to meet design constraints. The new control law takes the form

$$\mathbf{a}_p = \frac{\mu}{r^2}[\xi \cos\psi\,\mathbf{t} + (1 - 2\xi)\sin\psi\,\mathbf{n}] \tag{11.1}$$

where ξ is the control parameter.

The main advantage of using generalized logarithmic spirals over existing shape-based methods is the fact that the solution is dynamically intuitive. The constants to adjust are not arbitrary shape parameters, but constants of motion with a clear physical meaning. The conservation laws and the properties of the new spirals are closely related to Keplerian orbits. As a result, working in the spiral realm is similar to working with Keplerian solutions, and classical design techniques can be recovered and applied to the accelerated problem. This might help the mission analyst in the design process, leading to a "patched spirals" approach for low-thrust mission design: the boundary conditions define the values that the constants of motion must take, and different spiral or coast arcs can be defined easily. The continuity of the trajectory is guaranteed by matching the values of the constants of motion in the transition nodes.

Section 11.1 presents the design strategy. The first step is to characterize transfers between circular orbits. The second step consists in generalizing this method to the design of transfers between arbitrary orbits. Special attention is paid to the existence of solutions for a given transfer configuration. Periodic orbits are designed in Sect. 11.2. In Sect. 11.3 the flexibility of the design methodology is improved by introducing multiple nodes in the shaping sequence. Up to this point the analysis focuses on the planar case. Section 11.4 extends these techniques to the three-dimensional case. Finally, an example of application is presented in Sect. 11.5, consisting in the design of a mission to Ceres, including a flyby around Mars. In Appendix G we will explore a novel approach to shape-based trajectory design, recovering Seiffert's spherical to model orbital motion.

https://doi.org/10.1515/9783110559125-011

11.1 Orbit transfers

The control parameter provides a degree of freedom that can be adjusted as required to meet certain requirements. In addition, the transfer can be decomposed in a number of legs with different values of the control parameter, or even considering Keplerian arcs. The transition between two spiral arcs with different values of ξ is equivalent to changing the thrust magnitude and direction. No adjustments in the orbital velocity are required.

The control parameter only affects the value of the constant K_1. Changing the value of ξ adjusts the values of the constant of the generalized energy, K_1, but not the values of K_2. If K_2 needs to be changed along the transfer, coast arcs should be introduced.

11.1.1 Bitangent transfers

Consider the spiral transfer between two circular orbits of radii r_0 and $r_{\mathcal{F}}$. Along a spiral arc the evolution of the flight-direction angle ψ is monotonic, so if it is initially $\psi_0 = \pi/2$ (tangential departure) it is not possible to have $\psi_{\mathcal{F}} = \pi/2$ (tangential arrival) with one single spiral arc. Bitangent transfers are the composition of at least two continuous spiral arcs with different values of ξ. When $\psi = \pi/2$ the spiral is either a hyperbolic spiral at $r_{\min}$ or an elliptic spiral at $r_{\max}$. Parabolic spirals are not considered because of having a constant flight-direction angle.

The value of the constant K_2 corresponding to the first arc is solved initially from

$$K_{2,1} = r_0 v_0^2 \sin\psi_0 = 1$$

Since the control parameter does not affect the value of K_2 both spiral arcs share the same values of $K_2 \equiv K_{2,1} = K_{2,2} = 1$. The constant of the generalized energy on the first spiral arc takes the form

$$K_{1,1} = v_0^2 - \frac{2(1-\xi_1)}{r_0} = \frac{2\xi_1 - 1}{r_0}$$

because $v_0^2 = 1/r_0$. For transfers with $r_0 < r_{\mathcal{F}}$ the first arc is a hyperbolic spiral of Type II (with $r_0 = r_{\min}$) and the second arc is an elliptic spiral (with $r_{\max} = r_{\mathcal{F}}$). The condition $r_0 = r_{\min}$ is satisfied naturally by having imposed $K_2 = 1$:

$$r_{\min} = \frac{2\xi_1 - 1}{K_{1,1}} \implies r_{\min} = r_0$$

The radial distance at the transition point $\mathcal{A}$ is solved from the equation of the trajectory, Eq. (F.25):

$$\frac{r_{\mathcal{A}}}{r_0} = \frac{3 - 2\xi_1}{2(1-\xi_1) + \cos \ell_1 \theta_{\mathcal{A}}} \tag{11.2}$$

Similarly, the arrival spiral arc is defined by

$$K_{1,2} = v_{\mathcal{F}}^2 - \frac{2(1-\xi_2)}{r_{\mathcal{F}}} = \frac{2\xi_2 - 1}{r_{\mathcal{F}}}$$

and $K_{2,2} = 1$. From this equation it immediately follows that $r_{max} = r_{\mathcal{F}}$. The conditions at the transition point are obtained from the equation of the trajectory, Eq. (F.16):

$$\frac{r_{\mathcal{A}}}{r_{\mathcal{F}}} = \frac{3 - 2\xi_2}{2(1-\xi_2) + \cosh\{\ell_2[\theta_{\mathcal{A}} - (2n+1)\pi]\}} \tag{11.3}$$

The maximum radius occurs at $\theta_m = (2n+1)\pi$, where n is the number of revolutions. Dividing Eqs. (11.2) and (11.3) provides a relation between ξ_1 and ξ_2,

$$\frac{r_{\mathcal{F}}}{r_O} = \frac{3-2\xi_1}{3-2\xi_2}\left\{\frac{2(1-\xi_2) + \cosh\{\ell_2[\theta_{\mathcal{A}} - (2n+1)\pi]\}}{2(1-\xi_1) + \cos \ell_1\theta_{\mathcal{A}}}\right\} \tag{11.4}$$

In order for the two spiral arcs to be compatible, it must be

$$K_{1,2} = \frac{2\xi_2 - 1}{r_{\mathcal{F}}} = v_{\mathcal{A}}{}^2 - \frac{2(1-\xi_2)}{r_{\mathcal{A}}}$$

The velocity-matching condition at $\mathcal{A}$, $v_{\mathcal{A}1} = v_{\mathcal{A}2}$, follows from the integral of the generalized energy,

$$\frac{2\xi_2 - 1}{r_{\mathcal{F}}} = \frac{2\xi_1 - 1}{r_O} + \frac{2(\xi_2 - \xi_1)}{r_{\mathcal{A}}}$$

This condition yields an expression of ξ_2 as a function of ξ_1 and the boundary conditions,

$$\xi_2 = \frac{[(1-2\xi_1)r_{\mathcal{F}} - r_O]r_{\mathcal{A}} + 2\xi_1 r_O r_{\mathcal{F}}}{2r_O(r_{\mathcal{F}} - r_{\mathcal{A}})} \tag{11.5}$$

The problem of designing a bitangent transfer then reduces to solving for ξ_1 in Eq. (11.4). The value of the control parameter on the second arc, ξ_2, is given by Eq. (11.5). There is one degree of freedom in the solution. Under this formulation, it corresponds to the angular position of the node defining the transition point, $\theta_{\mathcal{A}}$. In addition, the number of revolutions can be adjusted by changing the values of n.

In a similar fashion, when $r_O > r_{\mathcal{F}}$ the equation to be solved for ξ_1 transforms into

$$\frac{r_{\mathcal{F}}}{r_O} = \frac{3-2\xi_1}{3-2\xi_2}\left\{\frac{2(1-\xi_2) + \cos\{\ell_2[\theta_{\mathcal{A}} - (2n+1)\pi]\}}{2(1-\xi_1) + \cosh \ell_1\theta_{\mathcal{A}}}\right\} \tag{11.6}$$

Recall that $\ell_i = |1 - 4(1-\xi_i)^2|^{1/2}$. Equation (11.5) still holds.

Figure 11.1 shows two examples of bitangent transfers with zero and one revolutions. The problem is based on an Earth to Mars transfer, where $r_{\mathcal{F}}/r_O = 1.527$. The transition point is selected arbitrarily. The departure spiral arc corresponds to a hyperbolic spiral of Type II in the raising regime, whereas the second arc corresponds to an elliptic spiral in the lowering regime. Their minimum and maximum radii are r_O and $r_{\mathcal{F}}$, respectively.

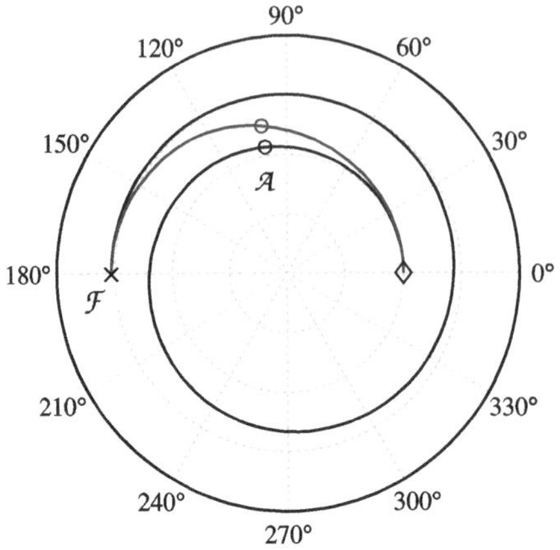

Fig. 11.1: Examples of bitangent transfers with 0 and 1 revolutions. The ∘ marks the transition point, fixed to $\theta_{\mathcal{A}} = 100$ deg. The spirals depart from ◊ and arrive at ×.

Low-thrust transfers are considered because they typically increase the mass delivered to the final orbit, by decreasing the propellant mass spent along the transfer. Figure 11.2 compares the fraction of mass that can be inserted into the final orbit using the Hohmann and the spiral transfers. The values of the specific impulse for the electric propulsion system and for the impulsive maneuvers are 2500 s and 250 s, respectively. In this example the zero-revolutions spiral transfer delivers over 50% of the initial mass, and the one-revolution case increases this percentage up to 75%. For the case of the Hohmann transfer the fraction of delivered mass falls to roughly 10% of the initial mass.

The time of flight for the different types of transfers is plotted in Fig. 11.3. The zero-revolutions spiral transfer is geometrically equivalent to the Hohmann transfer and the times of flight are comparable. Depending on the transition point, spiral transfers that are either slower or faster than the Hohmann transfer can be found. Increasing the number of revolutions can reduce the propellant consumption, but the time of flight

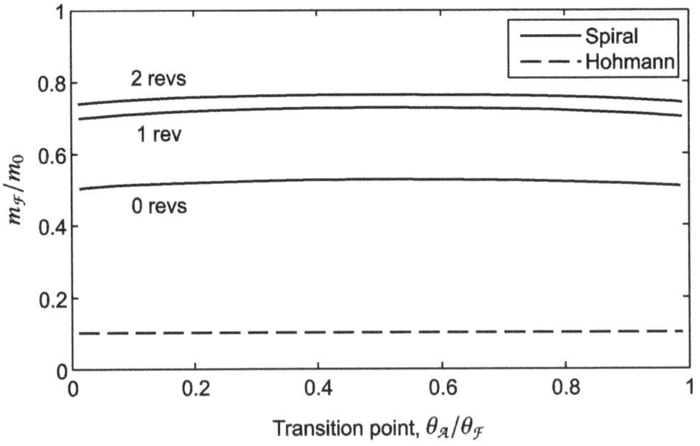

Fig. 11.2: Fraction of mass delivered to the final circular orbit.

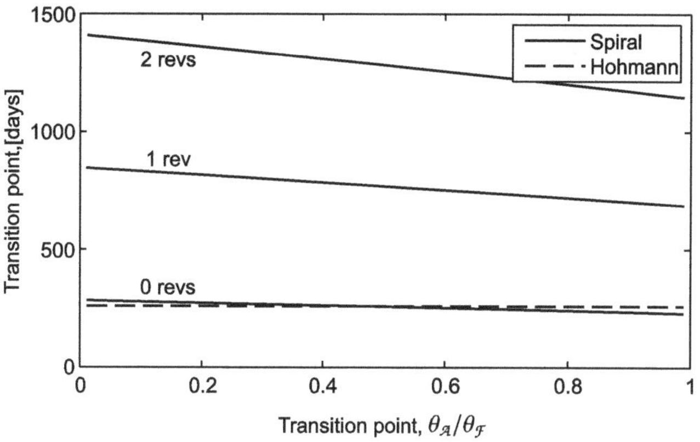

Fig. 11.3: Time of flight for Earth to Mars bitangent transfers.

grows significantly. In this example, the bitangent spiral transfer with one revolution increases the time of flight by a factor of three when compared with the Hohmann transfer, and for two revolutions the time of flight increases approximately by a factor six.

Figure 11.2 suggests that there is a particular position of the transition point $\theta_{\mathcal{A}}$ that maximizes the mass delivered into orbit. The zero-revolutions case is geometrically equivalent to the Hohmann transfer and can be compared directly in terms of mass delivered into orbit and time of flight. When considering the optimal transition point the spiral transfer delivers into Martian orbit 52.74% of the launch mass, whereas the Hohmann transfer can only deliver 10.12%. The time of flight for the Hohmann transfer is 259.38 d. The spiral transfer turns out to be faster, reducing the time of flight to 257.05 d. That means that the spiral transfer is more efficient both in terms of the mass delivered into orbit and in flight time.

The optimum spiral transfer is found by choosing adequately the transition point $\theta_{\mathcal{A}}$. Under the assumption that the specific impulse I_{sp} remains constant, the mass fraction is given by Tsiolkovsy's equation

$$\frac{m_{\mathcal{F}}}{m_0} = \exp\left(-\frac{\Delta v}{g_0 I_{\mathrm{sp}}}\right)$$

Here Δv stands for the time integral

$$\Delta v = \int_{t_0}^{t_{\mathcal{A}}} a_{p,1}\,\mathrm{d}t + \int_{t_{\mathcal{A}}}^{t_{\mathcal{F}}} a_{p,2}\,\mathrm{d}t$$

The integral of the thrust acceleration is decomposed in two arcs: the acceleration along the departure arc, $a_{p,1}$, is defined by ξ_1, while the acceleration along the second arc, $a_{p,2}$, is defined by ξ_2. The change in the thrust profile at $\theta_{\mathcal{A}}$ is the correction that the initial arc requires in order to meet the boundary conditions at arrival. Hence, the optimal solution is the one that requires the smallest correction, i.e., the transfer on which the difference $|a_{p,1} - a_{p,2}|$ at the transition point is minimum. The optimum transition point for maximizing the mass delivered into orbit is the one that minimizes the function

$$f(\theta_{\mathcal{A}}) = |a_{p,1} - a_{p,2}|$$

The magnitude of the thrust acceleration at $\theta_{\mathcal{A}}$ can be written

$$a_{p,i} = \frac{1}{{r_{\mathcal{A}}}^2}\sqrt{\xi_i^2 \cos^2\psi_{\mathcal{A}} + (1 - 2\xi_i)^2 \sin^2\psi_{\mathcal{A}}}$$

How the variables $r_{\mathcal{A}}$ and $\psi_{\mathcal{A}}$ relate to $\theta_{\mathcal{A}}$ depends on the type of spiral.

11.1.2 Transfers between arbitrary orbits. Introducing coast arcs

Changing the control parameter ξ only adjusts the value of K_1. The first integral (10.8) is not affected by the control parameter. If the values of K_2 on the departure and arrival spiral arcs are the same, no further corrections are required. However, in a general case $K_{2,1} \neq K_{2,2}$ and therefore the constant K_2 needs to be adjusted. A coast arc is introduced in order to connect the departure and arrival spiral arcs with the adequate values of K_2. The first spiral arc connects the departure point with a node $\mathcal{A}$. A Keplerian orbit connects $\mathcal{A}$ and $\mathcal{B}$, and the final thrust leg connects $\mathcal{B}$ and the final point, $\mathcal{F}$. If the difference between $K_{2,1}$ and $K_{2,2}$ is small the coast arc will be short too.

The transfer from an initial position and velocity vectors, $\mathbf{r}_0$ and $\mathbf{v}_0$, to a given state $\mathbf{r}_{\mathcal{F}}$ and $\mathbf{v}_{\mathcal{F}}$ imposes three constraints on the solution at $\theta_{\mathcal{F}}$. These are $r = r_{\mathcal{F}}$, $v = v_{\mathcal{F}}$, and $\psi = \psi_{\mathcal{F}}$. The control parameters on both spiral arcs ξ_1 and ξ_2 can be adjusted to solve the transfer, together with the orientation of the nodes $\mathcal{A}$ and $\mathcal{B}$, defined by $\theta_{\mathcal{A}}$ and $\theta_{\mathcal{B}}$. Having four variables for only three constraints there is a degree of freedom in the solution. The preferred choice is to use the position of the first node $\mathcal{A}$ as the free parameter and then solve for ξ_1, ξ_2, and the position of node $\mathcal{B}$.

The initial conditions define the values of the constants K_1 and K_2 on the first spiral arc:

$$K_{1,1} = v_0^2 - \frac{2(1-\xi_1)}{r_0} \qquad \text{and} \qquad K_{2,1} = r_0 v_0^2 \sin\psi_0$$

just like the arrival conditions define K_1 and K_2 on the final arc:

$$K_{1,2} = v_{\mathcal{F}}^2 - \frac{2(1-\xi_2)}{r_{\mathcal{F}}} \qquad \text{and} \qquad K_{2,2} = r_{\mathcal{F}} v_{\mathcal{F}}^2 \sin\psi_{\mathcal{F}}$$

The first arc can be propagated to $\mathcal{A}$ using the corresponding equation for the trajectory and for the flight-direction angle

$$r_{\mathcal{A}} = r(\theta_{\mathcal{A}}; K_{1,1}, K_{2,1}), \quad \text{and} \quad \sin\psi_{\mathcal{A}} = \frac{K_{2,1}}{2(1-\xi_1) + K_{1,1} r_{\mathcal{A}}}$$

The conditions at $\mathcal{A}$ provide the eccentricity, semimajor axis, and argument of periapsis of the Keplerian orbit:

$$e = \sqrt{K_{2,1}^2 + 1 - 2K_{2,1}\sin\psi_{\mathcal{A}}} \tag{11.7}$$

$$a = \frac{r_{\mathcal{A}}}{2\xi_1 - K_{1,1}r_{\mathcal{A}}} \tag{11.8}$$

$$\omega = \operatorname{atan2}(-\sin\theta_{\mathcal{A}} - K_{2,1}\cos(\theta_{\mathcal{A}} + \psi_{\mathcal{A}}), K_{2,1}\cos(\theta_{\mathcal{A}} + \psi_{\mathcal{A}}) - \sin\theta_{\mathcal{A}}) \tag{11.9}$$

The value of K_1 that a spiral will take if the thrust is switched on again at some point of the Keplerian orbit behaves as

$$K_1 = \frac{2a\xi - r}{ra}$$

The velocity-matching condition at $\mathcal{B}$ then yields

$$\frac{2a\xi_2 - r_{\mathcal{B}}}{r_{\mathcal{B}}a} = v_{\mathcal{F}}^2 - \frac{2(1-\xi_2)}{r_{\mathcal{F}}}$$

Recall that the semimajor axis and eccentricity have already been solved from the conditions at $\mathcal{A}$. The state at node $\mathcal{B}$ is solved analytically from Kepler's problem. The remaining two constraints are set on the values of the radius and flight-direction angle at $\mathcal{F}$. In sum, the variables $\{\xi_1, \xi_2, \theta_{\mathcal{B}}\}$ are solved from the system of nonlinear equations:

$$\begin{cases} \dfrac{2a\xi_2 - r_{\mathcal{B}}}{r_{\mathcal{B}}a} + \dfrac{2(1-\xi_2)}{r_{\mathcal{F}}} - v_{\mathcal{F}}^2 = 0 \\ r_{\mathcal{F}} - r(\theta_{\mathcal{F}}; K_{1,2}, K_{2,2}, r_{\mathcal{B}}, \theta_{\mathcal{B}}, \xi_2) = 0 \\ \psi_{\mathcal{F}} - \psi(\theta_{\mathcal{F}}; K_{1,2}, K_{2,2}, r_{\mathcal{B}}, \theta_{\mathcal{B}}, \xi_2) = 0 \end{cases} \tag{11.10}$$

The conditions at $\mathcal{B}$ have already been obtained by propagating the Keplerian arc.

Similarly, the evolution of K_2 on the coast leg renders

$$K_2 = \sqrt{1 + 2e\cos\vartheta + e^2} \tag{11.11}$$

where ϑ denotes the true anomaly. In Fig. 11.4 it is possible to study the evolution of the values of K_2 and the flight-direction angle depending on the position inside a Keplerian orbit. Spirals arriving/departing in the raising regime correspond to points in the Keplerian orbit that are between periapsis and apoapsis, whereas spirals in the lowering regime arrive at/depart from points between apoapsis and periapsis. In the interval $\vartheta \in (0, \pi)$, K_2 decreases toward apoapsis. Conversely, for $\vartheta \in (-\pi, 0)$ K_2 increases. This simple rule shows that if the first spiral arc arrives at $\mathcal{A}$ in the raising regime and the second arc departs from $\mathcal{B}$ in the raising regime too, then for $K_{2,2} > K_{2,1}$ the particle will travel $\Delta\vartheta > \pi$ along the Keplerian arc. This yields a *long* coast arc. The same discussion applies to the case where the spirals are in the lowering regime: when

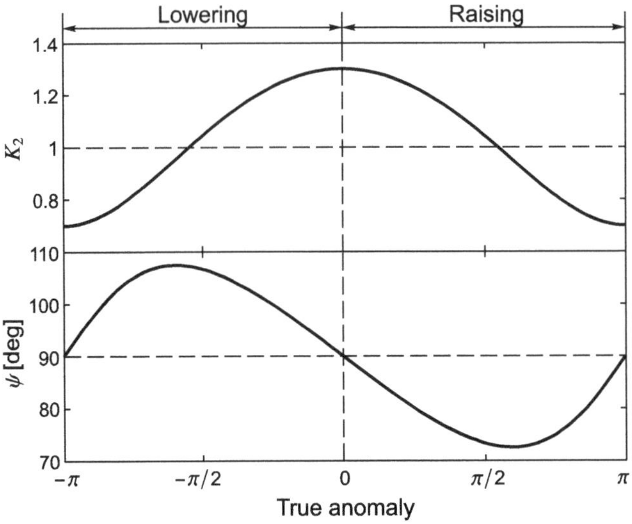

Fig. 11.4: Evolution of K_2 and ψ on a Keplerian orbit ($e = 0.3$).

$K_{2,2} < K_{2,1}$ the transfer includes a long coast arc, and a *short* coast arc otherwise. By virtue of Eq. (11.11) the condition $K_{2,2} \lessgtr K_{2,1}$ translates into:

$$\cos \vartheta_{\mathcal{F}} \lessgtr \frac{K_{2,1}^2 - \left(1 + e_{\mathcal{F}}^2\right)}{2e_{\mathcal{F}}}$$

This defines the points on the arrival orbit for which the constant $K_{2,2}$ is below/above $K_{2,1}$. In general, short transfers will be preferred because of involving shorter times of flight.

At $\mathcal{B}$ the true anomaly is simply

$$\vartheta_{\mathcal{B}} = \theta_{\mathcal{B}} - \omega$$

The position of the node $\mathcal{B}$ can be solved from the compatibility equation on K_2, which yields the relation

$$\cos \vartheta_{\mathcal{B}} = \frac{K_{2,2}^2 - (1 + e^2)}{2e} \tag{11.12}$$

There are two possible solutions to this equation depending on which quadrant the solution is in. As shown in Fig. 11.4, if $\vartheta_{\mathcal{B}} \in (0, \pi)$ the second spiral arc will be in the raising regime at $\mathcal{B}$. If $\vartheta_{\mathcal{B}} \in (\pi, 2\pi)$, then the second spiral departs in the lowering regime. The simplest criterion for selecting the adequate quadrant is:

$$r_{\mathcal{B}} < r_{\mathcal{F}} \implies \vartheta_{\mathcal{B}} \in (0, \pi)$$
$$r_{\mathcal{B}} > r_{\mathcal{F}} \implies \vartheta_{\mathcal{B}} \in (\pi, 2\pi)$$

It is valid for direct transfers from $\mathcal{B}$ to $\mathcal{F}$. In most practical applications the spirals are direct. If the spiral transfer is indirect then the previous criterion is inverted.

It is worth emphasizing that transfers to circular orbits will always be possible as long as the intermediate Keplerian orbit is closed. The target value of $K_{2,2}$ in this case is $K_{2,2} = 1$, which can be achieved by any Keplerian orbit.

The velocity-matching condition provides the relation

$$\xi_2 = \frac{\left[2a - r_{\mathcal{F}}\left(1 + av_{\mathcal{F}}^2\right)\right] r_{\mathcal{B}}}{2a(r_{\mathcal{B}} - r_{\mathcal{F}})} \tag{11.13}$$

Thanks to the previous expressions, the system of equations defined in Eq. (11.10) reduces to one single transcendental equation,

$$r_{\mathcal{F}} - r(\theta_{\mathcal{F}}; K_{1,2}, K_{2,2}, r_{\mathcal{B}}, \theta_{\mathcal{B}}, \xi_2) = 0 \tag{11.14}$$

to be solved for ξ_1. On each iteration the values of $\{\xi_2, \theta_{\mathcal{B}}\}$ are solved from Eqs. (11.13) and (11.12), respectively. Recall that $\theta_{\mathcal{B}} = \omega + \vartheta_{\mathcal{B}}$.

Figure 11.5 shows two examples of generic spiral transfers including a coast arc. The first example is the result of circularizing an elliptic orbit considering zero-revolutions spiral arcs. The switch point $\mathcal{A}$ can be adjusted as required. Two different solutions to the same transfer problem are presented. These are examples of long coast arcs, because the particle crosses the periapsis of the Keplerian orbit. The second example corresponds to an Earth to Ceres transfer. In this case the correction required on K_2 is small and therefore the coast arc is almost negligible. Different solutions are displayed corresponding to different positions of the transition point $\mathcal{A}$. In this example the coast arc is so small that it cannot be distinguished.

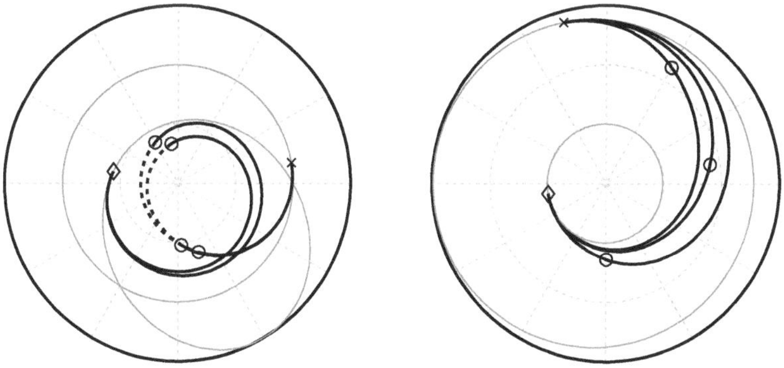

(a) Long coast arcs **(b)** Short coast arcs

Fig. 11.5: Examples of transfers between arbitrary orbits. The departure point is $\diamond$, the points $\mathcal{A}$ and $\mathcal{B}$ are denoted by $\circ$, and the arrival point is $\times$. The gray lines represent the departure and arrival orbits. The coast arcs are plotted using dashed lines.

11.1.3 Existence of solutions. The admissible region

The goal of this section is to find the set of values $K_{2,2} \in \mathbb{I}$ for which transfers from an initial state vector are possible. We refer to $\mathbb{I}$ as the admissible region and $K_{2,2} \in \mathbb{I}$ is a sufficient condition for existence. There is an interval $\mathbb{I}^*$ outside the admissible region ($\mathbb{I}^* \cap \mathbb{I} = \emptyset$) in which solutions exist under some very specific conditions. Therefore $K_{2,2} \in (\mathbb{I} \cup \mathbb{I}^*)$, is a necessary condition for solutions to exist.

The admissible region $\mathbb{I}$ is constructed by considering how much K_2 can change on the coast arc. If the value of $K_{2,2}$ is larger than the maximum value that K_2 can reach on the Keplerian orbit, then the transfer is unfeasible.

On a Keplerian orbit K_2 changes according to

$$K_2 = \sqrt{1 + e\,(e + 2\cos\vartheta)}$$

The maximum and minimum values of K_2 occur respectively at periapsis ($\vartheta = 0$) and at apoapsis ($\vartheta = \pi$), and are

$$K_{2,\max} = 1 + e \qquad \text{and} \qquad K_{2,\min} = 1 - e \tag{11.15}$$

This means that the values that K_2 will take on the coast leg are confined to $[1-e, 1+e]$. The eccentricity of the orbit depends on the selection of the control parameter ξ_1 and the transition point $\mathcal{A}$. Figure 11.6 builds the admissible region for an example set of initial conditions and one given value of $\theta_{\mathcal{A}}$. It is parameterized in terms of the control parameter on the first spiral arc, ξ_1. The upper branch is $K_{2,\max} = 1 + e$, and the lower branch is $K_{2,\min} = 1 - e$. The eccentricity is solved from Eq. (11.7). Notice that the number of revolutions along the first arc does have an effect on the admissible region, since it yields different values of $\sin\psi_{\mathcal{A}}$ and modifies e. Two conclusions can be derived from this figure: first, given a target value of $K_{2,2}$, if it falls in the forbidden region ($K_{2,2} \notin \mathbb{I}$), then the transfer is not feasible. Second, if $K_{2,2} \in \mathbb{I}$, it is possible to bound the range of values of ξ_1 where the solution is. The parameter ξ_p reads

$$\xi_p = 1 - \frac{r_0 v_0^2}{2}$$

and it defines the value of ξ_1 that makes the first spiral arc parabolic. That is, if $\xi_1 < \xi_p$ the spiral is elliptic, and for $\xi_1 > \xi_p$ the spiral is hyperbolic. The value of the control parameter that makes $K_{2,\max} = 2$ (or equivalently that makes $e = 1$) is denoted by ξ_1^*. There are two pairs of relevant points in the definition of the admissible region:

$$a^{\pm} = 1 \pm |K_{2,1} - 1|\,, \qquad b^{\pm} = 1 \pm \sqrt{K_{2,1}^2 + 1 - 2K_{2,1}\sin\psi_{\mathcal{A}}^0}$$

where $\psi_{\mathcal{A}}^0$ denotes the flight-direction angle at $\mathcal{A}$ propagated with $\xi_1 = 0$. The highest point in the region $\mathbb{I}^*$ reads

$$c = 2 + K_{2,1}$$

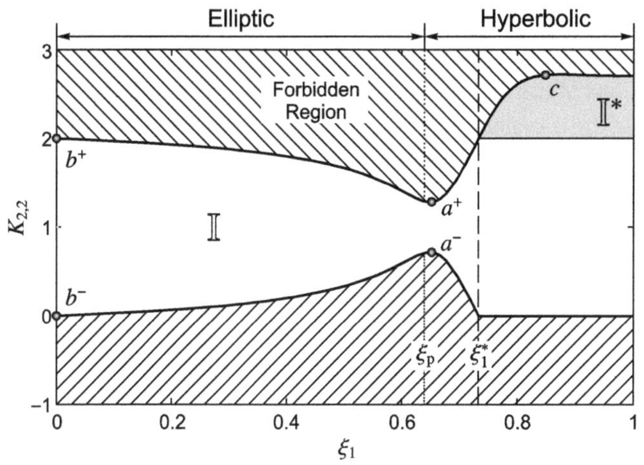

Fig. 11.6: Compatibility conditions in terms of the control parameter for an example transfer.

The difference between $\mathbb{I}$ and $\mathbb{I}^*$ is that for $K_{2,2} \in \mathbb{I}$ the Keplerian orbit is elliptic, whereas for $K_{2,2} \in \mathbb{I}^*$ the Keplerian orbit is hyperbolic. In the latter case, the true anomaly is restricted by the asymptotes of the hyperbola, i.e., $\cos \vartheta < 1/e$. In order to find solutions in $\mathbb{I}^*$ two additional constraints need to be satisfied:

- The coast arc must be a short arc, as defined from Fig. 11.4.
- Equation (11.12) must yield $\cos \vartheta_{\mathcal{B}} < 1/e$.

We emphasize that $K_{2,2} \in \mathbb{I}$ is a sufficient condition for the transfer to be feasible. If $K_{2,2} \in [b^-, a^-] \cup [a^+, b^+]$ the admissible region defines the intervals where adequate initial guesses in ξ_1 can be found. The discussion about the existence of solutions does not replace the procedure for computing the transfer; Equation (11.14) still needs to be solved for ξ_1.

In practice, the nonexistence of solutions relates to too short transfers that involve noticeable changes in the eccentricity. Increasing the number of revolutions changes the admissible region and can make the transfer possible.

11.1.4 Controlling the time of flight

In the previous sections it has been shown that there is a degree of freedom in the solution to the problem of finding a transfer orbit connecting two given state vectors. It can be adjusted in order to meet an additional constraint regarding the time of flight, $t = t_{\mathcal{F}}$. The degree of freedom is controlled by the position of the switch point $\mathcal{A}$, and its adequate value is solved from:

$$t_{\mathcal{F}} = t_1 + t_k + t_2$$

Here t_1 and t_2 are the times of flight for the first and second spiral arcs, respectively. The time spent on the Keplerian orbit is denoted t_k.

The number of revolutions of the first and second spiral arcs is an alternative way of modifying the geometry of the transfer. Consequently, it also affects the time of flight and introduces additional degrees of freedom in the solution.

It is important to note that the methods presented in the previous sections, which consider only two spiral arcs, are the simplest but not the only solutions. By introducing more spiral arcs (or even coast arcs) the degrees of freedom of the system can be increased arbitrarily. Introducing control nodes in the problem might provide a useful method for trajectory optimization. Similarly, if there are no constraints on the arrival velocity only one spiral arc and one coast arc will be needed.

Section 10.3.4 analyzed carefully the properties of the acceleration in Eq. (11.1). Closed-form expressions were given for locating the maximum acceleration reached along the transfer. With these results at hand, one can identify easily the transfers that violate limitations on the maximum admissible thrust.

11.2 Periodic orbits

The procedure for designing bitangent transfers can be extended to generate arbitrary periodic orbits. Because of the $\mathscr{T}$-symmetry of elliptic and hyperbolic spirals of Type II, any bitangent transfer can be closed to define a periodic orbit. When the spiral intersects a certain axis with $\psi = \pi/2$, the trajectory can be extended symmetrically to close the orbit. This procedure is then generalized to generate periodic orbits of different shapes.

11.2.1 Coaxial solutions

A symmetric extension of the bitangent transfers defines a periodic orbit. Having imposed a maximum and minimum radius, Eq. (11.4) or (11.17) can be solved to provide the generating bitangent spiral. The number of revolutions n can be replaced by $n = m/2$, where m controls the number of intersections with the horizontal axis before reaching the maximum/minimum radius. Periodic orbits require only two spiral arcs that are repeated based on the symmetry properties of the trajectory. The first, from θ_0 to $\theta_{\mathcal{A}}$, is propagated with $\xi = \xi_1$. For the second arc, spanning from $\theta_{\mathcal{A}}$ to $2m\pi - \theta_{\mathcal{A}}$, it is $\xi = \xi_2$. The final arc that closes the trajectory is defined again by ξ_1 and is propagated from $2(m+1)\pi - \theta_{\mathcal{A}}$ up to $2(m+1)\pi$.

Examples of coaxial periodic transfers are presented in Fig. 11.7. We emphasize that there is a degree of freedom in the construction process. It relates to the orientation of the transition point $\mathcal{A}$. Because of the symmetry of the spiral there are two symmetric transition points.

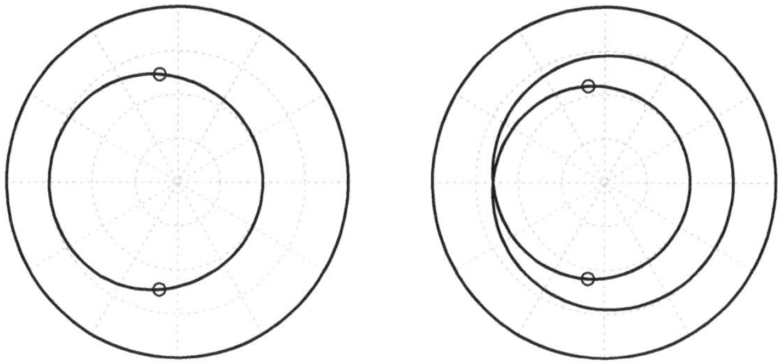

(a) $m = 0$ (b) $m = 1$

Fig. 11.7: Coaxial periodic orbits. The ∘ denotes the nodes where the control parameter ξ changes.

11.2.2 Generic periodic orbits

Relying on the symmetry properties of the generalized logarithmic spirals arbitrary periodic orbits can be generated. Numerically, the problem reduces to solving a bitangent spiral transfer where the orientation of the arrival point is adjusted depending on the periodicity conditions. Periodic orbits are constructed by combining elliptic and Type II hyperbolic spiral arcs; the exterior radius of the orbit corresponds to $r_{\max}$, whereas the interior radius defines $r_{\min}$. Only two arcs are required to build a periodic orbit; they are repeated sequentially according to the symmetries of the spirals.

Consider a periodic orbit with $s = 3, 4, 5, \ldots$ lobes. If the spiral departs from the maximum radius ($\theta = 0$) then the minimum radius will be reached at $\theta = \pi\Lambda(s)$. The parameter $\Lambda(s)$ depends on the number of lobes and determines the configuration of the periodic orbit. Typical values of $\Lambda(s)$ are $\Lambda(s) = (s - j)/s$, with $j = 1, 2, \ldots, s - 1$. Given a transition point $\theta_{\mathcal{A}}$ the method derived for solving coaxial bitangent transfers can be extended to compute the values of ξ_1 and ξ_2 defining a generic periodic orbit. When the periodic orbit intersects itself at least once it is called an interior periodic orbit ($j < s - 1$), and exterior otherwise ($j = s - 1$). Figure 11.8 depicts the construction of an interior periodic orbit with $s = 3$ and $\Lambda = (s - 1)/s = 2/3$.

The symmetry properties of the spirals guarantee that if the second spiral arc reaches a minimum $r_{\min} = r_{\text{in}}$ at $\theta_m = (s - 1)\pi/s$ then the resulting trajectory is a periodic orbit with s lobes. If $r_0 = r_{\text{in}}$ the equation to be solved for ξ_1 is

$$\frac{r_{\text{ex}}}{r_{\text{in}}} = \frac{3 - 2\xi_1}{3 - 2\xi_2} \left\{ \frac{2(1 - \xi_2) + \cosh[\ell_2(\theta_{\mathcal{A}} - \Lambda\pi)]}{2(1 - \xi_1) + \cos \ell_1 \theta_{\mathcal{A}}} \right\} \tag{11.16}$$

The value of ξ_2 is solved from the velocity-matching condition, Eq. (11.5). When the spiral is initially at the exterior radius it is $r_0 = r_{\text{ex}}$ and the equation becomes

$$\frac{r_{\text{in}}}{r_{\text{ex}}} = \frac{3 - 2\xi_1}{3 - 2\xi_2} \left\{ \frac{2(1 - \xi_2) + \cos[\ell_2(\theta_{\mathcal{A}} - \Lambda\pi)]}{2(1 - \xi_1) + \cosh \ell_1 \theta_{\mathcal{A}}} \right\} \tag{11.17}$$

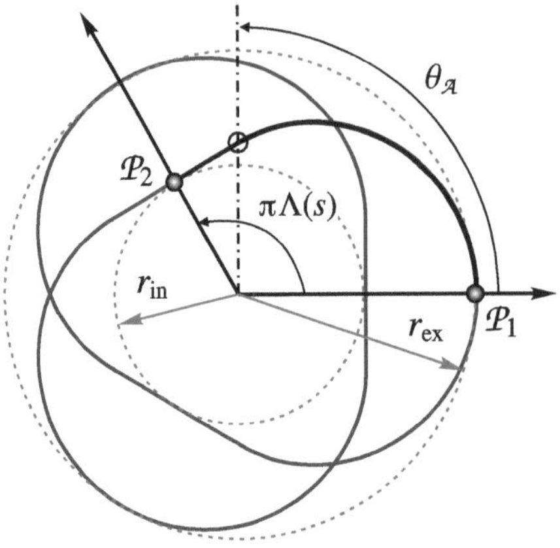

Fig. 11.8: Construction of an interior periodic orbit with $s = 3$.

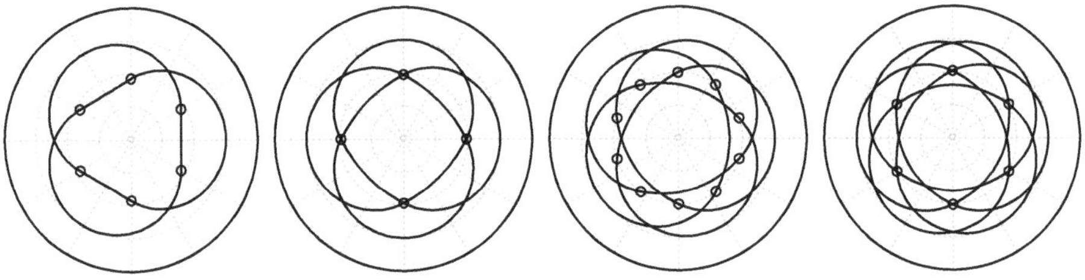

(a) $s = 3$ **(b)** $s = 4$ **(c)** $s = 5$ **(d)** $s = 6$

Fig. 11.9: Examples of periodic orbits with $j = 1$, $\theta_{\mathcal{A}} = 90$ deg, and different numbers of lobes. The ∘ denotes the nodes where the control parameter ξ changes.

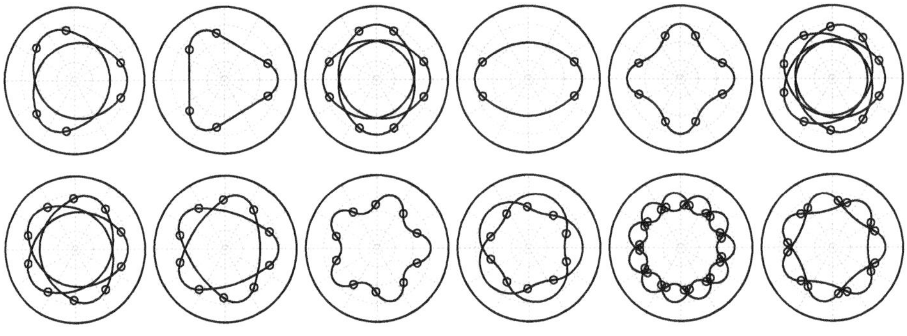

Fig. 11.10: Examples of periodic orbits for $\theta_{\mathcal{A}} = 20$ deg and different combinations of s and j.

Examples of interior periodic orbits with $\Lambda = (s - 1)/s$ are presented in Fig. 11.9. The circumference of radius r_{ex} is the exterior envelope of the orbits, and the circumference with $r = r_{in}$ is the interior envelope. Alternative configurations can be found by changing $\Lambda(s)$. Having fixed the number of lobes s there are four degrees of freedom in the definition of the periodic orbits: the interior and exterior radii, the transition point $\theta_{\mathcal{A}}$, and the integer parameter j in the definition of $\Lambda(s)$.

Figure 11.10 displays a number of example interior and exterior periodic orbits generated by combining two spiral arcs. The entire orbit is generated by symmetrically extending a bitangent transfer defined by two arcs.

11.3 Multinode transfers

We have shown that transfers between circular orbits require at least one transition node, whereas transfers between arbitrary state vectors require a coast arc. An arbitrary number of nodes can be introduced in order to increase the flexibility of the solution.

Consider the problem of finding transfers between circular orbits introducing $N \geq 1$ nodes. Let θ_n denote the orientation of the n-th node. There are $2N - 1$ degrees of freedom in the solution: the position of the nodes $(\theta_1, \theta_2, \theta_3, \ldots, \theta_N)$ and the control parameters of the interior arcs $(\xi_2, \xi_3, \ldots, \xi_{N-1})$. The arc connecting nodes $n - 1$ and n is labeled the n-th arc. With this, the initial arc connecting θ_0 and θ_1 is labeled "1". The final arc, connecting θ_N and $\theta_{\mathcal{F}}$, is denoted by $\mathcal{F}$. The problem consists in solving for ξ_1 given the position of the nodes and the values of the control parameter for the intermediate arcs.

The trajectory is propagated sequentially through the nodes

$$r_n = r(\theta_n; \xi_n)\,, \qquad n = 1, 2, \ldots, N$$

considering the initial conditions $(r_{n-1}, v_{n-1}, \theta_{n-1}, \psi_{n-1})$ defined by the previous node. The final arc needs to satisfy two conditions: first, the velocity-matching condition

$$v_N^2 = K_{1,\mathcal{F}} + \frac{2(1 - \xi_{\mathcal{F}})}{r_N} \implies \xi_{\mathcal{F}} = \frac{2r_{\mathcal{F}} - r_N\left[2 + r_{\mathcal{F}}\left(v_N^2 - v_{\mathcal{F}}^2\right)\right]}{2(r_{\mathcal{F}} - r_N)}$$

Second, the boundary condition

$$r_{\mathcal{F}} = r(\theta_{\mathcal{F}}; \xi_{\mathcal{F}})$$

Figure 11.11 shows two examples of multinode transfers between two circular orbits. Each figure presents two spiral transfers that depart/arrive from/at the same exact state vectors. The difference between the black and gray transfer is the position of the nodes and the values of the control parameters. It is worth emphasizing that the transitions at the nodes are continuous (no additional Δv's are required) and no correcting maneuvers are required at departure nor arrival.

Introducing multiple nodes in transfers between arbitrary state vectors is a simple procedure: the problem reduces to solving a thrust-coast-thrust transfer, but the boundary conditions are given by the forward/backward propagation of the additional spiral arcs. Consider that there are N_1 spiral arcs until the thrust is switched off, and N_2 arcs between the point where the thrust is switched on again and the arrival point.

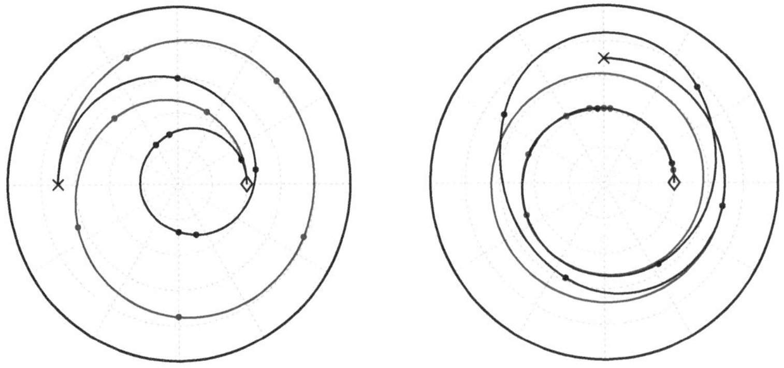

(a) 1 revolution (b) 2 revolutions

Fig. 11.11: Examples of bitangent transfers with $N = 7$. The nodes are denoted by •, ◊ is the departure point, and × is the arrival point.

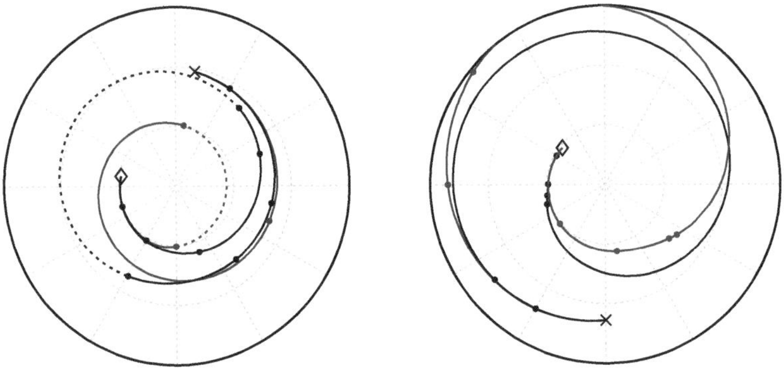

Fig. 11.12: Examples of multinode transfers between two arbitrary state vectors.

The initial conditions to be considered for solving Eq. (11.14) are defined by the final state after propagating the spirals $1, 2, \ldots, N_1 - 1$. The arrival conditions are obtained from the backward propagation of the $N_2, N_2 - 1, \ldots, 2$ arcs in the second leg. Figure 11.12 shows examples of transfers between two state vectors. The two trajectories plotted in each figure arrive at and depart from the same exact state vectors, and the only difference between them is the definition of the intermediate arcs. This technique can be generalized even more by introducing an arbitrary sequence of spiral and Keplerian arcs.

11.4 Three-dimensional motion

Let us now consider the more general case in which the transfer is no longer planar. The out-of-plane component of the motion, z, is different from zero. To prevent confusion the norm of vector **r** in the three-dimensional case will be written R, with $R^2 = r^2 + z^2$. The normalized ($\mu = 1$) and perturbed two-body problem obeys

$$\frac{\mathrm{d}^2\mathbf{r}}{\mathrm{d}t^2} + \frac{\mathbf{r}}{R^3} = \mathbf{a}_p \,, \qquad R = ||\mathbf{r}|| \tag{11.18}$$

Vector **r** can be represented by the set of cylindrical coordinates (r, θ, z). The climb angle is λ. Figure 11.13 depicts the geometry of the three-dimensional trajectory. The position of the subpoint $\mathcal{P}'$ is given by $\mathbf{r}_\parallel$, so that

$$\mathbf{r}(\theta) = \mathbf{r}_\parallel(\theta) + z(\theta)\,\mathbf{k}_\mathcal{I}$$

This means that $\mathbf{r}_\parallel$ describes the projected motion of the particle. The main hypothesis we will adopt is that $\mathbf{r}_\parallel(\theta)$ defines a generalized logarithmic spiral. It is called the *base spiral* (Roa and Peláez, 2016c). Under this notation it is $||\mathbf{r}_\parallel|| = r$.

Let $\mathcal{F} = \{\mathcal{P}; \mathbf{t}, \mathbf{n}, \mathbf{b}\}$ denote the Frenet–Serret frame, where **t** is the unit vector tangent to the trajectory, i.e.,

$$\mathbf{t} = \frac{\mathrm{d}\mathbf{r}}{\mathrm{d}s} = \frac{\mathbf{v}}{v}$$

Here s denotes an arclength along the curve. The unit vectors **n** and **b** are the normal and binormal unit vectors, and relate to each other and to **t** thanks to the Frenet–Serret formulas. In particular, the first of the Frenet–Serret formulas provides

$$\mathbf{n} = \frac{1}{\kappa}\frac{\mathrm{d}\mathbf{t}}{\mathrm{d}s}$$

where κ is the nonzero curvature of the trajectory, defined explicitly as

$$\kappa = \left|\left|\frac{\mathrm{d}\mathbf{t}}{\mathrm{d}s}\right|\right|$$

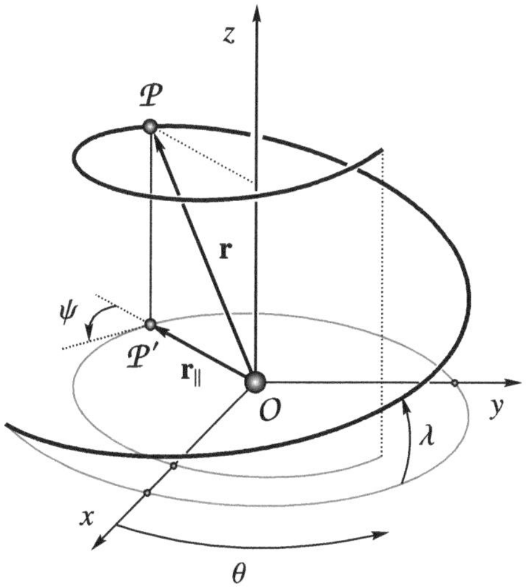

Fig. 11.13: Geometry of a 3D generalized logarithmic spiral.

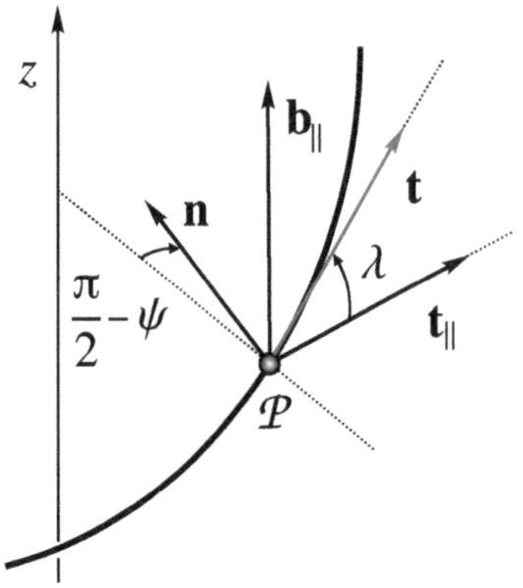

Fig. 11.14: The Frenet–Serret frame.

The binormal vector completes an orthonormal dextral reference frame,

$$\mathbf{b} = \mathbf{t} \times \mathbf{n}$$

Figure 11.14 shows the configuration of the Frenet–Serret frame. The auxiliary frame $\mathcal{F}_\parallel = \{\mathcal{P}; \mathbf{t}_\parallel, \mathbf{n}_\parallel, \mathbf{b}_\parallel\}$ is defined so that $(\mathbf{t}_\parallel \cdot \mathbf{t}) = \cos\lambda$ and $(\mathbf{t}_\parallel \cdot \mathbf{k}_\mathcal{J}) = 0$. It relates to the Frenet–Serret frame $\mathcal{F}$ by means of a rotation of magnitude λ about the normal vector.

The velocity, $\mathbf{v} \in \mathbb{R}^3$, can be decomposed in

$$\mathbf{v} = \mathbf{v}_\parallel + \mathbf{v}_z$$

where $\mathbf{v}_\parallel$ is the in-plane component of the velocity $(\mathbf{v}_\parallel \cdot \mathbf{k}_\mathcal{J} = 0)$ and the out-of-plane component is $\mathbf{v}_z = \dot{z}\,\mathbf{k}_\mathcal{J}$. From the geometry of the problem it follows that

$$\frac{\dot{z}}{v_\parallel} = \tan\lambda \implies \frac{\mathrm{d}z}{\mathrm{d}t} = v_\parallel \tan\lambda \tag{11.19}$$

Taking the time derivative of the out-of-plane velocity defines the transversal acceleration

$$\frac{\mathrm{d}^2 z}{\mathrm{d}t^2} = \tan\lambda\,\frac{\mathrm{d}v_\parallel}{\mathrm{d}t} + v_\parallel \sec^2\lambda \frac{\mathrm{d}\lambda}{\mathrm{d}t} \tag{11.20}$$

The time evolution of $\lambda(t)$ and its derivative are supposed to be known and depend on the shape assumed for the trajectory.

Under the hypothesis that $\mathbf{r}_\parallel$ defines a generalized logarithmic spiral, the equations of motion that govern the in-plane dynamics are those found in Sect. 11.1. When combined with Eqs. (11.19–11.20) it follows the system

$$\frac{\mathrm{d}v_\parallel}{\mathrm{d}t} = \frac{\xi - 1}{r^2}\cos\psi \tag{11.21}$$

$$\frac{\mathrm{d}r}{\mathrm{d}t} = +v_\parallel \cos\psi \tag{11.22}$$

$$\frac{\mathrm{d}\theta}{\mathrm{d}t} = +\frac{v_\parallel}{r}\sin\psi \tag{11.23}$$

$$\frac{d\psi}{dt} = \frac{\sin\psi}{r^2 v_\parallel}\left[2(1-\xi) - r v_\parallel^2\right] \tag{11.24}$$

$$\frac{dz}{dt} = v_\parallel \tan\lambda \tag{11.25}$$

$$\frac{d^2 z}{dt^2} = \frac{\xi - 1}{r^2}\cos\psi\tan\lambda + v_\parallel(1+\tan^2\lambda)\frac{d\lambda}{dt} \tag{11.26}$$

which needs to be integrated from the initial conditions:

$$t = 0: \quad v_\parallel(t_0) = v_{\parallel,0}\,, \quad r(t_0) = r_0\,, \quad \theta(t_0) = \theta_0$$
$$\psi(t_0) = \psi_0\,, \quad z(t_0) = z_0\,, \quad \dot z(t_0) = \dot z_0$$

In these equations ψ is the in-plane flight-direction angle, defined as the angle between the projections of the position and velocity vectors onto the $x-y$ plane, i.e.,

$$\cos\psi = \frac{(\mathbf{r}_\parallel \cdot \mathbf{v}_\parallel)}{r v_\parallel} \tag{11.27}$$

We devote the following lines to finding a similar expression for the absolute flight-direction angle, Ψ, between vectors $\mathbf{r}$ and $\mathbf{v}$.

Let $\{\mathbf{i}, \mathbf{j}, \mathbf{k}\}$ denote the orbital frame $\mathcal{L}$. The unit vectors defining this frame are

$$\mathbf{i} = \frac{\mathbf{r}}{R}\,, \qquad \mathbf{k} = \frac{\mathbf{r}\times\mathbf{v}}{||\mathbf{r}\times\mathbf{v}||}\,, \qquad \mathbf{j} = \mathbf{k}\times\mathbf{i}$$

Similarly, consider the orbital frame attached to the subpoint $\mathcal{P}'$, defined in terms of

$$\mathbf{i}_\parallel = \frac{\mathbf{r}_\parallel}{r}\,, \qquad \mathbf{k}_\parallel = \frac{\mathbf{r}_\parallel\times\mathbf{v}_\parallel}{||\mathbf{r}_\parallel\times\mathbf{v}_\parallel||}\,, \qquad \mathbf{j}_\parallel = \mathbf{k}_\parallel\times\mathbf{i}_\parallel$$

and denoted $\mathcal{L}'$. The Frenet–Serret frame referred to the projected spiral relates to frame $\mathcal{L}'$ by means of

$$\mathbf{t}_\parallel = \cos\psi\,\mathbf{i}_\parallel + \sin\psi\,\mathbf{j}_\parallel\,, \qquad \mathbf{n}_\parallel = -\sin\psi\,\mathbf{i}_\parallel + \cos\psi\,\mathbf{j}_\parallel$$

The vectors defining the Frenet–Serret frame attached to the trajectory admit an equivalent projection onto $\mathcal{L}$:

$$\mathbf{t} = \cos\Psi\,\mathbf{i} + \sin\Psi\,\mathbf{j}\,, \qquad \mathbf{n} = -\sin\Psi\,\mathbf{i} + \cos\Psi\,\mathbf{j}$$

where Ψ is the flight-direction angle defined by vectors $\mathbf{r}$ and $\mathbf{v}$,

$$\cos\Psi = \frac{(\mathbf{r}\cdot\mathbf{v})}{Rv} \tag{11.28}$$

We emphasize the difference between this equation and Eq. (11.27). It then follows

$$\begin{aligned}
\mathbf{t}_\parallel &= \cos\Psi\cos\lambda\,\mathbf{i} + \sin\Psi\cos\lambda\,\mathbf{j} - \sin\lambda\,\mathbf{k}\\
\mathbf{n}_\parallel &= -\sin\Psi\,\mathbf{i} + \cos\Psi\,\mathbf{j}\\
\mathbf{b}_\parallel &= \cos\Psi\sin\lambda\,\mathbf{i} + \sin\Psi\sin\lambda\,\mathbf{j} + \cos\lambda\,\mathbf{k}
\end{aligned}$$

The total velocity of the particle reads

$$\mathbf{v} = v_{\parallel} \cos\psi\, \mathbf{i}_{\parallel} + v_{\parallel} \sin\psi\, \mathbf{j}_{\parallel} + \dot{z}\, \mathbf{k}_{\parallel}$$

Provided that $\mathbf{r} = r\,\mathbf{i}_{\parallel} + z\,\mathbf{k}_{\parallel}$, Eq. (11.28) renders

$$\cos\Psi = \frac{r v_{\parallel} \cos\psi + z\dot{z}}{Rv}$$

11.4.1 The thrust acceleration

The equations of motion (11.21–11.26) show that the total acceleration governing the dynamics is

$$\frac{\mathrm{d}^2\mathbf{r}}{\mathrm{d}t^2} = \frac{\xi - 1}{r^2}[\cos\psi\,(\mathbf{t}_{\parallel} + \tan\lambda\,\mathbf{k}_{\mathcal{J}}) - 2\sin\psi\,\mathbf{n}] + v_{\parallel}\sec^2\lambda\frac{\mathrm{d}\lambda}{\mathrm{d}t}\,\mathbf{k}_{\mathcal{J}}$$

The gravitational acceleration is known to be

$$\mathbf{a}_g = -\frac{\mathbf{r}}{R^3} = -\frac{1}{R^3}\left[r(\cos\psi\,\mathbf{t}_{\parallel} - \sin\psi\,\mathbf{n}) + z\,\mathbf{k}_{\mathcal{J}}\right]$$

The acceleration due to the thrust decomposes in $\mathbf{a}_p = \mathbf{a}_{p,\parallel} + \mathbf{a}_{p,z}$. The in-plane component takes the form

$$\mathbf{a}_{p,\parallel} = \left(\frac{\xi - 1}{r^3} + \frac{1}{R^3}\right) r\cos\psi\,\mathbf{t}_{\parallel} + \left[\frac{2(1-\xi)}{r^3} - \frac{1}{R^3}\right] r\sin\psi\,\mathbf{n}$$

whereas the out-of-plane component of the acceleration reduces to

$$\mathbf{a}_{p,z} = \left[\frac{\xi - 1}{r^2}\cos\psi\tan\lambda + v_{\parallel}\frac{\mathrm{d}}{\mathrm{d}t}(\tan\lambda) + \frac{z}{R^3}\right]\mathbf{k}_{\mathcal{J}} \tag{11.29}$$

The climb-angle λ and its rate of change are known, given the shape of the trajectory. The previous expression is directly referred to $\tan\lambda$ in order to simplify the introduction of a steering law.

When the motion is confined to the plane it is $z = 0$ and $\lambda = 0$. This condition makes $r \equiv R$, and the perturbing acceleration reduces to

$$\mathbf{a}_p = \frac{1}{r^2}[\xi\cos\psi\,\mathbf{t} + (1 - 2\xi)\sin\psi\,\mathbf{n}] \qquad (z = 0, \lambda = 0)$$

and coincides with the acceleration defined in Eq. (11.1).

11.4.2 Modeling the out-of-plane motion

The following lines present different models for defining the transversal component of the motion.

11.4.2.1 Helices

The simplest law for the evolution of the climb-angle λ is

$$\lambda = \text{const.}$$

and of course its time derivative is null. The radial velocity has been defined in Eq. (11.22) as the projection of the velocity along the direction of $\mathbf{r}_{\parallel}$. Combining this expression with Eq. (11.19) yields

$$\frac{dz}{dr} = \frac{\tan\lambda}{\cos\psi} \tag{11.30}$$

The term $\cos\psi$ is given explicitly in Eq. (F.12). It is interesting to see that the sign of this expression depends on the regime of the spiral through the sign of $\cos\psi$. It becomes

$$\frac{dz}{dr} = \pm\frac{[2(1-\xi)+K_1 r]\tan\lambda}{\sqrt{[2(1-\xi)+K_1 r]^2 - K_2^2}} \tag{11.31}$$

This equation shows that the out-of-plane motion can be solved in closed form when the angle λ is constant, being:

$$z(r) - z_0 = \frac{K_2}{K_1}\tan\lambda(\cot\psi - \cot\psi_0) \tag{11.32}$$

Recall that

$$\tan\psi = \pm\frac{K_2}{\sqrt{[2(1-\xi)+K_1 r]^2 - K_2^2}}$$

The constant K_1 appears in the denominator of the solution in Eq. (11.32). In the limit case $K_1 \to 0$ this expression will become singular. But the singularity is avoidable as shown by the limit

$$\lim_{K_1\to 0} z(r) = z_0 + 2(r - r_0)(1-\xi)\tan\lambda\sec\psi_0$$

When $K_1 = 0$ the spiral is parabolic (logarithmic) and the flight-direction angle remains constant, $\psi = \psi_0$.

Figure 11.15 shows examples of different helicoidal trajectories depending on the type of the corresponding base spiral. By extension, we call elliptic helices those helices whose base spiral is elliptic. The same applies to parabolic and hyperbolic helices. Elliptic helices resemble spherical spirals, the poles corresponding to $r \to 0$ in the base spiral. Parabolic helices evolve on the surface of a cone.

The geometry of the problem is such that

$$(\mathbf{t}\cdot\mathbf{k}_{\mathcal{J}}) = \sin\lambda$$

Taking the derivative with respect to the arclength, s, provides

$$\left(\frac{d\mathbf{t}}{ds}\cdot\mathbf{k}_{\mathcal{J}}\right) = 0 \tag{11.33}$$

(a) Elliptic (b) Parabolic (c) Hyperbolic TI (d) Hyperbolic TII

Fig. 11.15: Helicoidal trajectories depending on the type of base spiral.

The derivative of **t** with respect to the arclength relates to **n** by means of the first of the Frenet–Serret formulas,

$$\frac{d\mathbf{t}}{ds} = \kappa\,\mathbf{n}$$

Therefore, Eq. (11.33) translates into

$$\kappa\,(\mathbf{n}\cdot\mathbf{k}_{\mathcal{I}}) = 0$$

so vector **n** is confined to the plane defined by $\mathbf{k}_{\mathcal{I}}$, provided that the helix has nonzero curvature. This proves that the normal vector of a generalized logarithmic helix **n** is always parallel to the $x - y$ plane.

From a geometrical perspective it is interesting to prove that the resulting curves satisfy the condition for curves being helices, given by Lancret's theorem. It states that a curve is called a helix if the ratio between the curvature, κ, and the torsion, τ, is constant.[1] The axis of the helix is defined by the unit vector $\mathbf{k}_{\mathcal{I}}$, and can be referred to the Frenet–Serret frame by virtue of

$$\mathbf{k}_{\mathcal{I}} = \sin\lambda\,\mathbf{t} + \cos\lambda\,\mathbf{b}$$

Differentiating with respect to the arclength s yields

$$0 = \sin\lambda\left(\frac{d\mathbf{t}}{ds}\right) + \cos\lambda\left(\frac{d\mathbf{b}}{ds}\right)$$

1 This theorem was first presented by the engineer Michel A. Lancret in *Mémoire sur les courbes à double courbure*, on April 26, 1802. But the formal proof of the theorem is attributed to Saint-Venant, published in his *Mémoire sur les lignes courbes non planes* in 1845. He worked with the concept of the rectifying plane (spanned by vectors **t** and **b**): by evaluating the relative inclination of the line of intersection between consecutive rectifying planes he proved that for a helix such lines are parallel. Moreover, they describe a cylinder. He then reduced the condition for the rectifying lines to be parallel to $d(\kappa/\tau) = 0$. We refer to Barros (1997) for a generalization of Lancret's theorem.

The Frenet–Serret formulas transform this expression into

$$\kappa(s)\sin\lambda\,\mathbf{n} - \tau(s)\cos\lambda\,\mathbf{n} = 0 \implies \frac{\kappa(s)}{\tau(s)} = \cot\lambda$$

so that the ratio κ/τ is indeed constant when $\lambda = \text{const}$.

The connection between Eq. (11.32) and the initial conditions (z_0, ψ_0) comes from solving for the constant of integration given the departure point. Without losing generality the constant of integration can be obtained from the conditions at apoapsis or periapsis when the spiral is either elliptic or hyperbolic of Type II. This provides a new form of the equation of the trajectory:

$$z(r) - z_m = \frac{K_2 \tan\lambda}{K_1 \tan\psi} \tag{11.34}$$

Here $z_m = z(r_{\max})$ or $z_m = z(r_{\min})$, depending on whether the helix is elliptic or hyperbolic of Type II.

This suggests a new concept of symmetry: a generalized helix is said to be $\mathscr{Z}$-symmetric if

$$z(\theta_m + \Delta) = -z(\theta_m - \Delta)$$

where $\Delta \geq 0$ denotes an arbitrarily large angular displacement. The component z depends on the polar angle through the relations $r = r(\theta)$ and $\psi = \psi(\theta)$. $\mathscr{Z}$-symmetric trajectories are invariant with respect to rotations of magnitude $\vartheta = n\pi$ (with $n = 1, 2, \ldots$) about the axis $\theta = \theta_m$ located in the $z = z_m$ plane. Clearly, a $\mathscr{Z}$-symmetric trajectory is $\mathscr{T}$-symmetric, because a maximum or minimum radius exists. Similarly, all $\mathscr{T}$-symmetric trajectories are $\mathscr{Z}$-symmetric because Eq. (11.34) will always change its sign when ψ reaches $\pi/2$.

Elliptic spirals are known to be bounded, so that $r \to 0$ when $\theta \to \pm\infty$. Similarly, Eq. (11.34) proves that the out-of-plane motion is also bounded:

$$\lim_{r\to 0} z(r) = z_m \pm \frac{\ell}{K_1}\tan\lambda$$

This result clearly shows the $\mathscr{Z}$-symmetry of the helix.

On the contrary, for the case of parabolic and hyperbolic spirals the particle escapes to infinity. This means $\lim_{r\to\infty} z(r) = \pm\infty$ depending on the type and regime of the helix. For hyperbolic helices of Type II there are two asymptotes that provide $z \to \pm\infty$.

11.4.2.2 Polynomial shape

Following the shape-based approach, the out-of-plane component of the motion can be modeled as

$$z(\theta) = \sum_{n=0}^{N} c_n \theta^n$$

The shape coefficients c_n are stored in the vector $\mathbf{c}_N \in \mathbb{R}^{N+1}$. The transversal velocity follows from the relation

$$\frac{\mathrm{d}z}{\mathrm{d}t} = \frac{\mathrm{d}z}{\mathrm{d}\theta}\frac{\mathrm{d}\theta}{\mathrm{d}t} = \frac{v_\parallel}{r}\sin\psi\sum_{n=1}^{N} nc_n\theta^{n-1} \tag{11.35}$$

having introduced Eq. (11.23).

The steering law $\lambda(\theta)$ can be solved by equating Eqs. (11.35) and (11.19):

$$\tan\lambda = \frac{\sin\psi}{r}\sum_{n=1}^{N} nc_n\theta^{n-1}$$

The transversal component of the thrust acceleration, given in Eq. (11.29), involves the time derivative of $\tan\lambda$:

$$\frac{\mathrm{d}}{\mathrm{d}t}(\tan\lambda) = \frac{\mathrm{d}}{\mathrm{d}t}\left(\frac{\sin\psi}{r}\right)\sum_{n=1}^{N} nc_n\theta^{n-1} + \frac{v_\parallel\sin^2\psi}{r^2}\sum_{n=2}^{N} n(n-1)c_n\theta^{n-2}$$

This solution requires an expression for $\mathrm{d}(\sin\psi/r)/\mathrm{d}t$, which takes the form

$$\frac{\mathrm{d}}{\mathrm{d}t}\left(\frac{\sin\psi}{r}\right) = -\frac{\sin\psi\cos\psi}{r^2 v_\parallel}\left(K_1 + v_\parallel^2\right) \tag{11.36}$$

11.4.2.3 Fourier series

Consider an alternative shape of the transversal component of the motion, modeled by a Fourier series:

$$z(\theta) = a_0 + \sum_{n=1}^{N}[a_n\cos(n\theta) + b_n\sin(n\theta)]$$

The out-of-plane velocity abides by

$$\frac{\mathrm{d}z}{\mathrm{d}t} = -\frac{v_\parallel}{r}\sin\psi\sum_{n=1}^{N}\{n[a_n\sin(n\theta) - b_n\cos(n\theta)]\}$$

Similarly, the steering law becomes

$$\tan\lambda = -\frac{\sin\psi}{r}\sum_{n=1}^{N}\{n[a_n\sin(n\theta) - b_n\cos(n\theta)]\}$$

Its time derivative, required for computing the out-of-plane acceleration, is

$$\begin{aligned}\frac{\mathrm{d}}{\mathrm{d}t}(\tan\lambda) = &-\frac{\mathrm{d}}{\mathrm{d}t}\left(\frac{\sin\psi}{r}\right)\sum_{n=1}^{N}\{n[a_n\sin(n\theta) - b_n\cos(n\theta)]\}\\ &-\frac{v_\parallel\sin^2\psi}{r^2}\sum_{n=1}^{N}\{n^2[a_n\cos(n\theta) + b_n\sin(n\theta)]\}\end{aligned}$$

The derivative that appears in the first term is given in Eq. (11.36).

11.4.3 Transfers between arbitrary orbits

The method for designing transfers between arbitrary orbits consists of two steps:
1. The target orbit is projected on the orbital plane of the departure orbit. The transfer between the departure orbit and the projection of the target orbit is solved using a generalized logarithmic spiral. This defines the in-plane components of the motion, the time of flight, the control parameter ξ, and the constants of motion K_1 and K_2.
2. The out-of-plane motion is designed to satisfy the boundary conditions at departure and arrival. If the polynomial law from the previous section is used, the problem reduces to adjusting the values of the coefficients c_n. A similar approach is followed when implementing the Fourier shape, by solving for a_n and b_n.

Given the initial conditions $(z_0, \dot{z}_0)$ two of the shape coefficients need to be adjusted in order for the trajectory to satisfy $z(r_0) = z_0$ and $\dot{z}(r_0) = \dot{z}_0$. These equations are linear in the coefficients c_i. Let $(z_f, \dot{z}_f)$ be the arrival conditions for the out-of-plane motion. The boundary conditions provide the relation $z_f = z(r_f)$, where r_f is the value of the cylindrical radius at the arrival point and $z(r)$ is the assumed law for the evolution of z. The conditions on the velocity render $\dot{z}_f = \dot{z}(r_f)$. The problem imposes a total of four constraints on the system: the trajectory is forced to depart from $(z_0, \dot{z}_0)$ and to arrive at $(z_f, \dot{z}_f)$. At least four shape coefficients are required.

11.4.3.1 Polynomial shape

One simple approach is to solve for the first four coefficients $n = 0, \dots, 3$, stored in $\mathbf{c}_3$. But different strategies can be adopted depending on the number of coefficients. The linear system of equations can be written in matrix form as

$$\mathbf{z} = \mathbf{M}\,\mathbf{c}_3 + \mathbf{N}\,\mathbf{c}_{4:N} \tag{11.37}$$

Here vector $\mathbf{z} = [z_0, z_f, \dot{z}_0, \dot{z}_f]^\top$ contains the boundary conditions. Vector $\mathbf{c}_{4:N}$ refers to the remaining coefficients not included in $\mathbf{c}_3$, and the matrices $\mathbf{M} \in \mathbb{R}^{4\times4}$ and $\mathbf{N} \in \mathbb{R}^{4\times(N-4)}$ depend on the shape function. Denoting M_n the $(n+1)$-th column of matrix $\mathbf{M}$, and N_m the $(m+1)$-th column of matrix $\mathbf{N}$, it follows that:

$$\mathsf{M}_n = \begin{bmatrix} \theta_0^n \\ \theta_f^n \\ n\theta_0^{n-1} C_0 \\ n\theta_f^{n-1} C_f \end{bmatrix} \quad \text{and} \quad \mathsf{N}_m = \begin{bmatrix} \theta_0^m \\ \theta_f^m \\ m\theta_0^{m-1} C_0 \\ m\theta_f^{m-1} C_f \end{bmatrix}$$

with $n = 0, \dots, 3$ and $m = 4, \dots, N$. For simplicity the coefficient $C = v_\parallel \sin\psi / r$ has been introduced.

The matrix equation (11.37) can be solved for $\mathbf{c}_3$ to provide

$$\mathbf{c}_3 = \mathbf{M}^{-1}(\mathbf{z} - \mathbf{N}\,\mathbf{c}_{4:N})$$

The determinant of matrix $\mathbf{M}$ is

$$\Delta = -C_0 C_f(\theta_f - \theta_0)^4$$

It will only vanish if $\theta_0 = \theta_f$ or $C_j = 0$. The former is not possible since θ always grows in time. The latter only occurs when $\psi \to 0$ (degenerate rectilinear orbit) or $r \to \infty$ (escape); both situations are limit cases that are omitted for consistency. Under these hypotheses matrix $\mathbf{M}$ is regular.

11.4.3.2 Fourier series

When considering the Fourier series representation the boundary conditions can be written

$$\mathbf{z} = \mathbf{M}\,\mathbf{c} + \mathbf{d}$$

Vector $\mathbf{c}$ groups the coefficients $[a_0, a_1, b_1, a_2]^\top$ and matrix $\mathbf{M}$ reads

$$\mathbf{M} = \begin{bmatrix} 1 & \cos\theta_0 & \sin\theta_0 & \cos 2\theta_0 \\ 1 & \cos\theta_f & \sin\theta_f & \cos 2\theta_f \\ 0 & -C_0\sin\theta_0 & C_0\cos\theta_0 & -2C_0\sin 2\theta_0 \\ 0 & -C_f\sin\theta_f & C_f\cos\theta_f & -2C_f\sin 2\theta_f \end{bmatrix}$$

The last vector, $\mathbf{d}$, refers to the remaining terms of the series and takes the form

$$\mathbf{d} = \begin{bmatrix} b_2\sin 2\theta_0 + \sum_{n=3}^{N}[a_n\cos(n\theta_0) + b_n\sin(n\theta_0)] \\ b_2\sin 2\theta_f + \sum_{n=3}^{N}[a_n\cos(n\theta_f) + b_n\sin(n\theta_f)] \\ 2C_0 b_2\cos 2\theta_0 - C_0\sum_{n=3}^{N} n[a_n\sin(n\theta_0) - b_n\cos(n\theta_0)] \\ 2C_f b_2\cos 2\theta_f - C_f\sum_{n=3}^{N} n[a_n\sin(n\theta_f) - b_n\cos(n\theta_f)] \end{bmatrix}$$

In this case the determinant of $\mathbf{M}$ is

$$\Delta = -\frac{C_0 C_f}{2}[\sin(3\theta_0 - \theta_f) + 6\sin(\theta_0 + \theta_f) - \sin(\theta_0 - 3\theta_f) - 4\sin(2\theta_f) - 4\sin(2\theta_0)]$$

meaning that the matrix is regular for $C_0 \neq 0$ and $C_f \neq 0$.

11.5 Applications

This section shows how the method presented in this chapter can be applied to the design of an interplanetary mission. In order to have a reference solution to compare with, we recovered an example that appears recursively in the literature (Sauer, 1997; McConaghy et al., 2003; Petropoulos and Longuski, 2004): a low-thrust transfer to rendezvous with Ceres including a flyby at Mars.

The problem is modeled in two different ways. First, a simplified ephemeris model is considered, following the examples in the literature. In this case the orbits of the Earth, Mars, and Ceres are assumed Keplerian and their inclination relative to the ecliptic plane is neglected. The second model considers the real ephemeris of the planets and Ceres.

11.5.1 Simplified model

The goal of this simplified example is to show how the generalized logarithmic spirals (GLS) perform when compared to other methods such as the exponential sinusoids, by adopting the same exact transfer configuration. The orbits are assumed to be contained in the ecliptic plane, as defined in Table 11.1. The specific impulse is assumed to be constant and equal to 3000 s. The spacecraft departs from the Earth on May 6, 2003 with $v_\infty = 1.6$ km/s at launch. The mass injected into orbit is 568 kg.

The Earth-Mars (EM) transfer leg decomposes in two spiral arcs connected by a coast arc. In the EM leg both spirals are hyperbolic of Type II (H-II). The spiral arcs are described in Table 11.2. In the Mars-Ceres (MC) leg the first spiral is also hyperbolic of Type II, whereas the second spiral is elliptic.

The performance of the GLS is compared with the solution provided by Petropoulos and Longuski (2004) and Sauer (1997) in Table 11.3. We also include the optimization using GALLOP (Sims and Flanagan, 1999; McConaghy et al., 2003) that Petropou-

Tab. 11.1: Reference orbits (MJD 52765).

	Units	Earth	Mars	Ceres
a	au	1.0000	1.5237	2.7656
e	–	0.0162	0.0936	0.0794
$\omega + \Omega$	deg	103.93	336.06	154.60
ϑ_0	deg	121.10	289.08	258.35

Tab. 11.2: Parameters defining the spiral segments.

	Type	ξ	K_1	K_2
EM-1	H-II	0.4936	0.0885	1.1021
EM-2	H-II	0.5181	0.0111	0.9684
MC-1	H-II	0.4983	0.0912	1.1371
MC-2	Ellip	0.4938	−0.0279	0.9291

Tab. 11.3: Comparison of the GLS solution with STOUR-LTGA, GALLOP, and Sauer's solution.

	Units	GLS	STOUR-LTGA*	GALLOP*	Sauer*
Launch v_∞	km/s	1.600	1.600	1.600	1.600
Mars flyby v_∞	km/s	1.590	1.435	1.920	–
Flyby altitude	km	562	5432	200	–
Propellant mass fraction	–	0.256	0.256	0.234	0.275
Arrival v_∞	km/s	0.000	0.237	0.000	0.000
TOF Earth-Mars	days	271	271	271	250
TOF Mars-Ceres	days	862	862	862 (739)	845
TOF total	days	1133	1133	1133 (1010)	1095

* Data taken from Petropoulos and Longuski (2004).

los and Longuski computed as a reference for their comparison. The launch v_∞ is equal in all cases. The configuration of the Mars flyby differs slightly to that predicted by STOUR-LTGA (the low-thrust version of the design tool created by Longuski, 1983), defining a faster and deeper flyby. The MC leg is solved in order for the arrival velocity to match the orbital velocity of Ceres, which yields a null v_∞ at arrival. The times of flight are taken from the solution by Petropoulos and Longuski (2004): 271 and 862 days for the EM and MC transfer legs, respectively. The solution found with the GLS yields the same propellant mass fraction as STOUR-LTGA, but requires no additional impulsive Δv for the final rendezvous with Ceres. Given the relatively low specific impulses of liquid propellants this final maneuver may increase significantly the propellant mass fraction. Note that Sauer used a more sophisticated model for the thruster and propellant expenditures, and he selected June 8, 2003, for the launch date. This is responsible for the higher propellant mass fraction of his solution.

The trajectory is plotted in Fig. 11.16. Both the EM and the MC legs include coast arcs to gain control over the final state. Right after launch there is a short thrust arc in order to increase the energy of the intermediate Keplerian orbit. The thrust is switched on again when the values of K_1 and K_2 are compatible with a feasible flyby about Mars. The flyby is designed to yield adequate values of the constants of motion. The MC leg is solved from Eq. (11.14) having selected the switch point that minimizes the propellant mass fraction. The configuration of the flyby is obtained with a genetic optimization algorithm.

The magnitude of the thrust required for the mission is plotted in Fig. 11.17. It is worth noting that the required thrust exhibits some peaks that exceed the limit thrust adopted in Petropoulos and Longuski (2004); they considered that the available thrust is 95 mN at 1 au and it decreases with the power law $1/r^2$. Despite the peaks of high thrust and due to the coast arcs the final result is a propellant mass fraction that is comparable to the optimized solution, as shown in Table 11.3.

The orientation of the thrust depends on the values of the control parameters (Table 11.2) and on the evolution of the flight-direction angle. Figure 11.18 shows how the

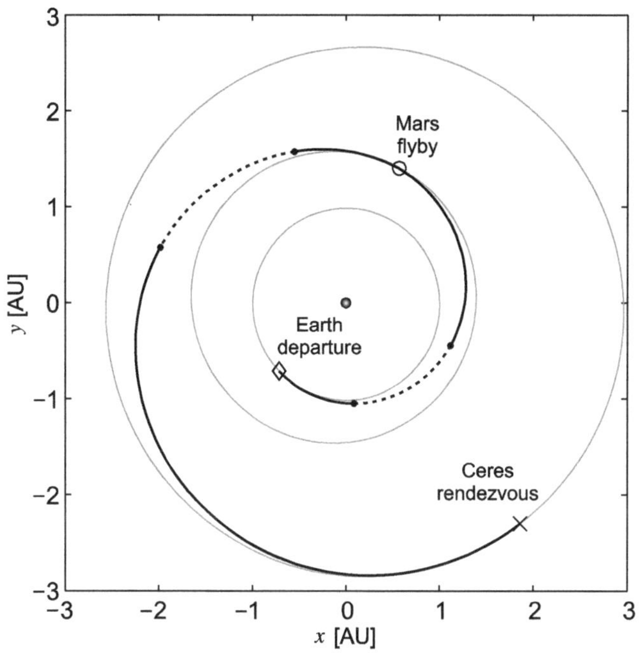

Fig. 11.16: Earth-Mars-Ceres spiral transfer. Solid lines correspond to the thrust arcs, whereas dashed lines are coast arcs. The black dots • represent the switch points.

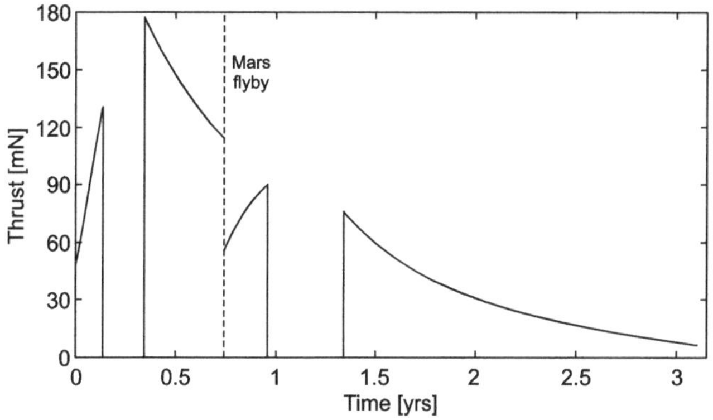

Fig. 11.17: Thrust profile for the Earth-Mars-Ceres transfer.

thrust angle evolves compared to the flight-direction angle. The orientation (and magnitude) of the thrust vector is adjusted in order to meet the constraints on the time of flight. In addition, in the MC leg the values of the control parameter are defined so that the final velocity matches that of Ceres. Along the final spiral arc, the magnitude of the thrust decreases and the thrust angle separates from the flight-direction angle for velocity matching.

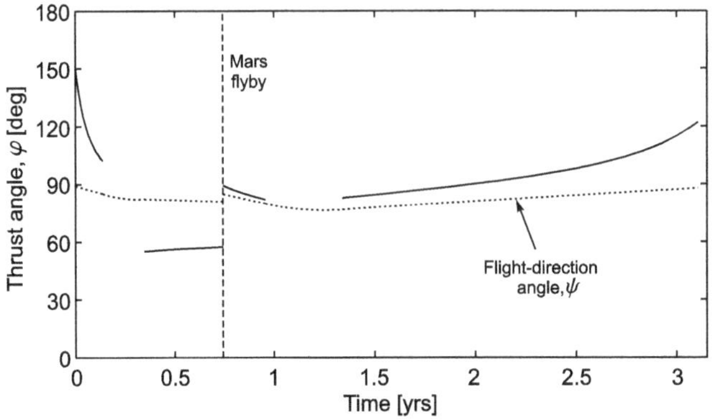

Fig. 11.18: Thrust angle compared to the flight-direction angle ψ. When $\varphi = \psi$ the thrust vector is directed along the velocity.

11.5.2 Real ephemeris

In the case of Mars, the low relative inclination (< 2 deg) will have little impact on the solution. However, the inclination of the orbit of Ceres is approximately 10 deg and the out-of-plane motion is more relevant.

The configuration of the planar transfer is obtained through a global optimization procedure using a genetic algorithm. The state vector is composed of the date of departure, the date of arrival at Mars, the date of arrival at Ceres, the launch v_∞, and the configuration of the flyby. The performance of the new solution can be analyzed in Table 11.4. The departure date changes to May 21, 2003, and the time of flight for each leg is reduced. The result is a propellant mass fraction for the in-plane motion of 23.42%. When accounting for the out-of-plane component of the thrust, this figure rises to 28.94%.

The shape of the trajectory is depicted in Fig. 11.19 and it is defined in the ecliptic ICRF/J2000 reference system. The orbit is projected in the xz-plane to show in more

Tab. 11.4: Definition of the three-dimensional transfer.

	Units	Value
Launch v_∞	km/s	1.995
Mars flyby v_∞	km/s	1.272
Flyby altitude	km	2058
Prop. mass fraction	–	0.289
Prop. mass fraction (in-plane)	–	0.234
Arrival v_∞	km/s	0.000
TOF Earth-Mars	days	246
TOF Mars-Ceres	days	846
TOF total	days	1092

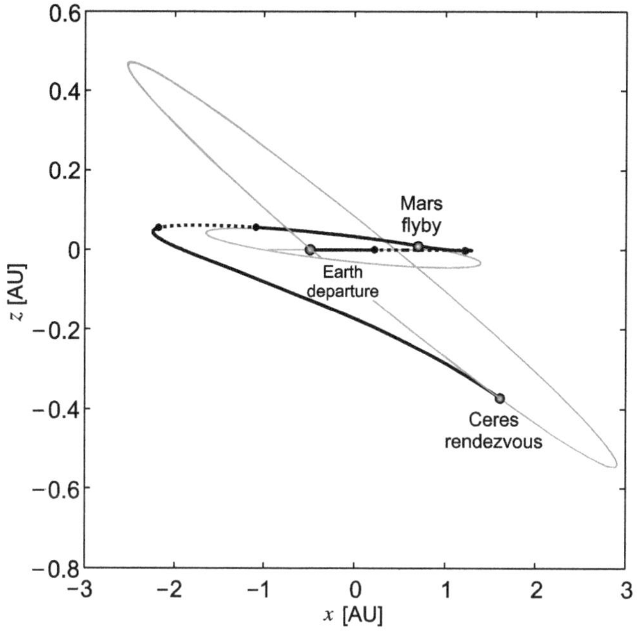

Fig. 11.19: Projection of the out-of-plane component of the motion.

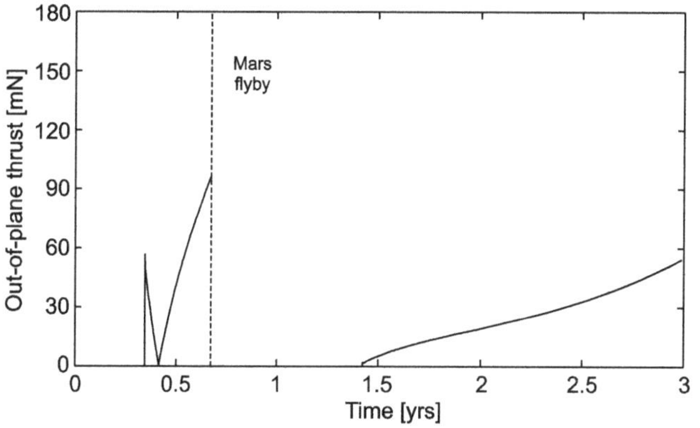

Fig. 11.20: Transversal component of the thrust.

detail the transversal component of the motion (the z-axis is exaggerated). Gray lines represent the orbits of the Earth, Mars, and Ceres. The arrival velocity at Ceres also matches the transversal component of the velocity of the minor planet.

The transversal component of the motion obeys a polynomial shaping law. The design procedure is based on the method described in Sect. 11.4.3.1, using only the first four coefficients (i.e., $\mathbf{c}_{4:N} = 0$). The evolution of the out-of-plane component of the thrust vector is plotted in Fig. 11.20. It is interesting to note that the first spiral arc and the coast arcs are coplanar (the transversal component of the thrust is only switched on in the second spiral arc). This is the simplest decomposition of the transfer, in which the correction of the out-of-plane motion occurs only in the last leg. Alternative strategies might be considered, like introducing an intermediate plane in which the Keplerian leg will be confined. The propellant expenditures associated to the out-of-plane dynamics are 18% of the total budget.

11.6 Conclusions

Thanks to the introduction of a control parameter the versatility of the family of generalized logarithmic spirals is improved. As a result, these new curves define a novel transfer strategy between arbitrary orbits. The existence of constants of motion of dynamical nature yields an intuitive design methodology, based on matching the values of the constants of motion along the spiral and coast arcs.

The feasibility conditions and design constraints reduce to simple compatibility equations. This not only simplifies the design problem to solving systems of algebraic equations, but it also allows one to define the conditions that ensure the existence of solutions. Additional degrees of freedom can be considered by introducing control nodes. The trajectory is decomposed in consecutive spirals or Keplerian arcs. The simple solutions obtained with single spiral arcs can then be generalized arbitrarily or even be handled by an optimizer. The distribution of the control nodes and the values of the control parameter on each spiral arc can be adjusted in order to minimize a certain cost function.

A fully three-dimensional transfer strategy is defined based on the family of planar spirals. It is based on a complete decoupling of the transversal dynamics, and the design procedure reduces to solving four linear equations. When applied to actual design problems, the solution provided by this method is in agreement with well-known optimized solutions.

12 Nonconservative extension of Keplerian integrals and new families of orbits

The invariance of the Lagrangian to time translations and rotations in Kepler's problem yields the conservation laws related to the energy and angular momentum. Noether's theorem reveals that these same symmetries furnish generalized forms of the first integrals in a special nonconservative case, which is a generalization of the acceleration introduced in Chap. 9. The system is perturbed by a biparametric acceleration with components along the tangential and normal directions. A similarity transformation reduces the biparametric disturbance to a simpler uniparametric forcing along the velocity vector. The solvability conditions of this new problem are discussed in the present chapter, and closed-form solutions for special cases are provided. Keplerian orbits and generalized logarithmic spirals appear naturally from this construction. After characterizing the orbits independently, a unified form of the solution is built based on the Weierstrass elliptic functions. Two new families of orbits are found in this chapter: the generalized cardioids, and the generalized sinusoidal spirals. The names come from the fact that in the parabolic case the orbits reduce to pure cardioids and sinusoidal spirals. These orbits approximate some instances of Schwarzschild geodesics. Finally, the connection with other known integrable systems in celestial mechanics is discussed (Roa, 2016a).

12.1 The role of first integrals

Finding first integrals is fundamental to characterizing a dynamical system. The motion is confined to submanifolds of lower dimensions on which the orbits evolve, providing an intuitive interpretation of the dynamics and reducing the complexity of the system. In addition, conserved quantities are good candidates when applying the second method of Lyapunov for stability analysis. Conservative systems related to central forces are typical examples of (Liouville) integrability, and provide useful analytic results. Hamiltonian systems have been widely analyzed in the classical and modern literature to determine adequate integrability conditions. The existence of first integrals under the action of small perturbations occupied Poincaré (1892, Chap. V) back in the nineteenth century. Emmy Noether (1918) established in her celebrated theorem that conservation laws can be understood as the system exhibiting dynamical symmetries. In a more general framework, Yoshida (1983a,b) analyzed the conditions that yield algebraic first integrals of generic systems. He relied on the Kowalevski exponents to characterize the singularity of the solutions and derived the necessary conditions for existence of first integrals exploiting similarity transformations.

Conservation laws are sensitive to perturbations and their generalization is not straightforward. For example, the Jacobi integral no longer holds when transforming

https://doi.org/10.1515/9783110559125-012

the circular restricted three-body problem to the elliptic case (Xia, 1993). Nevertheless, Contopoulos (1967) was able to find approximate conservation laws for orbits of small eccentricities. Szebehely and Giacaglia (1964) benefited from the similarities of the elliptic and the circular problems in order to define transformations between them. Hénon and Heiles (1964) deepened the study of the nature of conservation laws and reviewed the concepts of isolating and nonisolating integrals. Their study introduced a similarity transformation that embeds one of the constants of motion and transforms the original problem into a simplified one, reducing the degrees of freedom (Arnold et al., 2007, §3.2). Carpintero (2008) proposed a numerical method for finding the dimension of the manifold in which orbits evolve, i.e., the number of isolating integrals that the system admits.

The conditions for existence of integrals of motion under nonconservative perturbations received significant attention in the past due to their profound implications. Djukic and Vujanovic (1975) advanced on Noether's theorem and included nonconservative forces in the derivation. Relying on Hamilton's variational principle they not only extended Noether's theorem, but also its inverse form and the Noether–Bessel-Hagen and Killing equations. Later studies by Djukic and Sutela (1984) sought integrating factors that yield conservation laws upon integration. Examples of application of Noether's theorem to constrained nonconservative systems can be found in the work of Bahar and Kwatny (1987). Honein et al. (1991) arrived at a compact formulation using what was later called the neutral action method. Remarkable applications exploiting Noether's symmetries span from cosmology (Capozziello et al., 2009; Basilakos et al., 2011) to string theory (Beisert et al., 2008), field theory (Halpern et al., 1977), and fluid models (Narayan et al., 1987). In the book by Olver (2000, Chaps. 4 and 5) an exhaustive review of the connection between symmetries and conservation laws is provided within the framework of Lie algebras. We refer to Arnold et al. (2007, Chap. 3) for a formal derivation of Noether's theorem and a discussion of the connection between conservation laws and dynamical symmetries.

Integrals of motion are often useful for finding analytic or semianalytic solutions to a given problem. The acclaimed solution to the main satellite problem by Brouwer (1959) is a clear example of the decisive role of conserved quantities in deriving solutions in closed form. By perturbing the Delaunay elements, Brouwer and Hori (1961) solved the dynamics of a satellite subject to atmospheric drag and the oblateness of the primary. They proved the usefulness of canonical transformations even in the context of nonconservative problems. Whittaker (1917, pp. 81–82) approached the problem of a central force depending on powers of the radial distance, r^n, and found that there are only 14 values of n for which the problem can be integrated in closed form using elementary functions or elliptic integrals. Later, he discussed the solvability conditions for equations involving square roots of polynomials (Whittaker and Watson, 1927, p. 512). Broucke (1980) advanced on Whittaker's results and found six potentials that are a generalization of the integrable central forces discussed by the latter. These potentials include the referred to 14 values of n as particular cases. Numerical techniques for shaping the potential given the orbit solution were published by Carpintero and Aguilar (1998). Classical studies on the integrability of systems governed by central forces are based strongly on Newton's theorem of revolving orbits.[1] The problem of the orbital precession caused by central forces was recently recovered by Adkins and McDonnell (2007), who considered potentials involving both powers and logarithms of the radial distance, and the special case of the Yukawa potential (Yukawa, 1935). Chashchina and Silagadze (2008) relied on Hamilton's vector to simplify the analytic solutions found by Adkins and McDonnell (2007). More elaborated potentials have been explored for modeling the perihelion precession (Schmidt, 2008). The dynamics of a particle in Schwarzschild space-time can be regarded as orbital motion perturbed by an effective potential depending on inverse powers of the radial distance (Chandrasekhar, 1983, p. 102).

Potentials depending linearly on the radial distance appear recursively in the literature because they render constant radial accelerations, relevant for the design of spacecraft trajectories propelled by continuous-thrust systems. The pioneering work by Tsien (1953) provided the explicit solution to the problem in terms of elliptic integrals, as predicted by Whittaker (1917, p. 81). By means of a special change of variables, Izzo and Biscani (2015) arrived at an elegant solution in terms of the Weierstrass elliptic functions. These functions were also exploited by MacMillan (1908) when he solved the dynamics of a particle attracted by a central force decreasing with r^{-5}. Urrutxua et al. (2015a) solved the Tsien problem using the Dromo formulation, which models orbital motion with a regular set of elements (Chap. 4). The case of a constant radial force was approached by Akella and Broucke (2002) from an energy-driven perspective. They studied in detail the roots of the polynomial appearing in the denominator of the equation to integrate and connected their nature with the form of the solution. We refer again to San-Juan et al. (2012) for a detailed discussion on the integrability of the system.

Another relevant example of a system that admits an analytic solution is the Stark problem, governed by a constant acceleration fixed in the inertial frame. Lantoine and Russell (2011) provided the complete solution to the motion relying extensively on elliptic integrals and Jacobi elliptic functions. A compact form of the solution involv-

---

1 Section IX, Book I, of Newton's Principia is devoted to the motion of bodies in moveable orbits (*De Motu Corporum in Orbibus mobilibus, deq; motu Apsidum*, in the original Latin version). In particular, Thm. XIV states that "The difference of the forces, by which two bodies may be made to move equally, one in a quiescent, the other in the same orbit revolving, is in a triplicate ratio of their common altitudes inversely". Newton proved this theorem relying on elegant geometrical constructions. The motivation behind this result was the development of a theory for explaining the precession of the orbit of the Moon. A detailed discussion of this theorem can be found in the book by Chandrasekhar (1995, pp. 184–201).

ing the Weierstrass elliptic functions was later presented by Biscani and Izzo (2014), who also exploited this formalism for building a secular theory for the post-Newtonian model (Biscani and Carloni, 2012). The Stark problem provides a simplified model of radiation pressure. In general, the dynamics subject to this perturbation cannot be solved in closed form. An intuitive simplification that makes the problem integrable consists in assuming that the force due to the solar radiation pressure follows the direction of the Sun vector. The dynamics are equivalent to those governed by a Keplerian potential with a modified gravitational parameter.

In Chap. 9 it was proven that the problem defined by the perturbation

$$\mathbf{a}_p = \frac{\mu}{2r^2} \cos\psi\, \mathbf{t}$$

can be solved in closed form, and that there are two conservation laws. This acceleration was later extended in Chap. 10 to include a control parameter, while preserving the properties of the system. In the present chapter we approach this problem from a generic perspective, considering an even more general acceleration including two free parameters. Following this new approach, the problem is formulated in Sect. 12.2, where the biparametric acceleration is reduced to a uniparametric forcing thanks to a similarity transformation. The conservation laws hold in the more general case and characterize this type of system. The first integrals are obtained by exploiting known symmetries of Kepler's problem. Before solving the dynamics explicitly, we prove that there are four cases that can be solved in closed form using elementary or elliptic functions. The first two families are conic sections and generalized logarithmic spirals. Sections 12.5 and 12.6 introduce the two new families of orbits. Section 12.7 is a summary of the solutions, which are unified in Sect. 12.8 introducing the Weierstrass elliptic functions. Finally, Sect. 12.9 discusses the connection with known solutions to other problems, and the connection with Schwarzschild geodesics.

12.2 Dynamics

The two previous chapters dealt with the two-body problem

$$\frac{\mathrm{d}^2\mathbf{r}}{\mathrm{d}t^2} + \frac{\mu}{r^3}\mathbf{r} = \mathbf{a}_p \tag{12.1}$$

perturbed by the continuous acceleration defined in Eq. (10.2). This forcing term was defined as

$$\mathbf{a}_p = \frac{\mu}{r^2}[\xi \cos\psi\, \mathbf{t} + (1 - 2\xi)\sin\psi\, \mathbf{n}]$$

The coefficient $(1 - 2\xi)$ was chosen so that the integral of motion (9.16) still held. Replacing this term by a generic parameter η renders

$$\mathbf{a}_p = \frac{\mu}{r^2}(\xi \cos\psi\, \mathbf{t} + \eta \sin\psi\, \mathbf{n}) \tag{12.2}$$

This new form of the perturbation includes the Keplerian case as a particular instance, because making $\xi = \eta = 0$ cancels the perturbation. In addition, generalized logarithmic spirals are the solution with $\eta = (1 - 2\xi)$. The forcing parameter ξ controls the power exerted by the perturbation,

$$\frac{\mathrm{d}\mathcal{E}_k}{\mathrm{d}t} = \mathbf{a}_p \cdot \mathbf{v} = \frac{\xi\mu}{r^2} v \cos\psi$$

with $\mathcal{E}_k$ the Keplerian energy of the system. The second parameter η modulates the contribution of forces that do not perform any work.

For convenience we shall replace η by the parameter

$$\gamma = \frac{1+\eta}{1-\xi}$$

In this way, when the perturbation is added to the Keplerian acceleration it takes the simplified form

$$\frac{\mathrm{d}^2\mathbf{r}}{\mathrm{d}t^2} = -\frac{\mu}{r^2}(1-\xi)(\cos\psi\,\mathbf{t} - \gamma\sin\psi\,\mathbf{n}) \tag{12.3}$$

The term $\mu(1-\xi)$ can be regarded as a modified gravitational parameter μ^*. Consequently, when $\gamma = 1$ Eq. (12.3) reduces to Kepler's problem written in terms of μ^* instead of μ.

12.2.1 Similarity transformation

In Sect. 10.2 we discussed a similarity transformation that connects the solution to the original thrust (9.1) and the one including the control parameter ξ, Eq. (10.2). In the present section we will introduce a more general similarity transformation and discuss its properties in detail.

Consider a linear transformation $\mathscr{S}$ of the form

$$\mathscr{S} : (t, r, \theta, \dot{r}, \dot{\theta}) \mapsto (\tau, \rho, \theta, \rho', \theta')$$

defined explicitly by the positive constants α, β, and $\delta = \alpha/\beta$:

$$\tau = \frac{t}{\beta}, \qquad \rho = \frac{r}{\alpha}, \qquad \rho' = \frac{\dot{r}}{\delta}, \qquad \theta' = \beta\dot{\theta}$$

The constants α, β, and δ have units of length, time, and velocity, respectively. The symbol $\square'$ denotes derivatives with respect to τ, whereas $\dot{\square}$ is reserved for derivatives with respect to t. The scaling factor α can be seen as the ratio of a homothetic transformation that simply dilates or contracts the orbit. Similarly, β represents a time dilation or contraction. The velocity of the particle transforms into

$$\tilde{v} = \frac{v}{\delta}$$

In addition, β and δ are defined in terms of α by virtue of

$$\beta = \sqrt{\frac{\alpha^3}{\mu\gamma(1-\xi)}} \qquad \text{and} \qquad \delta = \frac{\alpha}{\beta} = \sqrt{\frac{\mu\gamma(1-\xi)}{\alpha}}$$

We assume $\gamma > 0$ and $\xi < 1$ for consistency.

Equation (12.3) then becomes

$$\frac{d^2\boldsymbol{\rho}}{d\tau^2} = -\frac{1}{\rho^2}\left(\frac{1}{\gamma}\cos\psi\,\mathbf{t} - \sin\psi\,\mathbf{n}\right) \tag{12.4}$$

which is equivalent to the normalized two-body problem

$$\frac{d^2\boldsymbol{\rho}}{d\tau^2} = -\frac{\boldsymbol{\rho}}{\rho^3} + \tilde{\mathbf{a}}_p \tag{12.5}$$

This result shows that $\mathscr{S}$ establishes, indeed, a similarity transformation between the two-body problem perturbed by the acceleration (12.2) and the simpler problem in Eq. (12.5), perturbed by the purely tangential acceleration:

$$\tilde{\mathbf{a}}_p = \frac{\gamma - 1}{\gamma\rho^2}\cos\psi\,\mathbf{t} \tag{12.6}$$

That is, $\mathscr{S}^{-1}$ transforms the solution to Eq. (12.4) into the solution to Eq. (12.3). It is worth mentioning that $\gamma = 2$ reduces the acceleration to

$$\tilde{\mathbf{a}}_p = \frac{\cos\psi}{2\rho^2}\,\mathbf{t}$$

which is none other than (9.1).

Using $\mathbf{q} = [\rho, \theta]^\top$ and $\mathbf{q}' = [\rho', \theta']^\top$ as the generalized coordinates and velocities, respectively, the dynamics of the problem abide by the Euler–Lagrange equations

$$\frac{d}{d\tau}\left(\frac{\partial\mathscr{L}}{\partial q_i'}\right) - \frac{\partial\mathscr{L}}{\partial q_i} = Q_i$$

The Lagrangian of the transformed system takes the form

$$\mathscr{L} = \frac{1}{2}\left(\rho'^2 + \rho^2\theta'^2\right) + \frac{1}{\rho} \tag{12.7}$$

and the generalized forces Q_i read

$$Q_\rho = \frac{(\gamma-1)}{\gamma}\left(\frac{\rho'}{\rho\tilde{v}}\right)^2 \qquad \text{and} \qquad Q_\theta = \frac{(\gamma-1)}{\gamma}\left(\frac{\rho'\theta'}{\tilde{v}^2}\right)$$

12.2.2 Integrals of motion and dynamical symmetries

Let us introduce an infinitesimal transformation $\mathscr{R}$:

$$\begin{aligned}\tau \to \tau^* &= \tau + \varepsilon f(\tau; q_i, q_i') \\ q_i \to q_i^* &= q_i + \varepsilon F_i(\tau; q_i, q_i')\end{aligned} \tag{12.8}$$

defined in terms of a small parameter $\varepsilon \ll 1$ and the generators F_i and f. For transformations that leave the action unchanged up to an exact differential,

$$\mathscr{L}\left(\tau^*; q_i^*, \frac{dq_i^*}{d\tau^*}\right) d\tau^* - \mathscr{L}\left(\tau; q_i, \frac{dq_i}{d\tau}\right) d\tau = \varepsilon\, d\Psi(\tau; q_i, q_i')$$

with $\Psi(\tau; q_i, q_i')$ a given gauge, Noether's theorem establishes that

$$\sum_i \left\{ \left(\frac{\partial \mathscr{L}}{\partial q_i'}\right) F_i + f\left[\mathscr{L} - \left(\frac{\partial \mathscr{L}}{\partial q_i'}\right) q_i'\right]\right\} - \Psi(\tau; q_i, q_i') = \Lambda \tag{12.9}$$

is a first integral of the problem. Here Λ is a certain constant of motion.

Since the perturbation in Eq. (12.6) is not conservative we shall focus on the extension of Noether's theorem to nonconservative systems by Djukic and Vujanovic (1975). It must be $F_i - q_i' f \neq 0$ for the conservation law to hold (Vujanovic et al., 1986). For the case of nonconservative systems, the generators F_i, f, and the gauge Ψ need to satisfy the following relation:

$$\sum_i \left\{ \left(\frac{\partial \mathscr{L}}{\partial q_i}\right) F_i + \left(\frac{\partial \mathscr{L}}{\partial q_i'}\right)\left(F_i' - q_i' f'\right) + Q_i\left(F_i - q_i' f\right)\right\} + f'\mathscr{L} + f\frac{\partial \mathscr{L}}{\partial \tau} = \Psi' \tag{12.10}$$

This equation and the condition $F_i - q_i' f \neq 0$ furnish the generalized Noether–Bessel-Hagen (NBH) equations (Trautman, 1967; Djukic and Vujanovic, 1975; Vujanovic et al., 1986). The NBH equations involve the full derivative of the gauge function and the generators with respect to τ, meaning that Eq. (12.10) depends on the partial derivatives of Ψ, F_i, and f with respect to time, the coordinates, and the velocities. By expanding the convective terms the NBH equations decompose in the system of Killing equations:

$$\begin{aligned}
&\mathscr{L}\frac{\partial f}{\partial q_j'} + \sum_i \frac{\partial \mathscr{L}}{\partial q_i'}\left(\frac{\partial F_i}{\partial q_j'} - q_i'\frac{\partial f}{\partial q_j'}\right) = \frac{\partial \Psi}{\partial q_j'} \\
&\frac{\partial}{\partial \tau}(f\mathscr{L} - \Psi) + \sum_i \left\{ \frac{\partial \mathscr{L}}{\partial q_i} F_i + \mathscr{L}\frac{\partial f}{\partial q_i} q_i' + Q_i(F_i - q_i' f) \right. \\
&\quad \left. + \frac{\partial \mathscr{L}}{\partial q_i'}\left[\frac{\partial F_i}{\partial \tau} - q_i'\frac{\partial f}{\partial \tau} + \sum_j\left(\frac{\partial F_i}{\partial q_j} q_j' - q_i' q_j' \frac{\partial f}{\partial q_j}\right)\right] - \frac{\partial \Psi}{\partial q_i} q_i' \right\} = 0
\end{aligned} \tag{12.11}$$

The system (12.11) is composed of three equations that can be solved for the generators F_ρ, F_θ, and f given a certain gauge. If the transformation defined in Eq. (12.8) satisfies the NBH equations, then the system admits the integral of motion (12.9).

12.2.2.1 Generalized equation of the energy

The Lagrangian in Eq. (12.7) is time-independent. Thus, the action is not affected by arbitrary time transformations. In the Keplerian case a simple time translation reveals the conservation of the energy. Motivated by this fact, we explore the generators

$$f = 1\,, \qquad F_\rho = 0\,, \qquad \text{and} \qquad F_\theta = 0$$

Clearly $F_i - q_i' f \neq 0$. Solving Killing equations (12.11) with the above generators the gauge function follows

$$\Psi = \frac{\gamma - 1}{\gamma r}$$

Provided that the NBH equations hold the system admits the integral of motion

$$\frac{\tilde{v}^2}{2} - \frac{1}{\gamma \rho} = -\Lambda \equiv \frac{\kappa_1}{2} \tag{12.12}$$

written in terms of the constant $\kappa_1 = -2\Lambda$. This term can be solved from the initial conditions

$$\kappa_1 = \tilde{v}_0^2 - \frac{2}{\gamma \rho_0}$$

When $\gamma = 1$ the perturbation (12.6) vanishes and Eq. (12.12) reduces to the normalized equation of the Keplerian energy. In fact, in this case κ_1 becomes twice the Keplerian energy of the system, $\kappa_1 = 2\mathcal{E}_k$. Moreover, the gauge vanishes and Eq. (12.9) furnishes the Hamiltonian of Kepler's problem. The integral of motion (12.12) is a generalization of the equation of the energy. Similarly, for $\gamma = 2$ Eq. (12.12) reduces to the generalized equation of the energy presented in Eq. (9.13).

12.2.2.2 Generalized equation of the angular momentum

In the unperturbed problem θ is an ignorable coordinate. Indeed, a simple translation in θ (a rotation) with $f = F_\rho = \Psi = 0$ and $F_\theta = 1$ yields the conservation of the angular momentum. In order to extend this first integral to the perturbed case, we consider the same generator $F_\theta = 1$. However, solving for the gauge and the remaining generators in Killing equations yields the nontrivial functions

$$\begin{aligned}
F_\rho &= \frac{\rho'}{\theta'}(1 - \tilde{v}^{\gamma-1}) \\
f &= \frac{1 - v^{\gamma-1}}{\theta'} + (1-\gamma)\rho^2\theta'\tilde{v}^{\gamma-3} \\
\Psi &= \frac{1}{2\rho\theta'}\Big\{\tilde{v}^2\left[\rho - (3-\gamma)\rho\tilde{v}^{\gamma-1}\right] + 2 - \tilde{v}^{\gamma-1}\left[2\gamma - (3-\gamma)\rho'^2\rho\right] \\
&\quad + \tilde{v}^{\gamma-3}\left[\rho(\rho^4\theta'^4 - \rho'^4) - 2(1-\gamma)\rho'^2\right]\Big\}
\end{aligned}$$

They satisfy $F_i - q_i' \neq 0$. Noether's theorem holds and Eq. (12.9) furnishes the integral of motion

$$\rho^2\tilde{v}^{\gamma-1}\theta' = \Lambda \equiv \kappa_2 \tag{12.13}$$

This first integral is the generalized form of the conservation of the angular momentum. Indeed, making $\gamma = 1$ Eq. (12.13) reduces to

$$\rho^2\theta' = \kappa_2$$

where κ_2 coincides with the angular momentum of the particle. In addition, the generators F_ρ and f vanish when γ equals unity.

The integral of motion (12.13) can be written:

$$\rho \tilde{v}^{\gamma} \sin \psi = \kappa_2$$

in terms of the coordinates intrinsic to the trajectory. It is easily verified that $\gamma = 2$ makes this equation coincide with the conservation law in Eq. (9.16). The fact that $\sin \psi \leq 1$ forces

$$\kappa_2 \leq \tilde{v}^{\gamma} \rho \implies \kappa_2^2 \leq \tilde{v}^{2\gamma} \rho^2$$

The second step is possible because all variables are positive. Combining this expression with Eq. (12.12) yields

$$\kappa_2^2 \leq \left(\kappa_1 + \frac{2}{\gamma \rho} \right)^{\gamma} \rho^2 \tag{12.14}$$

Depending on the values of γ, Eq. (12.14) may define upper or lower limits to the values that the radius ρ can reach. In general, this condition can be resorted to provide the polynomial constraint

$$P_{\text{nat}}(\rho) \geq 0$$

where $P_{\text{nat}}(\rho)$ is a polynomial of degree γ in ρ whose roots dictate the nature of the solutions. This inequality will be useful for defining the different families of solutions.

12.2.3 Properties of the similarity transformation

The main property of the similarity transformation $\mathscr{S}$ is that it does not change the type of solution, i.e., the sign of κ_1 is not altered. Applying the inverse similarity transformation $\mathscr{S}^{-1}$ to Eq. (12.12) yields

$$\kappa_1 = \frac{v^2}{\delta^2} - \frac{2\alpha}{\gamma r} = \frac{v^2 \alpha}{\mu \gamma (1 - \xi)} - \frac{2\alpha}{\gamma r} = \frac{\alpha}{\mu \gamma (1 - \xi)} \left[v^2 - \frac{2\mu}{r}(1 - \xi) \right]$$

and results in

$$K_1 = \delta^2 \kappa_1 = v^2 - \frac{2\mu}{r}(1 - \xi) = 2\left(\frac{v^2}{2} - \frac{\mu^*}{r} \right)$$

Since $\delta^2 > 0$ no matter the values of ξ or γ, the sign of K_1 is not affected by the transformation $\mathscr{S}$. If the solution to the original problem (12.1) is elliptic, the solution to the reduced problem (12.5) will be elliptic too, and vice-versa. The transformation reduces to a series of scaling factors affecting each variable independently.

The integral of the angular momentum transforms into

$$\kappa_2 = \frac{r^2 v^{\gamma - 1} \dot{\theta}}{\alpha \delta^{\gamma}} = \frac{K_2}{\alpha \delta^{\gamma}} \implies K_2 = \alpha \delta^{\gamma} \kappa_2$$

The constant κ_2 remains positive, although the scaling factor $\alpha \delta^{\gamma}$ can modify its value significantly.

The transformation is defined in terms of three parameters: α, ξ, and γ. For $\xi = \xi_\gamma$, with

$$\xi_\gamma = 1 - \frac{\alpha}{\gamma}$$

$\mathscr{S}$ reduces to the identity map. Choosing $\alpha = 1$ the special values of ξ that yield trivial transformations for $\gamma = 1, 2, 3$, and 4 are, respectively, $\xi_\gamma = 0, 1/2, 2/3$, and $3/4$. The similarity transformation can be understood from a different approach: solving the simplified problem is equivalent to solving the full problem but setting $\xi = \xi_\gamma$.

The polar angle relates to the radius by the differential equation

$$\frac{d\theta}{d\rho} = \frac{\tan\psi}{\rho} \tag{12.15}$$

The right-hand side of this equation can be written as a function of ρ alone thanks to

$$\tan\psi = \frac{n\kappa_2}{\sqrt{(\tilde{v}^\gamma\rho)^2 - \kappa_2^2}}$$

The parameter n is $n = +1$ for orbits in a raising regime ($\dot{r} > 0$), and $n = -1$ for a lowering regime ($\dot{r} < 0$). The velocity is solved from Eq. (12.12). Integrating Eq. (12.15) furnishes the solution $\theta(\rho)$, which can then be inverted to define the trajectory $\rho(\theta)$.

12.2.4 Solvability

The trajectory of the particle is obtained upon integration and inversion of Eq. (12.15). This equation can be written

$$\frac{d\theta}{d\rho} = \frac{n\kappa_2}{\sqrt{P_{\mathrm{sol}}(\rho)}} \tag{12.16}$$

where $P_{\mathrm{sol}}(\rho)$ is a polynomial in ρ, in particular

$$P_{\mathrm{sol}}(\rho) = \rho^2 P_{\mathrm{nat}}(\rho)$$

The roots of $P_{\mathrm{sol}}(\rho)$ determine the *form* of the solution and coincide with those of $P_{\mathrm{nat}}(\rho)$ (obviating trivial solutions). The integration of Eq. (12.16) depends on the factorization of $P_{\mathrm{sol}}(\rho)$. This polynomial expression can be expanded thanks to the binomial theorem:

$$P_{\mathrm{sol}}(\rho) = \sum_{k=0}^{\gamma} \binom{\gamma}{k} \frac{2^k}{\gamma^k} \rho^{4-k} \kappa_1^{\gamma-k} - \rho^2\kappa_2$$

with $\gamma \neq 0$ an integer. For $\gamma \leq 4$ the polynomial is of degree four: when $\gamma = 1$ or $\gamma = 2$ there are two trivial roots and Eq. (12.16) can be integrated using elementary functions; when $\gamma = 3$ or $\gamma = 4$ it yields elliptic integrals. Negative values of γ or positive values greater than four lead to a polynomial $P_{\mathrm{sol}}(\rho)$ with degree five or above. The solution can no longer be reduced to elementary functions or elliptic integrals

(Whittaker and Watson, 1927, p. 512), for it is given by hyperelliptic integrals. This special class of Abelian integrals can only be inverted in very specific situations (see Byrd and Friedman, 1954, pp. 252–271). Thus, we shall focus on the solutions to the cases $\gamma = 1, 2, 3$, and 4.

The following sections 12.3–12.6 present the corresponding families of orbits. For $\gamma = 1$ the solutions to the reduced problem are Keplerian orbits. For the case $\gamma = 2$ the solutions are generalized logarithmic spirals. The cases $\gamma = 3$ and $\gamma = 4$ yield two new families of orbits, *generalized cardioids* and *generalized sinusoidal spirals*, respectively (for $\kappa_1 = 0$ the orbits are cardioids and sinusoidal spirals).

12.3 Case $\gamma = 1$: Conic sections

For $\gamma = 1$ the integrals of motion (12.12) and (12.13) reduce to the equations of the energy and angular momentum, respectively. The condition on the radius given by Eq. (12.14) becomes

$$P_{\text{nat}}(\rho) \equiv \kappa_1\rho^2 + 2\rho - \kappa_2^2 \geq 0 \tag{12.17}$$

For the case $\kappa_1 < 0$ (elliptic solution) this translates into $\rho \in [\rho_{\min}, \rho_{\max}]$, where

$$\rho_{\min} = \frac{1}{(-\kappa_1)}\left(1 - \sqrt{1 + \kappa_1\kappa_2^2}\right)$$
$$\rho_{\max} = \frac{1}{(-\kappa_1)}\left(1 + \sqrt{1 + \kappa_1\kappa_2^2}\right)$$

These limits are none other than the periapsis and apoapsis radii, provided that κ_1 and κ_2 relate to the semimajor axis and eccentricity by means of:

$$\frac{1}{(-\kappa_1)} = \tilde{a} \qquad \text{and} \qquad \sqrt{1 + \kappa_1\kappa_2^2} = \tilde{e}$$

For $\kappa_1 = 0$ (parabolic case) the semimajor axis becomes infinite, and Eq. (12.17) has only one root corresponding to $\rho = \kappa_2^2/2$. Note that it must be $\kappa_2^2 > -1/\kappa_1$. Similarly, for hyperbolic orbits ($\kappa_1 > 0$) it is $\rho \geq \rho_{\min}$ as $\rho_{\max}$ becomes negative.

Thus, the solution is simply a conic section:

$$\rho(\theta) = \frac{\tilde{h}^2}{1 + \tilde{e}\cos(\theta - \theta_m)} = \frac{\kappa_2^2}{1 + \sqrt{1 + \kappa_1\kappa_2^2}\,\cos(\theta - \theta_m)} \tag{12.18}$$

The angle $\theta_m = \Omega + \omega$ defines the direction of the line of apses in the inertial frame, meaning that $\theta - \theta_m$ is the true anomaly. If $\kappa_1 < 0$ then $\rho(\theta_m) = \rho_{\min}$, and $\rho(\theta_m + \pi) = \rho_{\max}$. The velocity $\tilde{v}$ follows from the integral of the energy:

$$\tilde{v} = \sqrt{\kappa_1 + 2/\rho}$$

It is minimum at apoapsis and maximum at periapsis.

Applying the similarity transformation $\mathscr{S}^{-1}$ to the previous solution leads to the extended integral

$$\frac{K_1}{2} = \delta^2 \frac{\kappa_1}{2} = \frac{v^2}{2} - \frac{\mu}{r}(1 - \xi) = \frac{v^2}{2} - \frac{\mu^*}{r}$$

The factor μ^* is the modified gravitational parameter. These kinds of solutions arise from, for example, the effect of the solar radiation pressure directed along the Sun-line on a particle following a heliocentric orbit (McInnes, 2004, p. 121).

12.4 Case $\gamma = 2$: Generalized logarithmic spirals

The case $\gamma = 2$ yields the family of generalized logarithmic spirals. These orbits have been discussed extensively in Chaps. 9–11. The corresponding trajectories, properties, and applications can be found in the referenced chapters.

12.5 Case $\gamma = 3$: Generalized cardioids

The condition in Eq. (12.14), yields the polynomial inequality:

$$P_{\text{nat}}(\rho) \equiv \kappa_1^3 \rho^3 + 2\kappa_1^2 \rho^2 + \left(\frac{4\kappa_1}{3} - \kappa_2^2\right)\rho + \frac{8}{27} \geq 0 \tag{12.19}$$

The discriminant Δ of the polynomial $P_{\text{nat}}(\rho)$ predicts the nature of the roots. It is

$$\Delta = -4\kappa_1^3\kappa_2^4\left(3\kappa_1 - \kappa_2^2\right) \tag{12.20}$$

The intermediate value theorem shows that there is at least one real root. For the elliptic case ($\kappa_1 < 0$) it is $\Delta < 0$, meaning that the other two roots are complex conjugates. In the hyperbolic case ($\kappa_1 > 0$) the sign of the discriminant depends on the values of κ_2: if $\kappa_2^2 > 3\kappa_1$ it is $\Delta > 0$, and for $\kappa_2^2 < 3\kappa_1$ the discriminant is negative. This behavior yields two types of hyperbolic solutions.

12.5.1 Elliptic motion

The nature of elliptic motion is determined by the polynomial constraint in Eq. (12.19). The real root is given by

$$\rho_1 = \frac{\Lambda\left(\Lambda + 2\kappa_1^2\right) + 3\kappa_1^3\kappa_2^2}{3(-\kappa_1)^3\Lambda} \tag{12.21}$$

with

$$\Lambda = (-\kappa_1)\left\{3(-\kappa_1)\kappa_2^2\left[\sqrt{-3\kappa_1\left(\kappa_2^2 - 3\kappa_1\right)} + 3\kappa_1\right]\right\}^{1/3} \tag{12.22}$$

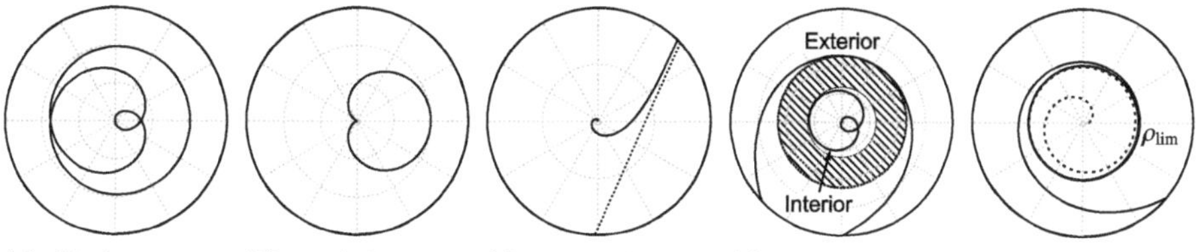

Fig. 12.1: Examples of generalized cardioids ($\gamma = 3$).

Equation (12.19) reduces to $\rho - \rho_1 \leq 0$ and we shall write

$$\rho_{\max} \equiv \rho_1$$

Thus, elliptic generalized cardioids never escape to infinity because they are bounded by $\rho_{\max}$. When in the raising regime they reach the apoapsis radius $\rho_{\max}$, then transition to the lowering regime and fall toward the origin. The velocity at apoapsis,

$$\tilde{v}_m = \sqrt{\kappa_1 + 2/(3\rho_{\max})}$$

is the minimum velocity in the cardioid.

Equation (12.15) can be integrated from the initial radius r_0 to the apoapsis of the cardioid, and the result provides the orientation of the line of apses:

$$\theta_m = \theta_0 + \frac{n\kappa_2}{\sqrt{AB}(-\kappa_1)^{3/2}} [2\,\mathrm{K}(k) - \mathrm{F}(\phi_0, k)]$$

where $\mathrm{K}(k)$ and $\mathrm{F}(\phi_0, k)$ are the complete and incomplete elliptic integrals of the first kind, respectively. Introducing the auxiliary term

$$\lambda_{ij} = \rho_i - \rho_j$$

their argument and modulus read

$$\phi_0 = \arccos\left[\frac{B\lambda_{10} - A\rho_0}{B\lambda_{10} + A\rho_0}\right], \quad k = \sqrt{\frac{\rho_{\max}^2 - (A - B)^2}{4AB}}$$

The previous definitions involve the auxiliary parameters:

$$A = \sqrt{(\rho_{\max} - b_1)^2 + a_1^2}, \qquad B = \sqrt{b_1^2 + a_1^2}$$

and

$$b_1 = \frac{\Lambda\left(\Lambda - 4\kappa_1^2\right) + 3\kappa_1^3\kappa_2^2}{6\kappa_1^3\Lambda}, \qquad a_1^2 = \frac{\left(\Lambda^2 - 3\kappa_1^3\kappa_2^2\right)^2}{12\kappa_1^6\Lambda^2}$$

Recall the definition of Λ in Eq. (12.22).

The equation of the trajectory is obtained by inverting the function $\theta(\rho)$, and results in:

$$\frac{\rho(\theta)}{\rho_{\max}} = \left\{1 + \frac{A}{B}\left[\frac{1 - \mathrm{cn}(v, k)}{1 + \mathrm{cn}(v, k)}\right]\right\}^{-1} \tag{12.23}$$

It is defined in terms of the Jacobi elliptic function $\mathrm{cn}(v, k)$. The anomaly v reads

$$v(\theta) = \frac{(-\kappa_1)^{3/2}}{\kappa_2}\sqrt{AB}\,(\theta - \theta_m)$$

Equation (12.23) is symmetric with respect to the apse line, $\rho(\theta_m + \Delta\theta) = \rho(\theta_m - \Delta\theta)$, as shown in Fig. 12.1a.

12.5.2 Parabolic motion: The cardioid

When the constant of the generalized energy κ_1 vanishes the condition in Eq. (12.14) translates into $\rho < \rho_{\max}$, where the maximum radius $\rho_{\max}$ takes the form

$$\rho_{\max} = \frac{8}{27\kappa_2^2}$$

Parabolic generalized cardioids, unlike logarithmic spirals or Keplerian parabolas, are bounded (they never escape the gravitational attraction of the central body).

The line of apses is defined by:

$$\theta_m = \theta_0 + n\left[\frac{\pi}{2} + \arcsin\left(1 - \frac{27}{4}\kappa_2^2\rho_0\right)\right]$$

The equation of the trajectory reveals that the orbit is in fact a pure cardioid:[2]

$$\frac{\rho(\theta)}{\rho_{\max}} = \frac{1}{2}[1 + \cos(\theta - \theta_m)] \tag{12.24}$$

This curve is symmetric with respect to θ_m. Figure 12.1b depicts the geometry of the solution.

12.5.3 Hyperbolic motion

The inequality in Eq. (12.14) determines the structure of the solutions and the sign of the discriminant (12.20) governs the nature of its roots. There are two types of hyperbolic generalized cardioids: for $\kappa_2 < \sqrt{3\kappa_1}$ the cardioids are of Type I, and for $\kappa_2 > \sqrt{3\kappa_1}$ the cardioids are of Type II.

---

2 The cardioid is a particular case of the limaçons, curves first studied by the amateur mathematician Étienne Pascal in the 17th century. The general form of the limaçon in polar coordinates is $r(\theta) = a + b\cos\theta$. Depending on the values of the coefficients the curve might reach the origin and form loops. It is worth noting that the inverse of a limaçon, $r(\theta) = (a + b\cos\theta)^{-1}$, results in a conic section. Limaçons with $a = b$ are considered part of the family of sinusoidal spirals.

12.5.3.1 Hyperbolic cardioids of Type I

For hyperbolic cardioids of Type I there is only one real root. The other two are complex conjugates. The real root is

$$\rho_3 = -\frac{\Lambda\kappa_1^2(2 + \Lambda\kappa_1) + 3\kappa_2^2}{3\kappa_1^3\Lambda} \tag{12.25}$$

having introduced the auxiliary parameter

$$\Lambda = \left\{\frac{3\kappa_2^2}{\kappa_1^{9/2}}\left[3\sqrt{\kappa_1} + \sqrt{3\left(3\kappa_1 - \kappa_2^2\right)}\right]\right\}^{1/3}$$

Provided that $\Lambda > 0$, Eq. (12.25) shows that $\rho_3 < 0$. Therefore, there are no limits to the values that ρ can take. As a consequence, the cardioid never transitions between regimes. If it is initially $\psi_0 < \pi/2$, it will always escape to infinity, and fall toward the origin for $\psi_0 > \pi/2$.

The equation of the trajectory for hyperbolic generalized cardioids of Type I is

$$\frac{\rho(\theta)}{\rho_3 A} = \frac{(A + B)\,\mathrm{sn}^2(v, k) + 2B[\mathrm{cn}(v, k) - 1]}{(A + B)^2\,\mathrm{sn}^2(v, k) - 4AB} \tag{12.26}$$

It is defined in terms of

$$A = \sqrt{b_1^2 + a_1^2}\,, \qquad B = \sqrt{(\rho_3 - b_1)^2 + a_1^2}$$

which require

$$b_1 = \frac{\Lambda\kappa_1^2(\Lambda\kappa_1 - 4) + 3\kappa_2^2}{6\kappa_1^3\Lambda}\,, \qquad a_1^2 = \frac{\left(\Lambda^2\kappa_1^3 - 3\kappa_2^2\right)^2}{6\Lambda^2\kappa_1^6}$$

There are no axes of symmetry. Therefore, the anomaly is referred directly to the initial conditions:

$$v(\theta) = \frac{\kappa_1^{3/2}}{\kappa_2}\sqrt{AB}\,(\theta - \theta_0) + \mathrm{F}(\phi_0, k)$$

The moduli of both the Jacobi elliptic functions and the elliptic integral, and the argument of the latter, are

$$k = \sqrt{\frac{(A + B)^2 - \rho_3^2}{4AB}}\,, \qquad \phi_0 = \arccos\left[\frac{A\lambda_{03} - B\rho_0}{A\lambda_{03} + B\rho_0}\right]$$

The cardioid approaches infinity along an asymptotic branch, with $\tilde{v} \to \tilde{v}_\infty$ as $\rho \to \infty$. The orientation of the asymptote follows from the limit

$$\theta_{\mathrm{as}} = \lim_{\rho\to\infty}\theta(\rho) = \theta_0 + \frac{n\kappa_2}{\kappa_1^{3/2}\sqrt{AB}}\left[\mathrm{F}(\phi_\infty, k) - \mathrm{F}(\phi_0, k)\right]$$

Here, the value of ϕ_∞,

$$\phi_\infty = \arccos\left(\frac{A - B}{A + B}\right)$$

is defined as $\phi_\infty = \lim_{\rho\to\infty}\phi$. An example of a hyperbolic cardioid of Type I with its corresponding asymptote is presented in Fig. 12.1c.

12.5.3.2 Hyperbolic cardioids of Type II

For hyperbolic cardioids of Type II the polynomial in Eq. (12.14) admits three distinct real roots, $\{\rho_1, \rho_2, \rho_3\}$, given by

$$\rho_{k+1} = \frac{2\kappa_2}{\sqrt{3\kappa_1^3}} \cos\left[\frac{\pi}{3}(2k+1) - \frac{1}{3}\arccos\left(\frac{\sqrt{3\kappa_1}}{\kappa_2}\right)\right] - \frac{2}{3\kappa_1}$$

with $k = 0, 1, 2$. The roots are then sorted so that $\rho_1 > \rho_2 > \rho_3$. Since

$$\rho_1\rho_2\rho_3 = -\frac{8}{27\kappa_1^3} < 0$$

then $\rho_3 < 0$ and also $\rho_1 > \rho_2 > 0$ for physical coherence. The polynomial constraint reads

$$(\rho - \rho_1)(\rho - \rho_2)(\rho - \rho_3) \geq 0$$

and holds for both $\rho > \rho_1$ and $\rho < \rho_2$. The integral of motion (12.12) shows that both situations are physically admissible, because $\tilde{v}^2(\rho_1) > 0$ and $\tilde{v}^2(\rho_2) > 0$. There are two families of solutions that lie outside the annulus $\rho \notin (\rho_2, \rho_1)$. When $\rho_0 < \rho_2$ the spirals are *interior*, whereas for $\rho > \rho_1$ they are *exterior* spirals. The geometry of the forbidden region can be analyzed in Fig. 12.1d. The particle cannot enter the barred annulus, whose limits coincide with the periapsis and apoapsis of the exterior and interior orbits, respectively.

The axis of symmetry of interior spirals is given by

$$\theta_m = \theta_0 + \frac{2n\kappa_2[\mathrm{K}(k) - \mathrm{F}(\phi_0, k)]}{\kappa_1^{3/2}\sqrt{\rho_1\lambda_{23}}}$$

in terms of the arguments:

$$\phi_0 = \arcsin\sqrt{\frac{\rho_0\lambda_{23}}{\rho_2\lambda_{03}}}, \qquad k = \sqrt{\frac{\rho_2\lambda_{13}}{\rho_1\lambda_{23}}}$$

The trajectory simplifies to:

$$\frac{\rho(\theta)}{\rho_3} = \frac{1}{1 + (\lambda_{32}/\rho_2)\,\mathrm{dc}^2(v, k)} \tag{12.27}$$

Here we made use of Glaisher's notation for the Jacobi elliptic functions,[3] so $\mathrm{dc}(v, k) = \mathrm{dn}(v, k)/\,\mathrm{cn}(v, k)$. The spiral anomaly v takes the form

$$v(\theta) = \frac{\kappa_1^{3/2}\sqrt{\rho_1\lambda_{23}}}{2\kappa_2}(\theta - \theta_m)$$

---

3 Glaisher's notation establishes that if p,q,r are any of the four letters s,c,d,n, then:

$$\mathrm{pq}(v, k) = \frac{\mathrm{pr}(v, k)}{\mathrm{qr}(v, k)} = \frac{1}{\mathrm{qp}(v, k)}$$

Under this notation repeated letters yield unity. See Appendix E for details.

The trajectory is symmetric with respect to the line of apses defined by θ_m.

For the case of exterior spirals, the largest root ρ_1 behaves as the periapsis. A cardioid initially in the lowering regime will reach ρ_1, then it will transition to the raising regime and escape to infinity. Equation (12.15) is integrated from the initial radius to the periapsis to provide the orientation of the line of apses:

$$\theta_m = \theta_0 - \frac{2n\kappa_2\,\mathrm{F}(\phi_0, k)}{\kappa_1^{3/2}\sqrt{\rho_1\lambda_{23}}} \tag{12.28}$$

with

$$\phi_0 = \arcsin\sqrt{\frac{\lambda_{23}\lambda_{01}}{\lambda_{13}\lambda_{02}}}, \qquad k = \sqrt{\frac{\rho_2\lambda_{13}}{\rho_1\lambda_{23}}}$$

the argument and modulus of the elliptic integral.

The trajectory of exterior spirals is obtained upon inversion of the equation for the polar angle,

$$\rho(\theta) = \frac{\rho_2\lambda_{13}\,\mathrm{sn}^2(v, k) - \rho_1\lambda_{23}}{\lambda_{13}\,\mathrm{sn}^2(v, k) - \lambda_{23}} \tag{12.29}$$

The anomaly is redefined as

$$v(\theta) = \frac{\kappa_1^{3/2}\sqrt{\rho_1\lambda_{23}}}{2\kappa_2}(\theta - \theta_m)$$

This variable is referred to the line of apses, given in Eq. (12.28). The form of the solution shows that hyperbolic generalized cardioids of Type II are symmetric with respect to θ_m.

Because of the symmetry of Eq. (12.29) the trajectory exhibits two symmetric asymptotes, defined by

$$\theta_{\mathrm{as}} = \theta_m \pm \frac{2\kappa_2\,\mathrm{F}(\phi_\infty, k)}{\kappa_1^{3/2}\sqrt{\rho_1\lambda_{23}}}$$

The argument ϕ_∞ reads

$$\phi_\infty = \arcsin\sqrt{\frac{\lambda_{23}}{\lambda_{13}}}$$

12.5.3.3 Transition between Type I and Type II hyperbolic cardioids

The limit case $\kappa_2 = \sqrt{3\kappa_1}$ defines the transition between hyperbolic cardioids of Types I and II. The discriminant Δ vanishes: the roots are all real and one is a multiple root, $\rho_{\mathrm{lim}} \equiv \rho_1 = \rho_2$. The region of forbidden motion degenerates into a circumference of radius ρ_{lim}. The roots take the form:

$$\rho_3 = -\frac{8}{3\kappa_1} < 0 \qquad \text{and} \qquad \rho_{\mathrm{lim}} \equiv \rho_1 = \rho_2 = \frac{1}{3\kappa_1} \tag{12.30}$$

The condition $(\rho - \rho_3)(\rho - \rho_{\mathrm{lim}})^2 \geq 0$ holds naturally for $\rho > 0$. When the cardioid reaches ρ_{lim} the velocity becomes $\tilde{v}_{\mathrm{lim}} = \sqrt{3\kappa_1} = \rho_{\mathrm{lim}}^{-1/2}$. It coincides with the local circular velocity. Moreover, from the integral of motion (12.13) one has

$$\kappa_2 = \rho_{\text{lim}}\tilde{v}_{\text{lim}}^3 \sin\psi_{\text{lim}} \implies \sin\psi_{\text{lim}} = 1$$

meaning that the orbit becomes circular as the particle approaches ρ_{lim}. When $\psi_{\text{lim}} = \pi/2$ the perturbing acceleration in Eq. (12.6) vanishes. As a result, the orbit degenerates into a circular Keplerian orbit. A cardioid with $\rho_0 < \rho_{\text{lim}}$ and in the raising regime will reach ρ_{lim} and degenerate into a circular orbit with radius ρ_{lim}. This phenomenon also appears in cardioids with $\rho_0 > \rho_{\text{lim}}$ and in the lowering regime.

The trajectory reduces to

$$\frac{\rho(\theta)}{\rho_{\text{lim}}} = \frac{\cosh\nu - 1}{\cosh\nu + 5/4} \tag{12.31}$$

which is written in terms of the anomaly

$$\nu(\theta) = \frac{\theta - \theta_0}{\sqrt{3}} + 2mn\,\text{arctanh}\left(3\sqrt{\frac{\rho_0}{8\rho_{\text{lim}} + \rho_0}}\right)$$

The integer $m = \text{sign}(1 - \rho_0/\rho_{\text{lim}})$ determines whether the particle is initially below $(m = +1)$ or above $(m = -1)$ the limit radius ρ_{lim}. The limit $\lim_{\theta\to\infty}\rho = \rho_{\text{lim}} = 1/(3\kappa_1)$ shows that the radius converges to ρ_{lim}. This limit only applies to the cases $m = n = +1$ and $m = n = -1$. When the particle is initially below ρ_{lim} and in the lowering regime, $\{m = +1, n = -1\}$, it falls toward the origin. In the opposite case, $\{m = -1, n = +1\}$, the cardioid approaches infinity along an asymptotic branch with

$$\theta_{\text{as}} = \theta_0 - 2\sqrt{3}\left[mn\,\text{arctanh}\left(3\sqrt{\frac{\rho_0}{8\rho_{\text{lim}} + \rho_0}}\right) + \text{arctanh}(3)\right]$$

Two example trajectories with $n = +1$ are plotted in Fig. 12.1e. The dashed line corresponds to $m = +1$ and the solid line to $m = -1$. The trajectories terminate/emanate from a circular orbit of radius ρ_{lim}.

12.6 Case $\gamma = 4$: Generalized sinusoidal spirals

Setting $\gamma = 4$ in Eq. (12.14) gives rise to the polynomial inequality

$$\left[4\left(\kappa_1^2\rho + \kappa_1 + \kappa_2\right)\rho + 1\right]\left[4\left(\kappa_1^2\rho + \kappa_1 - \kappa_2\right)\rho + 1\right] \geq 0 \tag{12.32}$$

which governs the subfamilies of solutions to the problem. The four roots of the polynomial are

$$\rho_{1,2} = +\frac{\kappa_2 - \kappa_1 \pm \sqrt{\kappa_2(\kappa_2 - 2\kappa_1)}}{2\kappa_1^2}$$

$$\rho_{3,4} = -\frac{\kappa_1 + \kappa_2 \mp \sqrt{\kappa_2(\kappa_2 + 2\kappa_1)}}{2\kappa_1^2}$$

and the discriminant of $P_{\text{nat}}(\rho)$ is

$$\Delta = \frac{\kappa_2^6}{\kappa_1^{20}}\left(\kappa_2^2 - 4\kappa_1^2\right) \tag{12.33}$$

The sign of the discriminant determines the nature of the four roots.

12.6.1 Elliptic motion

When $\kappa_1 < 0$ there are two subfamilies of elliptic sinusoidal spirals: Type I, with $\kappa_2 > -2\kappa_1$ ($\Delta > 0$), and Type II, with $\kappa_2 < -2\kappa_1$ ($\Delta < 0$). Both types are separated by the limit case $\kappa_2 = -2\kappa_1$ that makes $\Delta = 0$.

12.6.1.1 Elliptic sinusoidal spirals of Type I

For the case $\kappa_2 > -2\kappa_1$ the discriminant is positive and the four roots are real, with $\rho_1 > \rho_2 > \rho_3 > \rho_4$. Since $\rho_{3,4} < 0$, Eq. (12.32) reduces to

$$(\rho - \rho_1)(\rho - \rho_2) \geq 0\,, \tag{12.34}$$

meaning that it must be either $\rho > \rho_1$ or $\rho < \rho_2$. The integral of motion (12.12) reveals that only the latter case is physically possible, because $\tilde{v}^2(\rho_1) < 0$. Thus, ρ_2 is the apoapsis of the spiral:

$$\rho_{\max} \equiv \rho_2 = \frac{\kappa_2 - \kappa_1 - \sqrt{\kappa_2(\kappa_2 - 2\kappa_1)}}{2\kappa_1^2}$$

and $\rho \leq \rho_{\max}$. Equation (12.16) is then integrated from ρ_0 to $\rho_{\max}$ to define the orientation of the apoapsis,

$$\theta_m = \theta_0 - \frac{2n\kappa_2[\mathrm{F}(\phi_0, k) - \mathrm{K}(k)]}{\kappa_1^2\sqrt{\lambda_{13}\lambda_{24}}}$$

The arguments of the elliptic integrals are

$$\phi_0 = \arcsin\sqrt{\frac{\lambda_{24}\lambda_{03}}{\lambda_{23}\lambda_{04}}}\,, \qquad k = \sqrt{\frac{\lambda_{23}\lambda_{14}}{\lambda_{13}\lambda_{24}}}$$

Recall that $\lambda_{ij} = \rho_i - \rho_j$.

Elliptic sinusoidal spirals of Type I are defined by

$$\rho(\theta) = \frac{\rho_4\lambda_{23}\,\mathrm{cd}^2(v, k) - \rho_3\lambda_{24}}{\lambda_{23}\,\mathrm{cd}^2(v, k) - \lambda_{24}} \tag{12.35}$$

The spiral anomaly reads

$$v(\theta) = \frac{n\kappa_1^2}{2\kappa_2}(\theta - \theta_m)\sqrt{\lambda_{13}\lambda_{24}}$$

The trajectory is symmetric with respect to θ_m, which corresponds to the line of apses.

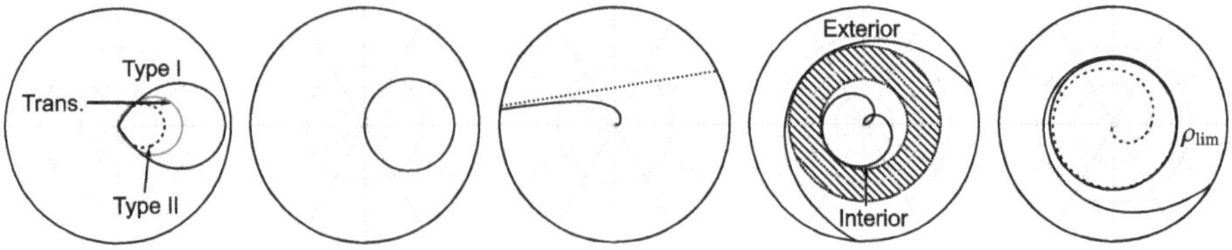

(a) Elliptic **(b)** Parabolic **(c)** Hyperbolic Type I **(d)** Hyperbolic Type II **(e)** Hyperbolic transition

Fig. 12.2: Examples of generalized sinusoidal spirals ($\gamma = 4$).

12.6.1.2 Elliptic sinusoidal spirals of Type II

When $\kappa_2 < -2\kappa_1$, two roots are real and the other two are complex conjugate. In this case the real roots are $\rho_{1,2}$ and the inequality (12.32) reduces again to Eq. (12.34), meaning that $\rho \leq \rho_2 \equiv \rho_{\max}$. However, the form of the solution is different from the trajectory described by Eq. (12.35), being

$$\rho(\theta) = \frac{\rho_1 B - \rho_2 A - (\rho_2 A + \rho_1 B)\,\mathrm{cn}(\nu, k)}{B - A - (A + B)\,\mathrm{cn}(\nu, k)} \tag{12.36}$$

The spiral anomaly is

$$\nu(\theta) = \frac{n\kappa_1^2}{\kappa_2}\sqrt{AB}\,(\theta - \theta_m)$$

It is referred to the orientation of the apse line, θ_m. This variable is given by

$$\theta_m = \theta_0 + \frac{n\kappa_2}{\kappa_1^2\sqrt{AB}}[2\,\mathrm{K}(k) - \mathrm{F}(\phi_0, k)]$$

considering the arguments:

$$\phi_0 = \arccos\left[\frac{A\lambda_{02} + B\lambda_{10}}{A\lambda_{02} - B\lambda_{10}}\right], \qquad k = \sqrt{\frac{(A + B)^2 - \lambda_{12}^2}{4AB}}$$

The coefficients A and B are defined in terms of

$$a_1^2 = -\frac{\kappa_2(\kappa_2 + 2\kappa_1)}{4\kappa_1^4} \quad \text{and} \quad b_1 = -\frac{\kappa_1 + \kappa_2}{2\kappa_1^2}$$

namely

$$A = \sqrt{(\rho_1 - b_1)^2 + a_1^2}, \quad B = \sqrt{(\rho_2 - b_1)^2 + a_1^2}$$

The fact that the Jacobi function $\mathrm{cn}(\nu, k)$ is symmetric proves that elliptic sinusoidal spirals of Type II are symmetric.

12.6.1.3 Transition between spirals of Types I and II

In this particular case of elliptic motion, $-2\kappa_1 = \kappa_2$, the roots of polynomial $P_{\mathrm{nat}}(\rho)$ are

$$\rho_1 = \frac{3 + 2\sqrt{2}}{\kappa_2}, \quad \rho_2 \equiv \rho_{\max} = \frac{3 - 2\sqrt{2}}{\kappa_2}, \quad \rho_{3,4} = -\frac{1}{\kappa_2}$$

These results simplify the definition of the line of apses to

$$\theta_m = \theta_0 + \sqrt{2}\, n \ln \left[\frac{\sqrt{2}(1 - \rho_0\kappa_2) + \sqrt{(3 - \rho_0\kappa_2)^2 - 8}}{1 + \rho_0\kappa_2} \right]$$

Introducing the spiral anomaly $\nu(\theta)$,

$$\nu(\theta) = \frac{\sqrt{2}}{2}(\theta - \theta_m)$$

the equation of the trajectory takes the form

$$\frac{\rho(\theta)}{\rho_{\max}} = \frac{5 - 4\sqrt{2}\cosh\nu + \cosh(2\nu)}{(3 - 2\sqrt{2})[3 - \cosh(2\nu)]} \tag{12.37}$$

The trajectory is symmetric with respect to the line of apses θ_m thanks to the symmetry of the hyperbolic cosine.

Figure 12.2a shows the three types of elliptic spirals. It is important to note that in all three cases the condition in Eq. (12.32) transforms into Eq. (12.34), equivalent to $\rho < \rho_{\max}$. As a result, there are no differences in their nature although the equations for the trajectory are different.

12.6.2 Parabolic motion: Sinusoidal spiral (off-center circle)

Making $\kappa_1 = 0$ the condition in Eq. (12.32) simplifies to

$$\rho \le \rho_{\max} = \frac{1}{4\kappa_2}$$

meaning that the spiral is bounded by a maximum radius $\rho_{\max}$. It is equivalent to the apoapsis of the spiral. Its orientation is given by

$$\theta_m = \theta_0 + n\left[\frac{\pi}{2} - \arcsin\left(\frac{\rho_0}{\rho_{\max}}\right)\right]$$

The trajectory reduces to a sinusoidal spiral,[4] and its definition can be directly related to θ_m:

$$\frac{\rho(\theta)}{\rho_{\max}} = \cos(\theta - \theta_m) \tag{12.38}$$

The spiral defined in Eq. (12.38) is symmetric with respect to the line of apses. The resulting orbit is a circle centered at $(\rho_{\max}/2, \theta_m)$ (Fig. 12.2b). Circles are indeed a special case of sinusoidal spirals.

4 It was the Scottish mathematician Colin Maclaurin the first to study sinusoidal spirals. In his "Tractatus de Curvarum Constructione & Mensura", published in *Philosophical Transactions* in 1717, he constructed this family of curves relying on the epicycloid. Their general form is $r^n = \cos(n\theta)$ and different values of n render different types of curves; $n = -2$ correspond to hyperbolas, $n = -1$ to straight lines, $n = -1/2$ to parabolas, $n = -1/3$ to Tschirnhausen cubics, $n = 1/3$ to Cayley's sextics, $n = 1/2$ to cardioids, $n = 1$ to circles, and $n = 2$ to lemniscates.

12.6.3 Hyperbolic motion

Given the discriminant in Eq. (12.33), for the case $\kappa_1 > 0$ the values of the constant κ_2 define two different types of hyperbolic sinusoidal spirals: spirals of Type I ($\kappa_2 < 2\kappa_1$), and spirals of Type II ($\kappa_2 > 2\kappa_1$).

12.6.3.1 Hyperbolic sinusoidal spirals of Type I

If $\kappa_2 < 2\kappa_1$, then $\rho_{1,2}$ are complex conjugates and $\rho_{3,4}$ are both real but negative. Therefore, the condition in Eq. (12.32) holds naturally for any radius and there are no limitations to the values of ρ. The particle can either fall to the origin or escape to infinity along an asymptotic branch. The equation of the trajectory is given by:

$$\rho(\theta) = \frac{\rho_3 B - \rho_4 A + (\rho_3 B + \rho_4 A)\,\mathrm{cn}(v, k)}{B - A + (A + B)\,\mathrm{cn}(v, k)} \tag{12.39}$$

where the spiral anomaly can be referred directly to the initial conditions:

$$v(\theta) = \frac{n\kappa_1^2}{\kappa_2}\sqrt{AB}\,(\theta - \theta_0) + \mathrm{F}(\phi_0, k)$$

This definition involves an elliptic integral of the first kind with argument and parameter:

$$\phi_0 = \arccos\left[\frac{A\lambda_{04} + B\lambda_{30}}{A\lambda_{04} - B\lambda_{30}}\right], \qquad k = \sqrt{\frac{(A + B)^2 - \lambda_{34}^2}{4AB}}$$

The coefficients A and B require the terms:

$$b_1 = -\frac{\kappa_1 + \kappa_2}{2\kappa_1^2}, \qquad a_1^2 = -\frac{\kappa_2(\kappa_2 + 2\kappa_1)}{4\kappa_1^4}$$

being

$$A = \sqrt{(\rho_3 - b_1)^2 + a_1^2}, \quad B = \sqrt{(\rho_4 - b_1)^2 + a_1^2}$$

The direction of the asymptote is defined by

$$\theta_{\mathrm{as}} = \theta_0 + \frac{n\kappa_2}{\kappa_1^2\sqrt{AB}}\left[\mathrm{F}(\phi_\infty, k) - \mathrm{F}(\phi_0, k)\right]$$

This definition involves the argument

$$\phi_\infty = \arccos\left(\frac{A - B}{A + B}\right)$$

The velocity of the particle when reaching infinity is $\tilde{v}_\infty = \sqrt{\kappa_1}$. Hyperbolic sinusoidal spirals of Type I are similar to the hyperbolic solutions of Type I with $\gamma = 2$ and $\gamma = 3$. Figure 12.2d depicts an example trajectory and the asymptote defined by θ_{as}.

12.6.3.2 Hyperbolic sinusoidal spirals of Type II

In this case the four roots are real and distinct, with $\rho_{3,4} < 0$. The two positive roots $\rho_{1,2}$ are physically valid, i.e., $\tilde{v}^2(\rho_1) > 0$ and $\tilde{v}^2(\rho_2) > 0$. This yields two situations in which the condition $(\rho - \rho_1)(\rho - \rho_2) \geq 0$ is satisfied: $\rho > \rho_1$ (exterior spirals) and $\rho < \rho_2$ (interior spirals).

Interior spirals take the form

$$\rho(\theta) = \frac{\rho_2 \lambda_{13} - \rho_1 \lambda_{23} \, \mathrm{sn}^2(v, k)}{\lambda_{13} - \lambda_{23} \, \mathrm{sn}^2(v, k)} \tag{12.40}$$

The spiral anomaly is

$$v(\theta) = \frac{\kappa_1^2}{2\kappa_2} \sqrt{\lambda_{13}\lambda_{24}} \, (\theta - \theta_m)$$

The orientation of the line of apses is solved from

$$\theta_m = \theta_0 + \frac{2n\kappa_2 \, \mathrm{F}(\phi_0, k)}{\kappa_1^2 \sqrt{\lambda_{13}\lambda_{24}}} \tag{12.41}$$

with

$$\phi_0 = \arcsin \sqrt{\frac{\lambda_{13}\lambda_{20}}{\lambda_{23}\lambda_{10}}}, \qquad k = \sqrt{\frac{\lambda_{23}\lambda_{14}}{\lambda_{13}\lambda_{24}}}$$

Interior hyperbolic spirals are bounded and their shape is similar to that of a limaçon.

The line of apses of an exterior spiral is defined by

$$\theta_m = \theta_0 - \frac{2n\kappa_2 \, \mathrm{F}(\phi_0, k)}{\kappa_1^2 \sqrt{\lambda_{13}\lambda_{24}}}$$

The modulus and the argument of the elliptic integral are

$$\phi_0 = \arcsin \sqrt{\frac{\lambda_{24}\lambda_{01}}{\lambda_{14}\lambda_{02}}}, \qquad k = \sqrt{\frac{\lambda_{23}\lambda_{14}}{\lambda_{13}\lambda_{24}}}$$

The trajectory becomes

$$\rho(\theta) = \frac{\rho_1 \lambda_{24} + \rho_2 \lambda_{41} \, \mathrm{sn}^2(v, k)}{\lambda_{24} + \lambda_{41} \, \mathrm{sn}^2(v, k)} \tag{12.42}$$

and it is symmetric with respect to θ_m. The geometry of the solution is similar to that of hyperbolic cardioids of Type II, mainly because of the existence of the forbidden region plotted in Fig. 12.2d. The asymptotes follow the direction of

$$\theta_{as} = \theta_m \pm \frac{2\kappa_2 \, \mathrm{F}(\phi_\infty, k)}{\kappa_1^2 \sqrt{\lambda_{13}\lambda_{24}}}, \quad \text{with} \quad \phi_\infty = \arcsin \sqrt{\frac{\lambda_{24}}{\lambda_{14}}}$$

12.6.3.3 Transition between Type I and Type II spirals

When $\kappa_2 = 2\kappa_1$ the radii ρ_1 and ρ_2 coincide, $\rho_1 = \rho_2 \equiv \rho_{\text{lim}}$, and become equal to the limit radius

$$\rho_{\text{lim}} = \frac{1}{2\kappa_1}$$

The equation of the trajectory is a particular case of Eq. (12.42), obtained with $\kappa_2 \to 2\kappa_1$:

$$\frac{\rho(\theta)}{\rho_{\text{lim}}} = \left[1 - \frac{8}{4 + m(\sinh \nu - 3\cosh \nu)}\right]^{-1} \tag{12.43}$$

The spiral anomaly can be referred to the initial conditions,

$$\nu(\theta) = \frac{nm}{\sqrt{2}}(\theta - \theta_0) + \ln\left[\frac{2(1 + 2\kappa_1\rho_0) + \sqrt{2 + 8\kappa_1\rho_0(3 + \kappa_1\rho_0)}}{m(1 - 2\kappa_1\rho_0)}\right]$$

avoiding additional parameters. Here $m = \text{sign}(1 - \rho_0/\rho_{\text{lim}})$ determines whether the spiral is below or over the limit radius ρ_{lim}. The asymptote follows from

$$\theta_{\text{as}} = \theta_0 - n\sqrt{2}\log(1 - \sqrt{2}/2)$$

Like in the case of hyperbolic cardioids, for $m = n = -1$ and $m = n = +1$ spirals of this type approach the circular orbit of radius ρ_{lim} asymptotically, i.e., $\lim_{\theta\to\infty} \rho = \rho_{\text{lim}}$. When approaching a circular orbit the perturbing acceleration vanishes and the spiral converges to a Keplerian orbit. See Fig. 12.2e for examples of hyperbolic sinusoidal spirals with $\{m = +1, n = +1\}$ (dashed) and $\{m = -1, n = +1\}$ (solid).

12.7 Summary

The solutions presented in the previous sections are summarized in Table 12.1, organized in terms of the values of γ. Each family is then divided into elliptic, parabolic, and hyperbolic orbits. The table includes references to the corresponding equations of the trajectories for convenience. The orbits are said to be bounded if the particle can never reach infinity, because $r < r_{\text{lim}}$.

12.8 Unified solution in Weierstrassian formalism

The orbits can be unified by introducing the Weierstrass elliptic functions. Indeed, Eq. (12.16) furnishes the integral expression

$$\theta(r) - \theta_0 = \int_{r_0}^{r} \frac{k\,\mathrm{d}s}{\sqrt{f(s)}} \tag{12.44}$$

with

$$f(s) \equiv P_{\text{sol}}(s) = a_0 s^4 + 4a_1 s^3 + 6a_2 s^2 + 4a_3 s + a_4 \tag{12.45}$$

Tab. 12.1: Summary of the families of solutions.

Family	Type	γ	κ_1	κ_2	Bounded	Trajectory
Conic sections	Elliptic	1	< 0	$> \sqrt{-1/\kappa_1}$	Y	Eq. (12.18)
	Parabolic	1	$= 0$	–	N	Eq. (12.18)
	Hyperb.	1	> 0	–	N	Eq. (12.18)
Generalized logarithmic spirals	Elliptic	2	< 0	< 1	Y	Eq. (9.30)
	Parabolic	2	$= 0$	≤ 1	N	Eq. (9.35)*
	Hyperbolic T-I	2	> 0	< 1	N	Eq. (9.43)
	Hyperb. T-II	2	> 0	> 1	N	Eq. (9.50)
	Hyperb. trans.	2	> 0	$= 1$	N	Eq. (9.57)
Generalized cardioids	Elliptic	3	< 0	< 1	Y	Eq. (12.23)
	Parabolic	3	$= 0$	≤ 1	Y	Eq. (12.24)†
	Hyperbolic T-I	3	> 0	$< \sqrt{3\kappa_1}$	N	Eq. (12.26)
	Hyperbolic T-II (int)	3	> 0	$> \sqrt{3\kappa_1}$	Y	Eq. (12.27)
	Hyperb. T-II (ext)	3	> 0	$> \sqrt{3\kappa_1}$	N	Eq. (12.29)
	Hyperb. trans.	3	> 0	$= \sqrt{3\kappa_1}$	Y/N	Eq. (12.31)
Generalized sinusoidal spirals	Elliptic T-I	4	< 0	$> -2\kappa_1$	Y	Eq. (12.35)
	Elliptic T-II	4	< 0	$< -2\kappa_1$	Y	Eq. (12.36)
	Elliptic trans.	4	< 0	$= -2\kappa_1$	Y	Eq. (12.37)
	Parabolic	4	$= 0$	–	Y	Eq. (12.38)‡
	Hyperb. T-I	4	> 0	$< 2\kappa_1$	N	Eq. (12.39)
	Hyperb. T-II (int)	4	> 0	$> 2\kappa_1$	Y	Eq. (12.40)
	Hyperb. T-II (ext)	4	> 0	$> 2\kappa_1$	N	Eq. (12.42)
	Hyperb. trans.	4	> 0	$= 2\kappa_1$	Y/N	Eq. (12.43)

* Logarithmic spiral.
† Cardioid.
‡ Sinusoidal spiral (off-center circle).

and $a_{0,1} \neq 0$. Introducing the auxiliary parameters

$$\vartheta = f'(r_0)/4 \qquad \text{and} \qquad \varphi = f''(r_0)/24$$

Eq. (12.44) can be inverted to provide the equation of the trajectory (see Appendix E and Whittaker and Watson, 1927, p. 454, for details),

$$r(\theta) - r_0 = \frac{1}{2[\wp(z) - \varphi]^2 - f(r_0)f^{(\mathrm{iv})}(r_0)/48} \times \left\{[\wp(z) - \varphi]2\vartheta - \wp'(z)\sqrt{f(r_0)} + f(r_0)f'''(r_0)/24\right\} \qquad (12.46)$$

The solution is written in terms of the Weierstrass elliptic function

$$\wp(z) \equiv \wp(z; g_2, g_3)$$

Tab. 12.2: Coefficients a_i of the polynomial $f(s)$ and factor k.

γ	a_0	a_1	a_2	a_3	a_4	k
1	κ_1	$1/2$	$-\kappa_2^2/6$	0	0	$n\kappa_2$
2	κ_1^2	$\kappa_1/2$	$\left(1-\kappa_2^2\right)/6$	0	0	$n\kappa_2$
3	$27\kappa_1^3$	$27\kappa_1^2/2$	$6\kappa_1 - 9\kappa_2^2/2$	2	0	$\sqrt{27}\, n\kappa_2$
4	$16\kappa_1^4$	$8\kappa_1^3$	$4\kappa_1^2 - 8\kappa_2^2/3$	$2\kappa_1$	1	$4n\kappa_2$

and its derivative $\wp'(z)$, where $z = (\theta - \theta_0)/k$ is the argument and the invariant lattices g_2 and g_3 read

$$g_2 = a_0a_4 - 4a_1a_3 + 3a_2^2$$
$$g_3 = a_0a_2a_4 + 2a_1a_2a_3 - a_2^3 - a_0a_3^2 - a_1^2a_4$$

The coefficients a_i and k are obtained by identifying Eq. (12.44) with Eq. (12.16) for different values of γ. They can be found in Table 12.2.

Symmetric spirals reach a minimum or maximum radius r_m, which is a root of $f(r)$. Thus, Eq. (12.46) can be simplified if referred to r_m instead of r_0:

$$r(\theta) - r_m = \frac{f'(r_m)/4}{\wp(z_m) - f''(z_m)/24} \tag{12.47}$$

This is the unified solution for all symmetric solutions, with $z_m = (\theta - \theta_m)/k$. Practical comments on the implementation of the Weierstrass elliptic functions can be found in Biscani and Izzo (2014). Although $\wp(z) = \wp(-z)$ the derivative $\wp'(z)$ is an odd function in z, $\wp'(-z) = -\wp'(z)$. Therefore, the integer n needs to be adjusted according to the regime of the spiral when solving Eq. (12.46): $n = 1$ for the raising regime, and $n = -1$ for the lowering regime.

12.9 Physical discussion of the solutions

Each family of solutions involves a fundamental curve: the case $\gamma = 1$ relates to conic sections, $\gamma = 2$ to logarithmic spirals, $\gamma = 3$ to cardioids, and $\gamma = 4$ to sinusoidal spirals. This section is devoted to analyzing the geometrical and dynamical connections between the solutions and other integrable systems.

12.9.1 Connection with Schwarzschild geodesics

The Schwarzschild metric is a solution to the Einstein field equations of the form

$$(\mathrm{d}s)^2 = \left(1 - \frac{2M}{r}\right)(\mathrm{d}t)^2 - \frac{(\mathrm{d}r)^2}{1 - 2M/r} - r^2(\mathrm{d}\phi)^2 - r^2\sin^2\phi\,(\mathrm{d}\theta)^2$$

written in natural units so that the speed of light and the gravitational constant equal unity, and the Schwarzschild radius reduces to $2M$. In this equation M is the mass of the central body, $\phi = \pi/2$ is the colatitude, and θ is the longitude.

The time-like geodesics are governed by the differential equation

$$\left(\frac{\mathrm{d}r}{\mathrm{d}\theta}\right)^2 = \frac{1}{L^2}(E^2 - 1)\,r^4 + \frac{2M}{L^2}\,r^3 - r^2 + 2Mr$$

where L is the angular momentum and E is a constant of motion related to the energy and defined by

$$E = \left(1 - \frac{2M}{r}\right)\frac{\mathrm{d}t_s}{\mathrm{d}t_p}$$

in terms of the proper time of the particle t_p and the Schwarzschild time t_s. Its solution is given by

$$\theta(r) - \theta_0 = \int_{r_0}^{r} \frac{nL\,\mathrm{d}s}{\sqrt{f(s)}}$$

and involves the integer $n = \pm 1$ that gives the raising/lowering regime of the solution. Here $f(s)$ is a quartic function defined like in Eq. (12.45). The previous equation is formally equivalent to Eq. (12.44). As a result, the Schwarzschild geodesics abide by Eqs. (12.46) or (12.47) depending on the selection of the roots, and are written in terms of the Weierstrass elliptic functions and identifying the coefficients

$$a_0 = E^2 - 1\,, \quad a_1 = M/2\,, \quad a_2 = -L^2/6\,,$$
$$a_3 = ML^2/2\,, \quad a_4 = 0\,, \quad k = nL$$

Compact forms of the Schwarzschild geodesics using the Weierstrass elliptic functions can be found in the literature (see for example Hagihara, 1930, §4). We refer to classical books like the one by Chandrasekhar (1983, Chap. 3) for an analysis of the structure of the solutions in the Schwarzschild metric.

The analytic solution to Schwarzschild geodesics involves elliptic functions, like generalized cardioids ($\gamma = 3$) and generalized sinusoidal spirals ($\gamma = 4$). If the values of κ_1 and κ_2 are adjusted in order for the roots of P_{sol} to coincide with the roots of $f(r)$ in the Schwarzschild metric, the solutions will be comparable. For example, when the polynomial $f(r)$ has three real roots and a repeated one the geodesics spiral toward or away from the limit radius

$$r_{\mathrm{lim}} = \frac{1}{2M} - \frac{L^2 + 4M^2}{2ML(2L + \sqrt{L^2 - 12M^2})}$$

If we make this radius coincide with the limit radius of a hyperbolic cardioid with $\kappa_2 = \sqrt{3}\kappa_1$ [Eq. (12.30)] it follows that

$$\kappa_1 = \frac{1}{3r_{\mathrm{lim}}} \quad \text{and} \quad \kappa_2 = \frac{1}{\sqrt{r_{\mathrm{lim}}}}$$

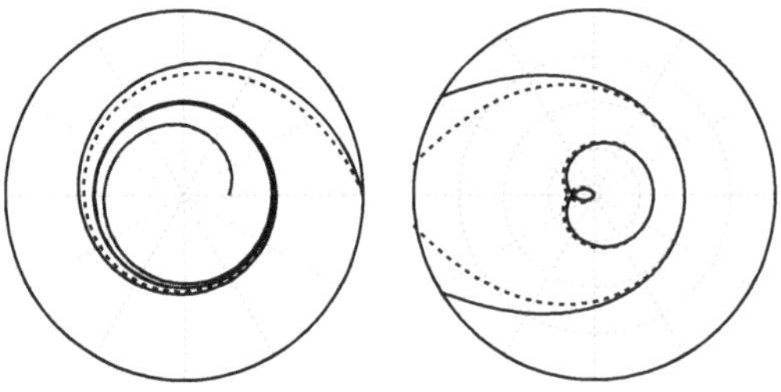

(a) $\gamma = 3$ (b) $\gamma = 4$

Fig. 12.3: Schwarzschild geodesic (solid line) compared to generalized cardioids and generalized sinusoidal spirals (dashed line).

Figure 12.3a compares hyperbolic generalized cardioids and Schwarzschild geodesics with the same limit radius. The interior solutions coincide almost exactly, whereas exterior solutions separate slightly. Similarly, when $f(r)$ admits three real and distinct roots the geodesics are comparable to interior and exterior hyperbolic solutions of Type II. For example, Fig. 12.3b shows interior and exterior hyperbolic sinusoidal spirals of Type II with the inner and outer radii equal to the limit radii of Schwarzschild geodesics. As in the previous example the interior solutions coincide, while the exterior solutions separate in time. In order for two completely different forces to yield a similar trajectory it suffices that the differential equation $\mathrm{d}r/\mathrm{d}\theta$ takes the same form. Even if the trajectory is similar the integrals of motion might not be comparable. As a result, the velocity along the orbit and the time of flight between two points will be different. Equation (12.15) shows that the radial motion is governed by the evolution of the flight-direction angle. For $\gamma = 3$ and $\gamma = 4$ the acceleration in Eq. (12.6) makes the radius evolve with the polar angle just like in the Schwarzschild metric. Consequently, the orbits may be comparable. However, since the integrals of motion (12.12) and (12.13) do not hold along the geodesics, the velocities do not necessarily match.

12.9.2 Newton's theorem of revolving orbits

Newton found that if the angular velocity of a particle following a Keplerian orbit is multiplied by a constant factor k, it is then possible to describe the dynamics by superposing a central force depending on an inverse cubic power of the radius. The additional perturbing terms depend only on the angular momentum of the original orbit and the value of k.

Consider a Keplerian orbit defined as

$$\rho(\theta) = \frac{\tilde{h}^2}{1 + \tilde{e}\cos\vartheta} \tag{12.48}$$

Here $\vartheta = \theta - \theta_m$ denotes the true anomaly. The radial motion of hyperbolic spirals of Type II – Eq. (9.50) – resembles a Keplerian hyperbola. The difference between both orbits comes from the angular motion, because they revolve with different angular velocities. Indeed, recovering Eq. (9.50), identifying $\kappa_2 = \tilde{e}$ and $\rho_{\min}(1 + \kappa_2) = \tilde{h}^2$, and calling the spiral anomaly $\beta = k(\theta - \theta_m)$, it is

$$\rho(\theta) = \frac{\tilde{h}^2}{1 + \tilde{e}\cos\beta}$$

When equated to Eq. (12.48), it follows a relation between the spiral (β) and the true (ϑ) anomalies:

$$\beta = k\vartheta$$

The factor k reads

$$k = \frac{\kappa_2}{\ell} = \frac{\kappa_2}{\sqrt{\kappa_2^2 - 1}}$$

Replacing ϑ by β/k in Eq. (12.48) and introducing the result in the equations of motion in polar coordinates gives rise to the radial acceleration that renders a hyperbolic generalized logarithmic spiral of Type II, namely

$$\tilde{a}_{r,2} = -\frac{1}{\rho^2} + \frac{\tilde{h}^2}{\rho^3}(1 - k^2) = \tilde{a}_{r,1} + \frac{\tilde{h}^2}{\rho^3}(1 - k^2)$$

This is in fact the same result predicted by Newton's theorem of revolving orbits.

The radial acceleration $\tilde{a}_{r,2}$ yields a hyperbolic generalized logarithmic spiral of Type II. This is a central force that preserves the angular momentum, but not the integral of motion in Eq. (12.13). Thus, a particle accelerated by $\tilde{a}_{r,2}$ describes the same trajectory as a particle accelerated by the perturbation in Eq. (12.6), but with different velocities. As a consequence, the times between two given points are also different. The acceleration derives from the specific potential

$$\mathcal{V}(\rho) = \mathcal{V}_k(\rho) + \Delta\mathcal{V}(\rho)$$

where $\mathcal{V}_k(\rho)$ denotes the Keplerian potential, and $\Delta\mathcal{V}(\rho)$ is the perturbing potential:

$$\Delta\mathcal{V}(\rho) = \frac{\tilde{h}^2}{2\rho^2}(1 - k^2)$$

12.9.3 Geometrical and physical relations

The inverse of a generic conic section $r(\theta) = a + b\cos\theta$ using one of its foci as the center of inversion defines a limaçon. In particular, the inverse of a parabola results in a cardioid. Let us recover the equation of the trajectory of a generalized parabolic cardioid (a true cardioid) from Eq. (12.24),

$$\rho(\theta) = \frac{\rho_{\max}}{2}[1 + \cos(\theta - \theta_m)]$$

Taking the cusp as the inversion center defines the inverse curve:

$$\frac{1}{\rho} = \frac{2}{\rho_{\max}}[1 + \cos(\theta - \theta_m)]^{-1}$$

Identifying the terms in this equation with the elements of a parabola it follows that the inverse of a generalized parabolic cardioid with apoapsis $\rho_{\max}$ is a Keplerian parabola with periapsis $1/\rho_{\max}$. The axis of symmetry remains invariant under inversion; the lines of apses coincide.

The subfamily of elliptic generalized logarithmic spirals is a generalized form of Cotes's spirals, more specifically of Poinsot's spirals:

$$\frac{1}{\rho} = a + b \cosh \nu$$

Cotes's spirals are known to be the solution to the motion immersed in a potential $\mathcal{V}(r) = -\mu/(2r^2)$ (Danby, 1992, p. 69).

The radial motion of interior Type II hyperbolic and Type I elliptic sinusoidal spirals has the same form, and is also equal to that of Type II hyperbolic generalized cardioids, except for the sorting of the terms.

It is worth noting that the dynamics along hyperbolic sinusoidal spirals with $\kappa_2 = 2\kappa_1$ are qualitatively similar to the motion under a central force decreasing with r^{-5}. Indeed, the orbits shown in Fig. 12.2e behave as the limit case $\gamma = 1$ discussed by MacMillan (1908, Fig. 4). Conversely, parabolic sinusoidal spirals (off-center circles) are also the solution to the motion under a central force proportional to r^{-5}.

12.10 Conclusions

The dynamical symmetries in Kepler's problem hold under a special nonconservative perturbation, which is a generalization of the thrust that rendered generalized logarithmic spirals. There are two integrals of motion that are extended forms of the equation of the energy and angular momentum. A similarity transformation reduces the original problem to a system perturbed by a tangential uniparametric forcing. It simplifies the dynamics significantly, for the integrability of the system is evaluated in terms of one unique parameter. The algebraic properties of the equations of motion dictate what values of the free parameter make the problem integrable in closed form.

The extended integrals of motion include the Keplerian ones as particular cases. The new conservation laws can be seen as generalizations of the original integrals. The new families of solutions are defined by fundamental curves in the zero-energy case, and there are geometrical transformations that relate different orbits. The orbits can be unified by introducing the Weierstrass elliptic functions. This approach simplifies the modeling of the system.

The solutions derived in this chapter are closely related to different physical problems. The fact that the magnitude of the acceleration decreases with $1/r^2$ makes it comparable with the perturbation due to solar radiation pressure. Moreover, the inverse similarity transformation converts Keplerian orbits into the conic sections obtained when the solar radiation pressure is directed along the radial direction. The structure of the solutions, governed by the roots of a polynomial, is similar in nature to the Schwarzschild geodesics. This is because under the considered perturbation and in the Schwarzschild metric the evolution of the radial distance takes the same form. Some of the solutions are comparable to the orbits derived from potentials depending on different powers of the radial distance. Although the trajectory may take the same form, the velocity will be different, in general.

13 Conclusions

Most problems in orbital mechanics are based on the same physical principles. Every effort toward advancing our understanding of the dynamical laws that govern motion in space will help in finding solutions to particular problems. In this sense, the present book shows how the improved description of the dynamics provided by the theory of regularization yields new results in different areas. These innovative techniques have been applied systematically to three challenging problems in modern astrodynamics: spacecraft relative motion, low-thrust mission design, and high-performance numerical propagation of orbits. The methods and techniques presented here might be extended to other areas of astronomy and engineering.

Special attention has been paid to the historical background of regularization. First of all, in order to pay tribute to the great minds who set the foundations of astrodynamics and, consequently, of spaceflight mechanics. But, hopefully, this review will also convince the reader of the power of regularization, by showing how different were the motivations that made astronomers and engineers derive new formulations, and how the same collection of techniques were useful in many scenarios. Computers have revolutionized science and engineering, enabling extremely complex numerical experiments that simulate physical processes with unprecedented accuracy. As a result, many disciplines have become eminently numerical. And regularization in celestial mechanics is one of these disciplines. Technical reports that Bond, Gottlieb, Stiefel, Velez, Janin, Baumgarte, etc. completed for both NASA and ESA (ESRO at the time of writing of many of them) in the 1960s, 1970s, and 1980s, were focused on the virtues of regularization for numerical integration. Regularization is undoubtedly well suited for numerical propagation of orbits, but there is also much more.

Regularization techniques can be applied to virtually any orbital problem. Conceptually, the main change is the use of different time variables. This is not only a matter of representation: the physical meaning of the variational equations, partial derivatives, relative states, etc. change because of using an alternative parameterization. This new idea led to the development of the theory of asynchronous relative motion, which provides a different approach to the concept of variational formulation. Its application to spacecraft relative motion constitutes a more efficient and accurate method for propagating the dynamics. Because it is a generic concept, the formulation admits any source of external perturbation. Other properties of regularized formulations can be leveraged too. For example, the universal solution to Kepler's problem can be extrapolated to derive the universal solution to relative motion.

The theory of asynchronous relative motion can be applied to virtually any problem that can be solved using variational methods. This extends the reach of the theory to other problems such as orbit determination, trajectory optimization, periodic orbit search, etc. In this book, we have derived a systematic formulation of the variational equations of orbital motion using regularization. Given the virtues of regularized tech-

https://doi.org/10.1515/9783110559125-013

niques for numerical propagation and the interest of the variational equations in many problems, allowing the user to explore the vicinity of the nominal trajectory with efficient methods is advantageous. The software tool PERFORM has specifically been adapted to the propagation of the partial derivatives of a given trajectory, in order to study its sensitivity to the initial conditions.

Orbital elements prove useful when integrating weakly perturbed problems. And low-thrust trajectories are good candidates, as even the name suggests. While seeking special sets of elements that would improve the numerical integration of these kinds of orbits, an entirely new family of analytic solutions with continuous thrust was found, called generalized logarithmic spirals. Looking for elements is closely related to finding conservation laws, and this was the key to deriving the new solution. The properties of generalized logarithmic spirals are surprisingly similar to Keplerian orbits. Even the spiral Lambert problem with these new orbits is almost equivalent to the classical ballistic problem. Further studies have revealed that this new family of orbits is, in fact, a subclass of a more general type of integrable system.

The new analytic solution obtained with regularized methods is not simply a mathematical curiosity. Based on it, a new strategy for low-thrust preliminary mission design has been derived. The advantages of the new methodology come from the unique dynamical properties of the solution. This is yet another proof of the importance of exploring new models, properties, and solutions when approaching a certain problem.

Regularization might also lead to more profound theoretical conclusions. The analysis of the topology underlying a special perturbation method yielded a new theory of stability. Although the theory is entirely constructed in an alternative phase space, not intuitive, it simplifies the characterization of complex systems that evolve into chaotic regimes. An indirect outcome is the discovery of a new Lyapunov-like indicator, which can be used to monitor the dynamical regimes of the orbits.

The orbit propagation tool PERFORM has been developed in an attempt to standardize the numerical experiments for evaluating the performance of orbital formulations. The software includes accurate force models that allow the user to integrate real-life problems and determine which combination of numerical integrator and dynamical formulation is best suited for the problem of interest. With the aid of sequential performance diagrams (SPD) the selection of the best combinations is simplified. In practice, investing some time in finding an efficient propagation strategy may save hours of computation when long simulation campaigns need to be conducted.

New formulations can be derived for specific applications. It is clear that today's challenges differ from those faced in the 1960s. But the principles are the same, and the basic concepts of regularization can still be applied to satisfy both present and future needs. For example, as missions incorporate more flybys, maintaining the accuracy of the propagations becomes an issue. In this context, a geometrical analysis of the problem showed how the propagation can be improved by relying on unconventional constructions. This book presents a new propagator specifically conceived for integrating hyperbolic orbits. It recovers the geometrical foundations of Minkowski space-time, and proves that hyperbolic geometry is more convenient than Euclidean for these kind of trajectories. The main advantage of the method is that the accuracy of the propagation is not affected by periapsis passage, which is a critical point during the integration phase. As shown when deriving this example formulation, the methods and techniques that make a regularized formulation are still powerful and worth considering. Their systematic application to generic problems might open a whole new world of possibilities.

Part III: **Appendices**

"The more I learn, the more I realize how much I don't know."
– Albert Einstein

A Hypercomplex numbers

Let $\mathfrak{a}_n$ denote an n-dimensional hypercomplex number, $\mathfrak{a}_n \in \mathbb{H}^n$. The field of hypercomplex numbers is denoted $\mathbb{H}^n$. The hypercomplex number $\mathfrak{a}_n$ can be written

$$\mathfrak{a}_n = a_0\mathrm{u}_0 + a_1\mathrm{u}_1 + a_2\mathrm{u}_2 + \ldots + a_{n-1}\mathrm{u}_{n-1}$$

where the coefficients $a_i \in \mathbb{R}$ are the components of $\mathfrak{a}_n$, and u_i (with $i = 0, \ldots, n-1$) are the generalized imaginary units, $\mathrm{u}_0 \equiv 1 \in \mathbb{R}$ and $\mathrm{u}_i \notin \mathbb{R}$ for $i \neq 0$. By analogy with vector algebra they are also referred to as versors, defining the n-dimensional basis $\{1, \mathrm{u}_1, \ldots, \mathrm{u}_{n-1}\}$ (Boccaletti et al., 2008, p. 5). The main rules in $\mathbb{H}^n$ are:

Equality:	$\mathfrak{a}_n = \mathfrak{b}_n$	$\Longleftrightarrow$	$a_i = b_i\,,$	$\forall\, i$
Addition:	$\mathfrak{c}_n = \mathfrak{a}_n + \mathfrak{b}_n$	$\Longleftrightarrow$	$c_i = a_i + b_i\,,$	$\forall\, i$
Null element w.r.t. sum:	$\mathfrak{a}_n + \mathfrak{z}_n = \mathfrak{a}_n$	$\Longleftrightarrow$	$z_i = 0\,,$	$\forall\, i$
Distributive w.r.t. sum:	$(\mathfrak{a}_n + \mathfrak{b}_n) + \mathfrak{c}_n$	$=$	$\mathfrak{a}_n + (\mathfrak{b}_n + \mathfrak{c}_n)$	

Multiplication of two hypercomplex numbers $\mathfrak{a}_n = \sum a_j\mathrm{u}_j$ and $\mathfrak{b}_n = \sum b_k\mathrm{u}_k$ is defined in terms of the product $(\mathrm{u}_j\,\mathrm{u}_k)$,

$$\mathfrak{a}_n\,\mathfrak{b}_n = \Big(\sum_j a_j\mathrm{u}_j\Big)\Big(\sum_k b_k\mathrm{u}_k\Big) = \sum_j\sum_k a_j b_k(\mathrm{u}_j\,\mathrm{u}_k)$$

with j, k integers spanning from 0 to $n-1$. Provided that $\mathbb{H}^n$ is closed under multiplication (Kantor et al., 1989, chap. 5), the product $(\mathrm{u}_j\,\mathrm{u}_k)$ is of the form

$$\mathrm{u}_j\,\mathrm{u}_k = \psi^0_{j,k} + \psi^1_{j,k}\mathrm{u}_1 + \psi^2_{j,k}\mathrm{u}_2 + \ldots + \psi^{n-1}_{j,k}\mathrm{u}_{n-1} \tag{A.1}$$

with $\psi^\ell_{j,k} \in \mathbb{R},\ \forall\, j, k, \ell$. That is, $(\mathrm{u}_j\,\mathrm{u}_k) \in \mathbb{H}^n$. The values for the coefficients $\psi^\ell_{j,k}$, given in the form of multiplication tables, define the nature of hypercomplex numbers.

Theorem 3. *Multiplication is commutative in* $\mathbb{H}^n$ *if and only if* $\psi^\ell_{j,k} = \psi^\ell_{k,j}$ *for all possible values of* j, k, ℓ.

Proof. Consider two hypercomplex numbers $\mathfrak{a}_n = \sum a_j\mathrm{u}_j$ and $\mathfrak{b}_n = \sum b_k\mathrm{u}_k$. The product $(\mathfrak{a}_n\,\mathfrak{b}_n)$ is commutative if $(\mathrm{u}_j\,\mathrm{u}_k)$ is commutative, provided that $a_j b_k = b_k a_j$. Multiplication yields

$$\mathrm{u}_j\,\mathrm{u}_k = \sum_\ell \psi^\ell_{j,k}\mathrm{u}_\ell\,, \qquad \mathrm{u}_k\,\mathrm{u}_j = \sum_\ell \psi^\ell_{k,j}\mathrm{u}_\ell$$

When equating term by term: $\mathrm{u}_j\,\mathrm{u}_k = \mathrm{u}_k\,\mathrm{u}_j \Longleftrightarrow \psi^\ell_{j,k} = \psi^\ell_{k,j}$. □

Theorem 4. *Multiplication is associative in* $\mathbb{H}^n$, $(\mathrm{u}_i\,\mathrm{u}_j)\,\mathrm{u}_k = \mathrm{u}_i(\mathrm{u}_j\,\mathrm{u}_k)$, *if and only if* $\sum_\ell \left(\psi^\ell_{i,j}\psi^m_{\ell,k} - \psi^\ell_{j,k}\psi^m_{i,\ell}\right) = 0$ *for all possible values of the coefficients* i, j, k, m.

https://doi.org/10.1515/9783110559125-014

Proof. Consider the products

$$(\mathfrak{u}_i\,\mathfrak{u}_j)\,\mathfrak{u}_k = \sum_\ell \psi^\ell_{i,j}\mathfrak{u}_\ell\,\mathfrak{u}_k = \sum_m\sum_\ell \psi^\ell_{i,j}\psi^m_{\ell,k}\mathfrak{u}_m$$

$$\mathfrak{u}_i(\mathfrak{u}_j\,\mathfrak{u}_k) = \sum_\ell \psi^\ell_{j,k}\mathfrak{u}_i\,\mathfrak{u}_\ell = \sum_m\sum_\ell \psi^\ell_{j,k}\psi^m_{i,\ell}\mathfrak{u}_m$$

Both expressions are equated term by term to provide $\sum_\ell \psi^\ell_{i,j}\psi^m_{\ell,k} = \sum_\ell \psi^\ell_{j,k}\psi^m_{i,\ell}$ for all values of the coefficients i, j, k, m. □

Multiplication in $\mathbb{H}^n$ is equipped with additional properties:
- Let $r \in \mathbb{R}$ and $\mathfrak{a}_n \in \mathbb{H}^n$. The product is $r\mathfrak{a}_n = r\sum a_i\mathfrak{u}_i = \sum ra_i\mathfrak{u}_i = \mathfrak{a}_n r$.
- Let $p, q \in \mathbb{R}$ and $\mathfrak{a}_n, \mathfrak{b}_n \in \mathbb{H}^n$. Then, it is $(pq)(\mathfrak{a}_n\mathfrak{b}_n) = (p\mathfrak{a}_n)(q\mathfrak{b}_n)$.
- $\mathfrak{c}_n(\mathfrak{a}_n + \mathfrak{b}_n) = \mathfrak{c}_n\,\mathfrak{a}_n + \mathfrak{c}_n\,\mathfrak{b}_n$ and $(\mathfrak{a}_n + \mathfrak{b}_n)\mathfrak{c}_n = \mathfrak{a}_n\,\mathfrak{c}_n + \mathfrak{b}_n\,\mathfrak{c}_n$.

Definition 7. A number $\mathfrak{p}_n \in \mathbb{H}^n$ is said to be a divisor of zero of $\mathfrak{s}_n \in \mathbb{H}^n$ if the equation $\mathfrak{s}_n\,\mathfrak{p}_n = 0$ is satisfied for $\mathfrak{p}_n, \mathfrak{s}_n \neq 0$.

The equation $\mathfrak{s}_n\,\mathfrak{p}_n = 0$ is expanded to provide,

$$\mathfrak{s}_n\,\mathfrak{p}_n = \sum_i\sum_j s_ip_j\mathfrak{u}_i\mathfrak{u}_j = \sum_i\sum_j\sum_k s_ip_j\psi^k_{i,j}\mathfrak{u}_k = 0$$

Each component of $\mathfrak{s}_n\,\mathfrak{p}_n$ in the basis $\{1, \mathfrak{u}_1, \ldots, \mathfrak{u}_{n-1}\}$ must cancel to comply with this result. This leads to a homogeneous system of n linear equations:

$$\sum_i\sum_j s_ip_j\psi^k_{i,j} = \sum_i\sum_j s_i\psi^k_{i,j}p_j = 0\,, \quad \text{with } k = 0, \ldots, n-1$$

There exists an equivalent expression in matrix form,

$$\mathbf{A}\,\mathbf{p} = 0\,, \quad \text{with } \mathrm{A}_{ij} = \sum_{\ell=0}^{n-1} s_\ell\psi^{i-1}_{\ell,j-1} \text{ and } \mathbf{p} = [p_0, \ldots, p_{n-1}]^\top \tag{A.2}$$

which provides the components of $\mathfrak{p}_n$. Such components are included in the kernel of **A**. Note that if matrix **A** is singular, i.e., $|\mathbf{A}| = 0$, there are infinitely many nontrivial solutions to the system. In that case $\mathfrak{p}_n$ is a divisor of zero of $\mathfrak{s}_n$. The hypercomplex numbers $\mathfrak{p}_n, \mathfrak{s}_n$ then verify

$$\mathfrak{p}_n = \frac{0}{\mathfrak{s}_n} \neq 0$$

The condition for $\mathfrak{p}_n$ being a divisor of zero deals both with its components and with the coefficients $\psi^\ell_{j,k}$ from the multiplication table.

A.1 Complex and hyperbolic numbers

Two-dimensional hypercomplex numbers of the form $\mathfrak{h} = x + y\,\mathfrak{u}_1$ are particularly interesting. These numbers correspond to the components $(x, y) \in \mathbb{R} \times \mathbb{R}$ in the basis

$\{1, \mathrm{u}_1\}$. From now on the generalized imaginary unit u_1 is simply written u. Multiplication reduces to

$$\mathrm{u}_j\,\mathrm{u}_k = \psi^0_{j,k} + \mathrm{u}\,\psi^1_{j,k}$$

These coefficients define the nature of the resulting hypercomplex numbers. In particular, the table

$$\begin{aligned} \psi^0_{00} &= 1\,, & \psi^1_{00} &= 0\,, & \psi^0_{10} &= 0\,, & \psi^1_{10} &= 1 \\ \psi^0_{01} &= 0\,, & \psi^1_{01} &= 1\,, & \psi^0_{11} &= -1\,, & \psi^1_{11} &= 0 \end{aligned} \tag{A.3}$$

defines the field of complex numbers $\mathbb{C}$,

$$\mathbb{C} = \{x + \mathrm{i}y \,|\, (x, y) \in \mathbb{R} \times \mathbb{R}\,,\ \mathrm{i}^2 = -1\}$$

where $\mathrm{i} \equiv \mathrm{u}$ is referred to as the imaginary unit. Since $\psi^\ell_{j,k} = \psi^\ell_{k,j}$, Thm. 3 shows that multiplication is commutative in $\mathbb{C}$. Let $\mathfrak{s}_2 \in \mathbb{C}$ be $\mathfrak{s}_2 = x + \mathrm{i}y$. Matrix $\mathbf{A}$ from Eq. (A.2) is such that $\|\mathbf{A}\| = x^2 + y^2$, that is nonzero for $\mathfrak{s}_2 \neq 0$. The kernel of $\mathbf{A}$ reduces to the null vector, and there are no divisors of zero in $\mathbb{C}$.

Now consider that $\psi^0_{11} = -1$ is replaced by $\psi^0_{11} = 1$ in Eq. (A.3); this defines the field of hyperbolic numbers $\mathbb{D}$,

$$\mathbb{D} = \{x + \mathrm{j}y \,|\, (x, y) \in \mathbb{R} \times \mathbb{R}\,,\ \mathrm{j}^2 = +1\}$$

where $\mathrm{j} \equiv \mathrm{u}$ is the hyperbolic imaginary unit, and $\mathrm{j} = \sqrt{+1} \notin \mathbb{R}$. Like in $\mathbb{C}$, multiplication in $\mathbb{D}$ is commutative. Let $\mathfrak{w}_2$ be the hyperbolic number $\mathfrak{w}_2 = x + \mathrm{j}y$. The determinant $\|\mathbf{A}\|$ is $\|\mathbf{A}\| = x^2 - y^2$, which vanishes for $y = \pm x$. Thus, hyperbolic numbers admit divisors of zero.

A.1.1 The modulus

Given a complex number $\mathfrak{z} \in \mathbb{C}$, there is a conjugate element in the field of complex numbers, $\mathfrak{z} =^\dagger x - \mathrm{i}y$, with $\mathfrak{z} \in^\dagger \mathbb{C}$. The modulus of $\mathfrak{z}$ is obtained from the definition,

$$|\mathfrak{z}|^2_{\mathbb{C}} = \mathfrak{z}\,\mathfrak{z}^\dagger = (x + \mathrm{i}y)(x - \mathrm{i}y) = x^2 - \mathrm{i}^2y^2 = x^2 + y^2$$

The locus $|\mathfrak{z}|_{\mathbb{C}} = \text{const.}$ results in a circumference on the (x, y) plane. This circumference is denoted by $\mathcal{c}^r$, with radius $r = |\mathfrak{z}|_{\mathbb{C}}$.

In a similar fashion, given a hyperbolic number $\mathfrak{w} \in \mathbb{D}$, its conjugate is a hyperbolic number $\mathfrak{w}^\dagger = x - \mathrm{j}y$. The definition of modulus in $\mathbb{D}$ yields

$$|\mathfrak{w}|^2_{\mathbb{D}} = \mathfrak{w}\,\mathfrak{w}^\dagger = (x + \mathrm{j}y)(x - \mathrm{j}y) = x^2 - \mathrm{j}^2y^2 = x^2 - y^2$$

Provided with this result, $x^2 - y^2 = \text{const.}$ defines the locus of the invariant modulus on the (x, y) plane. Geometrically, this invariant subspace is a rectangular (or equilateral) hyperbola written $\mathcal{h}^a$, where a is the semimajor axis. The right branch is $\mathcal{h}^a_+$, and $\mathcal{h}^a_-$ denotes the left branch. This hyperbola should not be confused with the osculating orbit.

A.1.2 The geometry of two-dimensional hypercomplex numbers

Two-dimensional hypercomplex numbers have been defined in terms of a particular basis $\{1, \mathrm{u}\}$. Given a pair of components (x, y), the two-dimensional hypercomplex number is written $\mathfrak{h} = x + \mathrm{u}\, y$. This interpretation is closely related to the definition of vectors: vector $\mathbf{v} = x\mathbf{i} + y\mathbf{j}$ is the compact expression of the pair (x, y) in the basis $\{\mathbf{i}, \mathbf{j}\}$.

The (x, y) plane in $\mathbb{C}$, with $x = \mathfrak{R}(\mathfrak{z})$ and $y = \mathfrak{I}(\mathfrak{z})$, is the Gauss–Argand plane. The norm of vector $\mathbf{v} = x\mathbf{i} + y\mathbf{j}$ in two-dimensional Euclidean geometry $\mathbb{E}^2$ is equivalent to the modulus of $\mathfrak{z} = x + \mathrm{i}y$,

$$\|\mathbf{v}\|^2_{\mathbb{E}^2} = x^2 + y^2 \equiv |\mathfrak{z}|^2_{\mathbb{C}}$$

The Gauss–Argand plane is conceived like the Cartesian plane, where Euclidean geometry applies. This is one of the major properties of complex numbers.

Just like how complex numbers comply with Euclidean $\mathbb{E}^2$ geometry, hyperbolic numbers relate to Minkowskian geometry (Boccaletti et al., 2008, chap. 4). The four-dimensional Minkowski space-time $\mathbb{M}$ is equipped with a metric tensor of signature $\langle +, -, -, - \rangle$. A detailed description of the metrics in Minkowskian space-time is given by Penrose (2004, chap. 18). The Minkowski plane, $\mathbb{R}^2_1$, is the two-dimensional representation of $\mathbb{M}$ and it is endowed with the inner product $\langle \mathbf{v}, \mathbf{v} \rangle_{\mathbb{R}^2_1} = x^2 - y^2$, with $\mathbf{v} = x\mathbf{i} + y\mathbf{j}$ in $\mathbb{R}^2_1$ (see Saloom and Tari, 2012; Yaglom, 1979, pp. 242–257).

Consider $\mathfrak{w} = x + \mathrm{j}y$ with $\mathfrak{w} \in \mathbb{D}$. Minkowskian geometry relates to the hyperbolic numbers in terms of the pseudonorm:

$$\|\mathbf{v}\|^2_{\mathbb{R}^2_1} = x^2 - y^2 \equiv |\mathfrak{w}|^2_{\mathbb{D}}$$

The geometrical representation of hyperbolic numbers on the Minkowski plane simplifies the description of hyperbolic geometries. Considering the metrics on the Minkowski plane the distance is preserved along rectangular hyperbolas. Hyperbolic angles and rotations appear naturally from this property, as explained in Sect. 5.1. The formulation of the orbital motion of a particle along a hyperbolic orbit benefits from this contrivance, due to a straight-forward definition of the hyperbolic anomaly (Sect. 5.1.2).

A.1.3 Angles and rotations

A complex number $\mathfrak{z} = x + \mathrm{i}y$ admits an alternative representation on the Gauss–Argand plane. Instead of the Cartesian form, the polar form is introduced,

$$\mathfrak{z} = r\,\mathrm{e}^{\mathrm{i}\vartheta}\,, \qquad \text{with } r = |\mathfrak{z}|_{\mathbb{C}}\,, \text{ and } \tan\vartheta = \frac{y}{x}$$

where r is the modulus and ϑ is the argument of $\mathfrak{z}$. Recall that under the mapping $\mathfrak{z} \mapsto \mathfrak{z}\,\mathrm{e}^{\mathrm{i}\vartheta_2}$ the modulus is invariant, $|\mathfrak{z}\,\mathrm{e}^{\mathrm{i}\vartheta_2}| = |\mathfrak{z}|\,|\mathrm{e}^{\mathrm{i}\vartheta_2}| = |\mathfrak{z}|$. This application defines a rotation along the circumference $\mathcal{c}^r$ of magnitude ϑ_2.

The polar form is related to the Cartesian form through Euler's identity,

$$\mathfrak{z} = r\,\mathrm{e}^{\mathrm{i}\vartheta} = r(\cos\vartheta + \mathrm{i}\sin\vartheta) = x + \mathrm{i}y$$

Its formal proof can be found in Ahlfors (1966, pp. 43–44).

The polar form of hyperbolic numbers on the Minkowski plane reads

$$\mathfrak{w} = s\,\mathrm{e}^{\mathrm{j}\gamma}$$

where s is the modulus and γ the hyperbolic argument. The form of the exponential function in $\mathbb{D}$ leads to the hyperbolic form of Euler's identity (Catoni et al., 2011, p. 16),

$$\mathrm{e}^{\mathrm{j}\gamma} = \cosh\gamma + \mathrm{j}\sinh\gamma$$

Considering this identity the hyperbolic components of $\mathfrak{w}$ on the Minkowski plane read

$$x = \mathfrak{R}(\mathfrak{w}) = s\,\cosh\gamma\,, \qquad y = \mathfrak{H}(\mathfrak{w}) = s\,\sinh\gamma$$

In this case, the modulus is invariant under the mapping $\mathfrak{w} \mapsto \mathfrak{w}\,\mathrm{e}^{\mathrm{j}\gamma_2}$. This application represents a hyperbolic rotation of magnitude γ_2 along $\hbar^s$. That is, circular rotations preserve the modulus in Euclidean geometry whereas hyperbolic rotations preserve the modulus in two-dimensional Minkowskian geometry. The sign criterion defines positive hyperbolic rotations on $\hbar^1_+$ from bottom to top, and from top to bottom on $\hbar^1_-$.

A.2 Quaternions

Quaternions are a particular instance of hypercomplex numbers. It is $\mathfrak{q} \in \mathbb{H}^4$, with $\mathfrak{q}$ a quaternion

$$\mathfrak{q} = q_0 + q_1\mathrm{u}_1 + q_2\mathrm{u}_2 + q_3\mathrm{u}_3$$

Quaternion algebra is endowed with the multiplication table:

$$\begin{aligned} \mathrm{u}_1^2 &= -1\,, \qquad & \mathrm{u}_1\,\mathrm{u}_2 &= \mathrm{u}_3 = -\mathrm{u}_2\,\mathrm{u}_1 \\ \mathrm{u}_2^2 &= -1\,, \qquad & \mathrm{u}_2\,\mathrm{u}_3 &= \mathrm{u}_1 = -\mathrm{u}_3\,\mathrm{u}_2 \\ \mathrm{u}_3^2 &= -1\,, \qquad & \mathrm{u}_3\,\mathrm{u}_1 &= \mathrm{u}_2 = -\mathrm{u}_1\,\mathrm{u}_3 \end{aligned}$$

The definition of multiplication leads to the properties:

- The product of quaternions is not commutative. Direct proof is obtained from the multiplication table, noting that $\mathrm{u}_j\,\mathrm{u}_k = -\mathrm{u}_k\,\mathrm{u}_j$ and then it is $\psi^\ell_{j,k} = -\psi^\ell_{k,j}$. This violates the condition from Thm. 1.
- The product of quaternions is associative. Note that $(\mathrm{u}_1\,\mathrm{u}_2)\,\mathrm{u}_3 = \mathrm{u}_3\,\mathrm{u}_3 = -1$, equal to $\mathrm{u}_1(\mathrm{u}_2\,\mathrm{u}_3) = \mathrm{u}_1\,\mathrm{u}_1 = -1$.
- Matrix **A** is regular, so there are no divisors of zero.

Quaternions appear recursively in many texts in mathematics, physics, and engineering, and their representation varies across disciplines. For instance, they are usually regarded as Pauli matrices, or as spinors.

A.2.1 Rotations in $\mathbb{R}^3$

Let $\mathcal{D} \subset \mathbb{H}^4$ be the subset of all pure quaternions, $\mathcal{D} = \{\mathfrak{q} \in \mathbb{H}^4 |\, q_0 = 0\}$. Any vector $\mathbf{r} \in \mathbb{R}^3$ admits an equivalent quaternionic representation in terms of $\mathfrak{r} \in \mathcal{D}$, given by the components of $\mathbf{r}$. A unit quaternion $\mathfrak{q}$ defines a rotation action $\mathfrak{r} \mapsto \mathfrak{r}_1$ by means of

$$\mathfrak{r}_1 = \mathfrak{q}\,\mathfrak{r}\,\mathfrak{q}^\dagger\,, \qquad \text{with } \mathfrak{q}\,\mathfrak{q}^\dagger = 1 \tag{A.4}$$

As a rotation, this transformation preserves the norm, i.e., $|\mathfrak{r}_1| = |\mathfrak{r}|$. This transformation is invertible,

$$\mathfrak{r} = \mathfrak{q}^\dagger\,\mathfrak{r}_1\,\mathfrak{q}$$

A.2.2 Quaternion dynamics

Theorem 5. *Let $\mathfrak{r}, \mathfrak{r}_1 \in \mathcal{D}$ be two pure quaternions defined by the components of vector $\mathbf{r}$ in a rotating and a fixed reference frame, respectively. The angular velocity of the rotating frame is $\boldsymbol{\omega}$. If quaternion $\mathfrak{q}$ defines the rotation $\mathfrak{r} \mapsto \mathfrak{r}_1$, the time evolution of $\mathfrak{q}$ is*

$$\dot{\mathfrak{q}} = \frac{1}{2}\,\mathfrak{q}\,(2\mathfrak{w})$$

where $2\mathfrak{w} \in \mathcal{D}$ is defined by the components of $\boldsymbol{\omega}$ in the rotating frame itself.

Proof. If vector $\mathbf{r}$ denotes the position of a particle, the time derivative of Eq. (A.4) yields the absolute velocity projected on the inertial reference, $\mathfrak{v}_1$,

$$\begin{aligned} \mathfrak{v}_1 &= \dot{\mathfrak{q}}\,\mathfrak{r}\,\mathfrak{q}^\dagger + \mathfrak{q}\,\mathfrak{r}\,\dot{\mathfrak{q}}^\dagger = \dot{\mathfrak{q}}\,\mathfrak{q}^\dagger\,\mathfrak{r}_1\,\mathfrak{q}\,\mathfrak{q}^\dagger + \mathfrak{q}\,\mathfrak{q}^\dagger\,\mathfrak{r}_1\,\mathfrak{q}\,\dot{\mathfrak{q}}^\dagger = \dot{\mathfrak{q}}\,\mathfrak{q}^\dagger\,\mathfrak{r}_1 + \mathfrak{r}_1\,\mathfrak{q}\,\dot{\mathfrak{q}}^\dagger \\ &= \mathfrak{w}_1\,\mathfrak{r}_1 + \mathfrak{r}_1\,\mathfrak{w}_1^\dagger \end{aligned} \tag{A.5}$$

where $\mathfrak{w}_1 = \dot{\mathfrak{q}}\,\mathfrak{q}^\dagger$ and the dot $\dot{\square}$ represents derivation with respect to time. From the condition $\mathfrak{q}\,\mathfrak{q}^\dagger = 1$ it follows that

$$\dot{\mathfrak{q}}\,\mathfrak{q}^\dagger + \mathfrak{q}\,\dot{\mathfrak{q}}^\dagger = 0 \implies \mathfrak{w}_1 = -\mathfrak{w}_1^\dagger$$

That is, $\mathfrak{w}_1 \in \mathcal{D}$. It is then possible to define an equivalent vector $\boldsymbol{\omega}_1/2$. The expression for the velocity then transforms into

$$\mathfrak{v}_1 = \mathfrak{w}_1\,\mathfrak{r}_1 - \mathfrak{r}_1\,\mathfrak{w}_1 \implies \mathbf{v}_1 = 2\left(\frac{\boldsymbol{\omega}_1}{2} \times \mathbf{r}_1\right) = \boldsymbol{\omega}_1 \times \mathbf{r}_1$$

The quaternionic product in Eq. (A.5) is equivalent to the cross product of the associated vector. Hence, quaternion $2\mathfrak{w}_1 \in \mathcal{D}$ is equivalent to the angular velocity vector of the rotating frame, projected onto the fixed frame. The components of the angular velocity in the rotating frame reduce to

$$2\mathfrak{w} = \mathfrak{q}^\dagger \, 2\mathfrak{w}_1 \, \mathfrak{q} = 2\mathfrak{q}^\dagger \, \dot{\mathfrak{q}}$$

The inverse equation yields the time evolution of quaternion $\mathfrak{q}$ in terms of the angular velocity $2\mathfrak{w}$,

$$\dot{\mathfrak{q}} = \frac{1}{2} \mathfrak{q} \, (2\mathfrak{w})$$

□

B Formulations in PERFORM

Chapter 6 explained the details of the propagator PERFORM. One of its main features is the fact that it incorporates a collection of orbital formulations. Table 6.2 listed all the methods available in PERFORM. Each of the methods was given a three-character identifier, and this appendix explains all the formulations in terms of their identifier.

COW: Cowell's method is the widely used propagator based on the use of Cartesian coordinates (x, y, z).

SCW: the stabilized Cowell method is sometimes referred to as s-Cowell, because it replaces the physical time t by a fictitious time s. The general stabilization method is explained in Sect. 2.3, leading to the equation of motion (2.11). The method is implemented following Janin (1974, eqs. 3.10 and 3.14).

DRO: the Dromo formulation has been explained in detail in Chap. 4. The evolution equations to be integrated are Eqs. (4.10–4.17).

TDR: TimeDromo is a modification of Dromo in which the independent variable is the physical time, instead of the ideal anomaly. Equations (4.11–4.17) are multiplied by $\zeta_3^3 s^2$, and Eq. (4.10) is no longer necessary. TimeDromo reduces the dimension of the system by one, and root-finding algorithms are no longer required for stopping the integration.

HDT: this formulation is the Minkowskian propagator defined in Chap. 5 including the time element [Eqs. (5.43–5.50)].

HDR: this is the version of the Minkowskian propagator not using the time element, Eqs. (5.44–5.50).

KS_: the transformation by Kustaanheimo and Stiefel (1965) leads to Eq. (3.19), which is integrated together with the Sundman transformation $\mathrm{d}t/\mathrm{d}s = r$ to provide the physical time and the vector $\mathbf{u}$.

KST: this version of the KS transformation includes a time element, defined as

$$\frac{\mathrm{d}\tau}{\mathrm{d}s} = \frac{1}{2h}\left\{1 + \left[\frac{4}{h}(\mathbf{u}\cdot\mathbf{u}')\,\mathbf{u}' + r\,\mathbf{u}\right]\cdot(\mathbf{L}^{\top}\mathbf{p})\right\}$$

with

$$h = \frac{(1 - 2\|\mathbf{u}'\|^2)}{r}$$

The physical time is retrieved by means of the equation

$$t = \tau - \frac{(\mathbf{u}\cdot\mathbf{u}')}{h}$$

KSR: instead of computing the value of the constant of the energy h at each integration step, it is possible to integrate the following equation

$$\frac{\mathrm{d}h}{\mathrm{d}s} = -2(\mathbf{L}^{\top}\mathbf{p})\cdot\mathbf{u}'$$

and treat this variable as an element. This is a redundant equation that is integrated together with (3.19) and the Sundman transformation.

https://doi.org/10.1515/9783110559125-015

KRT: this version of the KS regularization includes both the time element and the redundant equation.

SB_: the formulation originally developed by Sperling (1961) was later improved by Burdet (1967), and is given by Eq. (2.18). This formulation is called the Sperling–Burdet (SB) regularization in PERFORM.

SBT: given the form of the equations in the SB regularization, it is possible to use the time element defined by Janin (1974, eq. 3.10).

SBR: instead of computing the values of the eccentricity vector, the energy, and the radial distance at each integration step, it is possible to introduce a set of redundant equations to integrate these values directly. The equations can be found in Bond and Allman (1996, table 9.1, p. 154).

SRT: the redundant equations defined in the previous formulation are complemented with the time element by Janin (1974, eq. 3.10).

DEP: this is an element based formulation which uses a quaternion to model the orbital plane, and is conveniently summarized in Deprit (1975, table 1).

PAL: the method by Palacios and Calvo (1996) may be regarded as a version of the DEP.

MIL: the version of the Milankovitch elements implemented in PERFORM can be found in Rosengren and Scheeres (2014, proposition 3).

SSc: the entire Sect. 3.6 was devoted to the definition of KS elements, and the evolution equations were provided by Stiefel and Scheifele (1971, p. 91).

EQU: although defined originally by Cefola (1972), PERFORM implements the version of the equinoctial elements by Walker et al. (1985, eq. 9, taking into account the corresponding errata).

CLA: the traditional Gauss variational equations can be found in, for example, Battin (1999, p. 488, eq. 10.41).

BCP: the elements introduced by Burdet (1968, eqs. 306 and 307) are integrated following what he calls the *companion procedure* (eqs. 318–321 in the referenced paper). Except for the fact that they embedded the Jacobi integral in the formulation, this method is equivalent in practice to the BG14 method by Bond and Gottlieb (1989).

USM: the unified state model is summarized in Vittaldev et al. (2012, §3.5).

ED0: the remaining formulations are different versions of the methods presented by Baù et al. (2015). In particular, ED0 refers to the equations presented in Sect. 6.1 of the referenced paper, using the Sundman transformation instead of equation (6.1) in the paper. That is, the time is modeled with a time variable and not a time element.

ED1: refers to the same system of equations, using equation (6.1) in the paper.

ED2: this formulation features a constant time element, defined in equation (4.5) instead of (6.1) in the paper.

B0H: Baù et al. (2015, §C2) presented different versions of an improved form of the Burdet–Ferrándiz equations. The identifier B0H refers to the formulation in which the physical time is a state variable, and it obeys a first-order differential equation.

B1H: the physical time is given by a linear time element.

B2H: a constant time element replaces the linear time element.

B3H: the physical time is a state variable, and its evolution is governed by a second-order differential equation.

BXC: formulations B0C to B3C are the same as B0H to B3H; the only difference is the fact that the generalized angular momentum replaces the total energy as an element.

ST0: defined in Baù et al. (2015, §C2), this is a modification of the Stiefel–Fukushima formulation. In this case the physical time is integrated as a state variable.

ST1: a linear time element is propagated instead of the physical time.

ST2: the linear time element is replaced by a constant time element.

C Stumpff functions

Karl Stumpff (1947) introduced a special family of functions that appear recursively in astrodynamics. They were later called the Stumpff functions, and they are defined in terms of the series:

$$\mathfrak{C}_k(z) = \sum_{i=0}^{\infty} \frac{(-z)^i}{(2i+k)!}$$

The Stumpff functions are intimately related to universal variables (Everhart and Pitkin, 1983; Battin, 1999, chap. 4). They allow us to generalize the solution to Keplerian motion so the formulation is unique no matter the eccentricity of the orbit. In this work the argument of the Stumpff functions is $z = \omega^2 s^2$, with $\omega^2 = -2\mathcal{E}$.

When the orbital energy vanishes (in the parabolic case), it follows that $z = 0$ and the Stumpff functions reduce to

$$\mathfrak{C}_k(0) = \frac{1}{k!} \tag{C.1}$$

Theorem 6. *The Stumpff functions converge absolutely.*

Proof. Absolute convergence implies

$$\sum_{i=0}^{\infty} \left| \frac{(-z)^i}{(2i+k)!} \right| < \ell\,, \qquad \forall k \in \mathbb{N}$$

where ℓ is a finite number. This series verifies

$$\sum_{i=0}^{\infty} \left| \frac{(-z)^i}{(2i+k)!} \right| = \sum_{i=0}^{\infty} \left| \frac{z^i}{(2i+k)!} \right| < \sum_{i=0}^{\infty} \left| \frac{z^i}{(2i)!} \right|, \qquad \forall k \in \mathbb{N}$$

The problem is reduced to proving that

$$\sum_{i=0}^{\infty} \left| \frac{z^i}{(2i)!} \right| < \ell$$

It is possible to find a bounding value of ℓ when considering the following expression:

$$\sum_{i=0}^{\infty} \left| \frac{z^i}{(2i)!} \right| < \sum_{i=0}^{\infty} \left| \frac{z^i}{i!} \right| = \sum_{i=0}^{\infty} \frac{|z|^i}{i!} = \exp(|z|)$$

which bounds the series for all finite z. An elegant discussion of the growth rate of each term in the series can be found in Spivak (1994, p. 308). □

The first of the Stumpff functions admit simple closed-form expressions, as shown in Table C.1.

Increasing the degree k of the Stumpff functions yields:

$$\mathfrak{C}_{k+1}(z) = \sum_{i=0}^{\infty} \frac{(-z)^i}{(2i+k+1)!}\,, \quad \mathfrak{C}_{k+2}(z) = \sum_{i=0}^{\infty} \frac{(-z)^i}{(2i+k+2)!}$$

https://doi.org/10.1515/9783110559125-016

Tab. C.1: Explicit expressions for the first Stumpff functions $\mathfrak{C}_k(z)$, with $z = \omega^2 s^2$.

	$\omega^2 > 0$	$\omega^2 < 0$	$\omega^2 = 0$
$k = 0$	$\cos\sqrt{z}$	$\cosh\sqrt{-z}$	1
$k = 1$	$(\sin\sqrt{z})/\sqrt{z}$	$(\sinh\sqrt{-z})/\sqrt{-z}$	1
$k = 2$	$(1 - \cos\sqrt{z})/z$	$(1 - \cosh\sqrt{-z})/z$	1/2

From the latter it follows that

$$\mathfrak{C}_{k+2}(z) = \frac{1}{z}\sum_{i=0}^{\infty}\frac{(-1)^i z^{i+1}}{(2(i+1)+k)!} = \frac{1}{z}\left[\frac{1}{k!} - \mathfrak{C}_k(z)\right]$$

which provides the recurrence formula:

$$\mathfrak{C}_k(z) + z\,\mathfrak{C}_{k+2}(z) = \frac{1}{k!} \tag{C.2}$$

Techniques for computing the derivatives of the Stumpff functions can be found, for instance, in the book by Bond and Allman (1996, appx. E). Some useful relations are:

$$s\left[\partial_s \mathfrak{C}_k(z)\right] = \mathfrak{C}_{k-1}(z) - k\mathfrak{C}_k(z)\,, \qquad \partial_s \mathfrak{C}_k(z) = -\omega^2 s\,\mathfrak{C}^*_{k+2}(z) \tag{C.3}$$

$$\omega\left[\partial_\omega \mathfrak{C}_k(z)\right] = \mathfrak{C}_{k-1}(z) - k\mathfrak{C}_k(z)\,, \qquad \partial_\omega \mathfrak{C}_k(z) = -s^2\omega\mathfrak{C}^*_{k+2}(z) \tag{C.4}$$

where ∂_s and ∂_ω denote the partial derivatives with respect to s and ω, respectively. Note that Eqs. (C.3a) and (C.4b) are only valid for $k > 0$. The auxiliary term $\mathfrak{C}^*_{k+2}(z)$ corresponds to:

$$\mathfrak{C}^*_{k+2}(z) = \mathfrak{C}_{k+1}(z) - k\mathfrak{C}_{k+2}(z)$$

Danby (1992, p. 171) discusses in detail the computational aspects of handling the Stumpff functions. It is more convenient to compute the highest-degree functions via the series, and then to apply the recurrence formula – Eq. (C.2) – to obtain the remaining functions. If the high-degree functions were to be computed from the low degree ones, the recurrence relation may become singular for z. Usual applications of the Stumpff functions are restricted to $k \leq 3$, and do not consider higher degree functions (Danby, 1987; Sharaf and Sharaf, 1997). In this work, Stumpff functions up to $k = 5$ appear, and the required formulae are given explicitly in the following.

The numerical stability of the convergent series requires a detailed analysis. Let $S_k^i \in \mathbb{R}$ denote the i-th term of the series defining $\mathfrak{C}_k(z)$:

$$S_k^i(z) = \frac{(-z)^i}{(2i+k)!}$$

Since the Stumpff functions converge absolutely the factorial term compensates the power for i sufficiently large (Spivak, 1994, p. 308). Figure C.1 shows the evolution of $S_4^i(z)$ and $S_5^i(z)$ for different values of the argument z, and for increasing i. It is

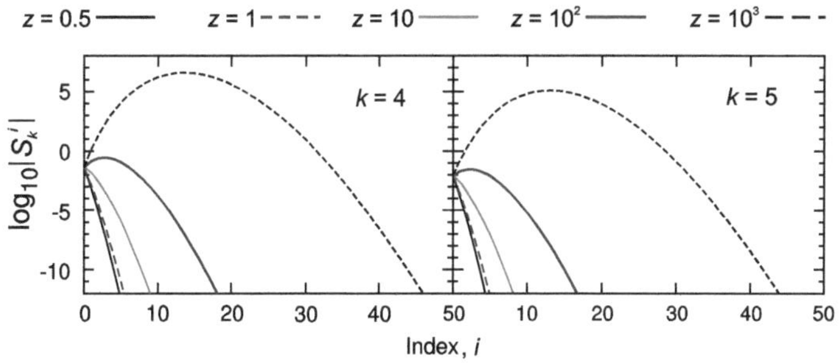

Fig. C.1: Analysis of the growth rate of the terms $S_k^i(z)$.

observed that the terms S_k^i experience changes of several orders of magnitude. As the argument grows the amplitude of the variation increases.

This phenomenon may lead to important losses in accuracy, provided that the least significant digits of the series are lost when added or subtracted from large quantities. Performing the computations in quadruple precision delays the appearance of these problems since round-off errors are reduced. However, as the computation advances the loss of accuracy will eventually appear at some point. To fully overcome this issue the argument of the Stumpff functions is reduced making use of the so-called half-angle relations:

$$\begin{aligned}
\mathfrak{C}_0(4z) &= 2[\mathfrak{C}_0(z)]^2 - 1\\
\mathfrak{C}_1(4z) &= \mathfrak{C}_0(z)\mathfrak{C}_1(z)\\
\mathfrak{C}_2(4z) &= \frac{1}{2}[\mathfrak{C}_1(z)]^2\\
\mathfrak{C}_3(4z) &= \frac{1}{4}[\mathfrak{C}_2(z) + \mathfrak{C}_0(z)\mathfrak{C}_3(z)]\\
\mathfrak{C}_4(4z) &= \frac{1}{8}\{[\mathfrak{C}_2(z)]^2 + 2\mathfrak{C}_4(z)\}\\
\mathfrak{C}_5(4z) &= \frac{1}{32}\{\mathfrak{C}_3(z) + 2\mathfrak{C}_5(z) + 2\mathfrak{C}_1(z)\mathfrak{C}_4(z)\}
\end{aligned}$$

Note that these expressions do not require the computation of higher order terms. They can be applied repeatedly to reduce the value of z below a certain threshold z_{crit}.

Figure C.2 shows a simple experiment to illustrate the previous discussion. The Stumpff function $\mathfrak{C}_0(z)$ is computed in three different ways and the results are compared to the exact solution $\mathfrak{C}_0(z) = \cos\sqrt{z}$. The function $\mathfrak{C}_0(z)$ is first computed by series. Then, it is computed by applying the recurrence relation in Eq. (C.2) starting at $\mathfrak{C}_4(z)$, and starting at $\mathfrak{C}_{10}(z)$. The goal of the experiment is to show how the error grows with the argument z, and how the recursive application of the half-angle formulae keeps the error bounded (setting $z_{\text{crit}} = 0.1$). Computations are performed both in double and quadruple precision, in order to evaluate the effect of truncation errors in the result. It is observed that the error grows exponentially with increasing z. How-

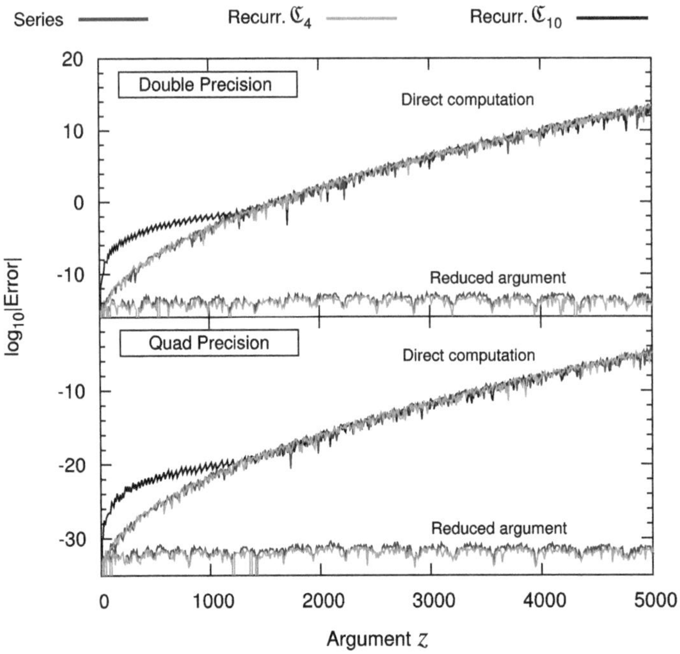

Fig. C.2: Error in computing the Stumpff function $\mathfrak{C}_0(z)$ from the definition by series, from the recurrence relation starting at $\mathfrak{C}_4(z)$, and from the recurrence relation starting at $\mathfrak{C}_{10}(z)$.

ever, when the argument is reduced before computing the series and the recurrence relations, the error exhibits no dependency on the magnitude of the argument. It remains under the tolerance set as the convergence criterion for computing the series. Performing the computations in quadruple precision yields the same error growth rate as in double precision. Although it is possible to obtain valid results up to higher values of z, numerical instabilities still appear. For small values of z there exists a noticeable difference between the solution obtained by recurrence from $\mathfrak{C}_{10}(z)$ and that from $\mathfrak{C}_4(z)$. As z grows, this difference dilutes due to truncation errors. The difference disappears when the argument is reduced below z_{crit}.

There are two possible ways of computing the Stumpff functions by series. First, the series can be truncated when a certain accuracy has been reached. In this case each term is computed sequentially. A possible way to compute the terms in the series is:

$$S_k^i = -\frac{z}{(2i+k)(2i+k-1)} S_k^{i-1} \,, \qquad \text{with } S_k^0 = \frac{1}{k!} \,.$$

Convergence will have been reached when $|S_k^i| < \varepsilon_{\text{tol}}$, where ε_{tol} denotes the tolerance. Second, the series can be truncated a priori and a nested expression for the Stumpff function can be constructed. Different forms of nesting the terms in the Stumpff functions can be found in the cited works.

D Inverse transformations

The present appendix contains the linearized form of the inverse of the series of transformations applied to the problem of relative motion in Chapters 7 and 8. Section D.1 focuses on the transformations involving the equinoctial elements. Next, Sect. D.2 contains the linearized transformation from Cartesian coordinates to Dromo elements. Thanks to this construction an arbitrary relative state vector can be easily transformed into the corresponding differences in the set of Dromo elements. Finally, the linearized form of the inverse KS transformation can be found in Sect. D.3.

D.1 Inverse transformations in equinoctial variables

D.1.1 Asynchronous case

Writing $\mathbf{A} \equiv \mathbf{J}_y$ for short, where $\mathbf{J}_y$ is defined in Eqs. (7.36) and (7.37), the nonzero terms of the inverse matrix $\mathbf{A}^{-1}$ read:

$$\mathrm{A}^{-1}_{1,1} = \frac{2s}{\eta^4}[2(u^2+\eta^2)+3s(s-1)] \qquad \mathrm{A}^{-1}_{1,2} = -\frac{2us^2}{\eta^4}$$

$$\mathrm{A}^{-1}_{2,2} = \frac{[s(1-s)-u^2]\sin\lambda - su\cos\lambda}{\eta^2} \qquad \mathrm{A}^{-1}_{1,4} = \frac{2u}{\eta}$$

$$\mathrm{A}^{-1}_{2,3} = \frac{q_2}{\eta^2}[k_1(q_1+\cos\lambda)+k_2(q_2+\sin\lambda)] \qquad \mathrm{A}^{-1}_{2,4} = -\tan\lambda$$

$$\mathrm{A}^{-1}_{2,6} = -\frac{q_2\eta}{s}(k_1\sin\lambda - k_2\cos\lambda) \qquad \mathrm{A}^{-1}_{2,5} = \frac{\eta}{s}(u\sin\lambda + 2s\cos\lambda)$$

$$\mathrm{A}^{-1}_{3,2} = \frac{[s(s-1)+u^2]\cos\lambda - su\sin\lambda}{\eta^2} \qquad \mathrm{A}^{-1}_{3,1} = \frac{3s^2\sin\lambda - 2su\cos\lambda}{\eta^2}$$

$$\mathrm{A}^{-1}_{3,5} = \frac{\eta}{s}(2q_2 + u\cos\lambda + 2\sin\lambda) \qquad \mathrm{A}^{-1}_{3,4} = \eta\sin\lambda$$

$$\mathrm{A}^{-1}_{4,3} = \frac{\ell^2}{2\eta^2}(q_2+\sin\lambda) \qquad \mathrm{A}^{-1}_{7,1} = -\frac{3s^2}{\eta^4}(s+1)(\tau-\tau_p) + \frac{u}{\eta}$$

$$\mathrm{A}^{-1}_{4,6} = \frac{\ell^2\eta\cos\lambda}{2s} \qquad \mathrm{A}^{-1}_{7,2} = \frac{3s^2u}{\eta^4}(\tau-\tau_p) - \frac{s}{\eta}$$

$$\mathrm{A}^{-1}_{5,6} = \frac{\ell^2\eta\sin\lambda}{2s} \qquad \mathrm{A}^{-1}_{7,4} = -\frac{3u}{\eta}(\tau-\tau_p) + 2r$$

$$\mathrm{A}^{-1}_{6,2} = \frac{1}{r} \qquad \mathrm{A}^{-1}_{7,5} = -\frac{3s}{\eta}(\tau-\tau_p)$$

https://doi.org/10.1515/9783110559125-017

$$A^{-1}_{7,7} = 1 \qquad A^{-1}_{7,6} = -r(k_1 \sin\lambda - k_2 \cos\lambda)$$

$$A^{-1}_{5,3} = -\frac{\ell^2}{2\eta^2}(q_1 + \cos\lambda) \qquad A^{-1}_{7,3} = \frac{1}{\eta}[k_1(q_1 + \cos\lambda) + k_2(q_2 + \sin\lambda)]$$

$$A^{-1}_{6,6} = \frac{\eta}{s}(k_1 \sin\lambda - k_2 \cos\lambda) \qquad A^{-1}_{6,3} = -\frac{1}{\eta^2}[k_1(q_1 + \cos\lambda) + k_2(q_2 + \sin\lambda)]$$

D.1.2 Synchronous case

As in the previous section we write $\mathbf{B} = \mathbf{J}_t$, where $\mathbf{J}_t$ can be found in Eq. (7.43). The nonzero terms in matrix $\mathbf{B}^{-1}$ take the form:

$$B^{-1}_{1,1} = \frac{2s^2(s+1)}{\eta^4}$$

$$B^{-1}_{1,2} = -\frac{2s^2 u}{\eta^4}$$

$$B^{-1}_{1,4} = \frac{2u}{\eta}$$

$$B^{-1}_{1,5} = \frac{2s}{\eta}$$

$$B^{-1}_{4,3} = \frac{\ell^2}{2\eta^2}(q_2 + \sin\lambda)$$

$$B^{-1}_{4,6} = \frac{\ell^2 \eta \cos\lambda}{2s}$$

$$B^{-1}_{5,3} = -\frac{\ell^2}{2\eta^2}(q_1 + \cos\lambda)$$

$$B^{-1}_{5,6} = \frac{\ell^2 \eta \sin\lambda}{2s}$$

$$B^{-1}_{6,5} = \frac{3s}{2\eta^2}(\tau - \tau_p)$$

$$B^{-1}_{3,5} = -\frac{3q_1 s}{\eta^2}(\tau - \tau_p) + \frac{\eta C_2}{s}$$

$$B^{-1}_{6,2} = -\frac{3s^2 u}{2\eta^5}(\tau - \tau_p) + \frac{s}{2\eta^2}$$

$$B^{-1}_{6,4} = \frac{3u}{2\eta^2}(\tau - \tau_p) - \frac{\eta}{s}$$

$$B^{-1}_{2,1} = \frac{3s^2 q_2}{\eta^5}(s+1)(\tau - \tau_p) + 3\cos\lambda + \left(\frac{s+1}{\eta^2} - \frac{2}{s}\right)C_1 + \frac{2u\sin\lambda}{s}$$

$$B^{-1}_{2,2} = -\frac{3q_2 s^2 u}{\eta^5}(\tau - \tau_p) - \frac{uC_1}{\eta^2}$$

$$B^{-1}_{2,4} = +\frac{3q_2 u}{\eta^2}(\tau - \tau_p) + \frac{\eta^3 \sin\lambda + \eta u C_1}{s^2}$$

$$B^{-1}_{2,5} = +\frac{3q_2 s}{\eta^2}(\tau - \tau_p) + \frac{\eta C_1}{s}$$

$$B^{-1}_{3,1} = -\frac{3s^2 q_1}{\eta^5}(s+1)(\tau - \tau_p) + 3\sin\lambda + \left(\frac{s+1}{\eta^2} - \frac{2}{s}\right)C_2 - \frac{2u\cos\lambda}{s}$$

$$B^{-1}_{3,2} = +\frac{3q_1 s^2 u}{\eta^5}(\tau - \tau_p) - \frac{uC_2}{\eta^2}$$

$$B^{-1}_{3,4} = -\frac{3q_1 u}{\eta^2}(\tau - \tau_p) - \frac{\eta^3 \cos\lambda - \eta u C_2}{s^2}$$

$$B^{-1}_{6,1} = \frac{3s^2}{2\eta^5}(s+1)(\tau - \tau_p) - \frac{u}{2\eta^2}$$

$$B^{-1}_{6,3} = -\frac{k_1(q_1 + \cos\lambda) + k_2(q_2 + \sin\lambda)}{2\eta^2}$$

$$B^{-1}_{6,6} = \frac{\eta(k_1 \sin\lambda - k_2 \cos\lambda)}{2s}$$

D.2 Cartesian to Dromo

This section presents the linear form of the transformation from Cartesian coordinates to Dromo elements. It completes the derivation presented in Sect. 8.1. The relative state vector at departure $\delta\mathbf{x}_0^\top = [\delta\mathbf{r}_0^\top, \delta\mathbf{v}_0^\top]$ is transformed to the differential Dromo elements $\delta\mathbf{q} = [\delta\zeta_1, \delta\zeta_2, \delta\zeta_3, \delta\eta_1, \delta\eta_2, \delta\eta_3, \delta\eta_4, \delta\sigma_0]^\top$ by the linear map

$$\delta\mathbf{q} = \mathbf{Q}\,\delta\mathbf{x}_0$$

In order to compute matrix $\mathbf{Q}$ there are two additional conditions that need to be accounted for. First, the condition $\mathfrak{n}\mathfrak{n}^\dagger = 1$ from Eq. (8.3). Second, the differential form of the angle β (between the osculating eccentricity vector and the $x_{\mathcal{P}}$-axis of the departure perifocal frame):

$$\delta\beta = \frac{-\zeta_2\,\delta\zeta_1 + \zeta_1\,\delta\zeta_2}{\sqrt{\zeta_1^2 + \zeta_2^2}}$$

Incorporating these conditions the transformation is given explicitly by

$$Q_{1,1} = +\frac{\zeta_1}{\zeta_2}Q_{2,1} = \zeta_1 s k, \qquad Q_{1,2} = +\frac{\zeta_1}{\zeta_2}Q_{2,2} = \zeta_1 \ell^2 f$$

$$Q_{1,4} = +\frac{\zeta_1}{\zeta_2}Q_{2,4} = \frac{\zeta_1 f}{\zeta_3^3}, \qquad Q_{1,5} = +\frac{\zeta_1}{\zeta_2}Q_{2,5} = \zeta_1 p$$

$$Q_{3,1} = -\frac{s}{u}Q_{3,2} = -\zeta_3^3 s, \qquad Q_{3,5} = -\frac{1}{s}$$

$$Q_{4,1} = -\frac{\eta_2}{\eta_1}Q_{5,1} = \frac{\eta_2}{\eta_4}Q_{6,1} = -\frac{\eta_2}{\eta_3}Q_{7,1} = \frac{s\eta_2 f}{2}$$

$$Q_{4,2} = -\frac{\eta_2}{\eta_1}Q_{5,2} = \frac{\eta_2}{\eta_4}Q_{6,2} = -\frac{\eta_2}{\eta_3}Q_{7,2} = -\frac{\eta_2 k}{2}$$

$$Q_{4,3} = -\frac{M_{43}^-}{M_{21}^-}Q_{6,3} = \frac{M_{43}^-}{M_{34}^+}Q_{5,3} = -\frac{M_{43}^-}{M_{12}^+}Q_{7,3} = -\frac{\zeta_3^2}{2}M_{43}^-$$

$$Q_{4,4} = -\frac{\eta_2}{\eta_1}Q_{5,4} = \frac{\eta_2}{\eta_4}Q_{6,4} = -\frac{\eta_2}{\eta_3}Q_{7,4} = \frac{\eta_2(1-s)}{\zeta_3(1-\ell^2)}$$

$$Q_{4,5} = -\frac{\eta_2}{\eta_1}Q_{5,5} = \frac{\eta_2}{\eta_4}Q_{6,5} = -\frac{\eta_2}{\eta_3}Q_{7,5} = \frac{\eta_2 u(1+s)}{s\zeta_3(1-\ell^2)}$$

$$Q_{4,6} = -\frac{N_{43}^-}{N_{21}^-}Q_{6,4} = \frac{N_{43}^-}{N_{34}^+}Q_{5,6} = -\frac{N_{43}^-}{N_{12}^+}Q_{6,4} = \frac{N_{43}^-}{\zeta_3 s}$$

$$Q_{8,1} = \frac{gu}{\zeta_3^3 s}, \qquad Q_{8,2} = \frac{s^2(1-s) - u^2(1+s)}{\zeta_3^3 s^2}$$

$$Q_{8,4} = \frac{g(1-s)}{s^2\zeta_3^6}, \qquad Q_{8,5} = \frac{gu(1+s)}{s^3\zeta_3^6}$$

with

$$f = \frac{\zeta_3^2 u}{\zeta_1^2 + \zeta_2^2}\,, \qquad k = (s - \ell^2)\frac{f}{u}\,, \qquad p = \frac{s^2 - \ell^2}{s_3\left(\zeta_1^2 + \zeta_2^2\right)}$$

$$N_{ij}^{\pm} = \eta_i \cos\sigma \pm \eta_j \sin\sigma\,, \qquad M_{ij}^{\pm} = uN_{ij}^{\pm} \pm sN_{ji}^{\mp}$$

Note that the transformation is singular for $e^2 = \zeta_1^2 + \zeta_2^2 = 0$, because of the indeterminacy of the perifocal frame and the eccentricity vector. Similarly, the transformation from polar to Cartesian coordinates becomes singular when the transformed vector vanishes.

D.3 Linear form of the Hopf fibration

This section deals with the linearization of the inverse form of the Hopf fibration. The inverse transformation has been defined in two different ways depending on the sign of the component x_0 – Eq. (3.22) –. This gives rise to two alternative definitions of the matrix $\mathbf{T}\colon\, \delta\mathbf{y} \mapsto \delta\mathbf{x}$,

$$\delta\mathbf{y}_0 = \begin{cases} \mathbf{T}^{+}(\mathbf{x}_0)\,\delta\mathbf{x}_0\,, & \text{if } x_0 \geq 0 \\ \mathbf{T}^{-}(\mathbf{x}_0)\,\delta\mathbf{x}_0\,, & \text{if } x_0 < 0 \end{cases}$$

The transformation is referred to $\delta\mathbf{x}_0^{\top} = [\delta\mathbf{r}_0^{\top}, \delta\dot{\mathbf{r}}_0^{\top}]$ meaning that the vanishing fourth component of the position and velocity vectors has been obviated. They do not contribute to $\delta\mathbf{y}_0^{\top} = [\delta\mathbf{u}_0^{\top}, \delta\mathbf{u}_0'^{\top}]$. When the first component of the initial state vector, x_0, is positive the transformation is defined by the matrix $\mathbf{T}^{+}(\mathbf{x}_0)$. Conversely, if the component x_0 is negative then it is more convenient to use the matrix $\mathbf{T}^{-}(\mathbf{x}_0)$. These two matrices read:

$$\mathbf{T}_0^{+} = c_0 \begin{bmatrix} 0, & 0, & 0 & \\ +2z_0, & +2\dfrac{y_0 z_0}{R}, & 2\dfrac{z_0^2}{R} - 4r_0, & [\mathbf{0}_{4\times 3}] \\ -2y_0, & 4r_0 - 2\dfrac{y_0^2}{R}, & -2\dfrac{y_0 z_0}{R}, & \\ -2R, & -2y_0\,, & -2z_0, & \\ -c_1\,, & 2r_0\dot{z}_0 - y_0\dfrac{c_1}{R}, & -2r_0\dot{y}_0 - z_0\dfrac{c_1}{R}, & \\ -R\dot{z}_0 - z_0\dot{x}_0, & -y_0\left(\dot{z}_0 + \dfrac{z_0\dot{x}_0}{R}\right), & 2\dot{x}_0 r_0 - z_0\dot{z}_0 - \dfrac{z_0^2\dot{x}_0}{R}, & [\mathbf{K}^{+}] \\ R\dot{y}_0 + y_0\dot{x}_0, & y_0\dot{y}_0 - 2\dot{x}_0 r_0 + \dfrac{y_0^2\dot{x}_0}{R}, & z_0\left(\dot{y}_0 + \dfrac{y_0\dot{x}_0}{R}\right), & \\ c_2 - \dot{x}_0 R, & y_0\dfrac{c_2}{R} - y_0\dot{x}_0 - 2\dot{y}_0 r_0, & z_0\dfrac{c_2}{R} - z_0\dot{x}_0 - 2\dot{z}_0 r_0, & \end{bmatrix}$$

$$\mathbf{T}_0^- = c_0 \begin{bmatrix} -2z_0, & 2\frac{y_0 z_0}{R}, & -4r_0 + 2\frac{z_0^2}{R}, & \\ 0, & 0, & 0 & \\ 2R, & -2y_0, & -2z_0 & [\mathbf{0}_{4\times 3}] \\ 2y_0, & 4r_0 - 2\frac{y_0^2}{R}, & -2\frac{y_0 z_0}{R}, & \\ R\dot{z}_0 - z_0\dot{x}_0, & -y_0\left(\dot{z}_0 - \frac{z_0\dot{x}_0}{R}\right), & \frac{z_0^2\dot{x}_0}{R} - 2r_0\dot{x}_0 - z_0\dot{z}_0, & \\ c_1, & 2\dot{z}_0 r_0 - y_0\frac{c_1}{R}, & -2\dot{y}_0 r_0 - z_0\frac{c_1}{R}, & [\mathbf{K}^-] \\ -R\dot{x}_0 - c_2, & y_0\dot{x}_0 - 2\dot{y}_0 r_0 + y_0\frac{c_2}{R}, & z_0\dot{x}_0 - 2\dot{z}_0 r_0 + z_0\frac{c_2}{R}, & \\ y_0\dot{x}_0 - R\dot{y}_0, & y_0\dot{y}_0 + 2\dot{x}_0 r_0 - \frac{y_0^2\dot{x}_0}{R}, & z_0\left(\dot{y}_0 - \frac{y_0\dot{x}_0}{R}\right), & \end{bmatrix}$$

having introduced the auxiliary variables:

$$c_0 = \frac{\sqrt{2}}{8}\frac{1}{r_0\sqrt{R}}, \qquad c_1 = y_0\dot{z}_0 - z_0\dot{y}_0$$
$$c_2 = y_0\dot{y}_0 + z_0\dot{z}_0, \qquad R = r_0 + |x_0|$$

and the submatrices

$$\mathbf{K}^+ = \begin{bmatrix} 0 & -2r_0 z_0 & +2r_0 y_0 \\ +2r_0 z_0 & 0 & -2r_0 R \\ -2r_0 y_0 & +2r_0 R & 0 \\ -2r_0 R & -2r_0 y_0 & -2r_0 z_0 \end{bmatrix}, \quad \mathbf{K}^- = \begin{bmatrix} -2r_0 z_0 & 0 & -2r_0 R \\ 0 & -2r_0 z_0 & +2r_0 y_0 \\ +2r_0 R & -2r_0 y_0 & -2r_0 z_0 \\ +2r_0 y_0 & +2r_0 R & 0 \end{bmatrix}$$

E Elliptic integrals and elliptic functions

Elliptic integrals are one of the many examples of geometrical problems leading to important advances in calculus. The English mathematician John Wallis is well known for introducing the symbol "∞" to denote infinity, and he was possibly the first to present a systematic treatment of the problem of rectifying curves (i.e., computing their arclength) in his *Arithmetica infinitorum* (Wallis, 1656). In the 17th century many authors dealt with the problem of computing the arclength of various transcendental curves. The ellipse was deeply studied, and Wallis and also Newton himself derived series solutions to rectify this conic section.

Hendrik van Heuraet in 1659 and Pierre de Fermat in 1660 showed independently that the problem of finding the arclength of a certain curve can be reformulated into the problem of computing an area. This means solving an integral. Following this same line of thought, Jakob Bernoulli published in 1694 the equation of the arclength of his *lemniscate*, which is given by the integral expression

$$\int_{s_0}^{s} \frac{\mathrm{d}t}{\sqrt{1-t^4}} \tag{E.1}$$

Giulio Fagnano was deeply interested in these types of integral expressions arising from transcendental curves, and tried to extend them to the ellipse. On the basis of Fagnano's work, it was Euler who derived a whole theory for the *elliptic integrals*. Later, Adrien-Marie Legendre made significant contributions to the theory and suggested alternative definitions and notation. Nowadays, the term elliptic integral refers to any function that admits the definition

$$\int R(s,t)\,\mathrm{d}t$$

in which $s^2(t)$ is a polynomial of degree three or four in t, and $R(s,t)$ is a rational function including at least an odd power of s. According to this general definition, Eq. (E.1) can be regarded as an elliptic integral.

Legendre culminated Euler's work and defined three fundamental functions:

$$\begin{aligned}
\mathrm{F}(\phi,k) &= \int_0^{\phi} \frac{\mathrm{d}\theta}{\sqrt{1-k^2\sin^2\theta}} = \int_0^{\sin\phi} \frac{\mathrm{d}t}{\sqrt{1-t^2}\sqrt{1-k^2t^2}} \\
\mathrm{E}(\phi,k) &= \int_0^{\phi} \sqrt{1-k^2\sin^2\theta}\,\mathrm{d}\theta = \int_0^{\sin\phi} \frac{\sqrt{1-k^2t^2}}{\sqrt{1-t^2}}\,\mathrm{d}t \\
\Pi(n;\phi,k) &= \int_0^{\phi} \frac{\mathrm{d}\theta}{(1-n\sin^2\theta)\sqrt{1-k^2\sin^2\theta}} = \int_0^{\sin\phi} \frac{\mathrm{d}t}{(1-nt^2)\sqrt{1-t^2}\sqrt{1-k^2t^2}}
\end{aligned}$$

https://doi.org/10.1515/9783110559125-018

They are the incomplete elliptic integrals of the first, second, and third kinds, respectively. Under this notation ϕ is the argument, k is the modulus, and n is the parameter of the elliptic integrals. When the argument takes the value $\phi = \pi/2$ these expressions reduce to the complete elliptic integrals of the first, second, and third kinds:

$$\begin{aligned} \mathrm{K}(k) &= \mathrm{F}(\pi/2, k) \\ \mathrm{E}(k) &= \mathrm{E}(\pi/2, k) \\ \Pi(n; k) &= \Pi(n; \pi/2, k) \end{aligned}$$

E.1 Properties and practical relations

Elliptic integrals appear recursively in the definition of the generalized logarithmic spirals introduced in Chap. 9. In this section we shall discuss some properties of practical nature. These properties are required to arrive at the solutions included in the referenced chapter. More details on the properties can be found in Olver (2010, §19).

A vanishing argument makes the elliptic integrals zero, i.e.,

$$\mathrm{F}(0, k) = \mathrm{E}(0, k) = \Pi(n; 0, k) = 0$$

Similarly, taking $k = 0$ simplifies the elliptic integrals to

$$\mathrm{F}(\phi, 0) = \mathrm{E}(\phi, 0) = \Pi(0; \phi, 0) = \phi \quad \text{and} \quad \Pi(1; \phi, 0) = \tan\phi$$

The incomplete elliptic integrals of the first and second kinds also satisfy the following identities

$$\mathrm{F}(\pi/2, 1) = \infty\,, \qquad \mathrm{E}(\pi/2, 1) = 1\,, \qquad \mathrm{E}(\phi, 1) = \sin\phi$$

E.1.1 Reciprocal-modulus transformation

The reciprocal-modulus transformation allows the user to confine the modulus of the elliptic integral to the interval $k < 1$. Given the modulus $k > 1$, the elliptic integrals can be referred to the reciprocal modulus $k_1 = 1/k < 1$ by means of the formulas:

$$\begin{aligned} \mathrm{F}(\phi, k_1) &= k\,\mathrm{F}(\beta, k) \\ \mathrm{E}(\phi, k_1) &= \frac{1}{k}[\mathrm{E}(\beta, k) - k'^2\,\mathrm{F}(\beta, k)] \\ \Pi(n; \phi, k_1) &= k\Pi(nk^2; \beta, k) \end{aligned}$$

The arguments ϕ and β relate by means of $\sin\beta = k_1 \sin\phi$.

E.1.2 Imaginary-argument transformation

When working in the complex plane it is sometimes useful to apply the following relations:

$$\begin{aligned} F(i\phi, k) &= i\,F(\psi, k') \\ E(i\phi, k) &= i\left[F(\psi, k') - E(\psi, k') + \tan\psi\sqrt{1 - k'^2 \sin^2\psi} \right] \\ \Pi(n; i\phi, k) &= \frac{i}{1-n}\left[F(\psi, k') - n\Pi(1-n; \psi, k') \right] \end{aligned}$$

These formulas involve the change of argument $\sinh\phi = \tan\psi$ and the modulus k is replaced by the complementary modulus, which satisfies

$$k^2 + k'^2 = 1$$

E.1.3 Imaginary-modulus transformation

Elliptic integrals involving a complex modulus verify the formulas

$$\begin{aligned} F(\phi, ik) &= \kappa'\,F(\theta, \kappa) \\ E(\phi, ik) &= \frac{1}{\kappa'}\left[E(\theta, \kappa) - \frac{\kappa^2 \sin\theta\cos\theta}{\sqrt{1 - \kappa^2\sin^2\theta}} \right] \\ \Pi(n; \phi, ik) &= \frac{\kappa'}{n_1}\left[\kappa^2\,F(\theta, \kappa) + n\kappa'^2\,\Pi(n_1; \theta, \kappa) \right] \end{aligned}$$

which involve the auxiliary variables

$$\kappa = \frac{k}{\sqrt{1+k^2}}, \qquad \kappa' = \frac{1}{\sqrt{1+k^2}}, \qquad n_1 = \frac{n+k^2}{1+k^2}$$

and also the arguments

$$\sin\theta = \frac{\sqrt{1+k^2}\,\sin\phi}{\sqrt{1+k^2\sin^2\phi}}$$

E.2 Implementation

E.2.1 Intrinsic functions

Most computing environments come with built-in implementations of elliptic integrals. However, the notation differs from one to another and migrating between them may not be straightforward. The following examples show how elliptic integrals are implemented in Matlab, Maple, and Mathematica.

In Abramowitz and Stegun (1964, p. 614, table 17.5), it is shown that

$$F(\pi/3, 0.8660254040) = 1.21259661$$

Note that the notation has been adapted to meet the standards of this chapter. In Maple (setting `Digits:=15`), this same result can be obtained evaluating the function

```
> evalf(EllipticF(sin(Pi/3),0.8660254040) );
                    1.21259661525498
```

The argument of the function `EllipticF` is $\sin\phi$, instead of ϕ. In Matlab, the functions `elliptic12` and `elliptic3`[1] provide efficient implementations of elliptic integrals. The present example can be solved with

```
>> elliptic12(pi/3,0.8660254040^2)
ans =
              1.21259661537869
```

The function takes $\phi = \pi/3$ as the argument, but the modulus needs to be squared. The convention in Mathematica is the same, with:

```
 In[1]:= EllipticF[Pi/3, 0.8660254040^2]
Out[1]:= 1.21259661537869
```

The differences between implementations of the incomplete elliptic integrals of the third kind reduce to the order in which the inputs are given. According to Abramowitz and Stegun (1964, p. 625, table 17.9):

$$\Pi(0.5; \pi/3, 0.8660254040) = 1.47906$$

The corresponding implementation in Maple yields

```
> evalf(EllipticPi(sin(Pi/3),0.5,0.8660254040) );
                    1.47906355878139
```

In Matlab it is

```
>> elliptic3(pi/3,0.8660254040^2,0.5)
ans =
              1.47906355895353
```

and in Mathematica it is written

```
 In[1]:= EllipticPi[0.5, Pi/3, 0.8660254040^2]
Out[1]:= 1.479063558953542
```

1 https://goo.gl/nsJFT1 and https://goo.gl/NiYTme. A wrapper function `elliptic123` is also available.

E.2.2 Approximation

When a fast evaluation of the elliptic integrals is required (for example, when the generalized logarithmic spirals are to be implemented in a search algorithm) it is often useful to compute them using a simplified approximation. For this reason, the following lines summarize the expressions suggested by Luke (1968):

$$\mathrm{F}(\phi, k) = \frac{1}{2N+1}\left[\phi + 2k^2 \sum_{j=1}^{N} \frac{1}{\sigma_j k^2} \arctan\left(\sigma_j \tan\phi\right)\right]$$

$$\mathrm{E}(\phi, k) = (2N+1)\phi - \frac{2}{2N+1}\sum_{j=1}^{N}\left[\frac{1}{\rho_j}\arctan(\rho_j \tan\phi)\tan^2\theta_j\right]$$

$$\Pi(n; \phi, k) = \frac{1}{2N+1}\left[A\left(1 - 2n\sum_{j=1}^{N}\frac{1}{k^2\sin^2\theta_j - n}\right)\right.$$
$$\left. + 2k^2\sum_{j=1}^{N}\frac{\arctan(\sigma_j\tan\phi)\sin^2\theta_j}{\sigma_j(k^2\sin^2\theta_j - n)}\right]$$

with

$$\sigma_j = \sqrt{1 - k^2\sin^2\theta_j}\,, \qquad \rho_j = \sqrt{1 - k^2\cos^2\theta_j}\,, \qquad \theta_j = \frac{j\pi}{2N+1}$$

and also

$$A = \begin{cases} \dfrac{1}{\sqrt{1-n}}\arctan\left(\sqrt{1-n}\tan\phi\right), & \text{if } n < 1 \\ \dfrac{1}{\sqrt{n-1}}\operatorname{arctanh}\left(\sqrt{n-1}\tan\phi\right), & \text{if } n \geq 1 \end{cases}$$

In most applications, $N = 10$ will suffice.

E.3 Jacobi elliptic functions

Inverting elliptic integrals is not an easy task, and occupied many mathematicians. Gauss approached the problem by the end of the 18th century, although he did not publish anything on the subject (Stillwell, 2010, p. 234). Later, Abel (1827) published some preliminary results that motivated Gustav Jakob Jacobi to complete his theory, presented in his book *Fundamenta nova theoriae functionum ellipticarum* in 1829.

Let u denote the incomplete elliptic integral of the first kind

$$u = \mathrm{F}(\phi, k) = \int_0^{\phi}\frac{\mathrm{d}\theta}{\sqrt{1 - k^2\sin^2\theta}}$$

The Jacobi amplitude is defined in terms of the inverse of the elliptic integral, and takes the form

$$\phi = \mathrm{F}^{-1}(u, k) = \mathrm{am}(u, k)$$

From this result, three important definitions follow:

$$
\begin{aligned}
\sin\phi &= \sin(\mathrm{am}(u,k)) &&= \mathrm{sn}(u,k)\\
\cos\phi &= \cos(\mathrm{am}(u,k)) &&= \mathrm{cn}(u,k)\\
\sqrt{1-k^2\sin^2\phi} &= \sqrt{1-k^2\sin^2(\mathrm{am}(u,k))} &&= \mathrm{dn}(u,k)
\end{aligned}
$$

The functions sn, cn, and dn are the Jacobi elliptic functions. This notation is actually due to Gudermann (1838), who also introduced a fourth function that is missing in most references (cf. p. 12):

$$\mathrm{tn}(u,k) = \frac{\mathrm{sn}(u,k)}{\mathrm{cn}(u,k)}$$

as it was replaced by $\mathrm{sc}(u,k)$. Glaisher (1879) worked with a larger collection of functions, unified under the following notation. Let p, q, r be any of the letters s, c, d, n. Then

$$\mathrm{pq}(u,k) = \frac{\mathrm{pr}(u,k)}{\mathrm{qr}(u,k)} = \frac{1}{\mathrm{qp}(u,k)}$$

The Jacobi elliptic functions satisfy two identities, similar in nature to the trigonometrical identity:

$$
\begin{aligned}
\mathrm{sn}^2(u,k) + \mathrm{cn}^2(u,k) &= 1\\
\mathrm{dn}^2(u,k) + k^2\,\mathrm{sn}^2(u,k) &= 1
\end{aligned}
$$

In fact, the elliptic functions relate to the trigonometric functions by means of

$$\mathrm{sn}(u,0) = \sin(u)\,,\quad \mathrm{cn}(u,0) = \cos(u)\,,\quad \mathrm{dn}(u,0) = 1$$

For $k = 1$ they reduce to the hyperbolic functions

$$\mathrm{sn}(u,1) = \tanh(u)\,,\quad \mathrm{cn}(u,1) = \mathrm{sech}(u)\,,\quad \mathrm{dn}(u,1) = \mathrm{sech}(u)$$

The Jacobi elliptic functions are periodic, with

$$
\begin{aligned}
\mathrm{sn}(u,k) &= \mathrm{sn}(u + 4m\,\mathrm{K}(k), k)\\
\mathrm{cn}(u,k) &= \mathrm{cn}(u + 4m\,\mathrm{K}(k), k)\\
\mathrm{dn}(u,k) &= \mathrm{dn}(u + 2m\,\mathrm{K}(k), k)
\end{aligned}
$$

and $m = 0, 1, 2, \dots$. That is, sn and cn have period $4\,\mathrm{K}(k)$, whereas dn is $2\,\mathrm{K}(k)$-periodic.

Another special function named after Jacobi is the Jacobi zeta function, $\mathrm{Z}(\phi,k)$. It is defined in terms of elliptic integrals of first and second kinds as

$$\mathrm{Z}(\phi,k) = \mathrm{E}(\phi,k) - \frac{\mathrm{E}(k)\,\mathrm{F}(\phi,k)}{\mathrm{K}(k)}$$

E.4 Weierstrass elliptic functions

Karl Weierstrass' first studies were in economy and laws, following the advice of his father, who worked in the Prussian administration. Motivated by the reading of Abel's work, he decided to abandon his studies after four years and to pursue a career in mathematics. He was given the chance to study at the Academy in Münster, where he enrolled in 1839. He was eager to enter the Academy because Christoph Gudermann was a lecturer there, and was teaching possibly the first course in elliptic functions. Gudermann discovered the mathematical talent of Weierstrass and motivated him to work on his own theories.

Most of Weierstrass' theory of elliptic function was developed in a series of unpublished works, collected as part of his lecture series. For practical reasons, we shall focus on the properties of the $\wp$ functions appearing in Chap. 12, and the connection of these functions with elliptic integrals.

According to Whittaker and Watson (1927, p. 454), it was Weierstrass' doctoral student Wilhelm Biermann who published the following result, attributing it to his advisor. Consider the problem of inverting the integral function

$$z = \int_a^x \frac{dt}{\sqrt{f(t)}} \tag{E.2}$$

where $f(x)$ is a quartic polynomial with no repeated roots, and a is not necessarily a zero of $f(x)$. This means solving for $x = x(z, a)$. Thanks to the $\wp$ functions, Eq. (E.2) can be inverted in terms of

$$x = a + \frac{48\wp'(z)\sqrt{f(a)} + bf'(a) + 2f(a)f'''(a)}{4b^2 - f(a)f^{\text{iv}}(a)}$$

having introduced the auxiliary coefficient $b = 24\wp(z) - f''(a)$. In this equation $f^n(x)$ refers to the n-th derivative of $f(x)$. It is easy to verify that, if $a = x_0$ is a root of $f(x)$, then the inverse of Eq. (E.2) reduces to

$$x = x_0 + \frac{f'(x_0)}{4[24\wp(z) - f''(x_0)]}$$

Writing

$$f(x) = a_0x^4 + 4a_1x^3 + 6a_2x^2 + 4a_3x + a_4$$

the invariant lattices

$$g_2 = a_0a_4 - 4a_1a_3 + 3a_2^2$$
$$g_3 = a_0a_2a_4 + 2a_1a_2a_3 - a_2^3 - a_0a_3^2 - a_1^2a_4$$

define the Weierstrass elliptic function

$$\wp(z) = \wp(z; g_2, g_3)$$

in the form that is expected by most software packages. For example, in Maple the $\wp$ and $\wp'$ functions are evaluated with

```
> WeierstrassP(.5, .2, .3);

    4.00317028652538

> WeierstrassPPrime(.5,.2,.3);

    -15.9846346069412
```

In Matlab they can be evaluated using MUPAD, and in Mathematica they are given by

```
In[1]:= WeierstrassP[.5, .2, .3]

Out[1]:= 4.00317028652538 + 0. i

In[2]:= WeierstrassPPrime[.5, .2, .3]

Out[2]:= -15.9846346069412 + 0. i
```

F Controlled generalized logarithmic spirals

This appendix presents the complete reformulation of the generalized logarithmic spirals including the control parameter ξ. Chapters 10–12 used this solution extensively, taking advantage of the similarity transformation that connects the original and the extended solutions. In the following lines the equations

$$\frac{\mathrm{d}v}{\mathrm{d}t} = \frac{\xi - 1}{r^2} \cos\psi \tag{F.1}$$

$$v\frac{\mathrm{d}}{\mathrm{d}t}(\psi + \theta) = \frac{2(1-\xi)}{r^2} \sin\psi \tag{F.2}$$

$$\frac{\mathrm{d}r}{\mathrm{d}t} = v\cos\psi \tag{F.3}$$

$$\frac{\mathrm{d}\theta}{\mathrm{d}t} = \frac{v}{r}\sin\psi \tag{F.4}$$

are integrated explicitly. The solution to the motion of a particle accelerated by

$$\mathbf{a}_p = \frac{1}{r^2}[\xi\cos\psi\,\mathbf{t} + (1-2\xi)\sin\psi\,\mathbf{n}] \tag{F.5}$$

is written here for convenience, instead of using the similarity transformation introduced in Chap. 10 and further developed in Chap. 12. This is an extension of the generalized logarithmic spirals.

Dividing Eqs. (F.1) and (F.3) provides

$$\frac{\mathrm{d}v}{\mathrm{d}r} = \frac{\xi - 1}{r^2 v} \implies v\,\mathrm{d}v = \frac{\xi - 1}{r^2}\,\mathrm{d}r$$

The resulting expression is an equation of separate variables that can be integrated easily to define an integral of motion:

$$v^2 - \frac{2}{r}(1-\xi) = K_1 \tag{F.6}$$

which introduces the control parameter ξ in the generalization of the equation of the energy provided in Chap. 9. The constant of motion K_1 is solved from the initial conditions

$$K_1 = v_0^2 - \frac{2}{r_0}(1-\xi) \tag{F.7}$$

If $\xi = 0$ the integral of motion (F.6) reduces to the equation of the Keplerian energy $\mathcal{E}_k$, with

$$K_1(\xi = 0) \equiv 2\mathcal{E}_k = v^2 - \frac{2}{r}$$

Dividing Eqs. (F.1) and (F.2) provides the relation

$$\frac{\mathrm{d}v}{v} = \frac{1}{2}\cot\psi\,(\mathrm{d}\psi + \mathrm{d}\theta)$$

https://doi.org/10.1515/9783110559125-019

By virtue of Eqs. (F.3–F.4) an equation of separate variables follows

$$2\frac{\mathrm{d}v}{v} + \cot\psi\,\mathrm{d}\psi + \frac{\mathrm{d}r}{r} = 0$$

that can be integrated analytically and defines a first integral:

$$2\ln v + \ln(\sin\psi) + \ln r_2 = \ln C \implies rv^2\sin\psi = K_2 \tag{F.8}$$

This result proves the $\mathscr{S}$-invariance of the integral of motion (9.16).

$$K_2 = r_0 v_0^{\gamma}\sin\psi_0 \tag{F.9}$$

In the following sections, the equations of motion are solved for the elliptic, parabolic, and hyperbolic cases.

F.1 Elliptic spirals

Elliptic spirals relate naturally to Keplerian ellipses; the particle never escapes the potential well of the central body. When propagated forward and backwards in time the spiral falls towards the origin. The trajectory is $\mathscr{T}$-symmetric and the axis of $\mathscr{T}$-symmetry is equivalent to the apse line. There is a natural transition from raising regime to lowering regime at the apoapsis of the spiral. They are defined by negative values of the constant K_1. This determines the values of ξ that yield elliptic spirals:

$$K_1 < 0 \iff \xi < 1 - \frac{r_0 v_0^2}{2}$$

From the integral of motion (F.8) and the condition $\sin\psi \le 1$ it must be that

$$\frac{K_2}{rv^{\gamma}} \le 1 \implies r\left[K_1 + \frac{2}{r}(1-\xi)\right]^{\gamma/2} \ge K_2$$

Thanks to the assumption $\gamma = 2$ the previous expression simplifies to

$$K_1 r + 2(1-\xi) \ge K_2$$

and shows that there is a maximum radius for the case of elliptic spirals ($K_1 < 0$):

$$r_{\max} = \frac{2(1-\xi) - K_2}{(-K_1)} \tag{F.10}$$

The condition on ξ that makes $K_1 < 0$ ensures $r_{\max} > 0$. The maximum radius can be seen as the apoapsis of the spiral. The velocity at $r_{\max}$ is

$$v_m = \sqrt{\frac{K_2}{r_{\max}}} = \sqrt{\frac{-K_1K_2}{2(1-\xi) - K_2}}$$

It is the minimum velocity that a particle can have on an elliptic spiral.

The flight-direction angle can be solved from the integral of motion (F.8) and results in

$$\sin\psi = \frac{K_2}{rv^\gamma} \equiv \frac{K_2}{rv^2} = \frac{K_2}{2(1-\xi) + K_1 r} \tag{F.11}$$

Only positive values of $\sin\psi$ will be considered. This restricts the solution to the case of prograde motions. Retrograde motions can be solved in an analogous way but are omitted for clarity. From Eq. (F.11) it follows that

$$\cos\psi = \pm\frac{\sqrt{[2(1-\xi) + K_1 r]^2 - K_2^2}}{2(1-\xi) + K_1 r} \tag{F.12}$$

The choice of the sign depends on the regime of the spiral. If the spiral is in the raising regime it is $\psi < \pi/2$ and therefore $\cos\psi > 0$. If the spiral is in the lowering regime it is $\cos\psi < 0$.

For $K_1 < 0$ Eq. (F.11) shows that K_2 is constrained to the open interval

$$0 < K_2 < 2(1-\xi)$$

The tangent of ψ is obtained by combining Eqs. (F.11) and (F.12),

$$\tan\psi = \pm\frac{K_2}{\sqrt{[2(1-\xi) + K_1 r]^2 - K_2^2}}$$

and introducing this result in the quotient of Eqs. (F.4) and (F.3) it follows that

$$\frac{\mathrm{d}\theta}{\mathrm{d}r} = \frac{\tan\psi}{r} \implies \mathrm{d}\theta = \pm\frac{K_2\,\mathrm{d}r}{r\sqrt{[2(1-\xi) + K_1 r]^2 - K_2^2}} \tag{F.13}$$

Integrating this equation determines the evolution of the polar angle θ as a function of the radial distance. This angle can be referred to as the axis of symmetry, defined by $\theta(r_{\max}) = \theta_m$, introducing the *spiral anomaly*:

$$\beta = \frac{\ell}{K_2}(\theta - \theta_m) \tag{F.14}$$

The spiral anomaly evolves according to

$$\beta(\theta) = \mp\left|\operatorname{arccosh}\left\{\frac{r_{\max}}{r} - \frac{2(1-\xi)}{K_2}\left(1 - \frac{r_{\max}}{r}\right)\right\}\right| \tag{F.15}$$

having introduced the auxiliary parameter

$$\ell = \sqrt{4(1-\xi)^2 - K_2^2}$$

The angle θ_m defines the orientation of the axis of symmetry, and can be solved directly from the initial conditions. The first sign corresponds to the raising regime and the second sign to the lowering regime.

The equation for the trajectory $r = r(\theta)$ is solved in terms of the spiral anomaly by inverting Eq. (F.15),

$$\frac{r(\theta)}{r_{\max}} = \frac{2(1-\xi) + K_2}{2(1-\xi) + K_2 \cosh \beta(\theta)} \tag{F.16}$$

The $\mathscr{T}$-symmetry of the trajectory is easily verified from this equation. This proves that introducing the control parameter ξ does not break the symmetries of the original family of solutions.

F.1.1 The time of flight

The time relates to the radial distance by means of the equation for the radial velocity, Eq. (F.3):

$$\frac{\mathrm{d}t}{\mathrm{d}r} = \frac{1}{v \cos \psi} = \pm \frac{\sqrt{r[2(1-\xi) + K_1 r]}}{\sqrt{\ell^2 + K_1 r [K_1 r + 4(1-\xi)]}} \tag{F.17}$$

This equation is integrated to provide the time of flight:

$$t(r) - t_m = \pm \frac{rv}{K_1} \sqrt{\frac{1 - \sin \psi}{1 + \sin \psi}} \pm \frac{2[2(1-\xi) k'^2 \Delta \Pi - K_2 \Delta \mathrm{E}] \sqrt{1-\xi}}{(-K_1)^{3/2} \sqrt{K_2}} \tag{F.18}$$

that is referred to the time of passage through the apoapsis $r_{\max}$, denoted t_m. The solution is given in terms of the complete and the incomplete elliptic integrals of the second, $\mathrm{E}(\phi, k)$, and third kinds, $\Pi(p; \phi, k)$, namely:

$$\Delta \mathrm{E} = \mathrm{E}(\phi, k) - \mathrm{E}(k) \,, \qquad \Delta \Pi = \Pi(p; \phi, k) - \Pi(p; k)$$

Their argument, modulus and parameter are, respectively:

$$\sin \phi = \frac{v_m}{v} \sqrt{\frac{2}{1 + \sin \psi}} \,, \qquad k = \sqrt{\frac{-K_1 r_{\max}}{4(1-\xi)}} \,, \qquad p = \frac{K_1 r_{\max}}{2K_2}$$

The complementary modulus k' is defined as $k' = \sqrt{1-k^2}$.

The time of apoapsis passage is solved from the initial conditions:

$$t_m - t_0 = \mp \frac{r_0 v_0}{K_1} \sqrt{\frac{1 - \sin \psi_0}{1 + \sin \psi_0}} \mp \frac{2[2(1-\xi) k'^2 \Delta \Pi_0 - K_2 \Delta \mathrm{E}_0] \sqrt{1-\xi}}{(-K_1)^{3/2} \sqrt{K_2}}$$

The signs in the expressions for the time of flight follow the sign criterion that has already been discussed: the first sign corresponds to the raising regime, and the second to the lowering regime.

F.2 Parabolic spirals

Parabolic spirals are equivalent to logarithmic spirals. For arbitrary values of ξ the velocity is no longer the local circular velocity, but a generalization in terms of the control parameter:

$$v(r) = \sqrt{\frac{2(1-\xi)}{r}}$$

This equation shows how changing the value of the control parameter yields logarithmic spirals but with different velocity profiles. This phenomenon was already discussed by Petropoulos et al. (1999), who found a parameterization of the thrust acceleration required by logarithmic spirals in terms of the orientation of the thrust vector. The value of the control parameter that makes the spiral logarithmic given a set of initial conditions is

$$\xi = 1 - \frac{r_0 v_0^2}{2}$$

Equation (F.11) governs the evolution of the flight-direction angle. Imposing $K_1 = 0$ in this equation yields

$$\sin\psi = \frac{K_2}{2(1-\xi)}$$

so the flight-direction angle remains constant during the propagation. The differential equation connecting the polar angle θ with the radial distance is simply

$$\frac{\mathrm{d}\theta}{\mathrm{d}r} = \frac{\tan\psi}{r} = \pm\frac{K_2}{r\ell}$$

The equation of the trajectory takes the form

$$\frac{r(\theta)}{r_0} = \mathrm{e}^{\pm\ell(\theta-\theta_0)/K_2} \tag{F.19}$$

Observe that

$$\pm\frac{\ell}{K_2} = \cot\psi$$

which is constant, and therefore Eq. (F.19) is no other than the equation of a logarithmic spiral. The thrust acceleration in this case is not tangential to the trajectory, because the normal component defined in Eq. (F.5) is not zero, in general. In fact, Eq. (10.10) shows that the thrust vector forms a constant angle with the radial direction.

As long as $K_1 = 0$, the trajectory will be a logarithmic spiral. If the flight-direction angle is fixed, then changing the value of the control parameter ξ changes the magnitude of the velocity on the spiral, although the trajectory remains the same.

F.2.1 The time of flight

Imposing the conditions $K_1 = 0$ in Eq. (F.17) results in a simple expression that can be integrated to provide the time of flight

$$t(r) - t_0 = \pm\frac{2\sqrt{2(1-\xi)}}{3\ell}\left(r^{3/2} - r_0^{3/2}\right) \tag{F.20}$$

This expression is referred directly to the initial conditions of the problem. The sign depends on the regime of the spiral as in the previous cases. It takes an infinite time to escape to infinity, meaning that the parabolic spirals follow a spiral branch.

F.3 Hyperbolic spirals

The family of hyperbolic generalized logarithmic spirals is defined by $K_1 > 0$. The integral of motion (F.6) sets no limits on the values that the radius can take. In fact, from this expression it follows that a particle traveling along a hyperbolic spiral reaches infinity with a finite, nonzero velocity:

$$\lim_{r\to\infty} v^2 = K_1$$

The constant K_1 is equivalent to the characteristic energy C_3 and readily provides the hyperbolic excess velocity. Once the initial conditions are fixed hyperbolic spirals appear for values of the control parameter satisfying:

$$\xi > 1 - \frac{r_0 v_0^2}{2}$$

Equation (F.11) sets a dynamical constraint that relates the radius with K_1 and K_2, because $\sin\psi \leq 1$. When $K_1 > 0$ this expression holds even for $K_2 > 2(1-\xi)$, unlike for elliptic spirals, but the radius must then satisfy

$$r \geq \frac{K_2 - 2(1-\xi)}{K_1}$$

This equation defines the minimum radius that the spiral can reach. This behavior suggests that there are two different subfamilies of hyperbolic spirals, as anticipated. Hyperbolic spirals of Type I correspond to $K_2 < 2(1-\xi)$. The previous constraint on the radius holds naturally so they escape to infinity if they are in the raising regime, and fall to the origin if in the lowering regime. Hyperbolic spirals of Type II ($K_2 > 2(1-\xi)$) exhibit a minimum radius where the spiral transitions from the lowering regime to the raising regime. These spirals describe two asymptotes. The two types of spirals are separated by the limit case $K_2 = 2(1-\xi)$.

F.3.1 Hyperbolic spirals of Type I

This subfamily of spirals corresponds to $K_1 > 0$ and $K_2 < 2(1-\xi)$. The polar angle is solved by integrating Eq. (F.13) and becomes:

$$\theta - \theta_0 = \pm\frac{K_2}{\ell}\ln\left\{\frac{r\sin\psi}{r_0\sin\psi_0}\left[\frac{2(1-\xi) - K_2\sin\psi_0 + \ell\,|\cos\psi_0|}{2(1-\xi) - K_2\sin\psi + \ell\,|\cos\psi|}\right]\right\}$$

The limit $r \to \infty$ determines the orientation of the asymptote:

$$\theta_{as} = \theta_0 \pm \frac{K_2}{\ell} \ln \left[\frac{K_2(\zeta - \ell - K_2 \sin\psi_0 + \ell\,|\cos\psi_0|)}{r_0 K_1 \zeta \sin\psi_0} \right] \tag{F.21}$$

having considered the auxiliary variable $\zeta = 2(1-\xi) + \ell$.

This type of spiral is not $\mathscr{T}$-symmetric so the definition of the spiral anomaly given in Eq. (F.14) cannot be applied directly. Redefining the angular parameter $\beta(\theta)$ as

$$\beta(\theta) = \pm \frac{\ell}{K_2}(\theta_{as} - \theta)$$

the equation of the trajectory becomes

$$r(\theta) = \frac{\zeta \ell^2 / K_1}{\sinh\frac{\beta}{2}\left[4\zeta(1-\xi)\sinh\frac{\beta}{2} + \left(\zeta^2 - K_2^2\right)\cosh\frac{\beta}{2}\right]} \tag{F.22}$$

The sign of $\beta(\theta)$ depends on the regime of the spiral. So does the value of the orientation of the asymptote given in Eq. (F.21). The regime of the spiral is easily determined from ψ_0. It is straightforward to verify the existence of an asymptote for $\theta = \theta_{as}$ ($\beta = 0$).

F.3.1.1 The time of flight

Inverting the equation of the radial velocity and integrating the resulting expression yields the time of flight,

$$t(r) = K_4 \pm \left\{ \frac{rv}{K_1}\sqrt{\frac{1+\sin\psi}{1-\sin\psi}} - \frac{2\{\mathrm{E} - (1-p)\Pi\}\sqrt{K_2(1-\xi)}}{K_1^{3/2}} \right\}$$

written in terms of a constant K_4. This constant can be easily solved from the previous equation particularized at the initial time. The solution is given in terms of the incomplete elliptic integrals of the second and third kinds, $\mathrm{E} = \mathrm{E}(\phi, k)$ and $\Pi = \Pi(p; \phi, k)$, being its argument:

$$\sin\phi = \sqrt{\frac{K_1 r \sin\psi}{pK_2(1-\sin\psi)}}$$

The modulus and the parameter of the elliptic integrals are:

$$k = \frac{1}{2}\sqrt{\frac{2(1-\xi)+K_2}{1-\xi}} \quad \text{and} \quad p = \frac{2(1-\xi)+K_2}{2K_2}$$

F.3.2 Hyperbolic spirals of Type II

For the case $K_1 > 0$ and $K_2 > 2(1-\xi)$ the condition $\sin\psi \le 1$ provides

$$r_{min} = \frac{K_2 - 2(1-\xi)}{K_1} \tag{F.23}$$

meaning that the spiral will never reach the origin. If the spiral is initially in the lowering regime, it will reach $r_{\min}$ and at this point it transitions to the raising regime. In this case ℓ takes the form

$$\ell = \sqrt{K_2^2 - 4(1-\xi)^2}$$

The polar angle is solved from Eq. (F.13) and yields:

$$\begin{aligned}\theta - \theta_m &= \pm\frac{K_2}{\ell}\left\{\frac{\pi}{2} + \arctan\left[\frac{2(1-\xi)[2(1-\xi)+K_1 r] - K_2^2}{\ell\sqrt{[2(1-\xi)+K_1 r]^2 - K_2^2}}\right]\right\} \\ &= \pm\frac{K_2}{\ell}\left\{\frac{\pi}{2} + \arctan\left[\frac{2(1-\xi) - K_2\sin\psi}{\ell\,|\cos\psi|}\right]\right\}\end{aligned} \tag{F.24}$$

The angle θ_m defines the direction of the axis of $\mathscr{T}$-symmetry. It can be solved from the initial conditions and the previous equation. Note that its definition depends on the regime of the spiral, because there are two possible $\mathscr{C}$-symmetric trajectories.

This equation can be inverted to provide the equation of the trajectory

$$\frac{r(\theta)}{r_{\min}} = \frac{2(1-\xi) + K_2}{2(1-\xi) + K_2\cos\beta} \tag{F.25}$$

having introduced the spiral anomaly:

$$\beta(\theta) = \frac{\ell}{K_2}(\theta - \theta_m)$$

Since the equation for the trajectory depends on the spiral anomaly by means of $\cos\beta$ the trajectory is $\mathscr{T}$-symmetric. The fact that there are two different values of β that cancel the denominator proves the existence of two distinct asymptotes. The asymptotes can be solved from the limit $r \to \infty$ in the equation for the polar angle and correspond to

$$\theta_{\text{as}} = \theta_m \pm \frac{K_2}{\ell}\left\{\frac{\pi}{2} + \arctan\left[\frac{2(1-\xi)}{\ell}\right]\right\}$$

The two asymptotes are given by the two different signs that appear in this equation. They are symmetric with respect to the apse line θ_m.

F.3.2.1 The time of flight

The time of flight is obtained following the same technique applied to elliptic, parabolic, and hyperbolic spirals of Type I. Integrating the inverse of the radial distance renders

$$\begin{aligned}t(r) - t_m = \mp\Bigg\{&\frac{[K_2 + 2(1-\xi)]K_2\,\mathrm{E} - K_1 r_{\min}[K_2\,\mathrm{F} + 2(1-\xi)\Pi]}{K_1\sqrt{K_1 K_2[K_2 + 2(1-\xi)]}} \\ &+ \frac{2(1-\xi)}{K_1^{3/2}}\operatorname{arcsinh}\left[\frac{\sqrt{2K_1 r(rv^2 - K_2)}}{2\sqrt{K_2 rv^2 + (rv^2 - K_2)(1-\xi)}}\right] - \frac{v}{K_1^2}\sqrt{r^2v^4 - K_2^2}\Bigg\}\end{aligned}$$

The solution is given in terms of the incomplete elliptic integrals of the first, $\mathrm{F} \equiv \mathrm{F}(\phi, k)$, second, $\mathrm{E} \equiv \mathrm{E}(\phi, k)$, and third kinds, $\Pi \equiv \Pi(p; \phi, k)$. The argument of the elliptic integrals is given by

$$\sin\phi = \frac{\sqrt{2(1-\xi)(1-\sin\psi)}}{k\sqrt{K_2 - 2(1-\xi)\sin\psi}}$$

and the modulus and parameter read:

$$k = \frac{2\sqrt{1-\xi}}{\sqrt{K_2 + 2(1-\xi)}} \qquad \text{and} \qquad p = \frac{2(1-\xi)}{K_2}$$

The time of flight is referred to t_m. It denotes the time of periapsis passage and is solved from the initial conditions:

$$t_m - t_0 = \pm\left\{\frac{[K_2 + 2(1-\xi)]K_2\,\mathrm{E}_0 - K_1 r_{\min}[K_2\,\mathrm{F}_0 + 2(1-\xi)\Pi_0]}{K_1\sqrt{K_1 K_2 [K_2 + 2(1-\xi)]}}\right.$$
$$+ \frac{2(1-\xi)}{K_1^{3/2}}\operatorname{arcsinh}\left[\frac{\sqrt{2K_1 r_0 (r_0 v_0^2 - K_2)}}{2\sqrt{K_2 r_0 v_0^2 + (r_0 v_0^2 - K_2)(1-\xi)}}\right]$$
$$\left. - \frac{v_0}{K_1^2}\sqrt{r_0^2 v_0^4 - K_2^2}\right\}$$

The regime of the spiral needs to be accounted for when solving this equation.

F.3.3 Limit case $K_2 = 2(1-\xi)$

Hyperbolic spirals of Type I are defined by $K_2 < 2(1-\xi)$ whereas hyperbolic spirals of Type II correspond to $K_2 > 2(1-\xi)$. In the limit case $K_2 \to 2(1-\xi)$, the minimum radius from Eq. (F.23) vanishes. This means that a spiral in the lowering regime will reach the origin but will not be able to transition to the raising regime and escape.

The equation for the evolution of the polar angle reduces to:

$$\theta(r) - \theta_0 = \mp\left(\sqrt{1 + \frac{4(1-\xi)}{K_1 r}} - \sqrt{1 + \frac{4(1-\xi)}{K_1 r_0}}\right)$$

There is only one asymptote whose definition depends on the initial regime of the spiral:

$$\theta_{as} - \theta_0 = \mp\left(1 - \sqrt{1 + \frac{4(1-\xi)}{K_1 r_0}}\right)$$

The spiral anomaly in this case is defined with respect to the direction of the asymptote,

$$\beta = \theta - \theta_{as}$$

Inverting the equation for the polar angle defines the trajectory:

$$r(\theta) = \frac{4(1-\xi)}{K_1\beta(\beta \mp 2)}, \qquad \text{with} \quad \beta = \theta - \theta_{\text{as}}$$

Recall that the first sign corresponds to the raising regime and the second to the lowering regime.

F.3.3.1 The time of flight

The time of flight can be directly related to the initial conditions, namely

$$t(r) - t_0 = \pm\frac{1}{K_1^{3/2}}\left\{\Xi - \Xi_0 + (1-\xi)\ln\left[\frac{r_0v_0^2 + 1 - \xi + \Xi_0}{rv^2 + 1 - \xi + \Xi}\right]\right\}$$

The auxiliary parameter $\Xi = \Xi(r)$ reads

$$\Xi(r) = v\sqrt{r[rv^2 + 2(1-\xi)]}$$

F.4 Osculating elements

The orbital elements defining the osculating orbit can be referred to the constants defining the spirals and the radial distance. In particular, the eccentricity reads

$$e(r) = \sqrt{K_2^2 + 1 - 2K_2 \sin\psi} \tag{F.26}$$

The angular momentum relates to the constant K_2: Roa et al. (2016a) connect Eq. (10.8) with the torque from the perturbed forces, and from that derivation it follows that

$$h(r) = K_2\sqrt{\frac{r}{2(1-\xi) + K_1 r}} \tag{F.27}$$

The semimajor axis is obtained by combining Eqs. (F.26) and (F.27), and results in

$$a(r) = \frac{h^2}{1-e^2} = \frac{r}{2\xi - K_1 r} \tag{F.28}$$

The semimajor axis always grows in the raising regime and decreases in the lowering regime, no matter the type of spiral.

G Dynamics in Seiffert's spherical spirals

Some 70 years after Carl G. J. Jacobi (1829) published his celebrated theory of elliptic functions, Alfred Seiffert (1896) equipped them with a beautiful geometric interpretation. He showed that the Jacobi elliptic functions render a three-dimensional curve confined to a spherical surface. The polar angle grows linearly with the arclength parameter and the curve connects the poles of the sphere. This defines what was later called a *Seiffert spiral*. More recently, Erdös (2000) recovered Seiffert's spirals to describe the motion of a plane circumnavigating the Earth with constant velocity. Chapters 9–11 proved the potential of using known curves for mission design. Following this line of thought, this chapter recovers Seiffert's spirals in order to characterize its dynamics from an orbital perspective.

G.1 Dynamics

Seiffert's spherical spiral is confined to a sphere of constant radius. In normalized variables which make the radius of the sphere equal unity, the spiral can be defined in cylindrical coordinates as:

$$r = \mathrm{sn}(u, k) \tag{G.1}$$

$$\theta = \theta_r + ku \tag{G.2}$$

$$z = \mathrm{cn}(u, k) \tag{G.3}$$

It is defined in terms of the Jacobi elliptic functions $\mathrm{sn}(u, k)$ and $\mathrm{cn}(u, k)$. The definition of these functions and some useful properties can be found in Appendix E. Here k is their modulus and u is the argument. In order to avoid singularities at the poles

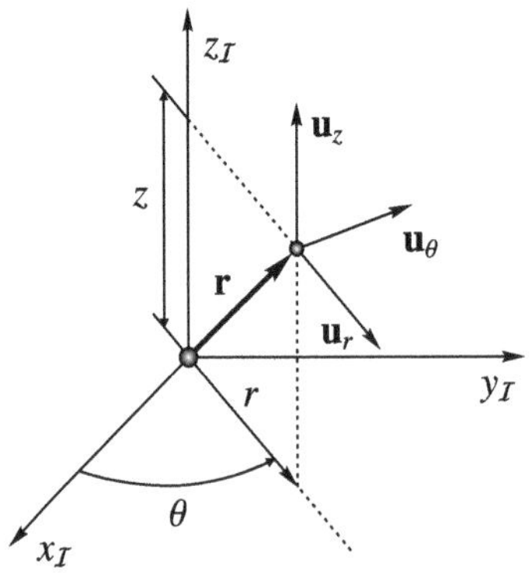

Fig. G.1: Geometry of the problem.

https://doi.org/10.1515/9783110559125-020

we adopt the sign criterion from Erdös (2000): the cylindrical radius r takes values in the interval $r \in [-1, 1]$ and the negative values of the radius ensure a smooth transition through the poles. He recovered these spirals and used them to explain the Jacobi elliptic functions from an academic perspective. The results presented here can be regarded as an extension of his analysis to orbital motion.

The angular variable u takes the form

$$u = nt + \phi_0$$

and shows that the angular velocity in Seiffert's spiral is constant. The coefficient n denotes the mean motion on the spiral, ϕ_0 controls the phasing, and θ_r determines the orientation of the spiral in the inertial space. The parameters $\{n, k, \phi_0, \theta_r\}$ define the trajectory.

Let $\mathcal{L} = \{\mathbf{u}_r, \mathbf{u}_\theta, \mathbf{u}_z\}$ be a rotating frame defined by the radial, circumferential, and normal unit vectors. The radius vector to the particle $\mathbf{r}$ is given by

$$\mathbf{r} = r\,\mathbf{u}_r + z\,\mathbf{u}_z$$

Figure G.1 depicts the construction of the problem referred to an inertial frame $\mathcal{I}$.

Geometrically, the spiral is confined to a spherical surface of radius $R^2 = r^2 + z^2 = 1$. The (positive) modulus of the Jacobi elliptic functions k controls the shape of the spiral. For $k > 1$ the out-of-plane component of the motion, z, never vanishes. The trajectory is confined to one of the hemispheres, the northern or the southern. For $k < 1$ the spiral is able to transition from the northern hemisphere to the southern hemisphere, and vice-versa. The limit case $k = 1$ is the asymptotic limit for which $z \to 0$ as $t \to \infty$.

Figure G.2 shows examples of orbits around Mars for different values of the modulus of the Jacobi elliptic functions. The case $k < 1$ provides a full coverage of the planet with the orbit going through the poles. For $k = 1$ the spiral only reaches the pole once and then approaches the equator. For $k > 1$ the orbit is confined to one single hemisphere. The vertical axis $z_{\mathcal{I}}$ coincides with the polar axis of the planet.

(a) $k < 1$ **(b)** $k = 1$ **(c)** $k > 1$

Fig. G.2: Seiffert's spherical spiral trajectories around Mars defined by different values of the modulus k.

The trajectory inherits the periodicity of the Jacobi elliptic functions: the spiral is periodic,

$$r(t + T_{sp}) = r(t) \quad \text{and} \quad z(t + T_{sp}) = z(t)$$

with period

$$T_{sp} = 4\,\mathrm{K}(k)/n$$

where $\mathrm{K}(k)$ denotes the complete elliptic integral of the first kind. Note that for $k = 1$ it is

$$\lim_{k\to 1} \mathrm{K}(k) = \infty \implies \lim_{k\to 1} T_{sp} = \infty$$

so it takes an infinite time to complete one revolution. A revolution in this case is considered to be complete when r and z take the same value again.

The velocity on the spiral can be obtained by differentiating Eqs. (G.1–G.3) with respect to time and yields

$$\dot{r} = +n\,\mathrm{cn}(u,k)\,\mathrm{dn}(u,k) \tag{G.4}$$

$$\dot{\theta} = kn \tag{G.5}$$

$$\dot{z} = -n\,\mathrm{sn}(u,k)\,\mathrm{dn}(u,k) \tag{G.6}$$

The angular velocity is simply kn. The initial position relates to the parameters of the spiral $\{n, \phi_0, k, \theta_r\}$ through

$$\begin{aligned} r_0 &= \mathrm{sn}(\phi_0, k) \\ \theta_0 &= \theta_r + k\phi_0 \\ z_0 &= \mathrm{cn}(\phi_0, k) \end{aligned}$$

and the initial velocity in cylindrical coordinates reads

$$\begin{aligned} \dot{r}_0 &= +n\,\mathrm{cn}(\phi_0,k)\,\mathrm{dn}(\phi_0,k) \\ \dot{\theta}_0 &= kn \\ \dot{z}_0 &= -n\,\mathrm{sn}(\phi_0,k)\,\mathrm{dn}(\phi_0,k) \end{aligned}$$

If the parameters $\{n, k, \phi_0, \theta_r\}$ are fixed the resulting Seiffert spiral is completely determined. The initial conditions must satisfy the previous relations. On the contrary, if the problem is initialized using the initial state vector the set of parameters defining the spiral need to be computed.

G.1.1 The accelerated two-body problem

The dynamics of a particle orbiting a central body and subject to a perturbing acceleration $\mathbf{a}_p$ abide by

$$\frac{\mathrm{d}^2\mathbf{r}}{\mathrm{d}t^2} + \frac{\mu}{R^3}\mathbf{r} = \mathbf{a}_p$$

In what follows the gravitational parameter μ is normalized to unity. Consider that the trajectory of the particle describes a Seiffert spherical spiral, so its motion is constrained to the surface of a sphere of unit radius, $||\mathbf{r}|| = 1$. The circular velocity on the sphere becomes unity with this normalization.

Under these assumptions the required acceleration for the particle to follow a Seiffert spherical spiral takes the form:

$$\mathbf{a}_p = a_{p,r}\,\mathbf{u}_r + a_{p,\theta}\,\mathbf{u}_\theta + a_{p,z}\,\mathbf{u}_z$$

and the components in frame $\mathcal{L}$ are

$$a_{p,r} = \operatorname{sn}(u,k)\left\{1 - n^2[1 + 2k^2\operatorname{cn}^2(u,k)]\right\} \tag{G.7}$$

$$a_{p,\theta} = 2n^2 k\operatorname{cn}(u,k)\operatorname{dn}(u,k) \tag{G.8}$$

$$a_{p,z} = \operatorname{cn}(u,k)\left\{1 - n^2[1 - 2k^2\operatorname{sn}^2(u,k)]\right\} \tag{G.9}$$

The magnitude of the acceleration due to the thrust reads

$$a_p = \sqrt{1 - 2n^2 + n^4[1 + 4k^2\operatorname{cn}^2(u,k)]}$$

It only depends on the time by means of the term $\operatorname{cn}^2(u,k)$. The maximum and minimum accelerations are, respectively

$$a_{p,\max} = \sqrt{1 - 2n^2 + n^4(1 + 4k^2)}$$

$$a_{p,\min} = |n^2 - 1|$$

The maxima and minima of the acceleration occur at

$$t_{a_p,\max} = \frac{1}{n}[2j\,\mathrm{K}(k) - \phi_0] \tag{G.10}$$

$$t_{a_p,\min} = \frac{1}{n}[(2j-1)\,\mathrm{K}(k) - \phi_0] \tag{G.11}$$

Here j is an integer that accounts for the periodicity of the function. The maximum acceleration is reached at the pole of the sphere.

When $n = 1$ the minimum acceleration vanishes. The minimum acceleration is obtained for $\operatorname{cn}(u,k) = 0$, which corresponds to the point where the trajectory crosses the equator ($r = 1$ and $z = 0$). In the limit case $k = 1$ the trajectory evolves from the north pole of the sphere to the equator following an asymptotic trajectory. Given the limit $\lim_{k\to 1}\mathrm{K}(k) = \infty$, the time to reach the maximum acceleration, defined in Eq. (G.10), will only be finite for $j = 0$. The minimum acceleration is reached in infinite time when propagated both forward and backward in time.

For $k = 0$ the acceleration reduces to

$$a_{p,r} = (1 - n^2)\sin u$$

$$a_{p,\theta} = 0$$

$$a_{p,z} = (1 - n^2)\cos u$$

and it vanishes for $n = 1$. This means that Seiffert's spiral with $k = 0$ and $n = 1$ is in fact a Keplerian orbit, as will be discussed in more detail in the following section. Introducing a positive parameter δ to quantify how much n separates from unity,

$$n = 1 \pm \delta$$

it follows that the low-thrust scenario corresponds to $k \ll 1$ and $\delta \ll 1$. The magnitude of the thrust in the tangential direction will be small compared to the radial acceleration. In addition, a direct consequence of $k \ll 1$ is that the angular velocity on the spiral will be small, because $\dot{\theta} = kn$.

G.1.2 Integrals of motion

The Keplerian energy of the system $\mathcal{E}_k$ evolves in time due to the power performed by the disturbing acceleration $\mathbf{a}_p$,

$$\frac{d\mathcal{E}_k}{dt} = \mathbf{a}_p \cdot \mathbf{v}$$

Considering the components of the velocity and the acceleration $\mathbf{a}_p$, given respectively in Eqs. (G.4–G.6) and (G.7–G.9), it is

$$\frac{d\mathcal{E}_k}{dt} = 0$$

so the Keplerian energy of the system is conserved. Moreover, the velocity on the spiral can be solved from Eqs. (G.4–G.6) and reduces to

$$v = n$$

provided that the angular velocity is constant and the motion is constrained to a spherical surface. In the definition of the Keplerian energy

$$\mathcal{E}_k = \frac{v^2}{2} - \frac{1}{r}$$

both terms are constant. So it is the semimajor axis,

$$a = \frac{1}{2 - n^2}$$

The acceleration due to the thrust can be rewritten in terms of the components of the motion as

$$a_{p,r}(r) = r\left\{1 + n^2[2k^2(r^2 - 1) - 1]\right\} \tag{G.12}$$

$$a_{p,\theta}(\theta) = 2n^2 k \,\mathrm{cn}\left(\frac{\theta - \theta_r}{k}, k\right) \mathrm{dn}\left(\frac{\theta - \theta_r}{k}, k\right) \tag{G.13}$$

$$a_{p,z}(z) = z\left\{1 + n^2[2k^2(1 - z^2) - 1]\right\} \tag{G.14}$$

The components of the acceleration derive from a perturbing potential

$$\mathbf{a}_p = -\nabla \mathcal{V}_p$$

This shows that the problem is conservative. The disturbing potential $\mathcal{V}_p$ is separable,

$$\mathcal{V}_p(r, \theta, z) = \mathcal{V}_r(r) + \mathcal{V}_\theta(\theta) + \mathcal{V}_z(z)$$

meaning that[1]

$$a_{p,r} = -\frac{\partial \mathcal{V}_r}{\partial r}, \quad a_{p,\theta} = -\operatorname{ns}\left(\frac{\theta - \theta_r}{k}, k\right)\frac{\partial \mathcal{V}_\theta}{\partial \theta}, \quad a_{p,z} = -\frac{\partial \mathcal{V}_z}{\partial z}$$

These three equations of separate variables can be integrated to provide

$$\begin{aligned}
\mathcal{V}_r(r) &= -\frac{r^2}{2}\left\{1 - n^2\left[1 - k^2(r^2 - 2)\right]\right\} \\
\mathcal{V}_\theta(\theta) &= -n^2 k^2 \operatorname{sn}^2\left(\frac{\theta - \theta_r}{k}\right) \\
\mathcal{V}_z(z) &= -\frac{z^2}{2}\left\{1 - n^2\left[1 + k^2(z^2 - 2)\right]\right\}
\end{aligned}$$

The total energy

$$\mathcal{E} = \mathcal{E}_k + \mathcal{V}_p = \frac{1}{2}(k^2 n^2 - 1) \tag{G.15}$$

is conserved and depends only on the angular velocity kn.

G.1.3 The osculating orbit

The angular momentum vector is solved from its definition $\mathbf{h} = \mathbf{r} \times \mathbf{v}$ and can be projected onto the rotating frame $\mathcal{L}$ as

$$\mathbf{h} = n\{k \operatorname{sn}(u, k)[\operatorname{sn}(u, k)\,\mathbf{u}_z - \operatorname{cn}(u, k)\,\mathbf{u}_r] + \operatorname{dn}(u, k)\,\mathbf{u}_\theta\}$$

Its magnitude is simply

$$h = n$$

which is constant. The equation of the angular momentum constitutes a first integral of the motion.

1 In order to find the equation for $a_{p,\theta}$ we transform Eq. (G.1) by means of Glaisher's rule:

$$\frac{1}{r} = \frac{1}{\operatorname{sn}(u, k)} = \operatorname{ns}(u, k)$$

and noting that $u = (\theta - \theta_r)/k$.

The eccentricity vector takes the form

$$\mathbf{e} = |n^2 - 1| \left[\operatorname{sn}(u, k)\, \mathbf{u}_r + \operatorname{cn}(u, k)\, \mathbf{u}_z \right] = |n^2 - 1|\, \mathbf{r}$$

Although the direction of the eccentricity vector changes in time its magnitude,

$$e = |n^2 - 1|$$

remains constant, meaning that the acceleration in Eqs. (G.7–G.9) does not affect the eccentricity of the osculating orbit. In addition, the apse line is directed along the radius vector $\mathbf{r}$. The particle is always at periapsis of the osculating orbit. Recall that $||\mathbf{r}|| = 1$.

The thrust profile $\mathbf{a}_p$ conserves the Keplerian (and the total) energy of the system so the osculating orbit remains unchanged. It is only rotated so that the eccentricity vector always follows the direction of the radius vector.

The inclination of the osculating orbit can be solved from the relation

$$\cos i = k \operatorname{sn}^2(u, k)$$

It oscillates with period $4\,\mathrm{K}(k)/n$ and the osculating orbit becomes polar ($\cos i = 0$) at

$$t_{\text{polar}} = \frac{1}{n}\left[2j\ \mathrm{K}(k) - \phi_0\right]$$

with $j = 0, 1, 2, \ldots$ This time coincides with the time when the spiral crosses the pole.

G.2 The geometry of Seiffert's spherical spirals

The equations for the trajectory have been defined in terms of the Jacobi elliptic functions $\operatorname{sn}(u, k)$ and $\operatorname{cn}(u, k)$. The geometry of the resulting Seiffert spiral is therefore defined by the properties of such functions. Different families of solutions are found depending on the values of the modulus k.

First note that $kn < 0$ defines retrograde motions ($\dot{\theta} < 0$). This problem is dynamically equivalent to the prograde case and is omitted from this analysis for convenience. In the limit case $k = 0$ Eq. (G.5) shows that the polar angle θ remains constant, so the trajectory degenerates into a circular (non-Keplerian) polar orbit on the spherical surface. The particle moves according to

$$r = \sin u \qquad \text{and} \qquad z = \cos u$$

The motion in the circular orbit is given as an explicit function of the time, provided that $u = nt + \phi_0$. Introducing $k = 0$ in the definition of the thrust acceleration will render

$$\mathbf{a}_p = (1 - n^2)(\sin u\, \mathbf{u}_r + \cos u\, \mathbf{u}_z) = (1 - n^2)\, \mathbf{r} \tag{G.16}$$

The acceleration opposes the gravitational attraction of the central body and modifies the velocity of the resulting orbit. In a Keplerian circular orbit with the same radius the velocity is normalized to unity, whereas for degenerate Seiffert's spirals it is $v = n$. When $n = 1$ the acceleration in Eq. (G.16) vanishes and the velocity matches the circular velocity of the Keplerian orbit.

The modulus of the Jacobi elliptic functions is typically assumed to be $k \in [0, 1]$. For $k < 1$ the functions $\mathrm{sn}(u, k)$ and $\mathrm{cn}(u, k)$ oscillate between -1 and $+1$. Since $r \in [-1, 1]$ by the sign convention that we adopted, the spiral crosses the poles. In addition, the fact that z changes its sign means that the particle can travel from the northern hemisphere to the southern hemisphere, and vice-versa. The angular velocity takes nonzero values and the polar angle evolves linearly in time.

When the modulus becomes unity the Jacobi elliptic functions transform into:

$$r = \mathrm{sn}(u, 1) \equiv \tanh u \qquad \text{and} \qquad z = \mathrm{cn}(u, 1) \equiv \mathrm{sech}\, u$$

At $u = 0$ it is $r = 0$ and $z = 1$, which corresponds to the north pole of the sphere. But for $t \to \pm\infty$ (with negative times meaning backward propagation) it is

$$\lim_{t\to\pm\infty} r = \pm 1 \qquad \text{and} \qquad \lim_{t\to\pm\infty} z = 0$$

The spiral reaches the pole once but it then approaches the equator asymptotically. The pole is reached at $t = -\phi_0/n$.

If the modulus lies in the interval $k > 1$ the Jacobi elliptic functions defining the trajectory transform into

$$\mathrm{sn}(u, k) = \frac{1}{k}\,\mathrm{sn}(ku, 1/k)$$
$$\mathrm{cn}(u, k) = \mathrm{dn}(ku, 1/k)$$

The out-of-plane component of the motion is now defined by the function $\mathrm{dn}(u, 1/k)$, with $1/k < 1$. This function does not change its sign. The resulting spiral is confined to one hemisphere and it cannot cross the equator.

It is remarkable that the two-body problem subject to the perturbing acceleration in Eqs. (G.7–G.9) admits an explicit solution in terms of the time. This is not the case even in the noncircular Kepler problem, where the time dependency is given by the solution to Kepler's equation. Tanguay (1960) and Petropoulos et al. (1999) found similar behaviors for special cases arising from the logarithmic spiral.

In the previous section it was shown that the acceleration will be small for $k \ll 1$ and $\delta \ll 1$, with $n = 1 \pm \delta$ and $\delta > 0$. The condition $n \sim 1$ is equivalent to the velocity on Seiffert's spiral being close to the local circular velocity. Since $k < 1$ the particle will travel from the north pole to the south pole and back, and $k \ll 1$ makes the angular velocity small.

G.3 Groundtracks

Let (λ, ψ) denote the latitude and longitude of the subsatellite point on the surface of the central body, and let ω_0 denote the angular velocity for the rotational motion of the central body.

Consider that θ at $t = 0$ defines the polar angle between the particle and the prime meridian at departure, i.e.,

$$\lambda_0 = \theta_r + k\phi_0$$

The longitude is then given by

$$\lambda(t) = \lambda_0 + (kn - \omega_0)t \tag{G.17}$$

The latitude of the subsatellite point is solved from

$$\psi(t) = \arctan\frac{z}{r} = \arctan\left[\frac{\mathrm{cn}(u,k)}{\mathrm{sn}(u,k)}\right] = \arctan\left[\,\mathrm{cs}(u,k)\right] \tag{G.18}$$

The last simplification follows from Glaisher's notation for the Jacobi elliptic functions (Appendix E).

Equation (G.17) shows that when $kn = \omega_0$ the angular velocity of the spiral matches that of the central body, so the longitude remains constant and equal to λ_0. The latitude vanishes for

$$u = 2j\ \mathrm{K}(k) \implies t = \frac{1}{n}[2j\ \mathrm{K}(k) - \phi_0]$$

meaning that the osculating orbit is a polar orbit precisely when the particle crosses the equator. In the limit case $k = 1$ the latitude given in Eq. (G.18) transforms into

$$\psi(t) = \arctan(\mathrm{csch}\, u)$$

This expression shows that the latitude becomes $\psi = \pi/2$ at $t = 0$, and the orbit becomes equatorial for $t \to \pm\infty$. This behavior was described in the previous section when studying the nature of the solutions. For $k > 1$ the latitude is

$$\psi(t) = \arctan\left[k\,\mathrm{ds}(ku, 1/k)\right]$$

Provided that $\mathrm{ds}(ku, 1/k) > 0 \Leftrightarrow ku > 0$, the motion is constrained to the northern hemisphere.

Seiffert's spirals can be used to design missions that observe the same longitude. Canceling the secular term in Eq. (G.17) will make $\lambda(t) = \lambda_0$. For this, the parameter k must satisfy

$$k = \omega_0/n$$

In order to control the maximum value of the acceleration we find that

$$a_{p,\max} = \sqrt{1 - 2n^2 + n^4[1 + 4(\omega_0/n)^2]}$$

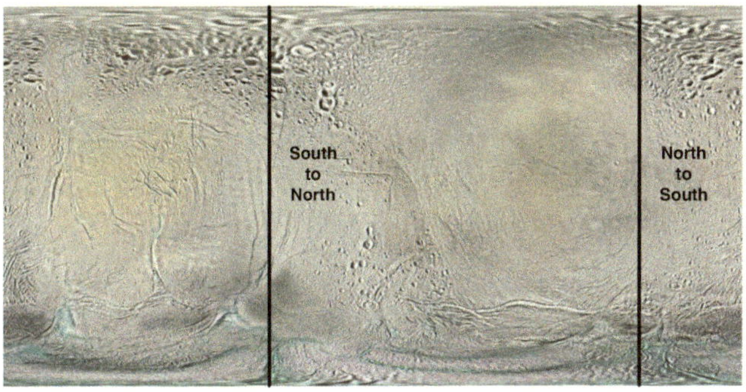

Fig. G.3: Groundtracks on the surface of Enceladus.

is minimum for

$$n_{\rm opt} = \sqrt{1 - 2\omega_0^2}$$

This is the optimal value of the orbital velocity, provided that it minimizes the thrust acceleration. In order to ensure that $k < 1$ it must be $n_{\rm opt} < \omega_0$. This condition yields a maximum height over the surface of the central body above which it is not possible to obtain constant groundtracks.

Figure G.3 shows a hypothetical example of an observation mission around Enceladus. Porco et al. (2006) found that its south pole is active. This region exhibits the highest temperature of the ice crust, and there are jets of icy particles emanating from the moon. The plume feeds Saturn's E-ring and, in October 28, 2015, the Cassini spacecraft passed through the geyser in order to analyze its composition. If an orbiter were placed in a polar orbit around Enceladus, it would go through the plume repeatedly and throw light on the mysterious origin and composition of the moon. Seiffert's spirals provide a very simple strategy for designing polar orbits and controlling their groundtrack. Figure G.3 shows a possible orbit, in which the probe is constantly observing the same parallel. This technique maximizes the observational time of important features in the moon's south pole. This kind of mission is also suitable for the use of cubesats deployed from the carrier, which serve as a communication relay for transmitting data back to Earth.

The spiral is polar, and the altitude of the orbit is 148 km. The south pole coverage of a 2.23-day orbit can be seen in Fig. G.4.

The required control acceleration is shown in Fig. G.5. The downside of this particular design is the high acceleration levels required for sustaining the spiral trajectory, although it serves as an illustrative example of the geometry of the spirals. The propellant mass fraction is approximately 6% of the initial mass.

Fig. G.4: Coverage of Enceladus' south pole using a Seiffert spiral.

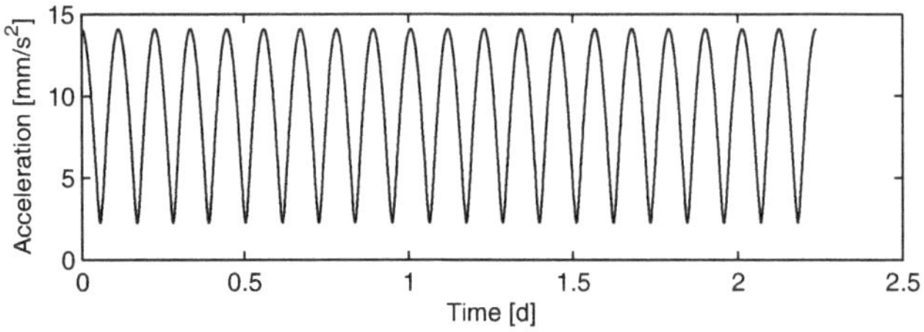

Fig. G.5: Control acceleration in Seiffert's spherical spiral.

G.4 Relative motion between Seiffert's spirals

The simple form of the solution to the motion along Seiffert's spirals invites one to model the relative dynamics of spacecraft following two different spirals. Let $\delta\mathbf{r}$ denote the relative position of the follower spacecraft with respect to the leader spacecraft, both describing a Seiffert spherical spiral. Assuming that the relative separation is small the first-order solution to relative motion reduces to:

$$\delta\mathbf{r} = \delta r\,\mathbf{u}_r + r\delta\theta\,\mathbf{u}_\theta + \delta z\,\mathbf{u}_z$$
$$\delta\dot{\mathbf{r}} = (\delta\dot{r} - r\dot{\theta}\delta\theta)\,\mathbf{u}_r + (\dot{r}\delta\theta + \delta r\dot{\theta} + r\delta\dot{\theta})\,\mathbf{u}_\theta + \delta\dot{z}\,\mathbf{u}_z$$

The orbit of the follower spacecraft can be constructed by considering that the parameters defining the orbit of the follower spacecraft read $\{n{+}\delta n,\ k{+}\delta k,\ \phi_0{+}\delta\phi_0,\ \theta_r{+}\delta\theta_r\}$, and the radius of the spiral is $1 + \delta R$. The set of differences $\delta\square$ are small to comply with the hypothesis of linear relative motion.

The components of the relative position vector take the form

$$\delta r = r\delta R - \frac{\delta k}{k'^2}\left[\frac{\dot{r}}{nk}\Lambda^-(u,k) - z^2kr\right] + \frac{\dot{r}}{n}(t\delta n + \delta\phi_0)$$

$$\delta\theta = \delta\theta_r + u\,\delta k + k(t\,\delta n + \delta\phi_0)$$

$$\delta z = z\delta R - \frac{\delta k}{k'^2}\left[\frac{\dot{z}}{nk}\Lambda^-(u,k) + zkr^2\right] + \frac{\dot{z}}{n}(t\,\delta n + \delta\phi_0)$$

We have introduced the auxiliary function

$$\Lambda^{\pm}(u,k) = \mathrm{Z}(u,k) + u\left[\frac{\mathrm{E}(k)}{\mathrm{K}(k)} \pm k'^2\right], \qquad k'^2 = 1 - k^2$$

defined in terms of Jacobi's zeta function $\mathrm{Z}(u,k)$ (Appendix E). It is worth noting that there are secular terms, growing linearly with t or u. The condition for canceling these terms and therefore defining closed orbits is

$$\delta k = \delta n = 0$$

Equation (G.15) shows that the previous condition translates into both orbits sharing the same value of the total energy. This is equivalent in practice to the well-known energy matching condition found in the unperturbed case (Alfriend et al., 2009, pp. 59–60).

The relative velocity is solved from

$$\begin{aligned}
\delta\dot{r} &= \dot{r}\,\delta R + A_n\,\delta n + A_k\,\delta k + A_{\phi_0}\,\delta\phi_0 \\
\delta\dot{\theta} &= n\,\delta k + k\,\delta n \\
\delta\dot{z} &= \dot{z}\,\delta R + B_n\,\delta n + B_k\,\delta k + B_{\phi_0}\,\delta\phi_0
\end{aligned}$$

given the coefficients

$$\begin{aligned}
A_n &= \frac{\dot{r}}{n} - \frac{rt}{nz^2}(\dot{r}^2 + k^2n^2z^4) \\
A_k &= -\frac{2kr^2\dot{r}}{k'^2} + \frac{r(\dot{r}^2 + k^2n^2z^4)}{nz^2k'^2k}\Lambda^-(u,k) \\
A_{\phi_0} &= -\frac{r}{nz^2}(\dot{r}^2 + k^2n^2z^4) \\
B_n &= \frac{\dot{z}}{n} - \frac{zt}{nz}(\dot{z}^2 - k^2n^2r^4) \\
B_k &= \frac{k\dot{z}(1-2r^2)}{k'^2} + \frac{z(\dot{z}^2 - k^2n^2r^4)}{nr^2kk'^2}\Lambda^+(u,k) \\
B_{\phi_0} &= -\frac{z}{nr^2}(\dot{z}^2 - k^2n^2r^4)
\end{aligned}$$

The secular terms appear in the coefficients corresponding to δn and δk. The energy matching condition ensures that the velocity is bounded too.

G.5 The significance of Seiffert's spirals

The orbital dynamics in Seiffert's spherical spirals are endowed with interesting properties. First, the problem is conservative. The perturbation that is required in order to describe this particular curve derives from an ordinary potential. Second, the orbital velocity is constant. This makes Seiffert's spirals similar to Keplerian circular orbits. In fact, the spirals will degenerate into a perfectly circular orbit under certain conditions. Third, the orbital frame and the osculating perifocal frame coincide. This is an interesting property that comes from the form of the perturbation, which makes the eccentricity vector parallel to the radius vector.

Thanks to having a fully analytic solution at hand it is possible to derive very simple strategies for planetary exploration. Although the thrust profile is not flexible and may exceed the limits of electric propulsion systems, the geometry and simplicity of the spirals is appealing.

List of Figures

https://doi.org/10.1515/9783110559125-202

Bibliography

Aarseth SJ. (1971) "Direct integration methods of the N-body problem". Astrophysics and Space Science 14(1):118–132

Aarseth SJ. (1999) "From NBODY1 to NBODY6: The growth of an industry". Publications of the Astronomical Society of the Pacific 111(765):1333–1346

Aarseth SJ. (2003) *Gravitational N-Body Simulations: Tools and Algorithms*. Cambridge University Press

Aarseth SJ., Zare K. (1974) "A regularization of the three-body problem". Celestial Mechanics 10(2):185–205

Abel NH. (1827) "Recherches sur les fonctions elliptiques.". Journal für die reine und angewandte Mathematik 2:101–181

Abramowitz M., Stegun IA. (1964) *Handbook of Mathematical Functions: With Formulas, Graphs, and Mathematical Tables* vol 55. Dover Publications

Adams JF. (1960) "On the non-existence of elements of Hopf invariant one". Annals of Mathematics pp 20–104

Adkins GS., McDonnell J. (2007) "Orbital precession due to central-force perturbations". Physical Review D 75(8):082,001

Ahlfors LV. (1966) *Complex analysis*. Krishna Prakashan Media

Akella MR., Broucke RA. (2002) "Anatomy of the constant radial thrust problem". Journal of Guidance, Control, and Dynamics 25(3):563–570

Alfriend K., Coffey S. (1984) "Elimination of the perigee in the satellite problem". Celestial Mechanics 32(2):163–172

Alfriend KT., Gim DW., Schaub H. (2000) "Gravitational perturbations, nonlinearity and circular orbit assumption effects on formation flying control strategies". In: AAS Guidance and Control Conference AAS 00-012

Alfriend KT., Vadali SR., Gurfil P., How J., Breger L. (2009) *Spacecraft formation flying: dynamics, control, and navigation*. Butterworth-Heinemann

Altobelli N., Kempf S., Landgraf M., Srama R., Dikarev V., Krüger H., Moragas-Klostermeyer G., Grün E. (2003) "Cassini between Venus and Earth: Detection of interstellar dust". Journal of Geophysical Research-Space 108(A10)

Amato D., Bombardelli C. (2014) "Mitigation of trajectory propagation error in spacecraft fly-bys". In: Key Topics in Orbit Propagation Applied to Space Situational Awareness p 82

Amato D., Bombardelli C., Baù G. (2016) "Efficient numerical propagation of planetary close encounters with regularized element methods". In: 6th International Conference on Astrodynamics Tools and Techniques (ICATT)

Arakida H., Fukushima T. (2000) "Long-term integration error of Kustaanheimo-Stiefel regularized orbital motion". The Astronomical Journal 120(6):3333

Archinal BA., Ahearn M., Bowell E., Conrad A., Consolmagno GJ., Courtin R., Fukushima T., Hestroffer D., Hilton JL., Krasinsky GA., et al. (2011) "Report of the IAU working group on cartographic coordinates and rotational elements: 2009". Celestial Mechanics and Dynamical Astronomy 109(2):101–135

Arnold VI., Kozlov VV., Neishtadt AI. (2007) *Mathematical aspects of classical and celestial mechanics* 3rd edn. Springer

Arora N., Russell RP. (2013) "A fast and robust multiple revolution Lambert algorithm using a cosine transformation". In: 2013 AAS/AIAA Astrodynamics Specialist Conference pp AAS13–728

Arora N., Russell RP., Strange N., Ottesen D. (2015) "Partial derivatives of the solution to the Lambert boundary value problem". Journal of Guidance, Control, and Dynamics 38(9):1563–1572

https://doi.org/10.1515/9783110559125-203

Astakhov SA., Farrelly D. (2004) "Capture and escape in the elliptic restricted three-body problem". Monthly Notices of the Royal Astronomical Society 354(4):971–979

Avanzini G. (2008) "A simple Lambert algorithm". Journal of Guidance, Control, and Dynamics 31(6):1587–1594

Avanzini G., Palmas A., Vellutini E. (2015) "Solution of low-thrust Lambert problem with perturbative expansions of equinoctial elements". Journal of Guidance, Control, and Dynamics 38(9):1585–1601

Bacon D., Sigurdsson S., Davies M. (1996) "Close approach during hard binary-binary scattering". Monthly Notices of the Royal Astronomical Society 281(3):830–846

Bacon RH. (1959) "Logarithmic spiral: An ideal trajectory for the interplanetary vehicle with engines of low sustained thrust". American Journal of Physics 27(3):164–165

Bahar LY., Kwatny HG. (1987) "Extension of Noether's theorem to constrained non-conservative dynamical systems". International Journal of Non-linear Mechanics 22(2):125–138

Baillaud B. (1876) "Exposition de la méthode de M. Gylden pour le développement des perturbations des comètes". In: Annales scientifiques de l'École Normale Supérieure vol 5 pp 355–398

Barros M. (1997) "General helices and a theorem of Lancret". Proceedings of the American Mathematical Society 125(5):1503–1509

Basilakos S., Tsamparlis M., Paliathanasis A. (2011) "Using the Noether symmetry approach to probe the nature of dark energy". Physical Review D 83(10):103,512

Bate RR., Mueller DD., White JE. (1971) *Fundamentals of astrodynamics*. Dover publications

Battin RH. (1977) "Lambert's problem revisited". AIAA Journal 15(5):707–713

Battin RH. (1999) *An introduction to the mathematics and methods of Astrodynamics*. AIAA

Battin RH., Vaughan RM. (1984) "An elegant Lambert algorithm". Journal of Guidance, Control, and Dynamics 7(6):662–670

Battin RH., Fill TJ., Shepperd SW. (1978) "A new transformation invariant in the orbital boundary-value problem". Journal of Guidance, Control, and Dynamics 1(1):50–55

Baù G., Bombardelli C. (2014) "Time elements for enhanced performance of the Dromo orbit propagator". The Astronomical Journal 148(3):43

Baù G., Urrutxua H., Peláez J. (2014) "Edromo: An accurate propagator for elliptical orbits in the perturbed two-body problem". Advances in the Astronautical Sciences 14-227(227)

Baù G., Bombardelli C., Peláez J., Lorenzini E. (2015) "Non-singular orbital elements for special perturbations in the two-body problem". Monthly Notices of the Royal Astronomical Society 454:2890–2908

Baumgarte J. (1972a) "Numerical stabilization of the differential equations of Keplerian motion". Celestial Mechanics 5(4):490–501

Baumgarte J. (1972b) "Stabilization of constraints and integrals of motion in dynamical systems". Computer Methods in Applied Mechanics and Engineering 1(1):1–16

Baumgarte J. (1976a) "A new time element for a general time transformation". Celestial Mechanics 14(1):121–127

Baumgarte J. (1976b) "Stabilized Kepler motion connected with analytic step adaptation". Celestial Mechanics 13(1):105–109

Baumgarte J., Stiefel E. (1974) "Examples of transformations improving the numerical accuracy of the integration of differential equations". In: Proceedings of the Conference on the Numerical Solution of Ordinary Differential Equations Springer pp 207–236

Beisert N., Ricci R., Tseytlin AA., Wolf M. (2008) "Dual superconformal symmetry from $AdS_5 \times S^5$ superstring integrability". Physical Review D 78(12):126,004

Belen'kii I. (1981) "A method of regularizing the equations of motion in the central force-field". Celestial Mechanics 23(1):9–32

Benney D. (1958) "Escape from a circular orbit using tangential thrust". Journal of Jet Propulsion 28(3):167–169

Berry MM. (2004) "A variable-step double-integration multi-step integrator". PhD thesis Virginia Polytechnic Institute and State University

Bertotti B., Iess L., Tortora P. (2003) "A test of general relativity using radio links with the Cassini spacecraft". Nature 425(6956):374–376

Bettis D., Szebehely V. (1971) "Treatment of close approaches in the numerical integration of the gravitational problem of N-bodies". Astrophysics and Space Science 14(1):133–150

Birkhoff GD. (1915) "The restricted problem of three bodies". Rendiconti del Circolo Matematico di Palermo 39(1):265–334

Biscani F., Carloni S. (2012) "A first-order secular theory for the post-Newtonian two-body problem with spin–I. The restricted case". Monthly Notices of the Royal Astronomical Society 428(3):2295–2310

Biscani F., Izzo D. (2014) "The Stark problem in the Weierstrassian formalism". Monthly Notices of the Royal Astronomical Society 439(1):810–822

Boccaletti D., Cannata R., Catoni V., Nichelatti E., Zampetti P. (2008) *The mathematics of Minkowski space-time: with an introduction to commutative hypercomplex numbers*. Birkhäuser Verlag AG, Basel, Boston, Berlin

Bohlin K. (1911) "Note sur le probléme des deux corps et sur une intégration nouvelle dans le probléme des trois corps". Bulletin Astronomique 28:113–119

Boltz FW. (1992) "Orbital motion under continuous tangential thrust". Journal of Guidance, Control, and Dynamics 15(6):1503–1507

Bombardelli C. (2014) "Analytical formulation of impulsive collision avoidance dynamics". Celestial Mechanics and Dynamical Astronomy 118:99–114

Bombardelli C., Hernando-Ayuso J. (2015) "Optimal impulsive collision avoidance in low earth orbit". Journal of Guidance, Control, and Dynamics 38(2):217–225

Bombardelli C., Baù G., Peláez J. (2011) "Asymptotic solution for the two-body problem with constant tangential thrust acceleration". Celestial Mechanics and Dynamical Astronomy 110(3):239–256

Bombardelli C., Gonzalo JL., Roa J. (2015) "Compact solution of circular orbit relative motion in curvilinear coordinates". In: 2015 AIAA/AAS Astrodynamics Specialist Conference 15-661

Bombardelli C., Gonzalo JL., Roa J. (2017) "Approximate solutions of non-linear circular orbit relative motion in curvilinear coordinates". Celestial Mechanics and Dynamical Astronomy 127(1):49–66

Bond V. (1990) "The elimination of secular terms from the differential equations of the BG14 element method". Tech. Rep. JSC-24351 (90-FM-3) NASA Johnson Space Center

Bond VR. (1982) "Error propagation in the numerical solutions of the differential equations of orbital mechanics". Celestial Mechanics 27(1):65–77

Bond VR., Allman MC. (1996) *Modern Astrodynamics: Fundamentals and perturbation methods*. Princeton University Press

Bond VR., Gottlieb RG. (1989) "A perturbed two-body element method utilizing the Jacobian integral". Tech. Rep. JSC-23495 (89-FM-3) NASA Johnson Space Center

Broucke R. (1980) "Notes on the central force r^n". Astrophysics and Space Science 72(1):33–53

Broucke R., Lass H., Ananda M. (1971) "Redundant variables in celestial mechanics". Astronomy & Astrophysics 13:390

Broucke RA. (2003) "Solution of the elliptic rendezvous problem with the time as independent variable". Journal of Guidance, Control, and Dynamics 26(4):615–621

Broucke RA., Cefola PJ. (1972) "On the equinoctial orbit elements". Celestial Mechanics 5(3):303–310

Brouwer D. (1959) "Solution of the problem of artificial satellite theory without drag". The Astronomical Journal 64(1274):378–396
Brouwer D., Clemence GM. (1961) "Methods of celestial mechanics". Methods of Celestial Mechanics, by D Brouwer and GM Clemence Academic Press, New York, 1961 1
Brouwer D., Hori G. (1961) "Theoretical evaluation of atmospheric drag effects in the motion of an artificial satellite". The Astronomical Journal 66:193
Brown RB., Gray A. (1967) "Vector cross products". Commentarii Mathematici Helvetici 42(1):222–236
Bulirsch R., Stoer J. (1966) "Numerical treatment of ordinary differential equations by extrapolation methods". Numerische Mathematik 8(1):1–13
Burdet CA. (1967) "Regularization of the two body problem". Journal of Applied Mathematics and Physics (ZAMP) 18(3):434–438
Burdet CA. (1968) "Theory of Kepler motion: the general perturbed two body problem". Journal of Applied Mathematics and Physics (ZAMP) 19(2):345–368
Burdet CA. (1969) "Le mouvement Keplerien et les oscillateurs harmoniques". Journal für die reine und angewandte Mathematik 238:71–84
Burkardt T., Danby J. (1983) "The solution of Kepler's equation, II". Celestial Mechanics and Dynamical Astronomy 31(3):317–328
Burrau C. (1913) "Numerische Berechnung eines Spezialfalles des Dreikörperproblems". Astronomische Nachrichten 195(6):113–118
Burton M., Buratti B., Matson D., Lebreton JP. (2001) "The Cassini/Huygens Venus and Earth flybys: An overview of operations and results". Journal of Geophysical Research-Space 106(A12):30,099–30,107
Byrd PF., Friedman MD. (1954) *Handbook of elliptic integrals for engineers and physicists*. Springer
Byrnes DV., Longuski JM., Aldrin B. (1993) "Cycler orbit between Earth and Mars". Journal of Spacecraft and Rockets 30(3):334–336
Capozziello S., Piedipalumbo E., Rubano C., Scudellaro P. (2009) "Noether symmetry approach in phantom quintessence cosmology". Physical Review D 80(10):104,030
Cariñena JF., López C., del Olmo MA., Santander M. (1991) "Conformal geometry of the Kepler orbit space". Celestial Mechanics and Dynamical Astronomy 52(4):307–343
Carpintero DD. (2008) "Finding how many isolating integrals of motion an orbit obeys". Monthly Notices of the Royal Astronomical Society 388(3):1293–1304
Carpintero DD., Aguilar LA. (1998) "Orbit classification in arbitrary 2D and 3D potentials". Monthly Notices of the Royal Astronomical Society 298(1):1–21
Carter TE. (1990) "New form for the optimal rendezvous equations near a Keplerian orbit". Journal of Guidance, Control, and Dynamics 13(1):183–186
Carter TE. (1996) "Closed-form solution of an idealized, optimal, highly eccentric hyperbolic rendezvous problem". Dynamics and Control 6(3):293–307
Carter TE. (1998) "State transition matrices for terminal rendezvous studies: brief survey and new example". Journal of Guidance, Control, and Dynamics 21(1):148–155
Casotto S. (1993) "Position and velocity perturbations in the orbital frame in terms of classical element perturbations". Celestial Mechanics and Dynamical Astronomy 55(3):209–221
Casotto S. (2014) "A nonsingular Keplerian differential state transition matrix". Advances in the Astronautical Sciences 152:167–183
Catoni F., Boccaletti D., Cannata R., Catoni V., Zampetti P. (2011) *Geometry of Minkowski space-time*. Springer
Cefola PJ. (1972) "Equinoctial orbit elements-application to artificial satellite orbits". In: AIAA/AAS Astrodynamics Conference pp AIAA 72–937

Chambers JE. (1999) "A hybrid symplectic integrator that permits close encounters between massive bodies". Monthly Notices of the Royal Astronomical Society 304(4):793–799

Chandrasekhar S. (1983) *The Mathematical Theory of Black Holes*. The International Series of Monographs on Physics Oxford University Press

Chandrasekhar S. (1995) *Newton's Principia for the common reader*. Oxford University Press

Chashchina O., Silagadze Z. (2008) "Remark on orbital precession due to central-force perturbations". Physical Review D 77(10):107,502

Cid R., Ferrer S., Elipe A. (1983) "Regularization and linearization of the equations of motion in central force-fields". Celestial Mechanics 31(1):73–80

Cid R., Ferrer S., Sein-Echaluce M. (1986) "On the radial intermediaries and the time transformation in satellite theory". Celestial Mechanics 38(2):191–205

Clohessy WH., Wiltshire RS. (1960) "Terminal guidance system for satellite rendezvous". Journal of the Aerospace Sciences 29(9):653–658

Colombo C., Vasile M., Radice G. (2009) "Semi-analytical solution for the optimal low-thrust deflection of near-earth objects". Journal of Guidance, Control, and Dynamics 32(3):796–809

Condurache D., Martinuşi V. (2007a) "A complete closed form vectorial solution to the Kepler problem". Meccanica 42(5):465–476

Condurache D., Martinuşi V. (2007b) "Kepler's problem in rotating reference frames part ii: relative orbital motion". Journal of Guidance, Control, and Dynamics 30(1):201–213

Condurache D., Martinuşi V. (2007c) "Relative spacecraft motion in a central force field". Journal of Guidance, Control, and Dynamics 30(3):873–876

Condurache D., Martinuşi V. (2009) "Exact solution to the relative orbital motion in eccentric orbits". Solar System Research 43(1):41–52

Condurache D., Martinuşi V. (2010) "Quaternionic exact solution to the relative orbital motion problem". Journal of Guidance, Control, and Dynamics 33(4):1035–1047

Contopoulos G. (1967) "Integrals of motion in the elliptic restricted three-body problem". The Astronomical Journal 72:669

Cowell PH., Crommelin ACD. (1908) "The orbit of Jupiter's eighth satellite". Monthly Notices of the Royal Astronomical Society 68(8):576–582

Cowell PH., Crommelin ACD. (1910) "Investigation of the motion of Halley's Comet from 1759 to 1910". Greenwich Observations in Astronomy, Magnetism and Meteorology 71:1–84

Cunningham LE. (1970) "On the computation of the spherical harmonic terms needed during the numerical integration of the orbital motion of an artificial satellite". Celestial Mechanics 2(2):207–216

D'Amico S., Montenbruck O. (2006) "Proximity operations of formation-flying spacecraft using an eccentricity/inclination vector separation". Journal of Guidance, Control, and Dynamics 29(3):554–563

Danby J. (1987) "The solution of Kepler's equation, III". Celestial Mechanics 40(3-4):303–312

Danby J. (1992) *Fundamentals of Celestial Mechanics* 2nd edn. Willman-Bell

Davtyan L., Mardoyan L., Pogosyan G., Sissakian A., Ter-Antonyan V. (1987) "Generalised KS transformation: from five-dimensional hydrogen atom to eight-dimensional isotrope oscillator". Journal of Physics A: Mathematical and General 20:6121–6125

De Pascale P., Vasile M. (2006) "Preliminary design of low-thrust multiple gravity-assist trajectories". Journal of Spacecraft and Rockets 43(5):1065–1076

De Vries JP. (1963) "Elliptic elements in terms of small increments of position and velocity components". AIAA Journal 1(11):2626–2629

Delaunay C. (1860) *Théorie du Mouvement de la Lune* vol 1. Didot

Deprit A. (1975) "Ideal elements for perturbed Keplerian motions". Journal of Research of the National Bureau of Standards 79:1–15

Deprit A. (1976) “Ideal frames for perturbed Keplerian motions”. Celestial Mechanics 13(2):253–263
Deprit A. (1981) “The elimination of the parallax in satellite theory”. Celestial Mechanics 24(2):111–153
Deprit A., Elipe A., Ferrer S. (1994) “Linearization: Laplace vs. Stiefel”. Celestial Mechanics and Dynamical Astronomy 58(2):151–201
Dickson LE. (1919) “On quaternions and their generalization and the history of the eight square theorem”. Annals of Mathematics pp 155–171
Djukic DDS., Vujanovic B. (1975) “Noether’s theory in classical nonconservative mechanics”. Acta Mechanica 23(1-2):17–27
Djukic DS., Sutela T. (1984) “Integrating factors and conservation laws for nonconservative dynamical systems”. International Journal of Non-linear Mechanics 19(4):331–339
Dormand JR., Prince PJ. (1980) “A family of embedded Runge-Kutta formulae”. Journal of Computational and Applied Mathematics 6(1):19–26
Dosopoulou F., Kalogera V. (2016) “Orbital evolution of mass-transferring eccentric binary systems I. phase-dependent evolution”. The Astrophysical Journal 825(1):70
Duncan MJ., Levison HF., Lee MH. (1998) “A multiple time step symplectic algorithm for integrating close encounters”. The Astronomical Journal 116(4):2067
Efroimsky M. (2002) “Equations for the Keplerian elements: hidden symmetry”. Preprint of the Institute of Mathematics and its Applications, University of Minnesota (1844)
Efroimsky M. (2005) “Gauge freedom in orbital mechanics”. Annals of the New York Academy of Sciences 1065(1):346–374
Efroimsky M., Goldreich P. (2003) “Gauge symmetry of the N-body problem in the Hamilton–Jacobi approach”. Journal of Mathematical Physics 44(12):5958–5977
Efroimsky M., Goldreich P. (2004) “Gauge freedom in the N-body problem of celestial mechanics”. Astronomy & Astrophysics 415(3):1187–1199
ElBialy MS. (2007) “The Kustaanheimo–Stiefel map, the Hopf fibration and the square root map on R3 and R4”. Journal of Mathematical Analysis and Applications 332(1):631–665
Erdös P. (2000) “Spiraling the Earth with CGJ Jacobi”. American Journal of Physics 68(10):888–895
Euler L. (1749) “Recherches sur la précession des équinoxes, et sur la nutation de l’axe de la Terre”. Histoire de l’Académie Royale des Sciences et Belles-lettres (Berlin) pp 289–325
Everhart E. (1976) “The evolution of comet orbits”. NASA Special Publication 393:445–464
Everhart E., Pitkin ET. (1983) “Universal variables in the two-body problem”. American Journal of Physics 51(8):712–717
Ferrándiz J. (1988) “A general canonical transformation increasing the number of variables with application to the two-body problem”. Celestial Mechanics 41(1-4):343–357
Ferrándiz JM., Ferrer S., Sein-Echaluce ML. (1987) “Generalized elliptic anomalies”. Celestial Mechanics 40(3-4):315–328
Ferraz-Mello S., Nesvorny D., Michtchenko T. (1998) “Chaos, diffusion, escape and permanence of resonant asteroids in gaps and groups”. Solar System Formation and Evolution 149:65
Fjelstad P. (1986) “Extending Special Relativity via the perplex numbers”. American Journal of Physics 54(5):416–422
Flanders H. (1989) *Differential Forms with Applications to the Physical Sciences* vol 11 2nd edn. Dover Publications
Flandro G. (1966) “Fast reconnaissance missions to the outer Solar System utilizing energy derived from the gravitational field of Jupiter”. Earth 108:6–1
Folta D., Quinn D. (1998) “A universal 3-D method for controlling the relative motion of multiple spacecraft in any orbit”. In: AIAA/AAS Astrodynamics Specialist Conference,(Boston, MA)
Fukushima T. (2003) “Efficient orbit integration by scaling for Kepler energy consistency”. The Astronomical Journal 126(2):1097

Fukushima T. (2004) "Efficient integration of highly eccentric orbits by quadruple scaling for Kustaanheimo-Stiefel regularization". The Astronomical Journal 128(6):3108

Fukushima T. (2005) "Efficient orbit integration by Kustaanheimo-Stiefel regularization using time element". The Astronomical Journal 129(3):1746

Funato Y., Hut P., McMillan S., Makino J. (1996) "Time-symmetrized Kustaanheimo-Stiefel regularization". The Astronomical Journal 112:1697

Gander W. (1985) "On Halley's iteration method". American Mathematical Monthly 92(2):131–134

Gauss KF. (1809) *Theoria Motus Corporum Coelestium in Sectionibus Conicis Solem Ambientium*. Little, Brown and Co., Boston

Gim DW., Alfriend KT. (2003) "State transition matrix of relative motion for the perturbed noncircular reference orbit". Journal of Guidance, Control, and Dynamics 26(6):956–971

Gim DW., Alfriend KT. (2005) "Satellite relative motion using differential equinoctial elements". Celestial Mechanics and Dynamical Astronomy 92(4):295–336

Glaisher JWL. (1879) "On definite integrals involving elliptic functions". J Proceedings of the Royal Society of London (1854-1905) 29:331–351

Glydén H. (1870) "Sur une méthode d'exprimer les perturbations d'une comète au moyen de seéries rapidement convergentes". Bulletin de l'Académie Impériale des Sciences de Saint-Petersbourg 14:195–231

Gondelach D., Noomen R. (2015) "Hodographic-shaping method for low-thrust interplanetary trajectory design". Journal of Spacecraft and Rockets 52(3):728–738

Gonzalo JL., Bombardelli C. (2014) "Asymptotic solution for the two body problem with radial perturbing acceleration". In: Advances in the Astronautical Sciences AAS/AIAA, Santa Fe, New Mexico, USA AAS 14-226

Gooding R. (1990) "A procedure for the solution of Lambert's orbital boundary-value problem". Celestial Mechanics and Dynamical Astronomy 48(2):145–165

Gudermann C. (1838) "Theorie der Modular-Functionen und der Modular-Integrale". Journal für die reine und angewandte Mathematik 18:1–54

Gurfil P. (2005a) "Euler parameters as nonsingular orbital elements in near-equatorial orbits". Journal of Guidance, Control, and Dynamics 28(5):1079–1084

Gurfil P. (2005b) "Relative motion between elliptic orbits: generalized boundedness conditions and optimal formation keeping". Journal of Guidance, Control, and Dynamics 28(4):761–767

Gurfil P. (2007) "Generalized solutions for relative spacecraft orbits under arbitrary perturbations". Acta Astronautica 60(2):61–78

Gurfil P., Kasdin N. (2003) "Canonical modelling of coorbital motion in Hill's problem using epicyclic orbital elements". Astronomy & Astrophysics 409:1135–1140

Gurfil P., Kasdin NJ. (2004) "Nonlinear modelling of spacecraft relative motion in the configuration space". Journal of Guidance, Control, and Dynamics 27(1):154–157

Gurfil P., Kholshevnikov KV. (2006) "Manifolds and metrics in the relative spacecraft motion problem". Journal of Guidance, Control, and Dynamics 29(4):1004–1010

Gurfil P., Klein I. (2006) "Mitigating the integration error in numerical simulations of Newtonian systems". International journal for numerical methods in engineering 68(2):267–297

Gurfil P., Lara M. (2013) "Motion near frozen orbits as a means for mitigating satellite relative drift". Celestial Mechanics and Dynamical Astronomy 116(3):213–227

Gurfil P., Lara M. (2014) "Satellite onboard orbit propagation using Deprit's radial intermediary". Celestial Mechanics and Dynamical Astronomy 120(2):217–232

Hagihara Y. (1930) "Theory of the relativistic trajectories in a gravitational field of Schwarzschild". Japanese Journal of Astronomy and Geophysics 8:67

Hairer E., Nørsett SP., Wanner G. (1991) *Solving ordinary differential equations I*. Springer

Hajduková Jr M. (2008) "Meteors in the IAU meteor data center on hyperbolic orbits". Earth, Moon, and Planets 102(1-4):67–71

Halpern M., Jevicki A., Senjanović P. (1977) "Field theories in terms of particle-string variables: Spin, internal symmetries, and arbitrary dimension". Physical Review D 16(8):2476

Hansen P. (1853) "Mémoire sur le calcul des perturbations qu'éprouvent les comètes". Comptes Rendus de l'Académie des Sciences de Paris 1:1–258

Hansen P. (1857) "Auseinandersetzung einer zweckmäßigen Methode zur Berechnung der absoluten Störungen der Kleinen Planeten, erste Abhandlung". Abhandlungen der mathematisch-physischen Classe der königlich sächsischen Gesellschaft der Wissenschaften, dritter Band, bei S Hirzel, Leipzig

Hedin AE. (1991) "Extension of the msis thermosphere model into the middle and lower atmosphere". Journal of Geophysical Research: Space Physics 96(A2):1159–1172

Heggie DC. (1974) "A global regularisation of the gravitational N-body problem". Celestial Mechanics 10(2):217–241

Hénon M., Heiles C. (1964) "The applicability of the third integral of motion: some numerical experiments". The Astronomical Journal 69:73

Herget P. (1962) "On the variation of arbitrary vectorial constants". The Astronomical Journal 67(1):16–18

Herrick S. (1953) "Icarus and the variation of parameters". The Astronomical Journal 58:156

Hill GW. (1878) "Researches in the lunar theory". American Journal of Mathematics 1(2):129–147

Hills J. (1981) "Comet showers and the steady-state infall of comets from the oort cloud". The Astronomical Journal 86:1730–1740

Hindmarsh AC. (1983) "ODEPACK, a systematized collection of ODE solvers". IMACS Transactions on Scientific Computation 1:55–64

Honein T., Chien N., Herrmann G. (1991) "On conservation laws for dissipative systems". Physics Letters A 155(4):223–224

Hopf H. (1931) "Über die Abbildungen der dreidimensionalen Sphäre auf die Kugelfläche". Mathematische Annalen 104(1):637–665

Hurwitz A. (1898) "Ueber die Composition der quadratischen Formen von belibig vielen Variablen". Nachrichten von der Gesellschaft der Wissenschaften zu Göttingen, Mathematisch-Physikalische Klasse 1898:309–316

Hut P., Bahcall JN. (1983) "Binary-single star scattering. I-numerical experiments for equal masses". The Astrophysical Journal 268:319–341

Ipatov SI. (1999) "Migration of trans-Neptunian objects to the Earth". Celestial Mechanics and Dynamical Astronomy 73(1-4):107–116

Izzo D. (2006) "Lambert's problem for exponential sinusoids". Journal of Guidance, Control, and Dynamics 29(5):1242–1245

Izzo D. (2014) "Revisiting lambert's problem". Celestial Mechanics and Dynamical Astronomy 121(1):1–15

Izzo D., Biscani F. (2015) "Explicit solution to the constant radial acceleration problem". Journal of Guidance, Control, and Dynamics 38(4):733–739

Izzo D., Becerra VM., Myatt DR., Nasuto SJ., Bishop JM. (2007) "Search space pruning and global optimisation of multiple gravity assist spacecraft trajectories". Journal of Global Optimization 38(2):283–296

Jacchia LG. (1977) "Thermospheric temperature, density, and composition: new models". SAO special report 375

Jacobi CGJ. (1829) *Fundamenta nova theoriae functionum ellipticarum*. Sumtibus fratrum Borntræger, Königsberg

Janin G. (1974) "Accurate computation of highly eccentric satellite orbits". Celestial Mechanics 10(4):451–467

Jochim E. (2012) "The significance of the Hansen ideal space frame". Astronomische Nachrichten 333(8):774–783

Kantor IL., Solodovnikov AS., Shenitzer A. (1989) *Hypercomplex numbers: an elementary introduction to algebras*. Springer-Verlag New York

Karlgaard CD., Lutze FH. (2003) "Second-order relative motion equations". Journal of Guidance, Control, and Dynamics 26(1):41–49

Kevorkian J., Cole JD. (1981) *Perturbation Methods in Applied Mathematics* vol 80. Springer New York

Kohlhase C., Penzo P. (1977) "Voyager mission description". Space Science Reviews 21(2):77–101

Kopeikin S., Efroimsky M., Kaplan G. (2011) *Relativistic celestial mechanics of the solar system*. John Wiley & Sons

Kozai Y. (1985) "Secular perturbations of resonant asteroids". Celestial Mechanics 36(1):47–69

Kriz J. (1978) "Variational formulae for solving perturbed boundary-value problems in Astrodynamics". Celestial Mechanics 18(4):371–382

Krogh FT. (1974) "Notes on a new package for ordinary differential equations". Tech. rep. Jet Propulsion Laboratory, Section 914, California Institute of Technology

Królikowska M. (2001) "A study of the original orbits of "hyperbolic"comets". Astronomy & Astrophysics 376:316–324

Królikowska M. (2006) "Non-gravitational effects in long-period comets and the size of the Oort cloud". Acta Astronomica 56:385–412

Kupi G., Amaro-Seoane P., Spurzem R. (2006) "Dynamics of compact object clusters: a post-Newtonian study". Monthly Notices of the Royal Astronomical Society: Letters 371(1):L45–L49

Kustaanheimo P., Stiefel E. (1965) "Perturbation theory of Kepler motion based on spinor regularization". Journal für die reine und angewandte Mathematik 218:204–219

Lancaster E., Devaney R., Blanchard R. (1966) "A note on Lambert's theorem.". Journal of Spacecraft and Rockets 3(9):1436–1438

Lancaster ER., Blanchard RC. (1968) "A unified form of Lambert's theorem". Tech. rep. NASA Goddard Space Flight Center

Landau DF., Longuski JM. (2007) "Guidance strategy for hyperbolic rendezvous". Journal of Guidance, Control, and Dynamics 30(4):1209–1213

Lantoine G., Russell RP. (2011) "Complete closed-form solutions of the Stark problem". Celestial Mechanics and Dynamical Astronomy 109(4):333–366

Lantoine G., Russell RP., Dargent T. (2012) "Using multicomplex variables for automatic computation of high-order derivatives". ACM Transactions on Mathematical Software (TOMS) 38(3):A16

Laplace PS. (1799) *Traité de Mécanique Céleste*. Paris, Duprat et Bachelier. Reprinted in Œuvres. Paris, Imprimerie royale, 1843.

Lara M. (2016) "Note on the ideal frame formulation". arXiv preprint arXiv:161208367

Lawden DF. (1954) "Fundamentals of space navigation". Journal of the British Interplanetary Society 13(2):87–101

Lawden DF. (1955) "Optimal programming of rocket thrust direction". Astronautica Acta 1(1):41–56

Lawden DF. (1963) *Optimal trajectories for space navigation*. Butterworths

Lee D., Cochran J., Jo J. (2007) "Solutions to the variational equations for relative motion of satellites". Journal of Guidance, Control, and Dynamics 30(3):669–678

Lega E., Guzzo M., Froeschlé C. (2011) "Detection of close encounters and resonances in three-body problems through Levi-Civita regularization". Monthly Notices of the Royal Astronomical Society 418(1):107–113

Levi-Civita T. (1903) "Sur les trajectoires singulières du problème restreint des trois corps". Comptes Rendus de l'Académie des Sciences de Paris 136:82–84

Levi-Civita T. (1904) “Traiettorie singolari ed urti nel problema ristretto dei tre corpi”. Annali di Matematica Pura ed Applicata 9(1):1–32

Levi-Civita T. (1906) “Sur la résolution qualitative du probleme restreint des trois corps”. Acta Mathematica 30(1):305–327

Levi-Civita T. (1920) “Sur la régularisation du probleme des trois corps”. Acta Mathematica 42(1):99–144

Lockwood EH. (1967) *A book of curves*. Cambridge University Press

London HS. (1963) “Second approximation to the solution of the rendezvous equations”. AIAA Journal 1(7):1691–1693

Longuski JM. (1983) “Galileo user's guide (phase 1), mission design system, satellite tour analysis and design subsystem, STOUR program set”. Tech. Rep. D-263 Jet Propulsion Laboratory, California Inst. of Technology Pasadena, CA

Longuski JM., Williams SN. (1991) “Automated design of gravity-assist trajectories to Mars and the outer planets”. Celestial Mechanics and Dynamical Astronomy 52(3):207–220

Luke YL. (1968) “Approximations for elliptic integrals”. Mathematics of Computation 22(103):627–634

Luo Q., Meng Z., Han C. (2011) “Solution algorithm of a quasi-Lambert's problem with fixed flight-direction angle constraint”. Celestial Mechanics and Dynamical Astronomy 109(4):409–427

MacMillan WD. (1908) “The motion of a particle attracted towards a fixed center by a force varying inversely as the fifth power of the distance”. American Journal of Mathematics 30(3):282–306

Martinuşi V., Gurfil P. (2011) “Solutions and periodicity of satellite relative motion under even zonal harmonics perturbations”. Celestial Mechanics and Dynamical Astronomy 111(4):387–414

McConaghy TT., Debban TJ., Petropoulos AE., Longuski JM. (2003) “Design and optimization of low-thrust trajectories with gravity assists”. Journal of Spacecraft and Rockets 40(3):380–387

McInnes CR. (2004) *Solar sailing: technology, dynamics and mission applications*. Springer Science & Business Media

Melton RG. (2000) “Time-explicit representation of relative motion between elliptical orbits”. Journal of Guidance, Control, and Dynamics 23(4):604–610

Mengali G., Quarta AA. (2009) “Escape from elliptic orbit using constant radial thrust”. Journal of Guidance, Control, and Dynamics 32(3):1018–1022

Mikkola S. (1983) “Encounters of binaries I. Equal energies”. Monthly Notices of the Royal Astronomical Society 203(4):1107–1121

Mikkola S. (1985) “A practical and regular formulation of the N-body equations”. Monthly Notices of the Royal Astronomical Society 215(2):171–177

Mikkola S., Aarseth SJ. (1989) “A chain regularization method for the few-body problem”. Celestial Mechanics and Dynamical Astronomy 47(4):375–390

Mikkola S., Aarseth SJ. (1993) “An implementation of N-body chain regularization”. Celestial Mechanics and Dynamical Astronomy 57(3):439–459

Mikkola S., Aarseth SJ. (1998) “An efficient integration method for binaries in N-body simulations”. New Astronomy 3(5):309–320

Mikkola S., Merritt D. (2008) “Implementing few-body algorithmic regularization with post-Newtonian terms”. The Astronomical Journal 135(6):2398

Mikkola S., Tanikawa K. (1999) “Algorithmic regularization of the few-body problem”. Monthly Notices of the Royal Astronomical Society 310(3):745–749

Milani A., Knežević Z. (1990) “Secular perturbation theory and computation of asteroid proper elements”. Celestial Mechanics and Dynamical Astronomy 49(4):347–411

Milani A., Carpino M., Hahn G., Nobili A. (1989) “Dynamics of planet-crossing asteroids: Classes of orbital behavior: Project SPACEGUARD”. Icarus 78(2):212–269

Milankovitch M. (1939) "Über die Verwendung vektorieller Bahnelemente in der Störungsrechnung". Bull Acad Sci Math Nat A 6:1–70

Miller J., Konopliv A., Antreasian P., Bordi J., Chesley S., Helfrich C., Owen W., Wang T., Williams B., Yeomans D., et al. (2002) "Determination of shape, gravity, and rotational state of asteroid 433 Eros". Icarus 155(1):3–17

Minkowski H. (1923) "Raum und Zeit". Das Relativitätsprinzip 18:54–66

Mittag-Leffler G. (1912) "Zur Biographie von Weierstrass". Acta Mathematica 35(1):29–65

Moser J. (1970) "Regularization of Kepler's problem and the averaging method on a manifold". Communications on Pure and Applied Mathematics 23(4):609–636

Motter A., Rosa M. (1998) "Hyperbolic calculus". Advances in Applied Clifford Algebras 8(1):109–128

Muinonen K., Virtanen J., Bowell E. (2001) "Collision probability for Earth-crossing asteroids using orbital ranging". Celestial Mechanics and Dynamical Astronomy 81(1-2):93–101

Murison MA. (1989) "On an efficient and accurate method to integrate restricted three-body orbits". The Astronomical Journal 97:1496–1509

Nacozy PE. (1969) "Hansen's method of partial anomalies: an application". The Astronomical Journal 74:544

Nacozy PE. (1977) "The intermediate anomaly". Celestial Mechanics 16(3):309–313

Nacozy PE. (1981) "Time elements in Keplerian orbital elements". Celestial Mechanics 23(2):173–198

Narayan R., Goldreich P., Goodman J. (1987) "Physics of modes in a differentially rotating system–analysis of the shearing sheet". Monthly Notices of the Royal Astronomical Society 228(1):1–41

Nayfeh AH., Balachandran B. (2004) *Applied nonlinear dynamics*. Wiley-VCH

Noether E. (1918) "Invariante Variationsprobleme". Nachr Akad Wiss Goett Math-Phys K1 II:235–257

Novak D., Vasile M. (2011) "Improved shaping approach to the preliminary design of low-thrust trajectories". Journal of Guidance, Control, and Dynamics 34(1):128–147

Olver FW. (2010) *NIST handbook of mathematical functions*. Cambridge University Press

Olver PJ. (2000) *Applications of Lie groups to differential equations* vol 107. Springer Science & Business Media

Palacios M., Calvo C. (1996) "Ideal frames and regularization in numerical orbit computation". The Journal of the Astronautical Sciences 44(1):63–77

Palacios M., Abad A., Elipe A. (1992) "An efficient numerical method for orbit computations". In: AAS/AIAA Astrodynamics Conference pp 265–274

Peláez J., Hedo JM., de Andrés PR. (2007) "A special perturbation method in orbital dynamics". Celestial Mechanics and Dynamical Astronomy 97(2):131–150

Pellegrini E., Russell RP., Vittaldev V. (2014) "F and g taylor series solutions to the stark and kepler problems with sundman transformations". Celestial Mechanics and Dynamical Astronomy 118

Penrose R. (2004) *The Road to Reality: A Complete Guide to the Laws of the Universe*. Jonathan Cape London

Perkins FM. (1959) "Flight mechanics of low-thrust spacecraft". Journal of the Aerospace Sciences 26(5):291–297

Peters CF. (1968) "Treatment of binary encounters in stellar dynamics". Bulletin of the Astronomical Institutes of the Netherlands 3:167

Peterson G., Campbell E., Hawkins A., Balbas J., Ivy S., Merkurjev E., Hall T., Rodriguez P., Gustafson E. (1991) "Relative performance of Lambert solvers 1: zero revolution methods". Advances in the Astronautical Sciences 136(1):1495–1510

Petropoulos AE. (2001) "A shape-based approach to automated, low-thrust, gravity-assist trajectory design". PhD thesis

Petropoulos AE., Longuski JM. (2004) "Shape-based algorithm for the automated design of low-thrust, gravity assist trajectories". Journal of Spacecraft and Rockets 41(5):787–796

Petropoulos AE., Sims JA. (2002) "A review of some exact solutions to the planar equations of motion of a thrusting spacecraft". In: 2nd International Symposium Low Thrust Trajectories, Toulouse, France

Petropoulos AE., Longuski JM., Vinh NX. (1999) "Shape-based analytic representations of low-thrust trajectories for gravity-assist applications". In: AAS/AIAA Astrodynamics Specialist Conference pp 563–581

Pinkham G. (1962) "Reference solution for low thrust trajectories". Journal of the American Rocket Society 32(5):775–776

Poincaré H. (1892) *Les méthodes nouvelles de la Mécanique Céleste* vol 1. Paris: Gauthier-Villars et Fils

Porco C., Helfenstein P., Thomas P., Ingersoll A., Wisdom J., West R., Neukum G., Denk T., Wagner R., Roatsch T., et al. (2006) "Cassini observes the active south pole of Enceladus". Science 311(5766):1393–1401

Porco CC., West RA., McEwen A., Del Genio AD., Ingersoll AP., Thomas P., Squyres S., Dones L., Murray CD., Johnson TV., et al. (2003) "Cassini imaging of Jupiter's atmosphere, satellites, and rings". Science 299(5612):1541–1547

Price GB. (1991) *An introduction to multicomplex spaces and functions.* Marcel Dekker, Inc.

Prussing JE., Coverstone-Carroll V. (1998) "Constant radial thrust acceleration redux". Journal of Guidance, Control, and Dynamics 21(3):516–518

Quarta AA., Mengali G. (2012) "New look to the constant radial acceleration problem". Journal of Guidance, Control, and Dynamics 35(3):919–929

Quinlan GD., Tremaine S. (1992) "On the reliability of gravitational N-body integrations". Monthly Notices of the Royal Astronomical Society 259(3):505–518

Richardson D., Mitchell J. (2003) "A third-order analytical solution for relative motion with a circular reference orbit". The Journal of the Astronautical Sciences 51(1):1–12

Roa J. (2016a) "Nonconservative extension of Keplerian integrals and a new class of integrable system". Monthly Notices of the Royal Astronomical Society 463(3):3204–3219

Roa J. (2016b) "Regularization in astrodynamics: applications to relative motion, low-thrust missions, and orbit propagation". PhD thesis Technical University of Madrid 10.13140/RG.2.2.27043.45605

Roa J., Handmer C. (2015) "Quantifying hazards: asteroid disruption in lunar distant retrograde orbits". ArXiv preprint:1505.03800

Roa J., Peláez J. (2014) "The elliptic rendezvous problem in Dromo formulation". Advances in the Astronautical Sciences 152:2531–2546

Roa J., Peláez J. (2015a) "Efficient trajectory propagation for orbit determination problems". In: 26th AAS/AIAA Astrodynamics Specialist Conference 15-730

Roa J., Peláez J. (2015b) "Frozen-anomaly transformation for the elliptic rendezvous problem". Celestial Mechanics and Dynamical Astronomy 121(1):61–81

Roa J., Peláez J. (2015c) "Generalized logarithmic spirals for low-thrust trajectory design". In: AAS/AIAA Astrodynamics Specialist Conference 15-729

Roa J., Peláez J. (2015d) "Orbit propagation in Minkowskian geometry". Celestial Mechanics and Dynamical Astronomy 123(1):13–43

Roa J., Peláez J. (2015e) "Orbit propagation in Minkowskian geometry". In: 25th Spaceflight Mechanics Meeting pp AAS 15–209

Roa J., Peláez J. (2015f) "Regularized formulations in relative motion". In: 25th Spaceflight Mechanics Meeting AAS 15-210

Roa J., Peláez J. (2016a) "Introducing a degree of freedom in the family of generalized logarithmic spirals". In: 26th Spaceflight Mechanics Meeting 16-317

Roa J., Peláez J. (2016b) "Spiral Lambert's problem with generalized logarithmic spirals". In: 26th Spaceflight Mechanics Meeting 16-316

Roa J., Peláez J. (2016c) "Three-dimensional generalized logarithmic spirals". In: 26th Spaceflight Mechanics Meeting 16-323

Roa J., Peláez J. (2017) "The theory of asynchronous relative motion ii: universal and regular solutions". Celestial Mechanics and Dynamical Astronomy 127(3):343–368

Roa J., Gómez-Mora JI., Peláez J. (2015a) "Error propagation in relative motion". In: 25th Spaceflight Mechanics Meeting pp 15–272

Roa J., Sanjurjo-Rivo M., Peláez J. (2015b) "Singularities in Dromo formulation. Analysis of deep flybys". Advances in Space Research 56(3):569–581

Roa J., Peláez J., Senent J. (2016a) "New analytic solution with continuous thrust: generalized logarithmic spirals". Journal of Guidance, Control, and Dynamics 39(10):2336–2351

Roa J., Peláez J., Senent J. (2016b) "Spiral Lambert's problem". Journal of Guidance, Control, and Dynamics 39(10):2250–2263

Roa J., Urrutxua H., Peláez J. (2016c) "Stability and chaos in Kustaanheimo-Stiefel space induced by the Hopf fibration". Monthly Notices of the Royal Astronomical Society 459(3):2444–2454

Rosengren AJ., Scheeres DJ. (2014) "On the Milankovitch orbital elements for perturbed Keplerian motion". Celestial Mechanics and Dynamical Astronomy 118:197–220

Roy A., Moran P. (1973) "Studies in the application of recurrence relations to special perturbation methods". Celestial Mechanics 7(2):236–255

Saha P. (2009) "Interpreting the Kustaanheimo–Stiefel transform in gravitational dynamics". Monthly Notices of the Royal Astronomical Society 400(1):228–231

Saloom A., Tari F. (2012) "Curves in the Minkowski plane and their contact with pseudo-circles". Geometriae Dedicata 159(1):109–124

San-Juan JF., López LM., Lara M. (2012) "On bounded satellite motion under constant radial propulsive acceleration". Mathematical Problems in Engineering 2012

Sanz-Serna JM. (1992) "Symplectic integrators for Hamiltonian problems: an overview". Acta numerica 1:243–286

Sarnecki A. (1988) "Minimal orbital dynamics". Acta Astronautica 17(8):881–891

Satō I. (1998) "A new anomaly of Keplerian motion". The Astronomical Journal 116(6):3038

Sauer CG. (1997) "Solar electric performance for Medlite and Delta Class planetary missions". American Astronautical Society pp AAS Paper 97–926

Schaub H. (2004) "Relative orbit geometry through classical orbit element differences". Journal of Guidance, Control, and Dynamics 27(5):839–848

Schmidt HJ. (2008) "Perihelion precession for modified newtonian gravity". Physical Review D 78(2):023,512

Schütze O., Vasile M., Junge O., Dellnitz M., Izzo D. (2009) "Designing optimal low-thrust gravity-assist trajectories using space pruning and a multi-objective approach". Engineering Optimization 41(2):155–181

Seiffert A. (1896) "Über eine neue geometrische Einführung in die Theorie der elliptischen Funktionen". Wissenschaftliche Beilage zum Jahresbericht der Städtischen Realschule zu Charlottenburg, Ostern

Senent JS. (2010) "Fast calculation of abort return trajectories for manned missions to the Moon". In: AIAA/AAS Astrodynamics Specialist Conference pp AAS 2010–8132

Sengupta P., Vadali SR., Alfriend KT. (2007) "Second-order state transition for relative motion near perturbed, elliptic orbits". Celestial Mechanics and Dynamical Astronomy 97(2):101–129

Sharaf M., Sharaf AA. (1997) "Closest approach in universal variables". Celestial Mechanics and Dynamical Astronomy 69(4):331–346

Shefer V. (2007) "Variational equations in parametric variables and transformation of their solutions". Cosmic Research 45(4):348–353

Sims JA., Flanagan SN. (1999) "Preliminary design of low-thrust interplanetary missions". In: AAS/AIAA Astrodynamics Specialist Conference, AAS Paper pp 99–338

Sims JA., Longuski JM., Staugler AJ. (1997) "V_∞ leveraging for interplanetary missions: Multiple-revolution orbit techniques". Journal of Guidance, Control, and Dynamics 20(3):409–415

Sobczyk G. (1995) "The hyperbolic number plane". College Mathematics Journal 26(4):268–280

Sperling HJ. (1961) "Computation of Keplerian conic sections". ARS Journal 31(5):660–661

Spivak M. (1994) *Calculus*. Publish or Perish

Stiefel EL., Scheifele G. (1971) *Linear and regular Celestial Mechanics*. Springer-Verlag, New York, Heidelberg, Berlin

Stillwell J. (2010) *Mathematics and Its History* 3rd edn. Springer-Verlag New York

Strömgren B. (1929) "Formeln zur Genäeherten Stöerungsrechnung in Bahnelementen". Publikationer og mindre Meddeler fra Københavns Observatorium 65:1–26

Stumpff K. (1947) "Neue Formeln und Hilfstafeln zur Ephemeridenrechnung". Astronomische Nachrichten 275(7-9):108–128

Sullivan J., Grimberg S., D'Amico S. (2017) "Comprehensive survey and assessment of spacecraft relative motion dynamics models". Journal of Guidance, Control, and Dynamics

Sun F. (1981) "A new treatment of Lambertian mechanics". Acta Astronautica 8(2):105–122

Sundman KF. (1913) "Mémoire sur le problème des trois corps". Acta Mathematica 36(1):105–179

Szebehely V. (1967) *Theory of orbits: the restricted problem of three bodies*. New York: Academic Press

Szebehely V., Bettis DG. (1971) "Recent developments of integrating the gravitational problem of N-bodies". Astrophysics and Space Science 13(2):365–376

Szebehely V., Giacaglia GE. (1964) "On the elliptic restricted problem of three bodies". The Astronomical Journal 69:230

Szebehely V., Peters CF. (1967) "Complete solution of a general problem of three bodies". The Astronomical Journal 72:876

Taheri E., Abdelkhalik O. (2012) "Shape based approximation of constrained low-thrust space trajectories using Fourier series". Journal of Spacecraft and Rockets 49(3):535–546

Taheri E., Abdelkhalik O. (2015a) "Fast initial trajectory design for low-thrust restricted-three-body problems". Journal of Guidance, Control, and Dynamics 38(11):2146–2160

Taheri E., Abdelkhalik O. (2015b) "Initial three-dimensional low-thrust trajectory design". Advances in Space Research 57(3):889–903

Tanguay AR. (1960) "Space maneuvers". Space Trajectories pp 163–199

Thiele TN. (1895) "Recherches numériques concernant des solutions périodiques d'un cas spécial du probléme des trois corps". Astronomische Nachrichten 138(1):1–10

Thomas F., Morbidelli A. (1996) "The Kozai resonance in the outer solar system and the dynamics of long-period comets". Celestial Mechanics and Dynamical Astronomy 64(3):209–229

Trautman A. (1967) "Noether equations and conservation laws". Communications in Mathematical Physics 6(4):248–261

Tschauner J., Hempel P. (1964) "Optimale Beschleunigeungsprogramme fur das Rendezvous-Manover". Astronautica Acta 10(5-6):296–307

Tschauner J., Hempel P. (1965) "Rendezvous zu einem in elliptischer Bahn umlaufenden Ziel". Astronautica Acta 11(2):104

Tsien H. (1953) "Take-off from satellite orbit". Journal of the American Rocket Society 23(4):233–236

Tsu T. (1959) "Interplanetary travel by solar sail". ARS Journal 29(6):422–427

Ulrych S. (2005) "Relativistic quantum physics with hyperbolic numbers". Physics Letters B 625(3):313–323

Urrutxua H. (2015) "High fidelity models for near-earth object dynamics". PhD thesis Technical University of Madrid
Urrutxua H., Peláez J. (2014) "A new regularization method for fast and accurate propagation of roto-translationally coupled asteroids". In: AAS/AIAA Astrodynamics Specialist Conference 14-4216
Urrutxua H., Morante D., Sanjurjo-Rivo M., Peláez J. (2015a) "Dromo formulation for planar motions: solution to the Tsien problem". Celestial Mechanics and Dynamical Astronomy 122(2):143–168
Urrutxua H., Sanjurjo-Rivo M., Peláez J. (2015b) "Dromo propagator revisited". Celestial Mechanics and Dynamical Astronomy 124(1):1–31
Vaddi SS., Vadali SR., Alfriend KT. (2003) "Formation flying: accommodating nonlinearity and eccentricity perturbations". Journal of Guidance, Control, and Dynamics 26(2):214–223
Vallado DA., Alfano S. (2014) "Curvilinear coordinate transformations for relative motion". Celestial Mechanics and Dynamical Astronomy 118(3):253–271
Vallado DA., Kelso TS. (2013) "Earth orientation parameter and space weather data for flight operation". Proceedings of the 23rd Spaceflight Mechanics Meeting pp 2679–2698
Valsecchi G., Milani A., Gronchi G., Chesley S. (2003) "Resonant returns to close approaches: Analytical theory". Astronomy & Astrophysics 408(3):1179–1196
Vasile M., De Pascale P. (2006) "Preliminary design of multiple gravity-assist trajectories". Journal of Spacecraft and Rockets 43(4):794–805
Vasile M., Schütze O., Junge O., Radice G., Dellnitz M., Izzo D. (2006) "Spiral trajectories in global optimisation of interplanetary and orbital transfers". Ariadna Study Report AO4919 5:4106
Vasile M., De Pascale P., Casotto S. (2007) "On the optimality of a shape-based approach based on pseudo-equinoctial elements". Acta Astronautica 61:286–297
Velez C. (1974) "Notions of analytic vs numerical stability as applied to the numerical calculation of orbits". Celestial Mechanics 10(4):405–422
Velez C. (1975) "Stabilization and real world satellite problem". Satellite dynamics pp 136–153
Velez C., Hilinski S. (1978) "Time transformations and Cowell's method". Celestial Mechanics 17(1):83–99
Velez CE. (1967) "Numerical integration of orbits in multirevolution steps". Tech. Rep. N67-39995 NASA Goddard Space Flight Center
Velte W. (1978) "Concerning the regularizing KS-transformation". Celestial Mechanics 17(4):395–403
Vitins M. (1978) "Keplerian motion and gyration". Celestial Mechanics 17(2):173–192
Vittaldev V., Mooij E., Naeije M. (2012) "Unified state model theory and application in astrodynamics". Celestial Mechanics and Dynamical Astronomy 112(3):253–282
Vivarelli MD. (1983) "The KS-transformation in hypercomplex form". Celestial Mechanics 29(1):45–50
Vivarelli MD. (1985) "The KS-transformation in hypercomplex form and the quantization of the negative-energy orbit manifold of the Kepler problem". Celestial Mechanics 36(4):349–364
Vivarelli MD. (1986) "The Kepler problem spinor regularization and pre-quantization of the negative-energy manifold". Meccanica 21(2):75–80
Vivarelli MD. (1987) "Geometrical and physical outlook on the cross product of two quaternions". Celestial Mechanics and Dynamical Astronomy 41(1):359–370
Vivarelli MD. (1994) "The KS-transformation revisited". Meccanica 29(1):15–26
Vujanovic B., Strauss A., Jones S. (1986) "On some conservation laws of conservative and non-conservative dynamic systems". International Journal of Non-linear Mechanics 21(6):489–499
Wagner WE., Velez CE. (1972) "Goddard Trajectory Determination Subsystem: Mathematical specifications". Tech. Rep. N72-30816 NASA Goddard Space Flight Center
Waite Jr JH., Niemann H., Yelle RV., Kasprzak WT., Cravens TE., Luhmann JG., McNutt RL., Ip WH., Gell D., De La Haye V., et al. (2005) "Ion neutral mass spectrometer results from the first flyby of Titan". Science 308(5724):982–986

Waldvogel J. (2006a) "Order and chaos in satellite encounters". In: Chaotic Worlds: From Order to Disorder in Gravitational N-Body Dynamical Systems Springer pp 231–251
Waldvogel J. (2006b) "Quaternions and the perturbed Kepler problem". Celestial Mechanics and Dynamical Astronomy 95(1):201–212
Waldvogel J. (2007) "Fundamentals of regularization in Celestial Mechanics". Tech. rep. ETH Zurich
Walker M., Ireland B., Owens J. (1985) "A set modified equinoctial orbit elements". Celestial Mechanics 36(4):409–419
Wall BJ., Conway BA. (2009) "Shape-based approach to low-thrust rendezvous trajectory design". Journal of Guidance, Control, and Dynamics 32(1):95–101
Wallis J. (1656) *Arithmetica Infinitorum*. Leon Lichfield Academiae Typographi, Imprensis Thomas Robinson
Wehry A., Mann I. (1999) "Identification of β-meteoroids from measurements of the dust detector onboard the Ulysses spacecraft". Astronomy & Astrophysics 341:296–303
Whittaker ET. (1917) *Treatise on analytical dynamics of particles and rigid bodies* 2nd edn. Cambridge University Press
Whittaker ET., Watson GN. (1927) *A course of modern analysis*. Cambridge University Press
Wisdom J., Holman M. (1991) "Symplectic maps for the n-body problem". The Astronomical Journal 102:1528–1538
Xia Z. (1993) "Arnold diffusion in the elliptic restricted three-body problem". Journal of Dynamics and Differential Equations 5(2):219–240
Yabushita S. (1979) "A statistical study of the evolution of the orbits of long-period comets". Monthly Notices of the Royal Astronomical Society 187:445–462
Yaglom IM. (1979) *A simple non-Euclidean geometry and its physical basis*. Springer-Verlag, New York, Heidelberg, Berlin
Yamanaka K., Ankersen F. (2002) "New state transition matrix for relative motion on an arbitrary elliptical orbit". Journal of Guidance, Control, and Dynamics 25(1):60–66
Yeomans D., Barriot JP., Dunham D., Farquhar R., Giorgini J., Helfrich C., Konopliv A., McAdams J., Miller J., Owen W., et al. (1997) "Estimating the mass of asteroid 253 Mathilde from tracking data during the NEAR flyby". Science 278(5346):2106–2109
Yoshida H. (1983a) "Necessary condition for the existence of algebraic first integrals I: Kowalevski's exponents". Celestial Mechanics 31(4):363–379
Yoshida H. (1983b) "Necessary condition for the existence of algebraic first integrals II: Condition for algebraic integrability". Celestial Mechanics 31(4):381–399
Yukawa H. (1935) "On the interaction of elementary particles. I". Proc Phys Math Soc Japan 17(0):48–57
Zare K., Szebehely V. (1975) "Time transformations in the extended phase-space". Celestial Mechanics 11(4):469–482
Zuiani F., Vasile M. (2015) "Extended analytical formulas for the perturbed Keplerian motion under a constant control acceleration". Celestial Mechanics and Dynamical Astronomy 121(3):275–300
Zuiani F., Vasile M., Palmas A., Avanzini G. (2012) "Direct transcription of low-thrust trajectories with finite trajectory elements". Acta Astronautica 72:108–120

Index

https://doi.org/10.1515/9783110559125-204

J

K

L

M

N

O

De Gruyter Studies in Mathematical Physics

Volume 41
Esra Russell, Oktay K. Pashaev
Oscillatory Models in General Relativity, 2017
ISBN 978-3-11-051495-7, e-ISBN (PDF) 978-3-11-051536-7,
e-ISBN (EPUB) 978-3-11-051522-0, Set-ISBN 978-3-11-051537-4

Volume 40
Joachim Schröter
Minkowski Space: The Spacetime of Special Relativity, 2017
ISBN 978-3-11-048457-1, e-ISBN (PDF) 978-3-11-048573-8,
e-ISBN (EPUB) 978-3-11-048461-8, Set-ISBN 978-3-11-048574-5

Volume 39
Vladimir K. Dobrev
Invariant Differential Operators: Quantum Groups, 2017
ISBN 978-3-11-043543-6, e-ISBN (PDF) 978-3-11-042770-7,
e-ISBN (EPUB) 978-3-11-042778-3, Set-ISBN 978-3-11-042771-4

Volume 38
Alexander N. Petrov, Sergei M. Kopeikin, Robert R. Lompay, Bayram Tekin
Metric theories of gravity: Perturbations and conservation laws, 2017
ISBN 978-3-11-035173-6, e-ISBN (PDF) 978-3-11-035178-1,
e-ISBN (EPUB) 978-3-11-038340-9, Set-ISBN 978-3-11-035179-8

Volume 37
Igor Olegovich Cherednikov, Frederik F. Van der Veken
Parton Densities in Quantum Chromodynamics: Gauge invariance, path-dependence and Wilson lines, 2016
ISBN 978-3-11-043939-7, e-ISBN (PDF) 978-3-11-043060-8,
e-ISBN (EPUB) 978-3-11-043068-4, Set-ISBN 978-3-11-043061-5

Volume 36
Alexander B. Borisov, Vladimir V. Zverev
Nonlinear Dynamics: Non-Integrable Systems and Chaotic Dynamics, 2016
ISBN 978-3-11-043938-0, e-ISBN (PDF) 978-3-11-043058-5,
e-ISBN (EPUB) 978-3-11-043067-7, Set-ISBN 978-3-11-043059-2

Bei Fragen zur Produktsicherheit wenden Sie sich bitte an:
If you have any questions regarding product safety,
please contact:

Walter de Gruyter GmbH
Genthiner Straße 13
10785 Berlin
productsafety@degruyterbrill.com